Lecture Notes in Computer Science   13878

Advanced Research in Computing and Software Science
Subline of Lecture Notes in Computer Science

More information about this series at https://link.springer.com/bookseries/558

Leszek Gąsieniec (Ed.)

SOFSEM 2023:
Theory and Practice
of Computer Science

48th International Conference on Current Trends
in Theory and Practice of Computer Science, SOFSEM 2023
Nový Smokovec, Slovakia, January 15–18, 2023
Proceedings

 Springer

Editor
Leszek Gąsieniec
University of Liverpool
Liverpool, UK

ISSN 0302-9743   ISSN 1611-3349   (electronic)
Lecture Notes in Computer Science
ISBN 978-3-031-23100-1   ISBN 978-3-031-23101-8   (eBook)
https://doi.org/10.1007/978-3-031-23101-8

This Springer imprint is published by the registered company Springer Nature Switzerland AG
The registered company address is: Gewerbestrasse 11, 6330 Cham, Switzerland

Preface

This volume contains the papers selected for presentation at SOFSEM 2023, the 48th International Conference on Current Trends in Theory and Practice of Computer Science, which was held during January 15–18, 2023, in Nový Smokovec, Slovakia. SOFSEM (originally the SOFtware SEMinar) is an annual international winter conference devoted to the theory and practice of computer science. Its aim is to present the latest developments in research for professionals from academia and industry working in leading areas of computer science. While being a well-established and fully international conference, SOFSEM also maintains the best of its original Winter School aspects, such as a high number of invited talks, in-depth coverage of selected research areas, and ample opportunity to discuss and exchange new ideas. The series of SOFSEM conferences was interrupted in 2022 due to the COVID-19 pandemic, but SOFSEM was held again this year, in scenic Nový Smokovec in the Tatra Mountains, Slovakia.

The renewed scope and format of SOFSEM is focused entirely on the original research and challenges in foundations of computer science including algorithms, AI-based methods, computational complexity, and formal models. The SOFSEM 2023 Program Committee was formed of 24 international experts supported by 55 external reviewers. Due to the new format and the post-pandemic sentiment in academia the event attracted only 43 full submissions; however, these were of high quality. The conference program consisted of 26 papers, where each accepted paper went through a thorough selection process supported by at least three external reviews. All reviews were single blind.

The SOFSEM 2023 Program Committee decided to split the prize for the Best Paper Award between two papers: "Balanced Substructures in Bicolored Graphs," by P. S. Ardra, R. Krithika, S. Saurabh, and R. Sharma and "On the Complexity of Scheduling Problems With a Fixed Number of Parallel Identical Machines," by K. Kahler and K. Jansen. Similarly, the Best Student Paper Award was split between two papers: "On the 2-Layer Window Width Minimization Problem," by M. Bekos, H. Förster, M. Kaufmann, S. Kobourov, M. Kryven, A. Kuckuk, and L. Schlipf and "Sequentially Swapping Tokens: Further on Graph Classes," by H. Kiya, Y. Okada, H. Ono, and Y. Otachi.

I would like to thank the invited speakers for their talks, including Věra Kůrková (Institute of Computer Science of the Czech Academy of Sciences, Czech Republic), "Some implications of high-dimensional geometry for classification by neural networks", Gerth Stølting Brodal (Aarhus University, Denmark), "Data Structure Design – Theory and Practice", and Sławomir Lasota (Uniwersytet Warszawski, Poland), "Ackermannian lower bound for the reachability problem of Petri nets". Many thanks go to the SOFSEM 2023 Program Committee members, the external reviewers, and the creators of EasyChair system. Special thanks go to the SOFSEM Steering Committee members Július Štuller and Jan van Leeuwen for their advice and support throughout the whole conference cycle.

January 2023   Leszek Gąsieniec

Organization

General Chair

Peter Gurský   Pavol Jozef Šafárik University in Košice, Slovakia

Program Committee

Amihood Amir	Bar-Ilan University, Israel, and Georgia Tech, USA
Michael Blondin	Université de Sherbrooke, Canada
Marek Chrobak	University of California, Riverside, USA
Paola Flocchini	University of Ottawa, Canada
Anna Gambin	Uniwersytet Warszawski, Poland
Robert Ganian	Vienna University of Technology, Austria
Leszek Gąsieniec (Chair)	University of Liverpool, UK
Davide Grossi	University of Groningen, Netherlands
Christoph Haase	University of Oxford, UK
Cezary Kaliszyk	University of Innsbruck, Austria
Ralf Klasing	CNRS and University of Bordeaux, France
Rastislav Královič	Comenius University in Bratislava, Slovakia
Giuseppe Liotta	University of Perugia, Italy
Paweł Parys	University of Warsaw, Poland
Vangelis Paschos	LAMSADE, Paris Dauphine University, France
Daniel Paulusma	Durham University, UK
Tomasz Radzik	King's College London, UK
David Sarne	Bar-Ilan University, Israel
Christian Scheideler	University of Paderborn, Germany
Paweł Sobociński	Tallinn University of Technology, Estonia
Paul Spirakis	University of Liverpool, UK
Grzegorz Stachowiak	University of Wrocław, Poland
Sebastian Wild	University of Liverpool, UK
Qin Xin	University of the Faroe Islands, Faroe Islands

Steering Committee

Barbara Catania	University of Genoa, Italy
Mirosław Kutyłowski	Wrocław University of Technology, Poland
Tiziana Margaria-Steffen	University of Limerick, Ireland
Branislav Rovan	Comenius University in Bratislava, Slovakia
Petr Šaloun	Technical University of Ostrava, Czech Republic
Július Štuller (Chair)	Czech Academy of Sciences, Czech Republic
Jan van Leeuwen	Utrecht University, The Netherlands

Additional Reviewers

Barloy, Corentin
Boneh, Itai
Bose, Prosenjit
Brakensiek, Joshua
Calamoneri, Tiziana
Chakraborty, Dibyayan
Chrzaszcz, Jacek
Chung, Neo Christopher
Czerwiński, Wojciech
Daming, Zhu
Deligkas, Argyrios
Di Lavore, Elena
Dojer, Norbert
Earnshaw, Matthew
Eiben, Eduard
Erlebach, Thomas
Fischer, Johannes
Fleischmann, Pamela
Foucaud, Florent
Fox, Kyle
Grüttemeier, Niels
Gupta, Siddharth
Hermelin, Danny
Itzhaki, Michael
Kindermann, Philipp
Klobas, Nina
Kobayashi, Yasuaki
Korchemna, Viktoriia

Kozłowski, Łukasz Paweł
Kurpicz, Florian
Kutner, David
Lubiw, Anna
Marcus, Shoshana
Montecchiani, Fabrizio
Niemiro, Wojciech
Niemyska, Wanda
Ochem, Pascal
Ordyniak, Sebastian
Ortali, Giacomo
Pajak, Dominik
Pardubska, Dana
Paszek, Jarosław
Patro, Subhasree
Polesiuk, Piotr
Przybylski, Bartłomiej
Raptopoulos, Christoforos
Raskin, Mikhail
Roditty, Liam
Román, Mario
Rzążewski, Paweł
Shabtay, Dvir
Skretas, George
Tappini, Alessandra
Walen, Tomasz
Wasa, Kunihiro

Contents

Robots and Strings

Graphs Problems and Optimisation

The Complexity of Finding Tangles

Oksana Firman[1]([✉])[iD], Philipp Kindermann[2][iD], Boris Klemz[1][iD],
Alexander Ravsky[3], Alexander Wolff[1][iD], and Johannes Zink[1][iD]

[1] Institut für Informatik, Universität Würzburg, Würzburg, Germany
{oksana.firman,boris.klemz,johannes.zink}@uni-wuerzburg.de
[2] Fachbereich IV – Informatikwissenschaften, Universität Trier, Trier, Germany
kindermann@uni-trier.de
[3] Pidstryhach Institute for Applied Problems of Mechanics and Mathematics,
National Academy of Sciences of Ukraine, Lviv, Ukraine
alexander.ravsky@uni-wuerzburg.de

Abstract. We study the following combinatorial problem. Given a set
of n y-monotone curves, which we call *wires*, a *tangle* determines the
order of the wires on a number of horizontal *layers* such that any two
consecutive layers differ only in swaps of neighboring wires. Given a
multiset L of *swaps* (that is, unordered pairs of wires) and an initial
order of the wires, a tangle *realizes* L if each pair of wires changes its
order exactly as many times as specified by L.

Deciding whether a given multiset of swaps admits a realizing tangle is
known to be NP-hard [Yamanaka et al., CCCG 2018]. We prove that this
problem remains NP-hard if every pair of wires swaps only a constant
number of times. On the positive side, we improve the runtime of a
previous exponential-time algorithm. We also show that the problem is
in NP and fixed-parameter tractable with respect to the number of wires.

Keywords: Tangle · NP-hard · Exponential-time algorithm · FPT

1 Introduction

This paper concerns the visualization of *chaotic attractors*, which occur in
(chaotic) dynamic systems. Such systems are considered in physics, celestial
mechanics, electronics, fractals theory, chemistry, biology, genetics, and population dynamics; see, for instance, [4], [13], and [6, p. 191]. Birman and Williams [3]
were the first to mention tangles as a way to describe the topological structure
of chaotic attractors. They investigated how the orbits of attractors are knotted. Later Mindlin et al. [9] characterized attractors using integer matrices that
contain numbers of swaps between the orbits.

Olszewski et al. [10] studied the problem of visualizing chaotic attractors.
Using $[n]$ as shorthand for $\{1, 2, \ldots, n\}$, define two permutations σ and τ of
$[n]$ to be *adjacent* if they differ only in transposing neighboring elements, that
is, for every $i \in [n]$, $\sigma(i) \in \{\tau(i)\} \cup \{\tau(i-1) \mid i > 1\} \cup \{\tau(i+1) \mid i < n\}$.

L. Gąsieniec (Ed.): SOFSEM 2023, LNCS 13878, pp. 3–17, 2023.
https://doi.org/10.1007/978-3-031-23101-8_1

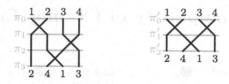

Fig. 1. Tangles T and T' of different heights realizing the list $L = \{(1,2),(1,4),(3,4)\}$.

For two adjacent permutations σ and τ, let $\operatorname{diff}(\sigma,\tau) = \{\{\sigma(i),\sigma(i+1)\} \mid i \in [n-1] \wedge \sigma(i) = \tau(i+1) \wedge \sigma(i+1) = \tau(i)\}$ be the set of neighboring transpositions in which σ and τ differ. Given a set of y-monotone curves called *wires* that hang off a horizontal line in a prescribed order π_0, and a multiset L (called *list*) of unordered pairs of wires (called *swaps*), the problem consists in finding a *tangle* realizing L, i.e., a sequence $\pi_0, \pi_1, \ldots, \pi_h$ of permutations of the wires such that (i) consecutive permutations are adjacent and (ii) $L = \bigcup_{i=0}^{h-1} \operatorname{diff}(\pi_i, \pi_{i+1})$.

For example, the list L in Fig. 1 admits a tangle realizing it. We call such a list *feasible*. The list $L' = L \cup \{(1,2)\}$, in contrast, is not feasible. Note that, if the start permutation π_0 is not given explicitly, we assume that $\pi_0 = \operatorname{id} = \langle 1, 2, \ldots, n \rangle$. In Fig. 2, the list L_n is feasible; it is specified by an $(n \times n)$-matrix. The gray horizontal bars correspond to the permutations (or *layers*).

Olszewski et al. gave an exponential-time algorithm for minimizing the *height* of a tangle, that is, the number of layers. They tested their algorithm on a benchmark set, which showed that instances with up to 18 swaps can be solved within seconds, but instances with more than 22 swaps can take several hours.

We [5] showed, by reduction from 3-PARTITION, that tangle-height minimization is NP-hard. We also presented two (exponential-time) algorithms, one for the general problem and one for *simple* lists, that is, lists where each swap occurs at most once. Using an extended benchmark set, we showed that in almost all cases our algorithm for the general problem is faster and more memory-efficient than the algorithm of Olszewski et al.

In an independent line of research, Yamanaka et al. [16] showed that the problem LADDER-LOTTERY REALIZATION is NP-hard. As it turns out, this problem is equivalent to deciding the feasibility of a list.

Sado and Igarashi [11] used the same optimization criterion for tangles in the setting where only the beginning and final permutation are given (but they can choose the swaps performed to get there). They used odd-even sort, a parallel variant of bubble sort, to compute tangles with at most one layer more than the minimum in $O(n^2)$ time. Wang [15] showed that there is always a height-optimal tangle where no swap occurs more than once. Bereg et al. [1,2] considered a similar problem. Given a final permutation, they showed how to minimize the number of bends or *moves* (which are maximal "diagonal" segments of the wires).

Notation and Conjecture. For n wires, a *list* $L = (l_{ij})$ of *order* n is a symmetric $n \times n$ matrix with entries in $\mathbb{N}_0$ and zero diagonal. Let $|L| = \sum_{i<j} l_{ij}$ be the *length* of L. A list $L' = (l'_{ij})$ is a *sublist* of L if $l'_{ij} \le l_{ij}$ for each $i, j \in [n]$.

$$L_n = \begin{pmatrix} 0 & 1 & 1 & \dots & 1 & 0 & 2 \\ 1 & 0 & 1 & \dots & 1 & 2 & 0 \\ 1 & 1 & 0 & \dots & 1 & 0 & 2 \\ \vdots & \vdots & \vdots & \ddots & \vdots & \vdots & \vdots \\ 1 & 1 & 1 & \dots & 0 & \mathbf{0} & \mathbf{2} \\ 0 & 2 & 0 & \dots & \mathbf{0} & 0 & n-1 \\ 2 & 0 & 2 & \dots & 2 & n-1 & 0 \end{pmatrix}$$

(The bold zeros and twos must be exchanged if n is even.)

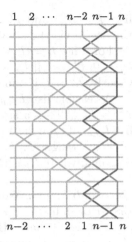

Fig. 2. A list L_n of order n (left) and a tangle realizing L_n (right). Entry (i,j) of L_n defines how often wires i and j must swap in the tangle. Here, $n = 7$.

If there is a pair $i, j \in [n]$ such that $l'_{ij} < l_{ij}$, then L' is a *strict* sublist of L. A list is *0–2* if all entries are zeros or twos; it is *even* (*odd*) if all non-zero entries are even (odd).

For a list to be feasible, it also has to fulfill the following property. We say that a list is *consistent* if the *final positions* of all wires form a permutation of $[n]$. For a wire i, its final position is its initial position (namely, i) minus one for each wire on its left that it swaps an odd number of times, plus one for each wire on its right that it swaps an odd number of times. We have shown that consistency is sufficient for the feasibility of odd lists [5]. Clearly, an even list is always consistent as, for any tangle realizing an even list, the initial permutation equals the final permutation.

For any list to be feasible, each triple of wires $i < j < k$ requires an i–j or a j–k swap if there is an i–k swap—otherwise wires i and k would be separated by wire j in any tangle. We call a list fulfilling this property *non-separable*. It is natural to ask whether this necessary condition is also sufficient. For odd lists, non-separability is implied by consistency (because consistency is sufficient for feasibility [5]). Although the NP-hardness reduction from Sect. 3 shows that a non-separable list can fail to be feasible even when it is consistent. For even lists, the following question remains.

Conjecture 1 [5]**.** *Every non-separable even list is feasible.*

In order to understood the structure of feasible lists better, we consider the following relation between them. Let $L = (l_{ij})$ be a feasible list. Consider the list L' that is identical to L except that it has two additional i–j swaps. We claim that if $l_{ij} > 0$ then the list L' is also feasible. Note that any tangle T that realizes L has a permutation π that supports the i–j swap. Directly after π, we can insert two i–j swaps into T. This yields a tangle that realizes L'. Given two lists $L = (l_{ij})$ and $L' = (l'_{ij})$, we write $L \to L'$ if the list L can be *extended* to the list L' via the above operation.

For a list $L = (l_{ij})$, let $1(L) = (l_{ij} \bmod 2)$ and let $2(L) = (l''_{ij})$ with $l''_{ij} = 0$ if $l_{ij} = 0$, $l''_{ij} = 1$ if l_{ij} is odd, and $l''_{ij} = 2$ otherwise. We call $2(L)$ the *type* of L. Clearly, $L' \to L$ if and only if $2(L') = 2(L)$ and $l'_{ij} \leq l_{ij}$ for each $i, j \in [n]$.

A feasible list $L_{\min}$ is *minimal* if there exists no feasible list $L^\star$ such that $L^\star \to L_{\min}$. Thus a list L is feasible, if and only if there exists a minimal feasible list $L_{\min}$ of type $2(L)$ such that $L_{\min} \to L$.

For a tangle T, let $L(T) = (l_{ij})$ be the symmetric $n \times n$ matrix with zero diagonal, where l_{ij} is the number of i–j swaps in T. Note that T realizes $L(T)$.

Our Contribution. We call the problem of testing the feasibility of a given list LIST-FEASIBILITY. As mentioned above, Yamanaka et al. [16] showed that this problem is NP-hard. However, in their reduction, for some swaps the number of occurrences is linear in the number of wires. We strengthen their result by showing that LIST-FEASIBILITY is NP-hard even if all swaps have constant multiplicity; see Sect. 3. Our reduction uses a variant of NOT-ALL-EQUAL 3-SAT (whereas Yamanaka et al. used 1-IN-3 3SAT).

We start the paper, however, by studying exact algorithms for the LIST-FEASIBILITY problem; see Sect. 2. We present an exponential-time algorithm with runtime $O\big((2|L|/n^2 + 1)^{n^2/2} \cdot n^3 \log |L|\big)$, where L is the given list of order n. The runtime is expressed in terms of the logarithmic cost model of computation. This improves our previous algorithm [5] (which actually computes a tangle of minimum height, if possible). That algorithm runs in $O((2|L|/n^2 + 1)^{n^2/2} \cdot \varphi^n \cdot n)$ time in the unit-cost model and in $O((2|L|/n^2 + 1)^{n^2/2} \cdot \varphi^n \cdot n \log |L|)$ time in the log-cost model, where $\varphi \approx 1.618$ is the golden ratio. Although we cannot characterize minimal feasible lists, we can bound their entries. Namely, we show that, in a minimal feasible list of order n, each swap occurs at most $n^2/4 + 1$ times. As a corollary, this yields that LIST-FEASIBILITY is in NP. Combined with our exponential-time algorithm, this also leads to a fixed-parameter tractable algorithm for testing feasibility (parameterized by the number of wires).

Finally, we disprove Conjecture 1; see Sect. 4. We could verify our counterexample (with 16 wires and 55 swaps of multiplicity 2) only by computer.

2 Exact Algorithms

We remind the reader of our algorithm for tangle-height minimization [5]; a dynamic program that runs in $O((2|L|/n^2 + 1)^{n^2/2} \cdot \varphi^n \cdot n)$ time in the unit-cost model; in the log-cost model the runtime increases by a factor of $O(\log |L|)$. We adjust this algorithm to the task of testing feasibility, which makes the algorithm simpler and faster. Then we will bound the entries of minimal feasible lists (defined above) and use this bound to turn our exact algorithm into a fixed-parameter algorithm where the parameter is the number of wires (i.e., n).

Theorem 1. *There is an algorithm that, given a list L of order n, tests whether L is feasible in $O((2|L|/n^2 + 1)^{n^2/2} \cdot n^3 \cdot \log |L|)$ time in the log-cost model.*

Proof. Let F be a Boolean table with one entry for each sublist L' of L such that $F(L') = \text{true}$ if and only if L' is feasible. This table can be filled by means of a dynamic programming recursion. The empty list is feasible. Let L' be a sublist of L with $|L'| \geq 1$ and assume that for each strict sublist of L', the corresponding entry in F has already been determined. A sublist $\tilde{L}$ of L is feasible if and only if there is a realizing tangle of $\tilde{L}$ of height $|\tilde{L}| + 1$. For each i–j swap in L', we check if there is a tangle realizing L' of height $|L'| + 1$ such that i–j is the last swap. If no such swap exists, then L' is infeasible, otherwise it is feasible. To perform the check for a particular i–j swap, we consider the strict sublist L'' of L' that is identical to L' except an i–j swap is missing. If $F(L'') = \text{true}$, we compute the final positions of i and j with respect to L''. The desired tangle exists if and only if these positions differ by exactly one.

The number of sublists of L is upper bounded by $(2|L|/n^2 + 1)^{n^2/2}$ [5]. For each sublist, we have to check $O(n^2)$ swaps. To check a swap, we have to compute the final positions of two wires, which can be done in $O(n \log |L|)$ time.   □

The following lemma follows from odd-even sort and is well-known [8].

Lemma 1. *For each integer $n \geq 2$ and each pair π, σ of permutations of $[n]$, we can construct in $O(n^2)$ time a tangle T of height at most $n + 1$ that starts with π, ends in σ, and whose list $L(T)$ is simple.*

Now we consider the following tangle shortening construction.

Example 1. Let T be a tangle and $T^* = (\pi_1, \ldots, \pi_h)$ be a subsequence of T containing the initial and the final permutations of T. By Lemma 1, we can augment T^* to a tangle T', overwriting each two consecutive elements π_k and π_{k+1} of T^* by a tangle T'_k which starts from π_k, ends at π_{k+1}, and whose list $L(T'_k) = (l'_{k,ij})$ is simple. Now let T_k be the subtangle of T that starts from π_k and ends at π_{k+1}. Let $L(T_k) = (l_{k,ij})$. The simplicity of the list $L(T'_k)$ implies that $l'_{k,ij} \leq l_{k,ij}$ for each $i, j \in [n]$. It follows that $l'_{ij} \leq \min\{l_{ij}, h - 1\}$ for each $i, j \in [n]$, where $L(T) = (l_{ij})$ and $L(T') = (l'_{ij})$. Since the tangles T' and T have common initial and final permutations, for each $i, j \in [n]$ the numbers l_{ij} and l'_{ij} have the same parity, that is, $1(T') = 1(T)$.

We want to upperbound the entries of a minimal feasible list. We first give a simple bound, which we then improve by a factor of 2 in Proposition 2 below.

Proposition 1. *If $L = (l_{ij})$ is a minimal feasible list of order n then $l_{ij} \leq \binom{n}{2} + 1$ for each $i, j \in [n]$.*

Proof. The list L is feasible, so there is a tangle T realizing L. We choose a subsequence T^* of T consisting of at most $h = \binom{n}{2} + 2$ permutations. To this end, we pick the initial and the final permutation of T and, for each pair $(i, j) \in [n]^2$ with $i < j$ and $l_{ij} \geq 1$, we pick a permutation that swaps i and j.

Let T' be the tangle that we construct from T using T^*, as described in the shortening construction, and let $L' = L(T') = (l'_{ij})$ be the list of T'. The construction of the tangle T' assures that, for any $i, j \in [n]$, if $l_{ij} > 0$, then $l'_{ij} > 0$. This, together with $1(L') = 1(L)$, yields $2(L') = 2(L)$. Hence, $L' \to L$. List L is minimal, so $L = L'$. Thus, $l_{ij} = l'_{ij} \le h - 1 \le \binom{n}{2} + 1$ for $i, j \in [n]$. □

Proposition 2. *If $L = (l_{ij})$ is a minimal feasible list of order n, then $l_{ij} \le n^2/4 + 1$ for each $i, j \in [n]$.*

Proof. Let T be a tangle (starting from id that realizes L). Given an i–j swap, we define its *span* to be $|i - j|$. Order the swaps in $1(L)$ according to decreasing span. We will color the swaps as follows. At each step we color in red the first non-colored swap i–j (with $i < j$) from the list. Since the tangle T realizes the list L, it contains a permutation π with $\pi(j) < \pi(i)$. Put π into an initially empty set P for later use. Let $k \in \{i + 1, \ldots, j - 1\}$ be an integer strictly between i and j. Since $\pi(j) < \pi(i)$, we have $\pi(k) < \pi(i)$ or $\pi(k) > \pi(j)$. We color in blue the i–k swap in the former case and the k–j swap in the latter case. In any case, if T_π is a tangle that starts from id and contains π, the list $L(T_\pi)$ contains the swap(s) we just colored in blue.

Let G be a graph with vertex set $[n]$ and edge set consisting of the red swaps. The coloring algorithm assures that the graph G is triangle-free, so, by Turán's theorem [14], it has at most $n^2/4$ edges.

Let T^* be a subsequence of the tangle T consisting of the initial permutation id, all permutations in P, and the final permutation of T. Let T' be the tangle that we construct from T using T^* as described in Example 1. Let $L' = L(T')$.

The construction of the tangle T' assures that if $1 \le i < j \le n$ and $l_{ij} > 0$ then $\pi(j) < \pi(i)$ in some selected permutation π. Since π is a member of the tangle T, $l'_{ij} > 0$. Moreover, since $1(L') = 1(L)$, $2(L') = 2(L)$. Hence $L' \to L$. The list L' is minimal, therefore $L' = L$. Since each entry of the list L' is at most $|P| + 2 - 1 \le n^2/4 + 1$, the same holds for the entries of the list L. □

Combining Proposition 2 and our exact algorithm yields a fixed-parameter tractable algorithm.

Theorem 2. *There is a fixed-parameter algorithm for* LIST-FEASIBILITY *with respect to the parameter n. Given a list L of order n, the algorithm tests whether L is feasible in $O\big((n/2)^{n^2} \cdot n^3 \log n + n^2 \log |L|\big)$ time.*

Proof. Given the list $L = (l_{ij})$, let $L' = (l'_{ij})$ with $l'_{ij} = \min\{l_{ij}, n^2/4 + 1\}$ for each $i, j \in [n]$. We use our exact algorithm described in the proof of Theorem 1 to check whether the list L' is feasible. Since our algorithm checks the feasibility of every sublist L'' of L', it suffices to combine this with checking whether $2(L'') = 2(L)$. If we find a feasible sublist L'' of the same type as L, then, by Proposition 2, L is feasible; otherwise, L is infeasible. Checking the type of L'' is easy. The runtime for this check is dominated by the runtime for checking the feasibility of L''. Constructing the list L' takes $O(n^2 \log |L|)$ time. Note that $|L'| \le \binom{n}{2} \cdot (n^2/4 + 1) \le (n^4 - 4n^2)/8$. Plugging this into the runtime

$O\big((2|L'|/n^2 + 1)^{n^2/2} \cdot n^3 \log |L'|\big)$ of our exact algorithm (Theorem 1) yields a total runtime of $O\big((n/2)^{n^2} \cdot n^3 \log n + n^2 \log |L|\big)$.

3   Complexity

Yamanaka et al. [16] showed that LIST-FEASIBILITY is NP-hard. In their reduction, however, some swaps have multiplicity $\Theta(n)$. In this section, we show that LIST-FEASIBILITY is NP-hard even if all swaps have multiplicity at most 8. We reduce from POSITIVE NAE 3-SAT DIFF, a variant of NOT-ALL-EQUAL 3-SAT. Recall that in NOT-ALL-EQUAL 3-SAT one is given a Boolean formula in conjunctive normal form with three literals per clause and the task is to decide whether there exists a variable assignment such that in no clause all three literals have the same truth value. By Schaefer's dichotomy theorem [12], NOT-ALL-EQUAL 3-SAT is NP-hard even if no negative literals are admitted. In POSITIVE NAE 3-SAT DIFF, additionally each clause contains three different variables. We show that this variant is NP-hard, too.

Lemma 2. POSITIVE NAE 3-SAT DIFF *is NP-hard.*

Proof. We show NP-hardness of POSITIVE NAE 3-SAT DIFF by reduction from NOT-ALL-EQUAL 3-SAT. Let $\Phi = c_1 \wedge c_2 \wedge \cdots \wedge c_m$ be an instance of NOT-ALL-EQUAL 3-SAT with variables $v_1, v_2, \ldots, v_n$. First we show how to get rid of negative variables and then of multiple occurrences of the same variable in a clause.

We create an instance Φ' of POSITIVE NAE 3-SAT DIFF as follows. For every variable v_i, we introduce two new variables x_i and y_i. We replace each occurrence of v_i by x_i and each occurrence of $\neg v_i$ by y_i. We need to force y_i to be $\neg x_i$. To this end, we introduce the clause $(x_i \vee y_i \vee y_i)$.

Now, we introduce three additional variables a, b, and d that form the clause $(a \vee b \vee d)$. Let $c = (x \vee x \vee y)$ be a clause that contains two occurrences of the same variable. We replace c by three clauses $(x \vee y \vee a)$, $(x \vee y \vee b)$, $(x \vee y \vee d)$. Since at least one of the variables a, b, or d has to be true and at least one has to be false, x and y cannot have the same assignment, i.e., $x = \neg y$. Hence, Φ' is satisfiable if and only if Φ is. Clearly, the size of Φ' is polynomial in the size of Φ.   □

Our main result is as follows.

Theorem 3. LIST-FEASIBILITY *is NP-complete even if every pair of wires has at most eight swaps.*

We split our proof into several parts. First, we introduce some notation, then we give the intuition behind our reduction. Next, we explain variable and clause gadgets in more detail. Finally, we show the correctness of the reduction.

Notation. We label the wires by their index in the initial permutation of a tangle. In particular, for a wire ε, its neighbor to the right is wire $\varepsilon + 1$. If a wire μ is to the left of some other wire ν, then we write $\mu < \nu$. If all wires in a set M are to the left of all wires in a set N, then we write $M < N$.

Setup. Given an instance $F = d_1 \wedge \cdots \wedge d_m$ of POSITIVE NAE 3-SAT DIFF with variables $w_1, \ldots, w_n$, we construct in polynomial time a list L of swaps such that there is a tangle T realizing L if and only if F is a yes-instance.

In L, we have two inner wires λ and $\lambda' = \lambda + 1$ that swap eight times. This yields two types of loops (see Fig. 3): four λ'–λ loops, where λ' is on the left and λ is on the right side, and three λ–λ' loops with λ on the left and λ' on the right side. Notice that we consider only *closed* loops, which are bounded by swaps between λ and λ'. In the following, we construct variable and clause gadgets. Each variable gadget will contain a specific wire that represents the variable, and each clause gadget will contain a specific wire that represents the clause. The corresponding variable and clause wires swap in one of the four λ'–λ loops. We call the first two λ'–λ loops *true-loops*, and the last two λ'–λ loops *false-loops*. If the corresponding variable is true, then the variable wire swaps with the corresponding clause wires in a true-loop, otherwise in a false-loop.

Apart from λ and λ', our list L contains (many) other wires, which we split into groups. For every $i \in [n]$, we introduce sets V_i and V_i' of wires that together form the gadget for variable w_i of F. These sets are ordered (initially) $V_n < V_{n-1} < \cdots < V_1 < \lambda < \lambda' < V_1' < V_2' < \cdots < V_n'$; the order of the wires inside these sets will be detailed in the next two paragraphs. Let $V = V_1 \cup V_2 \cup \cdots \cup V_n$ and $V' = V_1' \cup V_2' \cup \cdots \cup V_n'$. Similarly, for every $j \in [m]$, we introduce a set C_j of wires that contains a *clause wire* c_j and three sets of wires D_j^1, D_j^2, and D_j^3 that represent occurrences of variables in a clause d_j of F. The wires in C_j are ordered $D_j^3 < D_j^2 < D_j^1 < c_j$. Together, the wires in $C = C_1 \cup C_2 \cup \cdots \cup C_m$ represent the clause gadgets; they are ordered $V < C_m < C_{m-1} < \cdots < C_1 < \lambda$. Additionally, our list L contains a set $E = \{\varphi_1, \ldots, \varphi_7\}$ of wires that will make our construction rigid enough. The order of all wires in L is $V < C < \lambda < \lambda' < E < V'$. Now we present our gadgets in more detail.

Variable Gadget. For each variable w_i of F, $i \in [n]$, we introduce two sets of wires V_i and V_i'. Each V_i' contains a *variable wire* v_i that has four swaps with λ and no swaps with λ'. Therefore, v_i intersects at least one and at most two λ'–λ loops. In order to prevent v_i from intersecting both a true- and a false-loop, we introduce two wires $\alpha_i \in V_i$ and $\alpha_i' \in V_i'$ with $\alpha_i < \lambda < \lambda' < \alpha_i' < v_i$; see Fig. 3. These wires neither swap with v_i nor with each other, but they have two swaps with both λ and λ'. We want to force α_i and α_i' to have the two true-loops on their right and the two false-loops on their left, or vice versa. This will ensure that v_i cannot reach both a true- and a false-loop.

To this end, we introduce, for $j \in [5]$, a β_i-*wire* $\beta_{i,j} \in V_i$ and a β_i'-*wire* $\beta_{i,j}' \in V_i'$. These are ordered $\beta_{i,5} < \beta_{i,4} < \cdots < \beta_{i,1} < \alpha_i$ and $\alpha_i' < \beta_{i,1}' < \beta_{i,2}' < \cdots < \beta_{i,5}' < v_i$. Every pair of β_i-wires as well as every pair of β_i'-wires swaps exactly once. Neither β_i- nor β_i'-wires swap with α_i or α_i'. Each β_i'-wire

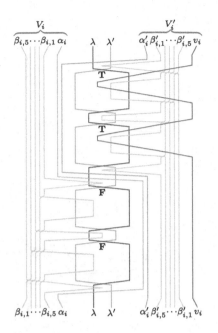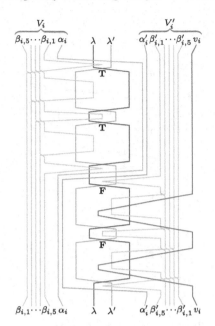

Fig. 3. A variable gadget with a variable wire v_i that corresponds to the variable that is true (left) or false (right). The λ-λ' loops are labeled **T** for true and **F** for false.

has four swaps with v_i. Moreover, $\beta_{i,1}, \beta_{i,3}, \beta_{i,5}, \beta'_{i,2}, \beta'_{i,4}$ swap with λ twice. Symmetrically, $\beta_{i,2}, \beta_{i,4}, \beta'_{i,1}, \beta'_{i,3}, \beta'_{i,5}$ swap with λ' twice; see Fig. 3.

We use the β_i- and β'_i-wires to fix the minimum number of λ'-λ loops that are on the left of α_i and on the right of α'_i, respectively. Note that, together with λ and λ', the β_i- and β'_i-wires have the same rigid structure as the wires shown in Fig. 2.

Observation 1 [5]. *The tangle in Fig. 2 realizes the list L_n specified there; all tangles that realize L_n have the same order of swaps along each wire.*

This means that there is a unique order of swaps between the β_i-wires and λ or λ', i.e., for $j \in [4]$, every pair of $\beta_{i,j+1}$-λ swaps (or $\beta_{i,j+1}$-λ' swaps, depending on the parity of j) can be done only after the pair of $\beta_{i,j}$-λ' swaps (or $\beta_{i,j}$-λ swaps, respectively). We have the same rigid structure on the right side with β'_i-wires. Hence, there are at least two λ'-λ loops to the left of α_i and at least two to the right of α'_i. Since α_i and α'_i do not swap, there cannot be a λ'-λ loop that appears simultaneously on both sides.

Note that the λ-λ' swaps that belong to the same side have to be consecutive, otherwise α_i or α'_i would need to swap more than twice with λ and λ'. Thus, there are only two ways to order the swaps among the wires α_i, α'_i, λ, λ'; the order is either α'_i-λ', α'_i-λ, four times λ-λ', α'_i-λ, α'_i-λ', α_i-λ, α_i-λ', four times λ-λ', α_i-λ', α_i-λ (see Fig. 3(left)) or the reverse (see Fig. 3(right)). It is easy to see

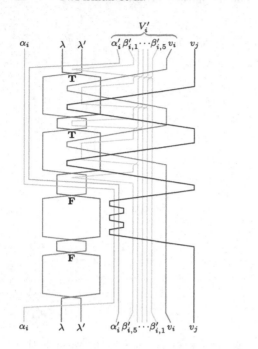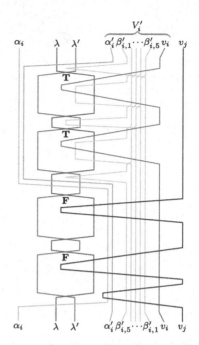

Fig. 4. A realization of swaps between the variable wire v_j and all wires that belong to the variable gadget corresponding to the variable w_i. On the left the variables w_i and w_j are both true, and on the right w_i is true, whereas w_j is false.

that in the first case v_i can reach only the first two $\lambda'-\lambda$ loops (the true-loops), and in the second case only the last two (the false-loops).

To avoid that the gadget for variable w_i restricts the proper functioning of the gadget for some variable w_j with $j > i$, we add the following swaps to L: for any $j > i$, α_j and α_j' swap with both V_i and V_i' twice, the β_j-wires swap with α_i' and V_i twice, and, symmetrically, the β_j'-wires swap with α_i and V_i' twice, v_j swaps with α_i and all wires in V_i' six times. We briefly explain these multiplicities. Wires from V_j and $V_j' \setminus \{v_j\}$ swap their partners twice so that they reach the corresponding $\lambda-\lambda'$ or $\lambda'-\lambda$ loops and go back. None of the wires from V_i or V_i' is restricted in which loop to intersect. Considering the wire v_j, note that it has to reach the $\lambda'-\lambda$ loops twice. For simplicity and in order not to have any conflicts with the β_i'-wires, we introduce exactly six swaps with α_i and all wires in V_i', see Fig. 4.

Clause Gadget. For every clause d_j from F, $j \in [m]$, we introduce a set of wires C_j. It contains the clause wire c_j that has eight swaps with λ'. We want to force each c_j to appear in all $\lambda'-\lambda$ loops. To this end, we use the set E with the seven φ-*wires* $\varphi_1, \ldots, \varphi_7$ ordered $\varphi_1 < \cdots < \varphi_7$. They create a rigid structure similar to the one of the β_i-wires. Each pair of φ-wires swaps exactly once. For each $k \in [7]$, if k is odd, then φ_k swaps twice with λ and twice with c_j for every

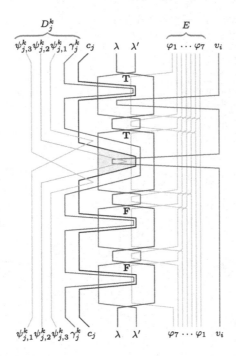

Fig. 5. A gadget for clause c_j showing only one of the three variables, namely v_i. The region shaded in yellow is the arm of c_j that is protected from other variables by γ_j^k. (Color figure online)

$j \in [m]$. If k is even, then φ_k swaps twice with λ'. Since c_j does not swap with λ, each pair of swaps between c_j and a φ-wire with odd index appears inside a λ'–λ loop. Due to the rigid structure, each of these pairs of swaps occurs in a different λ'–λ loop; see Fig. 5.

If a variable w_i belongs to a clause d_j, then L contains two v_i–c_j swaps. Since every clause has exactly three different positive variables, we want to force variable wires that belong to the same clause to swap with the corresponding clause wire in different λ'–λ loops. This way, every clause contains at least one true and at least one false variable if F is satisfiable.

We call a part of a clause wire c_j that is inside a λ'–λ loop—i.e., a λ'–c_j loop—an *arm of the clause c_j*. We want to "*protect*" the arm that is intersected by a variable wire from other variable wires. To this end, for every occurrence $k \in [3]$ of a variable in d_j, we introduce four more wires. The wire γ_j^k will protect the arm of c_j that the variable wire of the k-th variable of d_j intersects. Below we detail how to realize this protection. For now, just note that, in order not to restrict the choice of the λ'–λ loop, γ_j^k swaps twice with φ_ℓ for every odd $\ell \in [7]$. Similarly to c_j, the wire γ_j^k has eight swaps with λ' and appears once in every λ'–λ loop. Additionally, γ_j^k has two swaps with c_j.

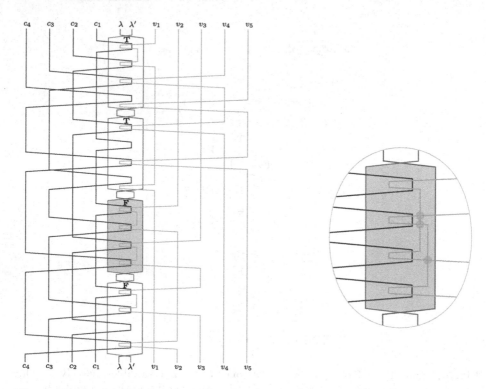

Fig. 6. Tangle obtained from the satisfiable formula $F = (w_1 \lor w_2 \lor w_3) \land (w_1 \lor w_3 \lor w_4) \land (w_2 \lor w_3 \lor w_4) \land (w_2 \lor w_3 \lor w_5)$. Here, w_1, w_4 and w_5 are set to true, whereas w_2 and w_3 are set to false. We show only λ, λ', and all variable and clause wires. Inset: problems that occur if variable wires swap with clause wires in a different order.

We force γ_j^k to protect the correct arm in the following way. Consider the λ'–λ loop where an arm of c_j swaps with a variable wire v_i. We want the order of swaps along λ' inside this loop to be fixed as follows: λ' first swaps with γ_j^k, then twice with c_j, and then again with γ_j^k. This would prevent all variable wires that do not swap with γ_j^k from reaching the arm of c_j. To achieve this, we introduce three ψ_j^k-wires $\psi_{j,1}^k, \psi_{j,2}^k, \psi_{j,3}^k$ with $\psi_{j,3}^k < \psi_{j,2}^k < \psi_{j,1}^k < \gamma_j^k$.

The ψ_j^k-wires also have the rigid structure similar to the one that β_i-wires have, so that there is a unique order of swaps along each ψ_j^k-wire. Each pair of ψ_j^k-wires swaps exactly once, $\psi_{j,1}^k$ and $\psi_{j,3}^k$ have two swaps with c_j, and $\psi_{j,2}^k$ has two swaps with λ' and v_i. Note that no ψ_j^k-wire swaps with γ_j^k. Also, since $\psi_{j,2}^k$ does not swap with c_j, the $\psi_{j,2}^k$–v_i swaps can appear only inside the λ'–c_j loop that contains the arm of c_j we want to protect from other variable wires. Since c_j has to swap with $\psi_{j,1}^k$ before and with $\psi_{j,3}^k$ after the $\psi_{j,2}^k$–λ' swaps, and since there are only two swaps between γ_j^k and c_j, there is no way for any variable wire except for v_i to reach the arm of c_j without also intersecting γ_j^k; see Fig. 5.

Finally, we consider the behavior of wires from different clause gadgets among each other and with respect to wires from variable gadgets. For every $\ell > k$ and for every $j \in [m]$, the wires c_j and γ_j^ℓ have eight swaps and the ψ_j^ℓ-wires have two swaps with all wires in C_j. Since all wires in V are to the left of all wires in C, each wire in C swaps twice with all wires in V and, for $i \in [n]$, with α_i'. Finally, all α- and α'-wires swap twice with each φ-wire.

Note that the order of the arms of the clause wires inside a λ'–λ loop cannot be chosen arbitrarily. If a variable wire intersects more than one clause wire, the arms of these clause wires occur consecutively, as for v_2 and v_3 in the shaded region in Fig. 6. If we had an interleaving pattern of variable wires (see inset), say v_2 first intersects c_1, then v_3 intersects c_2, then v_2 intersects c_3, and finally v_3 intersects c_4, then v_2 and v_3 would have to swap at least three times within the same λ'–λ loop. However, we have reserved only eight swaps for each pair of variable wires—two for each of the four λ'–λ loops.

Correctness. Clearly, if F is satisfiable, then there is a tangle obtained from F as described above that realizes the list L, so L is feasible; see Fig. 6 for an example. On the other hand, if there is a tangle that realizes the list L that we obtain from the reduction, then F is satisfiable. This follows from the rigid structure of a tangle that realizes L. The only flexibility is in which type of loop (true or false) a variable wire swaps with the corresponding clause wire. As described above, a tangle exists if, for each clause, the corresponding three variable wires swap with the clause wire in three different loops (at least one of which is a true-loop and at least one is a false-loop). In this case, the position of the variable wires yields a truth assignment satisfying F.

Membership in NP. To show that LIST-FEASIBILITY is in NP, we proceed as indicated in the introduction. Given a list $L = (l_{ij})$, we guess a list $L' = (l'_{ij})$ with $2(L) = 2(L')$ and $l'_{ij} \leq \min\{l_{ij}, n^2/4+1\}$ together with a permutation of its $O(n^4)$ swaps. Then we can efficiently test whether we can apply the swaps in this order to id $= \langle 1, 2, \ldots, n \rangle$. If yes, then the list L' is feasible (and, due to $L' \to L$, a witness for the feasibility of L), otherwise we discard it. By Proposition 2, L is feasible if and only if such a list L' exists and is feasible.

4   Counterexample to Conjecture 1

Recall that Conjecture 1 claims that every non-separable *even* list is feasible. We showed that all non-separable 0–2 lists up to $n = 8$ wires are feasible [5].

We now construct a family $(L_m^*)_{m \geq 1}$ of non-separable 0–2 lists such that L_m has 2^m wires and is not feasible for $m \geq 4$. We number the wires from 0 to $2^m - 1$. There is no swap between two wires $i < j$ in L_m^* if each 1 in the binary representation of i also belongs to the binary representation of j, that is, the bitwise OR of i and j equals j; otherwise, there are two swaps between i and j. E.g., for $m = 4$, wire $1 = 0001_2$ swaps twice with wire $2 = 0010_2$, but doesn't swap with wire $3 = 0011_2$.

Each list L_m^* is clearly non-separable: assume that there exists a swap between two wires $i = (i_1 i_2 \ldots i_m)_2$ and $k = (k_1 k_2 \ldots k_m)_2$ with $k > i+1$. Then there has to be some index a with $i_a = 1$ and $k_a = 0$. Consider any $j = (j_1 j_2 \ldots j_m)_2$ with $i < j < k$. By construction of L_m^*, if $j_a = 0$, then there are two swaps between i and j; if $j_a = 1$, then there are two swaps between j and k.

We confirmed by computer experiments (the Java source code is available on github [7]) that L_4^* – and hence all L_m^* with $m \geq 4$ – are infeasible. The list L_m^* has $\frac{1}{2} \sum_{r=1}^{m} 3^{r-1} 2^{m-r} (2^{m-r} - 1)$ swaps of multiplicity 2, so L_4^* has 55 distinct swaps. The full list L_4^* in list form and in matrix form is given below. Unfortunately, we could not (yet) find a combinatorial proof that the non-separable 0–2 list L_4^* is not feasible.

Theorem 4. *Conjecture 1 is false.*

$$L_4^* = 2\,\{(1,2),(1,4),(1,6),(1,8),(1,10),(1,12),(1,14),(2,4),(2,5),(2,8),(2,9),$$
$$(2,12),(2,13),(3,4),(3,5),(3,6),(3,8),(3,9),(3,10),(3,12),(3,13),$$
$$(3,14),(4,8),(4,9),(4,10),(4,11),(5,6),(5,8),(5,9),(5,10),(5,11),$$
$$(5,12),(5,14),(6,8),(6,9),(6,10),(6,11),(6,12),(6,13),(7,8),(7,9),$$
$$(7,10),(7,11),(7,12),(7,13),(7,14),(9,10),(9,12),(9,14),(10,12),$$
$$(10,13),(11,12),(11,13),(11,14),(13,14)\}$$

$$L_4^* = \begin{pmatrix}
0 & 0 & 0 & 0 & 0 & 0 & 0 & 0 & 0 & 0 & 0 & 0 & 0 & 0 & 0 & 0 \\
0 & 0 & 2 & 0 & 2 & 0 & 2 & 0 & 2 & 0 & 2 & 0 & 2 & 0 & 2 & 0 \\
0 & 0 & 0 & 0 & 2 & 2 & 0 & 0 & 2 & 2 & 0 & 0 & 2 & 2 & 0 & 0 \\
0 & 0 & 0 & 0 & 2 & 2 & 2 & 0 & 2 & 2 & 2 & 0 & 2 & 2 & 2 & 0 \\
0 & 0 & 0 & 0 & 0 & 0 & 0 & 0 & 2 & 2 & 2 & 2 & 0 & 0 & 0 & 0 \\
0 & 0 & 0 & 0 & 0 & 0 & 2 & 0 & 2 & 2 & 2 & 2 & 2 & 0 & 2 & 0 \\
0 & 0 & 0 & 0 & 0 & 0 & 0 & 0 & 2 & 2 & 2 & 2 & 2 & 2 & 0 & 0 \\
0 & 0 & 0 & 0 & 0 & 0 & 0 & 0 & 2 & 2 & 2 & 2 & 2 & 2 & 2 & 0 \\
0 & 0 & 0 & 0 & 0 & 0 & 0 & 0 & 0 & 0 & 0 & 0 & 0 & 0 & 0 & 0 \\
0 & 0 & 0 & 0 & 0 & 0 & 0 & 0 & 0 & 0 & 2 & 0 & 2 & 0 & 2 & 0 \\
0 & 0 & 0 & 0 & 0 & 0 & 0 & 0 & 0 & 0 & 0 & 0 & 2 & 2 & 0 & 0 \\
0 & 0 & 0 & 0 & 0 & 0 & 0 & 0 & 0 & 0 & 0 & 0 & 2 & 2 & 2 & 0 \\
0 & 0 & 0 & 0 & 0 & 0 & 0 & 0 & 0 & 0 & 0 & 0 & 0 & 0 & 0 & 0 \\
0 & 0 & 0 & 0 & 0 & 0 & 0 & 0 & 0 & 0 & 0 & 0 & 0 & 0 & 2 & 0 \\
0 & 0 & 0 & 0 & 0 & 0 & 0 & 0 & 0 & 0 & 0 & 0 & 0 & 0 & 0 & 0 \\
0 & 0 & 0 & 0 & 0 & 0 & 0 & 0 & 0 & 0 & 0 & 0 & 0 & 0 & 0 & 0
\end{pmatrix}$$

Acknowledgments. We thank Stefan Felsner for discussions about the complexity of LIST-FEASIBILITY.

References

1. Bereg, S., Holroyd, A., Nachmanson, L., Pupyrev, S.: Representing permutations with few moves. SIAM J. Discrete Math. **30**(4), 1950–1977 (2016). https://arxiv.org/abs/1508.03674, https://doi.org/10.1137/15M1036105
2. Bereg, S., Holroyd, A.E., Nachmanson, L., Pupyrev, S.: Drawing permutations with few corners. In: Wismath, S., Wolff, A. (eds.) GD 2013. LNCS, vol. 8242, pp. 484–495. Springer, Cham (2013). https://doi.org/10.1007/978-3-319-03841-4_42
3. Birman, J.S., Williams, R.F.: Knotted periodic orbits in dynamical systems—I: Lorenz's equation. Topology **22**(1), 47–82 (1983). https://doi.org/10.1016/0040-9383(83)90045-9
4. Crutchfield, J.P., Farmer, J.D., Packard, N.H., Shaw, R.S.: Chaos. Sci. Am. **254**(12), 46–57 (1986)
5. Firman, O., Kindermann, P., Ravsky, A., Wolff, A., Zink, J.: Computing height-optimal tangles faster. In: Archambault, D., Tóth, C.D. (eds.) GD 2019. LNCS, vol. 11904, pp. 203–215. Springer, Cham (2019). https://doi.org/10.1007/978-3-030-35802-0_16
6. Kauffman, S.A.: The Origins of Order: Self-Organization and Selection in Evolution. Oxford University Press, Oxford (1993)
7. Kindermann, P., Zink, J.: Java and Python code for computing tangles. Github repository (2022). https://github.com/PhKindermann/chaotic-attractors
8. Knuth, D.E.: The Art of Computer Programming, Volume III: Sorting and Searching, 2nd edn. Addison-Wesley, Boston (1998). https://www.worldcat.org/oclc/312994415
9. Mindlin, G., Hou, X.-J., Gilmore, R., Solari, H., Tufillaro, N.B.: Classification of strange attractors by integers. Phys. Rev. Lett. **64**, 2350–2353 (1990). https://doi.org/10.1103/PhysRevLett.64.2350
10. Olszewski, M., et al.: Visualizing the template of a chaotic attractor. In: Biedl, T., Kerren, A. (eds.) GD 2018. LNCS, vol. 11282, pp. 106–119. Springer, Cham (2018). https://doi.org/10.1007/978-3-030-04414-5_8
11. Sado, K., Igarashi, Y.: A function for evaluating the computing time of a bubbling system. Theor. Comput. Sci. **54**, 315–324 (1987). https://doi.org/10.1016/0304-3975(87)90136-8
12. Schaefer, T.J.: The complexity of satisfiability problems. In: Proceedings of 10th Annual ACM Symposium on Theory Computing (STOC'78), pp. 216–226 (1978). https://doi.org/10.1145/800133.804350
13. Stewart, I.: Does God Play Dice?: The New Mathematics of Chaos. Penguin (1997)
14. Turán, P.: On an external problem in graph theory. Mat. Fiz. Lapok **48**, 436–452 (1941)
15. Wang, D.C., Novel routing schemes for IC layout part I: two-layer channel routing. In: Proceedings of 28th ACM/IEEE Design Automation Conference (DAC 1991), pp. 49–53 (1991). https://doi.org/10.1145/127601.127626
16. Yamanaka, K., Horiyama, T., Uno, T., Wasa, K.: Ladder-lottery realization. In: Proceedings of 30th Canadian Conference on Computational Geometry (CCCG), pp. 61–67 (2018). http://www.cs.umanitoba.ca/~cccg2018/papers/session2A-p3.pdf

A Spectral Algorithm for Finding Maximum Cliques in Dense Random Intersection Graphs

Filippos Christodoulou[1](✉)[iD], Sotiris Nikoletseas[2,3][iD],
Christoforos Raptopoulos[3][iD], and Paul G. Spirakis[2,4][iD]

[1] Gran Sasso Science Institute, L'Aquila, Italy
`filippos.christodoulou@gssi.it`
[2] Computer Engineering and Informatics Department,
University of Patras, Patras, Greece
`{nikole,raptopox}@ceid.upatras.gr`
[3] Computer Technology Institute & Press Diophantus (CTI), Patras, Greece
[4] Department of Computer Science, University of Liverpool, Liverpool, UK
`P.Spirakis@liverpool.ac.uk`

Abstract. In a random intersection graph $G_{n,m,p}$, each of n vertices selects a random subset of a set of m labels by including each label independently with probability p and edges are drawn between vertices that have at least one label in common. Among other applications, such graphs have been used to model social networks, in which individuals correspond to vertices and various features (e.g. ideas, interests) correspond to labels; individuals sharing at least one common feature are connected and this is abstracted by edges in random intersection graphs. In this paper, we consider the problem of finding maximum cliques when the input graph is $G_{n,m,p}$. Current algorithms for this problem are successful with high probability only for relatively sparse instances, leaving the dense case mostly unexplored. We present a spectral algorithm for finding large cliques that processes vertices according to respective values in the second largest eigenvector of the adjacency matrix of induced subgraphs of the input graph corresponding to common neighbors of small cliques. Leveraging on the Single Label Clique Theorem from [16], we were able to construct random instances, without the need to externally plant a large clique in the input graph. In particular, we used label choices to determine the maximum clique and then concealed label information by just giving the adjacency matrix of $G_{n,m,p}$ as input to the algorithm. Our experimental evaluation showed that our spectral algorithm clearly outperforms existing polynomial time algorithms, both with respect to the failure probability and the approximation guarantee metrics, especially in the dense regime, thus suggesting that spectral properties of

Christoforos Raptopoulos was supported by the Hellenic Foundation for Research and Innovation (H.F.R.I.) under the "2nd Call for H.F.R.I. Research Projects to support Post-Doctoral Researchers" (Project Number: 704).
Paul Spirakis was supported by the NeST initiative of the EEE and CS of the University of Liverpool and by the EPSRC grant EP/P02002X/1.

L. Gąsieniec (Ed.): SOFSEM 2023, LNCS 13878, pp. 18–32, 2023.
https://doi.org/10.1007/978-3-031-23101-8_2

random intersection graphs may be also used to construct efficient algorithms for other NP-hard graph theoretical problems as well.

Keywords: Random intersection graphs · Maximum cliques · Heuristics

1 Introduction

A *clique* in an undirected graph G is a subset of vertices any two of which are connected by an edge. The problem of finding the maximum clique in an arbitrary graph is fundamental in Theoretical Computer Science and appears in many different settings. As an example, consider a social network where vertices represent people and edges represent mutual acquaintance. Finding a maximum clique in this network corresponds to finding the largest subset of people who all know each other. More generally, the analysis of large networks in order to identify communities, clusters, and other latent structure has come to the fore-front of much research. The Internet, social networks, bibliographic databases, energy distribution networks, and global networks of economies are some of the examples motivating the development of the field.

From a computational complexity point of view, it is well known that determining the size of the largest clique of an arbitrary graph of n vertices is NP-complete [13]. This fact is further strengthened in [11], showing that, if k is the size of the maximum clique, then the clique problem cannot be solved in time $n^{o(k)}$, unless the exponential time hypothesis fails. Additionally, there are several results on hardness of approximation which suggest that there can be no approximation algorithm with an approximation ratio significantly less than linear (see e.g. [10]).

The intractability of the maximum clique problem for arbitrary graphs lead researchers to the study of the problem for appropriately generated random graphs. In particular, for Erdős-Rényi random graphs $G_{n,\frac{1}{2}}$ (i.e. random graphs of n vertices, in which each edge appears independently with probability $\frac{1}{2}$), there are several greedy algorithms that find a clique of size about $\log_2 n$ with high probability (whp, i.e. with probability that tends to 1 as n goes to infinity), see e.g. [9,14]. Since the clique number of $G_{n,\frac{1}{2}}$ is asymptotically equal to $2\log_2 n$ with high probability, these algorithms approximate the clique number by a factor of 2. It has been conjectured that finding a clique of size $(1+\Theta(1))\log_2 n$, in a random graph instance $G_{n,\frac{1}{2}}$, in which we have planted a randomly chosen clique of size $n^{0.49}$, with at least constant probability, would require techniques beyond the current limits of complexity theory. This conjecture seems to identify a certain bottleneck for the problem; finding the maximum clique in the case where the planted clique has size at least $\sqrt{n}$ can be done in polynomial time by using spectral properties of the adjacency matrix of the graph (see [1]).

In this paper, we consider random instances of the random intersection graphs model (introduced in [12,19]) as input graphs. In this model, denoted by $\mathcal{G}_{n,m,p}$, each one of m labels is chosen independently with probability p by each one of n vertices, and there are edges between any vertices with overlaps in the labels chosen. One of the most interesting results regarding this model is that, when the number of labels is sufficiently large (in particular, when $m = n^\alpha, \alpha \geq 3$) the random intersection graphs model is equivalent to the Erdős-Rényi random graphs model (in the sense that the total variation distance between the two spaces tends to 0; see [7,18]). Random intersection graphs are relevant to and capture quite nicely social networking. Indeed, a social network is a structure made of nodes (individuals or organizations) tied by one or more specific types of interdependency, such as values, visions, financial exchange, friends, conflicts, web links etc. Social network analysis views social relationships in terms of nodes and ties. Nodes are the individual actors within the networks and ties are the relationships between the actors. Other applications include oblivious resource sharing in a (general) distributed setting, efficient and secure communication in sensor networks [15], interactions of mobile agents traversing the web etc. For recent research related to random intersection graphs we refer the interested reader to the surveys [3,4] and references therein.

1.1 Previous Work on Maximum Cliques in Random Intersection Graphs

In [19], the authors used the first moment probabilistic method to provide a lower bound on the clique number of random instance of $\mathcal{G}_{n,m,p}$ in the case where mp^2 tends to a constant as $n \to \infty$. In [16,17] this range of values was considerably extended and a precise characterization of maximum cliques was given in the case where $m = n^\alpha, \alpha < 1$ and $p = O(m^{-1/2})$. In particular, the Single Label Clique Theorem was proved, indicating that, with high probability any clique Q of size $|Q| \sim np$ in a random instance of $\mathcal{G}_{n,m,p}$ (and thus also the maximum clique) is formed by a single label. However, these structural results are existential and thus do not lead to algorithms for finding the maximum clique. It is worth noting that, the equivalence results between the random intersection graphs model and the Erdős-Rényi random graphs model for large number of vertices suggest that the problem of finding a maximum clique in a random instance of $\mathcal{G}_{n,m,p}$ in this range of values should not be any easier in the former that it is in the latter. On the other hand, in the range of values $m = n^\alpha, \alpha < 1$ and $p = O(m^{-2/3})$, greedy algorithms for finding large cliques in random intersection graphs were presented in the work [5]. The first algorithm in that paper, referred as GREEDY-CLIQUE, finds a clique of the optimal order in a random instance of $\mathcal{G}_{n,m,p}$ with high probability, in the case where the asymptotic degree distribution is a power-law with exponent within $(1,2)$. The algorithm considers vertices in decreasing order of degree and greedily constructs a clique by extending by a vertex if and only if the latter is connected to all other vertices already included; it can be implemented to run in expected time $O(n^2)$. In the same paper [5], in the case where the input graph is a random instance of $\mathcal{G}_{n,m,p}$ with bounded degree variance, a second greedy algorithm, named MONO-CLIQUE, was suggested, which can be implemented to run in expected time $O(n)$. The main idea of this algorithm is to try and construct a large clique directly by considering common neighbours of endpoints of every edge in the graph. The pseudocodes for GREEDY-CLIQUE and MONO-CLIQUE can be found in Appendixes 7.1 and 7.2 respectively.

In [2] a more general greedy algorithm was presented, namely the Maximum-Clique Algorithm, which constructs a large clique by considering the common neighborhood of vertex subsets of fixed size k (i.e. independent of n) and checking whether it forms a clique. From the cliques found in this way, it takes the largest ones in order to cover the graph. This algorithm finds maximum cliques whp for a wider range of parameters of the model (but still within the sparse regime) than both algorithms GREEDY-CLIQUE and MONO-CLIQUE, at the cost of larger running time. In particular, the Maximum-Clique Algorithm outputs a maximum clique in a random instance of $\mathcal{G}_{n,m,p}$ with $m = n^\alpha, \alpha < 1$ and $\ln^2 n/n \leq p = O(m^{-2/3})$, with high probability. Since in this paper we consider metrics regarding the ability of an algorithm to find large cliques (namely failure probability and approximation guarantee), we use the Maximum-Clique Algorithm as a benchmark in relation to which we evaluate our spectral algorithm. In fact, we use a slightly more efficient version where we directly exclude k-subsets of vertices that are not complete, in order to significantly reduce the $\binom{n}{k}$ factor corresponding to the number of all k-sets in the running time. The pseudocode of the benchmark algorithm is shown in Appendix 7.3. Different pruning ideas for reducing the running time of greedy algorithms for finding large cliques through the reduction of the size of the input graph, have been considered in [8].

2 Our Contribution

In this paper we consider the problem of finding maximum cliques when the input graph is $G_{n,m,p}$. We present a spectral algorithm for finding large cliques that processes vertices according to respective values in the second largest eigenvector of the adjacency matrix of carefully selected induced subgraphs of the input graph created by common neighborhoods of small (constant size) k-cliques. Because of the computation of the spectral decomposition, the running time of our algorithm is larger than greedy algorithms in the relevant literature, but it succeeds with higher probability in finding large cliques. In particular, we compared our algorithm with the most efficient version of the Maximum-Clique Algorithm from [2]. Leveraging on the Single Label Clique Theorem from [16], we were able to avoid the construction of artificial input graph instances with known planted large cliques. In particular, we used label choices to determine the maximum clique and then concealed label information by just giving the adjacency matrix of $G_{n,m,p}$ as input to the algorithm. Our experimental evaluation showed that, as we move from sparse instances to denser ones, both metrics regarding the failure probability of our algorithm as well as the approximation guarantee (when the maximum clique is not found) are much better than the corresponding values for the Maximum-Clique algorithm. This difference is especially highlighted as we move from sparser instances to denser ones, in which there is no guarantee that greedy algorithms will succeed (but the Single Label Clique Theorem still holds) and also as the size of the k-cliques used for creating induced subgraphs increases (yet remains a relatively small constant, e.g. $k = 6, 7, 8$). We believe that our current paper suggests that spectral properties of random intersection graphs may be used to construct efficient algorithms for other NP-hard graph theoretical problems as well.

3   Definitions, Notation and Useful Results

Given an undirected graph G, we denote by $V(G)$ and $E(G)$ the set of vertices and the set of edges respectively. Edges of G will be denoted as 2-sets; two vertices v, u are connected in G if and only if $\{u, v\} \in E(G)$. For any vertex $v \in V(G)$, we denote by $N(v) = N_G(v)$ the set of neighbours of v in G, namely $N(v) \stackrel{\text{def}}{=} \{u \in V(G) : \{u, v\} \in E(G)\}$. In addition, for any subset of vertices $S \subseteq V$, we denote by $N(S)$ the set of vertices having at least one neighbor in S. We denote by $\deg(v) = |N(v)|$ the degree of v. For any subset $S \subseteq V$, we denote by $G[S]$ the induced subgraph of G on S, namely $G[S] = (S, \{\{u, v\} \in E(G) : v, u \in S\})$. Given an arbitrary ordering of the vertices, say $v_1, v_2, \ldots, v_{|V|}$, the adjacency matrix A_G of G is an $|V| \times |V|$ matrix where $A_G[i, j] = 1$ if $\{v_i, v_j\} \in E(G)$ and $A_G[i, j] = 0$ otherwise. An eigenvector of $\mathbf{A}_G$ with corresponding eigenvalue λ is a vector $\mathbf{x}$ for which $\mathbf{A}_G \mathbf{x} = \lambda \mathbf{x}$. Since by definition $\mathbf{A}_G$ is symmetric, it has $|V|$ real eigenvalues $\lambda_1 \geq \lambda_2 \geq \cdots \geq \lambda_{|V|}$, with orthogonal corresponding eigenvectors $\mathbf{x}^{(1)}, \mathbf{x}^{(2)}, \ldots, \mathbf{x}^{(|V|)}$.

The formal definition of the random intersection graphs model is as follows:

Definition 1 (Random Intersection Graph - $\mathcal{G}_{n,m,p}$ [12,19]). *Consider a universe $\mathcal{M} = \{1, 2, \ldots, m\}$ of labels and a set of n vertices V. Assign independently to each vertex $v \in V$ a subset S_v of $\mathcal{M}$, choosing each element $\ell \in \mathcal{M}$ independently with probability p and draw an edge between two vertices $v \neq u$ if and only if $S_v \cap S_u \neq \emptyset$. The resulting graph is an instance $G_{n,m,p}$ of the random intersection graphs model.*

In this model we also denote by L_ℓ the set of vertices that have chosen label $\ell \in \mathcal{M}$. Given $G_{n,m,p}$, we refer to $\{L_\ell, \ell \in \mathcal{M}\}$ as its *label representation*. Furthermore, the bipartite graph with vertex set $V \cup \mathcal{M}$ and edge set $\{(v, \ell) : \ell \in S_v\} = \{(v, \ell) : v \in L_\ell\}$ is the *bipartite random graph* $B_{n,m,p}$ *associated to* $G_{n,m,p}$. Notice that the associated bipartite graph is uniquely defined by the label representation.

Given a graph G, a *clique* is a set of vertices every two of which are connected by an edge; the size of the maximum clique in G is its *clique number*. Notice that, by definition, for any ℓ the set of vertices within L_ℓ forms a clique in $G_{n,m,p}$. Furthermore, the expected size of such a clique is $\mathbb{E}[|L_\ell|] = np$. Observe that edges of cliques in $G_{n,m,p}$ may be formed by different labels. However, when the number of labels is smaller than the number of vertices, the following theorem states that, under mild conditions, with high probability, in any large enough clique of $G_{n,m,p}$, edges are formed by a single label.

Theorem 1 (Single Label Clique Theorem [17]). *Let $G_{n,m,p}$ be a random instance of the random intersection graphs model with $m = n^\alpha, 0 < \alpha < 1$ and $mp^2 = O(1)$. Then whp, any clique Q of size $|Q| \sim np$ in $G_{n,m,p}$ is formed by a single label. In particular, the maximum clique is formed by a single label.*

Leveraging on the above theorem, in our experiments we avoid the artificial construction of graph instances with planted cliques. In particular, during the construction of the random intersection graph, we use its label representation to find a set L_ℓ of maximum cardinality; by the above theorem, this will correspond to a maximum clique, and its size will be the clique number of $G_{n,m,p}$ with high probability. Subsequently, we hide the label representation and give the constructed graph $G_{n,m,p}$ as input to the algorithms that we consider in our experimental evaluation (i.e. just the vertex and edge sets).

3.1 Range of Values for m, n, p

It follows from the definition of the model that the edges in $G_{n,m,p}$ are not independent. In particular, the (unconditioned) probability that a specific edge exists is $1 - (1 - p^2)^m$. Therefore, when $mp^2 = o(1)$, the expected number of edges of $G_{n,m,p}$ is $(1 + o(1))\binom{n}{2}mp^2$. For the range of values $p = O(m^{-2/3}), m = n^\alpha, \alpha < 1$, where the Maximum-Clique Algorithm of [2] is guaranteed to output a large enough clique whp, this becomes $O(n^2 m^{-1/3}) = O(n^{2-\alpha/3})$. On the other end, when $mp^2 = \omega(1)$ then the graph is almost complete. In view of this, we will refer to the range of values $m = n^\alpha, \alpha < 1, mp^2 = \Omega(m^{-2/3})$ as the dense regime, noting that the Single Label Clique Theorem continues to hold in this range of values.

4 The Spectral Algorithm

We can now give the details of *Spectral-Max-Clique* algorithm. Inspired by the algorithm in [1], our algorithm takes as input the graph $G_{n,m,p}$, the size k of a witness k-clique (i.e. a small clique that is assumed to belong to the maximum clique) and a parameter t, which is a lower bound on the maximum clique size in $G_{n,m,p}$ (recall that, by Theorem 1, when $m = n^\alpha, \alpha < 1$, any clique Q with size $|Q| \sim np$ in $G_{n,m,p}$ is formed only by a single label; since np is the expected size of $\mathbb{E}[|L_\ell|]$, for any $\ell \in \mathcal{M}$, we set $t = np$). The main difference between our algorithm and the algorithm of [1] is a kind of preprocessing on the input graph, which is done at step 3 of the algorithm; in particular, since the input graph $G_{n,m,p}$ has many large cliques of size np (in fact, by Theorem 1, it has exactly m whp), we work on the induced graph H which has fewer (ideally exactly one) large cliques, namely the ones including S.

At the beginning of the execution, we initialize an empty set M, which at the end of the execution will be the output of the algorithm (the maximum clique of the graph). The algorithm enters a for-loop to be repeated as many times as the number of subsets $S \subseteq V$ of size k in the $G_{n,m,p}$. At every iteration of the for-loop, we construct the induced graph H which contains all the vertices of the subset S of the original graph $G_{n,m,p}$ as well as all the neighbors of the vertices in subset S (namely $H = G[S \cup N(S)]$). We then find the adjacency matrix $\mathbf{A}_H$ of H and we find the eigenvector corresponding to the second largest eigenvalue, namely $\mathbf{x}^{(2)}$; the latter can be done in polynomial time. The algorithm then sorts the vertices of H in decreasing order of the absolute values of the corresponding coordinates in the second eigenvector $\mathbf{x}^{(2)}$, where equalities are broken arbitrarily. Subsequently, we consider only the first t vertices in this ordering and store them in an empty set W. We then define an empty set Q, where the clique (not necessarily the maximum) will be added. Afterward, for every vertex v in H, the algorithm checks whether v has at least $3np/4$ neighbors in the set W and exactly $|Q|$ neighbors in the set Q. If the two conditions are true, v is included in the set Q. In the end, we check if the size of the newly added clique Q is largest from the size of the existing clique in set M ($|Q| \geq |M|$) and finally the maximum clique M is returned by the algorithm. The main heuristic idea why this algorithm works as intended is that, most of the time, the second eigenvector $\mathbf{x}^{(2)}$ of $\mathbf{A}_H$ can be used to find a big portion of the largest clique; intuitively this happens because the maximum clique will be by far the largest most dense induced subgraph of the graph, and this will be depicted in the (absolute) values of the corresponding positions of the second eigenvector (in the extreme case where the n-vertex graph consists only of a k-sized clique Q, the only positions where an eigenvector corresponding to the second largest eigenvalue can have non-zero elements is on the positions corresponding to the vertices of Q); see also [1] for a theoretical explanation why this heuristic works in the planted clique model. Therefore, since the algorithm checks all of the subsets of V of size k, in some step it will reach a subset S, which belongs to the maximum clique M. Our experimental evaluation shows that at this iteration our algorithm succeeds in finding the largest clique of the graph in most cases. The pseudocode of our algorithm is shown below.

Algorithm 1. Spectral-Max-Clique

Input: Random instance of $\mathcal{G}_{n,m,p}$, parameters $k \in \mathbb{N}$, $t = np$
Output: Clique M of $G_{n,m,p}$
 1: $M = \emptyset$;
 2: **foreach** subset $S \subseteq V, |S| = k$ **do**
 3:     Construct the induced graph $H = G[S \cup N(S)]$;
 4:     Compute the eigenvector $\mathbf{x}^{(2)}$ corresponding to the second largest eigenvalue of $\mathbf{A}_H$;
 5:     Sort the vertices of H in decreasing order of the absolute values of corresponding coordinates in $\mathbf{x}^{(2)}$;
 6:     Let W be the first t vertices in this ordering;
 7:     Set $Q = \emptyset$;
 8:     **foreach** $v \in H$ **do**
 9:         **if** v has at least $3np/4$ neighbors in W and $|Q|$ neighbors in Q **then**
10:             $Q = Q \cup \{v\}$;
11:         **end if**

```
12:     end for
13:     if |Q| > |M| then
14:         Set M = Q;
15:     end if
16: end for
17: return M
```

4.1 Running Time of Our Algorithm

We note that the outer for-loop of our algorithm runs for $\binom{n}{k}$ times. Furthermore, for a given k-set S, steps 3, 4 and 5 take $O(n^3)$ time, with step 4 regarding spectral decomposition being the most expensive (in theory, spectral decomposition can be done more efficiently in $O(n^{2.4})$, but here we use the time complexity of most practical implementations). Finally, it is easy to see that the for-loop in steps 8 to 12 runs in $O(n^3)$ time, while all other steps are either direct assignments, definition of easily checked conditions and thus take $O(1)$ time. Overall, the running time of our algorithm is $\binom{n}{k} \cdot O(n^3)$. Clearly the most time consuming factor is the number of times that the outer for-loop is running in order to find a good enough starting k-set S. A similar situation arises also in the algorithm Maximum clique (see Appendix 7.3), whose running time is $\binom{n}{k} \cdot O(n^2)$. To allow for the algorithms considered and evaluated in our paper to run for larger values of k in the experimental evaluation, we assume, without loss of generality, that a suitable k-set is known from the start, thus avoiding the $\binom{n}{k}$ factor in the running time. This is where Theorem 1 becomes useful, since a suitable k-set S can be any subset of a (single label) maximum clique.

5 Experimental Evaluation

This section is devoted to the presentation of our experimental results regarding the comparison of the algorithms *Spectral-Max-Clique* and *Maximum-Clique* with respect to two metrics, namely failure probability and approximation guarantee. In particular, the failure probability is defined as the probability that an algorithm fails to find the maximum clique; in our experimental evaluation this probability is approximated by the fraction of the number of independent instances of random intersection graphs where an algorithm fails to find a maximum clique. The approximation guarantee is defined as the fraction of the clique found by an algorithm over the size of a maximum clique; in our experimental evaluation this is approximated by the average of the corresponding fractions achieved by an algorithm for various independent instances of random intersection graphs.

The number of $G_{n,m,p}$ graph instances that have been given as input to the algorithms for small values of k ($k = 1, 2, 3$) were 2000. However, it is worth noting that, as we increase the value of k, the computational resources required also increases, because the dependency on k becomes more prevalent even for straightforward steps of the algorithm 1 (as for example in step 3 for constructing the induced subgraph H). In particular, the number of independent graph instances used for $k = 4, 5, 6, 7, 8$ were 1600, 1400, 800, 700, 500 respectively.

In this section we provide our experimental results between the two algorithms for finding the maximum clique that we have considered and secondly we show an approximation guarantee (denoted by *fraction* variable in the Figs. 4, 5 and 6 of the output maximum clique found by each algorithm over the maximum clique of the graph. We present a comparison between *Spectral-Max-Clique* and *Maximum-Clique* algorithm. Further experimental results for different values of k and α can be found in the full version of our paper [6].

Our computing platform is a machine with AMD Ryzen Threadripper 3970X at 3.7 GHz, 32 cores, 256 GB RAM, GPU with 2x NVIDIA GeForce RTX 3080 10GB and running Ubuntu Linux version 20.04.2 LTS. The code has been written in Python 3. The code for repeating the experiments is available here.

The goal of our experimental evaluation is to verify whether the *Spectral-Max-Clique* algorithm performs better, meaning that it succeeds to find the maximum clique in a $G_{n,m,p}$, in comparison with the *Maximum-Clique*. As we already mentioned, this happens for dense graphs. In our experiments, we ran both algorithms for different instances of $G_{n,m,p}$. Specifically, we use three different values of parameter α, with $\alpha = 1/3, 2/3, 1$ and for the number of nodes (n), we set n equal to 1000 and 3000. Regarding parameter p, we cover a different range of values for each experiment in order to test sparse instances as well as dense ones. It is important to explain how we understand that the output of each algorithm is actually the maximum clique in the graph. We take a $G_{n,m,p}$ instance with the above parameters and find the heaviest label (i.e. the label with the largest number of vertices), call it $\ell_{\max}$. By Theorem 1, the set of vertices in $L_{\ell_{\max}}$ form a maximum clique in $G_{n,m,p}$ whp, so in this way there is no need to externally plant a known large enough clique in the graph. Then, we run the algorithms on this instance, but we give only the graph as input (i.e. the algorithm is unaware of the specific label choices). We say that the algorithms fail if they do not find a clique at least as large as $|L_{\ell_{\max}}|$. Note that this is a strict condition, namely, even finding a clique of size $|L_{\ell_{\max}}| - 1$ is considered a failure. In each case, we gradually increase the selection probability p, in order to highlight that the failure probability curve of *Spectral-Max-Clique* is much lower than the failure probability curve of *Maximum-Clique*, especially when the input graphs become denser.

It is worth noting that, the selection of a correct starting set S of k vertices in Step 1 of *Spectral-Max-Clique* pseudocode, implies a multiplicative $\Theta(n^k)$ factor on the running time of our algorithm. Even though for constant k this remains polynomially bounded, in order to allow our experiments to run for large values of n and k, we have assumed that the initial set of k vertices is always chosen from those in the maximum clique.

The Figs. 1,2 and 3 show the failure probability of each algorithm, when p increases, meaning that the $G_{n,m,p}$ becomes denser. These experiments show that for smaller values of k and p, the two algorithms perform in a similar manner;

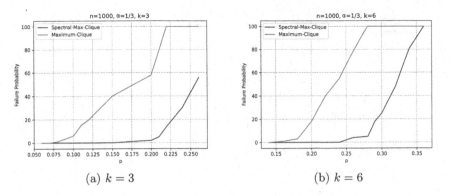

(a) $k = 3$ (b) $k = 6$

Fig. 1. Failure probability curves for $\alpha = 1/3$, $n = 1000$ and $k = 3, 6$.

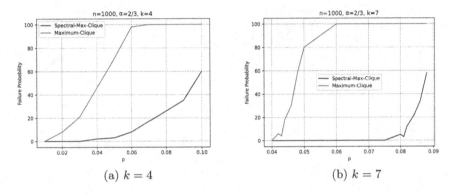

(a) $k = 4$ (b) $k = 7$

Fig. 2. Failure probability curves for $\alpha = 2/3$, $n = 1000$ and $k = 4, 7$.

they both find successfully the maximum clique in the graph. This is true for all the different values of parameter a. Indeed, as it is demonstrated in Fig. 2b, when $k = 7$, the *Spectral-Max-Clique* algorithm has failure probability close to 0 for the smaller values of p, while *Maximum-Clique* fails to find the maximum clique in almost all the cases, with failure probability close to 100%. One more example is Fig. 1b, where the failure probability of *Maximum-Clique* starts at $p \approx 0.17$ and increases as the graph gets denser, and fails in all of the cases to find the maximum clique when $p \approx 0.27$. On the other hand, the probability of failure of *Spectral-Max-Clique* begins when $p \approx 0.25$ and fails in all cases when $p \approx 0.36$. From all the figures it is obvious that the failure probability of the spectral algorithm also increases but slower than the failure probability of *Maximum-Clique*.

It is also interesting to demonstrate how far the resulting clique of each algorithm is from the maximum clique of the graph. For that reason, we ran experiments in the cases where both algorithms fail. In particular, Figs. 4,5 and 6 show the curves of the average of the fraction of the clique size found by the algorithms over the maximum clique size of the input graph. By studying these

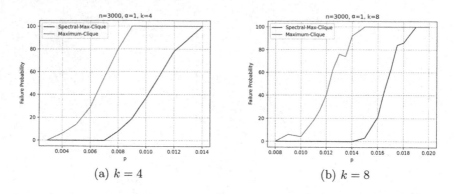

Fig. 3. Failure probability curves for $\alpha = 1$, $n = 3000$ and $k = 4, 8$.

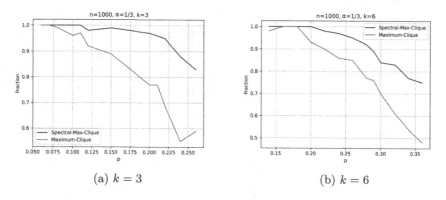

Fig. 4. Approximation guarantee curves for $\alpha = 1/3$, $n = 1000$ and $k = 3, 6$.

figures, we can observe that for all the different values of parameters α and k, the output clique of *Spectral-Max-Clique* algorithm is closer to the maximum clique of the graph with respect to the output clique of *Maximum-Clique* algorithm; the size of the clique that the *Spectral-Max-Clique* algorithm finds is closer to the size of the maximum clique of the graph. For instance, for the case $a = 1/3$ and $k = 3$, Fig. 4a, when $p \approx 0.250$ and the graph is denser, the approximation guarantee for *Spectral-Max-Clique* is $fraction \approx 0.82$ while for *Maximum-Clique* is $fraction \approx 0.57$. One more apparent example is Fig. 6b, when $a = 1$ and $k = 8$. In this case, $fraction \approx 0.6$ for *Spectral-Max-Clique* algorithm, although for *Maximum-Clique*, $fraction \approx 0.15$. Therefore, the former algorithm finds more than half of the maximum clique while the latter fails to find approximately 85% of the maximum clique of the graph.

It should be noted that, as the value of α gets closer to 0, we were only able to run our experiments for smaller values of n, because random graph instances are denser and choosing the right k-clique S that leads to the maximum clique is more time consuming.

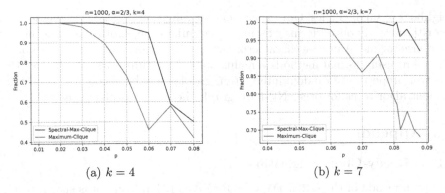

Fig. 5. Approximation guarantee curves for $\alpha = 2/3$, $n = 1000$ and $k = 4, 7$.

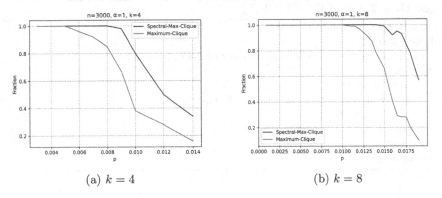

Fig. 6. Approximation guarantee curves for $\alpha = 1$, $n = 3000$ and $k = 4, 8$.

We conclude that for the cases when the graph gets more dense or when parameter k gets larger, *Spectral-Max-Clique* has a lower failure probability as well as it succeeds to find a larger portion of the maximum clique of the graph. The spectral algorithm performs better in dense instances, while the other algorithms for dense graphs do not perform well, meaning that they fail to find the maximum clique for each instance of the graph. Hence, spectral algorithm works for a larger interval of p than the other algorithms.

6   Conclusions

In this paper, we considered the problem of finding maximum cliques when the input graph is a random instance of the random intersection graphs model. Current algorithms for this problem are successful with high probability only for relatively sparse instances, leaving the dense case mostly unexplored. We presented a spectral algorithm for finding large cliques that processes vertices according to respective values in the second largest eigenvector of the adjacency matrix of induced subgraphs of the input graph corresponding to common neighbors of small cliques. Our experimental evaluation showed that our spectral algorithm clearly outperforms existing polynomial time algorithms, especially in the dense regime. A precise characterization of the performance guarantees of our algorithm using formal methods remains open for future work. We believe that spectral properties of random intersection graphs may be also used to construct efficient algorithms for other NP-hard graph theoretical problems as well.

7   Appendix

7.1   Greedy-Clique Algorithm

The pseudocode of the GREEDY-CLIQUE Algorithm from [5] is shown below.

Algorithm 2. GREEDY-CLIQUE [5]

Input: Random instance G of $\mathcal{G}_{n,m,p}$
Output: Clique Q
1: Let $v_1, \ldots, v_2$ the vertices of G in order of decreasing degree;
2: $Q = \emptyset$;
3: **for** $i = 1$ to n **do**
4:     **if** v_i is adjacent to each vertex in Q **then**
5:         $Q = Q \cup \{v_i\}$;
6:     **end if**
7: **end for**
8: **return** Q

7.2   Mono-Clique Algorithm

The pseudocode of the MONO-CLIQUE Algorithm from [5] is shown below.

7.3   Maximum-Clique Algorithm

The pseudocode of the Maximum-Clique Algorithm from [2] is shown below.

Algorithm 3. MONO-CLIQUE [5]

Input: Random instance G of $\mathcal{G}_{n,m,p}$
Output: Clique Q
1: **for** $\{u, v\} \in E(G)$ **do**
2:     $D(\{u, v\}) = |N(u) \cap N(v)|$;
3: **end for**
4: **for** $\{u, v\} \in E(G)$ in order of decreasing $D(\{u, v\})$ **do**
5:     $S = N(u) \cap N(v)$;
6:     **if** S is a clique **then**
7:         **return** $Q = S \cup \{u, v\}$
8:     **end if**
9: **end for**
10: **return** any vertex $v \in V(G)$

Algorithm 4. Maximum-Clique [2]

Input: Random instance G of $\mathcal{G}_{n,m,p}$ and (fixed) parameter $k \in \mathbb{N}$
Output: Clique Q of G
1: $\mathcal{L} = \emptyset$;
2: **for** $U_k = \{v_1, \dots, v_k\} \subseteq V$ such that $G[U_k]$ is complete **do**
3:    **if** $\exists L \in \mathcal{L} : U_k \subseteq L$ **then**
4:        **continue** to the next U_k;
5:    **end if**
6:    Let $Z = Z(U_k) := \cap_{k=1}^{i} N(v_i)$;
7:    **if** $G[Z]$ is complete **then**
8:        $\mathcal{L} = \mathcal{L} \cup \{Z\}$;
9:    **end if**
10: **end for**
11: $\mathcal{M} = \emptyset$;
12: $Y = \emptyset$;
13: **for** $Z \in \mathcal{L}$ in decreasing order $|Z|$ **do**
14:    **if** $E(G[Z]) \not\subseteq Y$ **then**
15:        $Y = Y \cup E(G[Z])$;
16:        $\mathcal{M} = \mathcal{M} \cup \{Z\}$;
17:    **end if**
18: **end for**
19: Sort $\mathcal{M}$ by decreasing order;
20: **return** largest $Q \subset \mathcal{M}$

References

1. Alon, N., Krivelevich, M., Sudakov, B.: Finding a large hidden clique in a random graph. Random Struct. Algor. **13**, 457–466 (1998)
2. Behrisch, M., Taraz, A.: Efficiently covering complex networks with cliques of similar vertices. Theor. Comput. Sci. **355**(1), 37–47 (2006)
3. Bloznelis, M., Godehardt, E., Jaworski, J., Kurauskas, V., Rybarczyk, K.: Recent progress in complex network analysis: properties of random intersection graphs. In: Lausen, B., Krolak-Schwerdt, S., Böhmer, M. (eds.) Data Science, Learning by Latent Structures, and Knowledge Discovery. SCDAKO, pp. 79–88. Springer, Heidelberg (2015). https://doi.org/10.1007/978-3-662-44983-7_7
4. Bloznelis, M., Godehardt, E., Jaworski, J., Kurauskas, V., Rybarczyk, K.: Recent progress in complex network analysis: models of random intersection graphs. In: Lausen, B., Krolak-Schwerdt, S., Böhmer, M. (eds.) Data Science, Learning by Latent Structures, and Knowledge Discovery. SCDAKO, pp. 69–78. Springer, Heidelberg (2015). https://doi.org/10.1007/978-3-662-44983-7_6
5. Bloznelis, M., Kurauskas, V.: Large cliques in sparse random intersection graphs. Electr. J. Comb. **24**(2), P2.5 (2017)
6. Christodoulou, F., Nikoletseas, S., Raptopoulos, C., Spirakis, P.: A spectral algorithm for finding maximum cliques in dense random intersection graphs (2022). https://doi.org/10.48550/ARXIV.2210.02121, https://arxiv.org/abs/2210.02121
7. Fill, J.A., Sheinerman, E.R., Singer-Cohen, K.B.: Random intersection graphs when $m = \omega(n)$: an equivalence theorem relating the evolution of the $g(n, m, p)$ and $g(n, p)$ models. Random Struct. Algor. **16**(2), 156–176 (2000)

8. Friedrich, T., Hercher, C.: On the kernel size of clique cover reductions for random intersection graphs. J. Discrete Algorithms **34**, 128–136 (2015)
9. Grimmett, G.R., McDiarmid, C.: On coloring random graphs. Math. Proc. Cambridge Philos. Soc. **77**, 313–324 (1975)
10. Håstad, J.: Clique is hard to approximate within $n^{1-\varepsilon}$. Acta Math. **182**, 105–142 (1999)
11. Jianer, C., Xiuzhen, H., Iyad, A.K., Ge, X.: Strong computational lower bounds via parameterized complexity. J. Comput. Syst. Sci. **72**(8), 1346–1367 (2006)
12. Karoński, M., Scheinerman, E.R., Singer-Cohen, K.B.: On random intersection graphs: the subgraph problem. Comb. Probab. Comput. **8**, 131–159 (1999)
13. Karp, R.M.: Reducibility among combinatorial problems. In: Complexity of computer computations, pp. 85–103. Plenum Press (1972)
14. Karp, R.M.: Probabilistic analysis of some combinatorial search problems. In: Algorithms and Complexity: New Directions and Recent Results, pp. 85–103. Academic Press (1976)
15. Nikoletseas, S.E., Raptopoulos, C.L., Spirakis, P.G.: Communication and security in random intersection graphs models. In: 12th IEEE International Symposium on a World of Wireless, Mobile and Multimedia Networks (WOWMOM), pp. 1–6 (2011)
16. Nikoletseas, S.E., Raptopoulos, C.L., Spirakis, P.G.: Maximum cliques in graphs with small intersection number and random intersection graphs. In: Proceedings of the 37th International Symposium on Mathematical Foundations of Computer Science (MFCS), pp. 728–739 (2012)
17. Nikoletseas, S.E., Raptopoulos, C.L., Spirakis, P.G.: Maximum cliques in graphs with small intersection number and random intersection graphs. Comput. Sci. Rev. **39**, 100353 (2021)
18. Rybarczyk, K.: Equivalence of a random intersection graph and $g(n,p)$. Random Struct. Algor. **38**(1–2), 205–234 (2011)
19. Singer-Cohen, K.B.: Random intersection graphs. Ph.D. thesis, John Hopkins University (1995)

Solving Cut-Problems in Quadratic Time for Graphs with Bounded Treewidth

Hauke Brinkop$^{(\boxtimes)}$ and Klaus Jansen

Kiel University, Kiel, Germany
{hab,kj}@informatik.uni-kiel.de

Abstract. In the problem (Unweighted) Max-Cut we are given a graph $G = (V, E)$ and asked for a set $S \subseteq V$ such that the number of edges from S to $V \setminus S$ is maximal. In this paper we consider an even harder problem: (Weighted) Max-Bisection. Here we are given an undirected graph $G = (V, E)$ and a weight function $w \colon E \to \mathbb{Q}_{>0}$ and the task is to find a set $S \subseteq V$ such that (i) the sum of the weights of edges from S is maximal; and (ii) S contains $\lceil \frac{n}{2} \rceil$ vertices (where $n = |V|$). We design a framework that allows to solve this problem in time $\mathcal{O}(2^t n^2)$ if a tree decomposition of width t is given as part of the input. This improves the previously best running time for Max-Bisection of Hanaka, Kobayashi, and Sone [9] by a factor t^2. Under common hardness assumptions, neither the dependence on t in the exponent nor the dependence on n can be reduced [7,9,16]. Our framework can be applied to other cut problems like Min-Edge-Expansion, Sparsest-Cut, Densest-Cut, β-Balanced-Min-Cut, and Min-Bisection. It also works in the setting with arbitrary weights and directed edges.

1 Introduction

Unweighted Max-Cut is one of Karp's 21 NP-complete problems [13]; given a graph $G = (V, E)$ one is asked for a set $S \subseteq V$ such that the number of edges from S to $V \setminus S$ is maximal. Formally, a *cut* is determined by a set of vertices $S \subseteq V$ of a graph. The *size of a cut* is given by the number of edges from S to $V \setminus S$. We denote these edges as ∂S and, for the sake of shortness, if $S = \{v\}$ for some v, we write ∂v instead of $\partial S = \partial \{v\}$. If the graph is weighted, the size of the cut is given by the sum of the edge weights $w(\partial S) := \sum_{e \in \partial S} w(e)$ instead of their number $|\partial S|$. In this paper we consider different cut problems for directed and weighted graphs[1], more precisely Max-Cut, β-Balanced-Min-Cut, Max-Bisection, Min-Bisection, Min-Edge-Expansion, (uniform) Sparsest-Cut, and Densest Cut. See Table 1 for precise formulations of these problems.

---

[1] The undirected and unweighted versions can easily be modelled as directed and weighted by setting each edge weight to 1 and by replacing each undirected edge between vertices v_1 and v_2 by two directed edges, $v_1 v_2$ and $v_2 v_1$.

This work was partially supported by DFG Project "Fein-granulare Komplexität und Algorithmen für Scheduling und Packungen", JA 612 /25-1.

L. Gąsieniec (Ed.): SOFSEM 2023, LNCS 13878, pp. 33–46, 2023.
https://doi.org/10.1007/978-3-031-23101-8_3

Observe that Densest-Cut and Sparsest-Cut can easily be reduced on each other in time $\mathcal{O}(n^2)$ (where $n = |V|$) using the complementary graph [3]; however, this reduction might change the treewidth and the corresponding decomposition, hence we have to consider both problems individually. We want to point out that if negative edge weights are allowed, as they are in our algorithm, Max-Bisection and Min-Bisection coincide. This does not hold for Max-Cut and its corresponding minimization variant; Min-Cut is solvable in polynomial time. We call those more general variants of the problem, where we get rid of the non-negativity restrictions, Max-Bisection', β-Balanced-Min-Cut', Min-Edge-Expansion', Sparsest-Cut', and Densest-Cut'. As we will see, our framework is able to solve these more general variants of the problems.

Table 1. Problems that we solve in quadratic time.

Name	Weights	Objective
Max-Cut	Arbitrary	$\max\limits_{S \subseteq V} w(\partial S)$
β-Balanced-Min-Cut [8]	Non-negative	$\min\limits_{\substack{S \subseteq V \\ \beta \cdot |V| \leq |S| \leq (1-\beta) \cdot |V|}} w(\partial S)$
Max-Bisection [11]	Non-negative	$\max\limits_{\substack{S \subseteq V \\ \left| |S| - |V \setminus S| \right| \leq 1}} w(\partial S)$
Min-Bisection [11]	Non-negative	$\min\limits_{\substack{S \subseteq V \\ \left| |S| - |V \setminus S| \right| \leq 1}} w(\partial S)$
Min-Edge-Expansion [17]	Non-negative	$\min\limits_{\substack{\emptyset \neq S \subseteq V \\ |S| \leq |V \setminus S|}} \dfrac{w(\partial S)}{|S|}$
Sparsest-Cut [3]	Non-negative	$\min\limits_{\emptyset \neq S \subsetneq V} \dfrac{w(\partial S)}{|S| \cdot |V \setminus S|}$
Densest-Cut [3]	Non-negative	$\max\limits_{\emptyset \neq S \subsetneq V} \dfrac{w(\partial S)}{|S| \cdot |V \setminus S|}$

Many graph problems are in FPT if parametrized by treewidth. This holds especially for the problems mentioned above [3,7,9,12]. The corresponding algorithms usually assume that a tree decomposition with $\mathcal{O}(n)$ (or a similar bound like $\mathcal{O}(nt)$) nodes of width t is given as part of the input; we will later see why that is a reasonable assumption and what we can do if a tree decomposition is not given. We show that in the above setting, all aforementioned problems can be solved in time $\mathcal{O}(2^t n^2)$.

2 Related Work

Jansen et al. [11] proposed an algorithm for Max-Bisection and Min-Bisection that runs in time $\mathcal{O}(2^t n^3)$ for a graph with n vertices, given a tree decomposition of width t with $\mathcal{O}(n)$ nodes[2]. They transform the tree decomposition into a so-called *nice tree decomposition* [15] and then formulate a dynamic program over the nodes of the tree decomposition. The bottleneck of their analysis are nodes that have more than one child, the so-called *join nodes*. There might be $\Omega(n)$ join nodes and for each the dynamic program might take time $\Omega(2^t n^2)$ to compute all the entries. Eiben, Lokshtanov, and Mouawad [7] have been able to improve the running time[3] in its dependence on n by balancing the tree decomposition and recognizing that the entries in join nodes can be computed via $(\max, +)$-convolution; this yields a running time of $\mathcal{O}(8^t t^5 n^2 \log n)$. Hanaka, Kobayashi, and Sone [9] proved that the algorithm of Jansen et al. does in fact run in time $\mathcal{O}(2^t(nt)^2)$, using a clever idea to improve the analysis: while for a single join node computing all entries of the dynamic program might take $\Omega(2^t n^2)$, the overall time for all join nodes altogether is $\mathcal{O}(2^t(nt)^2)$.

Lokshtanov, Marx, and Saurabh [16] proved that Max-Cut (without a tree decomposition given as part of the input) cannot be solved in time $\mathcal{O}((2 - \varepsilon)^t \operatorname{poly} n)$ for any $\varepsilon > 0$ assuming the *Strong Exponential Time Hypothesis (SETH)* [5,10]. It is not hard to see that this result can be extended to the case where a tree decomposition is part of the input. By adding isolated vertices, this result can also be applied to Max-Bisection and Min-Bisection [9]. Eiben, Lokshtanov, and Mouawad [7] proved that Min-Bisection (and hence Max-Bisection') cannot be solved in truly subquadratic time, that is $\mathcal{O}(n^{2-\varepsilon})$ for some $\varepsilon > 0$, even if a tree decomposition of width 1 is given as part of the input, unless $(\min, +)$-convolution can be solved in truly subquadratic time, which is considered unlikely [6].

Given a tree decomposition of width t with $\mathcal{O}(nt)$ nodes, in time $\mathcal{O}(2^t n^3)$ the problems Sparsest-Cut [3], Densest-Cut [3], and Min-Edge-Expansion [12] can be solved. To our knowledge, those are the best running times achieved so far.

Our Contribution and Organization of this Paper. In Sect. 4 we show how to improve the running time by a factor t^2 for Max-Bisection. In Sect. 5 we then generalize this to a framework which can be used to solve different cut problems in time $\mathcal{O}(2^t n^2)$ (compare Table 1). Some problems (like Sparsest Cut) are improved by a factor n, which is substantial when t is small. The instantiations of our framework together with the corresponding correctness proofs are omited due to space restrictions. We refer to the extended version [4] of the paper.

---

[2] The upper bound on the number of nodes occurs only implicitly in their work within the analysis of their algorithm's running time.

[3] To be precise, they consider Min-Bisection; however, in their paper as well as in [11], all arguments work for both problems.

3   Preliminaries

Notation. For tuples we write $(a, b) \oplus (c, d) = (a + c, b + d)$ (and analogously define $\ominus$). We write π to denote the projection on tuples, that is: for a tuple t and an index i, $\pi_i(t)$ is the i-th component of t.

For a set M and a number k we write $\binom{M}{k} = \{M' \subseteq M : |M| = k\}$. For a graph edge from v_1 to v_2 we write $v_1 v_2$ for both, directed and undirected graphs. A rooted tree $T = (V, r, E)$ is a graph (V, E) that is connected, has no circles, and where $r \in V$.

We use n as abbreviation for $|V|$. By $\bar{\mathbb{Q}} := \mathbb{Q} \,\dot\cup\, \{\infty, -\infty\}$ we denote the rational numbers with positive and negative infinity.

As already mentioned in the previous sections, we make use of *tree decompositions*, which are defined as follows:

Definition 1 (Tree decomposition). Let $G = (V, E)$ be an undirected graph, I be a finite set and $X = (X_i)_{i \in I}$ be a family of sets such that for any $i \in I$ one has $X_i \subseteq V$. Moreover, let $T = (I, r, H)$ be a rooted tree with root $r \in I$. Then (I, r, X, H) is called a *tree decomposition* of G if and only if T has the following properties:

(i) **Node coverage:** Every vertex occurs in some X_i for some $i \in I$ (and no further vertices occur): $\bigcup_{i \in I} X_i = V$;

(ii) **Edge coverage:** for every edge $v_1 v_2 \in E$ there is a node $i \in I$ such that both, v_1 and v_2, are contained in X_i;

(iii) **Coherence:** for every vertex $v \in V$ the subgraph $T - \{i \in I : v \notin X_i\}$ is connected.

By convention the nodes of the graph G are called *vertices* while the nodes of the tree are just called *nodes*. For a node $i \in I$ the set X_i is called a *bag*. The *width* of a decomposition is the largest cardinality of any of its bags minus 1. The minimum width among all decompositions of G is called the *treewidth* of G. If the node set of a decomposition is sufficiently small, more precisely if $|I| \leq 4 \cdot (|V| + 1)$, we call the decomposition *small*. A tree decomposition of a directed graph is a tree decomposition of the underlying graph.

As the approaches mentioned before, our approach also makes use of a specific kind of tree decompositions, which are of a very simple structure:

Definition 2 (Nice Tree Decomposition). Let $G = (V, E)$ be an undirected graph and (I, r, X, H) a tree decomposition of G. We call (I, r, X, H) a *nice tree decomposition* if and only if for any $i \in I$ the node i is of one of the following forms:

(i) **Leaf node:** i has no child node in T, that is i is a *leaf* of the tree T;

(ii) **Forget node:** i has exactly one child node $j \in I$ in T and $X_i \,\dot\cup\, \{v\} = X_j$ for some $v \in X_j$, that is i *forgets* a vertex from X_j;

(iii) **Introduce node:** i has exactly one child node $j \in I$ in T and $X_i = X_j \,\dot\cup\, \{v\}$ for some $v \in V \setminus X_j$, that is i *introduces* a new vertex $v \in V \setminus X_j$; or

(iv) **Join node:** i has exactly two child nodes $j \in I$, $k \in I$, $j \neq k$, in T such that $X_i = X_j = X_k$, that is *joining* two branches of the tree T.   ⌋

The conversion of a tree decomposition into a nice tree decomposition can be done in time $\mathcal{O}(nt^2)$ as long as the number of nodes of the decomposition is at most linear in the number of vertices.

Lemma 3 ([15, **Lemma 13.1.3, p. 150**]). Given a small tree decomposition of a graph G with width t one can find a nice tree decomposition of G with width t and with at most $4n$ nodes in $\mathcal{O}(nt^2)$ time, where n is the number of vertices of G.   ⌋

It is reasonable to assume that given tree decompositions are small for the following reason: No matter how the tree decomposition is constructed, it is always possible to incorporate the following mechanism without asymptotically increasing the running time: If a node j with parent i has $X_j \subseteq X_i$, merge those nodes.

We claim that we now can only have n edges. This is because for every node j with parent i, we have $X_j \not\subseteq X_i$; this means, at least one vertex has to disappear when going from j up to i. Since by Coherence every vertex can disappear at most once[4], this upper bounds the number of edges by n and hence the number of nodes by $n + 1$.

Note that this also allows us to easily extend our approach to the case where a tree decomposition with $\mathcal{O}(nt)$ nodes is given as part of the input: simply apply the procedure described above to reduce the number of nodes.

4   Max-Bisection: From $\mathcal{O}(2^t n^3)$ to $\mathcal{O}(2^t n^2)$

In this section we focus on our idea on how the running time of the algorithm of Jansen et al. [11] for Max-Bisection can be improved to $\mathcal{O}(2^t n^2)$, incorporating the idea of Hanaka, Kobayashi, and Sone [9].

The algorithm of Jansen et al. [11] is a subroutine used in their PTAS for the Max-Bisection problem on planar graphs. Their approach uses Baker's technique (see [1]) where the idea is to solve the problem for k-outerplanar graphs (instead of general planar graphs), for a k depending only on the approximation factor, and then combining the results. Note that k-outerplanar graphs have a treewidth in $\mathcal{O}(k)$ [14]. For those k-outerplanar graphs, the problem is solved exactly using the aforementioned subroutine. Since – as opposed to the general case – tree decompositions for k-outerplanar graphs can be computed in time $\mathcal{O}(kn)$ [14], this subroutine gets tree decomposition as part of its input; otherwise the running time of the subroutine would be dominated by the computation of the decomposition.

Let us now focus on the subroutine. We will traverse the nice tree decomposition bottom up in the algorithm of Jansen et al., hence the following notations come in handy: For a node i the set Y_i contains all the vertices appearing in

---

[4] It might disappear on multiple leaf-root-paths; however, the node at which a vertex disappears, is the same on each of those leaf-root-paths.

bags associated with nodes below i. Moreover, we write $F_i := Y_i \setminus X_i$ to describe the set of vertices that "have been forgotten" somewhere below i, that is, that they have appeared in bag of some node j below i, but are not contained in X_i. Due to Definition 1 (iii), those vertices can never reoccur in any bag above i.

The algorithm of Jansen et al. uses a dynamic program to compute

$$\mathcal{B}_i \colon \{0, \ldots, |Y_i|\} \times 2^{X_i} \to \bar{\mathbb{Q}}_{>0}$$

$$\mathcal{B}_i(\ell, S) = \max_{\substack{\widehat{S} \subseteq Y_i \\ |\widehat{S}| = \ell \\ S \subseteq \widehat{S}}} w(\partial \widehat{S} \cap Y_i^2), \tag{1}$$

given a small nice tree decomposition of a weighted, undirected graph $G = (V, E)$ with weight function $w \colon E \to \mathbb{Q}_{>0}$. The idea is that for a node i the entry $\mathcal{B}_i(\ell, S)$ is the size of the largest possible cut that consists of ℓ vertices from Y_i and includes the set $S \subseteq X_i$. As the table might have preimages (ℓ, S) where there does not exist be a cut meeting the requirements above, ∞ and $-\infty$ have to be used to deal with those – we omit further details.

For the root r of the tree decomposition we can compute our optimal objective value by iterating over all entries of $\mathcal{B}_r$ and picking the best value where the number of vertices is in the feasible interval for Max-Bisection.

The dynamic program traverses the decomposition bottom up. The bottleneck of the running time comes from the time spent at join nodes – the values for each node of any different type can be computed in time $\mathcal{O}(2^t n)$ using a simple DP. For a join node i with left child j and right child k, they use the following recurrence to compute $\mathcal{B}_i$:

$$\mathcal{B}_i(\ell, S) = \max_{\substack{|S| \leq \ell_1 \leq |V| \\ |S| \leq \ell_2 \leq |V| \\ \ell_1 + \ell_2 - |S| = \ell}} \left(\mathcal{B}_j(\ell_1, S) + \mathcal{B}_k(\ell_2, S) - w(\partial S \cap (S \times (X_i \setminus S))) \right) \tag{2}$$

We omit the details on how the necessary values of w are computed in their case; it suffices to see that a rough analysis of the equation above, assuming that we are given the value of the w expression, is $\mathcal{O}(n^2)$ per entry for a single join node. It is also easy to see that there are indeed instances where the computation of an entry for (ℓ, S) takes $\Theta(n^2)$ time. This yields an overall running time of $\mathcal{O}(2^t n^3)$ as stated by Jansen et al. [11].

Hanaka, Kobayashi, and Sone [9] provided an improved analysis for the algorithm of Jansen et al. [11]. They defined ν_i to be the sum of all $|X_j|$ for nodes j that are below i. It is not hard to see that the running time for the computation of a single entry of a join node i with left child j and right child k can be done in time $\mathcal{O}(\nu_j \nu_k)$. Using a labeling argument they then proved that

$$\sum_{i \, : \, \text{join node with children } j,k} \nu_j \nu_k \leq (nt)^2.$$

Their idea is that after labeling the vertices in all bags (possibly giving the same vertex different labels for different bags), every pair of those labels can occur at at most one join node. The consequence of the above statement is that the worst case for join nodes cannot occur too often; overall, all entries of all join nodes can be computed in time $\mathcal{O}(2^t(nt)^2)$.

Our approach is now to reformulate the recurrence by something that can be thought of as an index shift; for each node i we define a table

$$\Gamma_i\colon \{0,\ldots,|F_i|\} \times 2^{X_i} \to \mathbb{Q}$$
$$\Gamma_i(\ell, S) := \max_{\widehat{S}\in\binom{F_i}{\ell}} w(\partial(S \cup \widehat{S}) \cap Y_i^2). \tag{3}$$

In comparison to (1), there are two differences.

1. The indices have a different meaning: In $\Gamma_i(\ell, S)$ we store the value of the best cut (with respect to f) that consists of the set $S \subseteq X_i$ and ℓ further vertices that *occur in bags below i, but not in X_i*.
2. The table's size is now $\mathcal{O}(|F_i|2^t)$; this is not only smaller but also for every entry (ℓ, S) there is a cut consisting of ℓ vertices from F_i and the vertices from S (hence we do not have to consider those special cases of undefinedness as it had to be done in [11]).

For the modified recursion, the join nodes are still the bottleneck; their recurrence is[5]

$$\Gamma_i(\ell, S) = \left(\max_{\substack{0\le\ell_1\le|F_j|\\0\le\ell_2\le|F_k|\\\ell_2+\ell_1=\ell}} \left(\Gamma_j(\ell_1, S) + \Gamma_k(\ell_2, S)\right) \right) - w(\partial S \cap Y_i^2).$$

The key observation is that the running time is dominated by computing the max expression[6], which depends linearly on $|F_j| \cdot |F_k| = |F_j \times F_k|$. We can now show that all of those occurring Cartesian products are disjunct:

Proposition 4. For each pair $(v_1, v_2) \in V^2$ there is at most one join node i with left child j and right child k such that $(v_1, v_2) \in F_j \times F_k$.

Proof. Proof by contradiction. Assume there was a join node $i' \ne i$ with left child j' and right child k' such that $(v_1, v_2) \in F_{j'} \times F_{k'}$. Then, by Coherence, either i' is below i or i is below i'. We assume without loss of generality that i' is below i. Moreover, we assume without loss of generality that i' is somewhere in the *left* subtree of i. As $(v_1, v_2) \in F_j \times F_k$ by assumption, we have in particular $v_2 \in F_k$, and, additionally taking into account that $F_j \cap F_k = \emptyset$ by Coherence, $v_1 \notin F_k$. Since i' is in the left subtree of i, we also have $F_{j'} \subseteq F_{i'} \subseteq F_j$, hence $v_1 \notin F_{j'}$. This is a contradiction to $(v_1, v_2) \in F_{j'} \times F_{k'}$. □

---

[5] Since we only did an index shift, we can reuse the recurrence from [11] by applying the same shift to it.

[6] Note that ∂S can only take on $\mathcal{O}(2^t)$ different values at some fixed node i. We use a simple DP to precompute the $w(\cdot)$ terms efficiently (for a fixed node in time $\mathcal{O}(2^t n)$).

Using this statement we can now deduce that

$$\sum_{i \,:\, \text{join node with children } j,k} |F_j \times F_k|$$

$$= \left| \bigcup_{i \,:\, \text{join node with children } j,k} (F_j \times F_k) \right| \leq |V^2| = n^2. \quad (4)$$

We can thus deduce the overall running time of computing all entries for all join nodes is $\mathcal{O}(2^t n^2)$, as we need time $\mathcal{O}(2^t |F_j \times F_k|)$ for a single join node.

As the running time for the other node types obviously remain unchanged, this yields an algorithm with overall running time $\mathcal{O}(2^t n^2)$ for Max-Bisection.

5   Our Framework

In this section we discuss how we can generalize the idea from the previous section to other cut problems. More precisely, we present a framework that can solve several cut-problems (for directed, arbitrarily-weighted graphs $G = (V, E)$ with weight function $w \colon E \to \mathbb{Q}$) in time $\mathcal{O}(2^t n^2)$ if a small tree decomposition of width t is given as part of the input. Without loss of generality we assume that this small tree decomposition is also a nice tree decomposition (if not, we could simply use Lemma 3 to convert it accordingly in sufficiently small time).

The main obstacle is finding an abstraction of the algorithm for Max-Bisection that maintains the running time, but also allows us tackle all the listed problems. Especially extracting the formal arguments hidden implicitly in existing algorithms turned out to be a non-trivial task.

For our framework, we assume that we are given an objective function $f \colon \mathbb{N}_0 \times \mathbb{Q} \to \bar{\mathbb{Q}}$ that is either monotonic or antitonic[7] in its second argument, and a validator function $\Lambda \colon \mathbb{N}_0 \to \{\text{true}, \text{false}\}$. We use a function of arity 2 to be able to not only model Max-Bisection and similar, but also e.g. Sparsest-Cut, where the objective depends on the size of the cut *and* the number of vertices selected. The validator function is needed e.g. for Max-Bisection, as we somehow have to tell the framework which entries correspond to feasible solutions and which are infeasible; for Max-Bisection we would set $\Lambda(x) := (|x - (n - x)| \leq 1)$. We assume that both, f and Λ, can be evaluated in time $\mathcal{O}(1)$.

Our task is now to compute an element of all *possible preimages* (in the sense: there exists a corresponding cut) $\left\{ \big(|S|, w(\partial S) \big) : S \subseteq V, \Lambda(|S|) \right\}$ of the objective function f that maximizes f. We can reformulate this task in a more elegant way by introducing the total quasiorder $(\sqsubseteq) \subseteq (\{0, \dots, n\} \times \mathbb{Q}) \,\dot\cup\, \{\bot\})^2$ defined by

$$a \sqsubseteq b \iff a = \bot \vee \big(a \neq \bot \neq b \wedge f(a) \leq f(b) \big).$$

The intuition behind that quasiorder is as follows: if we compare two values $a, b \neq \bot$, then $a \sqsubseteq b$ iff $f(a) \leq f(b)$, that is, we compare (non-$\bot$) values by their image under f.

----

[7] $x \leq y \implies f(a, x) \geq f(a, y)$.

The idea of the new symbol $\bot$ is to represent the case where there is no feasible solution and hence no possible preimage to f as we have to deal with that case, too.

Our task is now to compute (where $\bigsqcup$ is the supremum operator regarding $\sqsubseteq$).

$$\Phi = \bigsqcup_{\substack{S \subseteq V \\ \Lambda(\overline{|S|})}} \left(|S|, w(\partial S) \right). \tag{5}$$

By definition, $\bot$ is the smallest element of our order, so if there is a feasible solution, the result cannot be $\bot$.

From a strict mathematical perspective, Eq. 5 is incorrect as in general there is no such thing as a unique supremum for a total quasiorder (there might be multiple possible preimages of f taking on the optimal value). Taking this into account would make the description of our approach way more complicated, as we would need to reason about equivalence classes and eventually give a recurrence to compute a representant of the class of element optimizing the objective function. Thus, we identify elements and their corresponding equivalence class (set of possible preimages that have the same objective value) in this paper.

For the sake of shortness, for a subset of edges $M \subseteq E$ we write $w_i(M) := w(M \cap Y_i^2)$. We set up the dynamic program similar to the one for Max-Bisection, that is for a node i we define

$$\Gamma_i \colon \{0, \ldots, |F_i|\} \times 2^{X_i} \to \{0, \ldots, n\} \times \mathbb{Q}$$

$$\Gamma_i(\ell, S) := \bigsqcup_{\widehat{S} \in \binom{F_i}{\ell}} \left(\ell + |S|, w_i(\partial(S \cup \widehat{S})) \right). \tag{6}$$

In comparison to Eq. 3 there are two differences.

1. An entry is no longer just the size of the corresponding cut, but a 2-tuple consisting of the number of vertices selected and the size of the cut.
2. Instead of storing values for the largest cut, as we did for Max-Bisection, we store the tuple that maximizes the function f.

We can now rewrite Eq. 5 in terms of Γ (recall that r is the root of the given nice tree decomposition):

Lemma 5.

$$\Phi = \bigsqcup_{\substack{S \subseteq X_r \\ 0 \leq \ell \leq |F_r| \\ \Lambda(\ell + |S|)}} \Gamma_r(\ell, S) \tag{7}$$

Proof (Omitted due to space restrictions, we refer to the extended version [4]).

It is easy to see that, if we are given the values $\Gamma_r(\ell, S)$ for all $0 \leq \ell \leq |F_r|$ and all $S \subseteq X_r$, we can compute Φ in time $\mathcal{O}(2^t n)$. We claim that we can compute the table Γ for all nodes together in overall time $\mathcal{O}(2^t n^2)$, implying that Φ can be computed in time $\mathcal{O}(2^t n^2)$. To see this, we now show that we can use a dynamic program to compute the table Γ and eventually Γ_r in the desired time. Therefor, we first set up recurrences for each node type that we can use to efficiently compute the value for a node i of this type, given that we already know all the values *below* i.

For our approach there is an important property of $\sqsubseteq$: Basically, we are able to move the addition with a constant tuple outside of the supremum operator, if the elements all have the same first component.

Lemma 6. For any a, any finite set $M \subseteq \{a\} \times \mathbb{Q}$ and any z it holds that

$$z \oplus \bigsqcup_{x \in M} x = \bigsqcup_{x \in M} (z \oplus x)$$

Proof. As M is finite, it suffices to show[8] this property for the binary supremum $\sqcup$. Let $z = (b, w)$ and $(a, x) \in M$, $(a, y) \in M$. If $x = y$ or $f(a, x) = f(a, y)$, the property is trivial. Thus we may assume without loss of generality that $x > y$ and $f(a, x) \neq f(a, y)$. We now have two cases, depending on $h \mapsto f(a, h)$. The first case is that $h \mapsto f(a, h)$ monotonic. Then our assumption implies that $f(a, x) > f(a, y)$ and hence:

$$(b, z) \oplus ((a, x) \sqcup (a, y)) = (b, z) \oplus (a, x) = (a+b, x+z) = (a+b, x+z) \sqcup (a+b, y+z)$$

The last step follows from the monotony, as $x + z \geq y + z$. The second case, where $h \mapsto f(a, h)$ is antitonic, can be proven analogously.

This property is absolutely crucial as it gives us some (necessary) freedom for transformations of Eq. 6. Intuitively, this lemma tells us that the optimization process works no different than it does e.g. Max-Cut or Max-Bisection; if we fix the number of vertices we choose, given a set of cuts to pick from the cut maximizing f is either the cut of largest size or the cut of smallest size.

Recurrences

We now set up a recurrence, depending on the node type, to compute the table Γ by traversing the tree decomposition in a bottom-up fashion. In this section we give the intuition behind the recurrences step by step. For correctness proofs of the equations (which are very technical) we refer to the extended version [4] of the paper.

---

[8] For the reader not familiar with order theory: The binary supremum is an associative and commutative map. If for every pair of elements there is a supremum, that is, a smallest element that is larger than both elements of the pair, then so it does for any finite set M. This can be shown by a simple inductive argument using the aforementioned associativity/commutativity.

Leaf Node. Let i be a leaf node. Then $F_i = \emptyset$. Thus, Γ_i is only defined for $\ell = 0$ and $S \subseteq X_i$. If $S = \emptyset$, there are no edges in the cut, hence $\Gamma_i(0, \emptyset) = (0, 0)$. If $S = \{v\}$ consists of a single node, we can simply go through all its edges, that is set

$$\Gamma_i(0, \{v\}) = \left(1, \sum_{\substack{v' \in X_i \\ vv' \in E}} w(vv')\right). \tag{8}$$

Now let $S = S' \,\dot\cup\, \{v\}$ where $S' \neq \emptyset$. We will now argue how we can compute the value $\Gamma_i(0, S)$ given the value $\Gamma_i(0, S')$. In the situation considered for $\Gamma_i(0, S')$ we have $v \notin S'$. If we move v into the selection and are able to track and compute the changes, we can also compute $\Gamma_i(0, S)$. After moving v into the selection, there might be edges from S' to $\{v\}$ (which all have been in the cut before); those edges are no more in the cut for S. Also, there might also be some new edges in the cut, more precisely all edges from v to $X_i \setminus S$. There are no more new edges in the cut and no other edges are removed from the cut. This yields

$$\Gamma_i(0, \underbrace{\overset{\substack{\neq\emptyset \\ \downarrow}}{S'} \,\dot\cup\, \{v\}}_{S}) = \Gamma_i(0, S') \oplus \Big(\overset{\substack{|S|-|S'|=1, \text{ we now add the new vertex } v \\ \downarrow}}{1}, \underbrace{\sum_{\substack{v' \in X_i \setminus S' \\ vv' \in E}} w(vv')}_{\text{edges from } v \text{ to } X_i \setminus S'} - \underbrace{\sum_{\substack{v' \in S' \\ v'v \in E}} w(v'v)}_{\text{edges from } S' \text{ to } v}\Big). \tag{9}$$

Forget Node. Let i be a forget node with child j. Then there is $v \in V$ such that $X_i \,\dot\cup\, \{v\} = X_j$. Now, for the computation of the entries $\Gamma_i(\ell, S)$ we only have to deal with one question: is it better to include v into the selection or not? As $v \in F_i$, this question is only relevant if $\ell \geq 1$; if $\ell = 0$, then $\Gamma_i(\ell, S) = \Gamma_j(\ell, S)$. Now let $\ell \geq 1$. If we included v into our selection, we have to include $\ell - 1$ further vertices from F_j; if we do not include v, we have to include ℓ further from F_j. We now simply pick the better result of both options. Overall, this yields the following recurrence:

$$\Gamma_i(\ell, S) = \begin{cases} \Gamma_j(\ell - 1, S \,\dot\cup\, \{v\}) \sqcup \Gamma_j(\ell, S) & \ell \geq 1 \\ \Gamma_j(\ell, S) & \ell = 0 \end{cases}. \tag{10}$$

Introduce Node. Let i be an introduce node with child j. Then there is $v \in V$ such that $X_i = X_j \,\dot\cup\, \{v\}$. When computing $\Gamma_i(\ell, S)$ we have to make a case distinction whether v is in S or not.

In the first case $v \in S$ we rely on $\Gamma_j(\ell, S \setminus \{v\})$ to compute the value. As we consider a selection of S and ℓ further vertices of F_i as opposed to the entry $\Gamma_j(\ell, S \setminus \{v\})$ which only considers a selection of $S \setminus \{v\}$ and ℓ further vertices of $F_j = F_i$, we have to add 1 to the first component of $\Gamma_j(\ell, S \setminus \{v\})$ to account for the additional vertex v. For the second component, the size of the cut, we only have to add the weight of all edges from v to $X_j \setminus S$. Note that there are no weights that we need to subtract; while there might be edges between v and S, those are not considered in the computation of $\Gamma_j(\ell, S \setminus \{v\})$ as the vertex v

does not appear in Y_j (due to Coherence and Edge Coverage). Thus, for $v \in S$ we get

we now add the new vertex v to the selection $S \setminus \{v\}$

$$\Gamma_i(\ell, S) = \Big(\overbrace{1}, \underbrace{w_i(\partial v \cap (\{v\} \times (X_i \setminus S)))} \Big) \oplus \Gamma_j(\ell, S \setminus \{v\}). \qquad (11)$$

edges from v to $X_i \setminus S$; we will later discuss how to compute this value

Now let us consider the case $v \notin S$. The argumentation is very similar. As $v \notin S$, there is no new vertex we add to the selection. However, there are possibly new edges and that change the value of the cut: the edges from S to $\{v\}$. This gives us the following recurrence for the case $v \notin S$:

$$\Gamma_i(\ell, S) = \Big(0, \underbrace{w_i(\partial S \cap (X_i \times \{v\}))} \Big) \oplus \Gamma_j(\ell, S). \qquad (12)$$

edges from S to $\{v\}$; we will later discuss how to compute this value

In both cases there is a (case-dependent) additive term that has to b eadded to an entry of Γ_j. Observe that this additive term is independent of ℓ. As it turns out, this additive term (depending only on S) can be rewritten as

$$\pi_2 \big(\Gamma_i(\ell, S) \ominus \Gamma_j(\ell, S) \big) \qquad (13)$$

for any arbitrary, but fixed ℓ; for details we refer to the extended version [4]. Thus, we can do the following: for $\ell = 0$ we need to explicitly compute the additive term for all $S \subseteq X_i$. For all $\ell \geq 1$ we can simply use $\pi_2(\Gamma_i(\ell-1, S) \ominus \Gamma_i(\ell-1, S))$ for all $S \subseteq X_i$.

Join Node. Let i be a join node with left child j and right child k. Then $X_i = X_j = X_k$. The value $\Gamma_i(\ell, S)$ can then be interpreted as the tuple maximizing f over all selections $\widehat{S} \subseteq Y_i$ where there are ℓ_1 vertices from F_j and ℓ_2 from F_k for every $0 \leq \ell_1 \leq |F_j|$, $0 \leq \ell_2 \leq |F_k|$ where $\ell_1 + \ell_1 = \ell$. This is closely related to

$$\Gamma_i(\ell, S) = \bigsqcup_{\substack{0 \leq \ell_1 \leq |F_j| \\ 0 \leq \ell_2 \leq |F_k| \\ \ell_2 + \ell_1 = \ell}} (\Gamma_j(\ell_1, S) \oplus \Gamma_k(\ell_2, S));$$

it is not hard to see that this equation almost computes the desired tuple, with the exception that each vertex in S and each edge of $E \cap X_i^2$ that is in the cut is counted twice. Thus, by subtracting the doubly counted vertices/edge weights and applying Lemma 6 we get

$$\Gamma_i(\ell, S) = \Big(\bigsqcup_{\substack{0 \leq \ell_1 \leq |F_j| \\ 0 \leq \ell_2 \leq |F_k| \\ \ell_2 + \ell_1 = \ell}} (\Gamma_j(\ell_1, S) \oplus \Gamma_k(\ell_2, S)) \Big) \ominus \Big(|S|, w_i(\partial S) \Big). \qquad (14)$$

The running time analysis is analogous to the analysis of the algorithm for Max-Bisection we presented before. We thus skip this part here (due to space restrictions); if interested, the reader can find a detailed analysis of every case in the extended version [4].

6   Conclusion

We showed that – given a small tree decomposition of width t – many cut problems can be solved in time $\mathcal{O}(2^t n^2)$ using our framework. To our knowledge, the running times achieved by our framework are better than the previously known algorithms for the considered problems. Moreover, this running time is unlikely to be improved significantly (improvements by factors $\operatorname{poly} t$ and/or $\operatorname{poly} \log n$ are not excluded) in general: An algorithm (solving Min-Bisection) that runs in time $\mathcal{O}(n^{2-\varepsilon} f(t))$ for some f and some $\varepsilon > 0$ would imply an algorithm of running time $\mathcal{O}(n^{2-\delta})$ for some $\delta > 0$ for (min, +)-convolution [7], which is considered unlikely [6]. An algorithm (solving Max-Cut) cannot have a running time $\mathcal{O}((2 - \varepsilon)^t \operatorname{poly} n)$ for some $\varepsilon > 0$ unless SETH fails [5,9,10, 16]. However, there might be problems that can be solved using our framework that we have not considered yet. Moreover, it might be possible to generalize the framework (with possibly worse running time) to e.g. be able to also cover connectivity problems (see [2]).

References

1. Baker, B.S.: Approximation algorithms for NP-complete problems on planar graphs. J. ACM **41**(1), 153–180 (1994). https://doi.org/10.1145/174644.174650. ISSN 0004-5411
2. Bodlaender, H.L., Cygan, M., Kratsch, S., Nederlof, J.: Deterministic single exponential time algorithms for connectivity problems parameterized by treewidth. Inf. Comput. **243**, 86–111 (2015). https://doi.org/10.1016/j.ic.2014.12.008
3. Bonsma, P.S., Broersma, H., Patel, V., Pyatkin, A.V.: The complexity of finding uniform sparsest cuts in various graph classes. J. Discrete Algorithms **14**, 136–149 (2012). https://doi.org/10.1016/j.jda.2011.12.008
4. Brinkop, H., Jansen, K.: Solving cut-problems in quadratic time for graphs with bounded treewidth. https://arxiv.org/abs/2101.00694
5. Calabro, C., Impagliazzo, R., Paturi, R.: The complexity of satisfiability of small depth circuits. In: Chen, J., Fomin, F.V. (eds.) IWPEC 2009. LNCS, vol. 5917, pp. 75–85. Springer, Heidelberg (2009). https://doi.org/10.1007/978-3-642-11269-0_6
6. Cygan, M., Mucha, M., Węgrzycki, K., Włodarczyk, M.: On problems equivalent to (min,+)-convolution. ACM Trans. Algorithms **15**(1), 1–25 (2019). https://doi.org/10.1145/3293465. ISSN 1549-6325
7. Eiben, E., Lokshtanov, D., Mouawad, A.E.: Bisection of bounded treewidth graphs by convolutions. J. Comput. Syst. Sci. **119**, 125–132 (2021). https://doi.org/10.1016/j.jcss.2021.02.002
8. Feige, U., Yahalom, O.: On the complexity of finding balanced oneway cuts. Inf. Process. Lett. **87**(1), 1–5 (2003). https://doi.org/10.1016/S0020-0190(03)00251-5
9. Hanaka, T., Kobayashi, Y., Sone, T.: A (probably) optimal algorithm for bisection on bounded-treewidth graphs. Theor. Comput. Sci. **873**, 38–46 (2021). https://doi.org/10.1016/j.tcs.2021.04.023
10. Impagliazzo, R., Paturi, R.: On the complexity of k-sat. J. Comput. Syst. Sci. **62**(2), 367–375 (2001). https://doi.org/10.1006/jcss.2000.1727
11. Jansen, K., Karpinski, M., Lingas, A., Seidel, E.: Polynomial time approximation schemes for max-bisection on planar and geometric graphs. SIAM J. Comput. **35**(1), 110–119 (2005). https://doi.org/10.1137/S009753970139567X

12. Javadi, R., Nikabadi, A.: On the parameterized complexity of sparsest cut and small-set expansion problems. Computing Research Repository, abs/1910.12353 (2019). http://arxiv.org/abs/1910.12353

13. Karp, R.M.: Reducibility among Combinatorial Problems. In: Miller, R.E., Thatcher, J.W., Bohlinger, J.D. (eds.) Complexity of Computer Computations, pp. 85–103. Springer, Boston (1972). https://doi.org/10.1007/978-1-4684-2001-2_9

14. Katsikarelis, I.: Computing bounded-width tree and branch decompositions of k-outerplanar graphs. Computing Research Repository, abs/1301.5896 (2013). http://arxiv.org/abs/1301.5896

15. Kloks, T. (ed.): Treewidth, Computations and Approximations. LNCS, vol. 842. Springer, Heidelberg (1994). https://doi.org/10.1007/BFb0045375

16. Lokshtanov, D., Marx, D., Saurabh, S.: Known algorithms on graphs of bounded treewidth are probably optimal. ACM Trans. Algorithms **14**(2), 13:1–13:30 (2018). https://doi.org/10.1145/3170442

17. Mahoney, M.: Lecture: Overview of graph partitioning. https://www.stat.berkeley.edu/~mmahoney/s15-stat260-cs294/Lectures/lecture05-05feb15.pdf. Accessed 04 Jan 2022

More Effort Towards Multiagent Knapsack

Sushmita Gupta[1], Pallavi Jain[2], and Sanjay Seetharaman[1(✉)]

[1] The Institute of Mathematical Sciences, HBNI, Chennai, India
{sushmitagupta,sanjays}@imsc.res.in
[2] IIT Jodhpur, Jodhpur, India
pallavi@iitj.ac.in

Abstract. In this paper, we study two multiagent variants of the knapsack problem. Fluschnik et al. [AAAI 2019] studied the model in which each agent expresses its preference by assigning a utility to every item. They studied three preference aggregation rules for finding a subset (knapsack) of items: individually best, diverse, and Nash welfare-based. Informally, diversity is achieved by satisfying as many agents as possible. Motivated by the application of aggregation operators in multiwinner elections, we extend the study from diverse aggregation rule to Median and Best scoring functions. We study the computational and parameterized complexity of the problem with respect to some natural parameters, namely, the number of agents, the number of items, and the distance from an easy instance. We also study the complexity of the problem under domain restrictions. Furthermore, we present significantly faster parameterized algorithms with respect to the number of agents for the diverse aggregation rule.

Keywords: Social choice · Voting · Complexity

1 Introduction

Knapsack is a paradigmatic problem in the area of optimization research, and its versatility in modeling situations with dual objectives/criterion is well established [28]. Unsurprisingly, it has been generalized and extended to incorporate additional constraints; and encoding *preferences* is a move in that direction. This type of modeling allows us to address the need for mechanisms to facilitate complex decision-making processes involving multiple agents with competing objectives while dealing with limited resources. We will use *multiagent knapsack* to refer to this setting. One natural application of multiagent knapsack would be in the area of participatory budgeting (PB, in short), a democratic process in which city residents decide on how to spend the municipal budget.

In the context of multiagent knapsack, approval ballot/voting is among the most natural models. It fits quite naturally with the motivation behind PB, in which for a given set of projects, every voter approves a set of projects that s/he would like to be executed. However, *utilitarian voting* (variously called score

L. Gąsieniec (Ed.): SOFSEM 2023, LNCS 13878, pp. 47–62, 2023.
https://doi.org/10.1007/978-3-031-23101-8_4

or range voting) is a more enriched ballot model, where every voter expresses his/her preferences via a utility function that assigns a numerical value to each alternative on the ballot (approval ballot is a special case as 0 utility denotes disapproval). This model applies to PB very well in situations where residents do not really have the motivation to reject any project. For example, consider the proposal to build different sporting facilities in a city. One might like a sport over another but would not really have any objections against building facilities for *any* sport. So, instead of disapproving a project, residents could be asked to give some numerical value (called utility) to every project, depending on how much they value that project. In a realistic scenario, citizens would be asked to "rate" the proposed projects by a number between 1 to 10, with 10 being the highest. A larger (smaller) range can be conveniently chosen if the number of proposed projects is reasonably high (low).

Recently, this model has been studied by Fluschnik et al. [18], analysing the problem from a computational viewpoint; and Aziz and Lee [5] who considered the axiomatic properties. Formally, we define the problem as follows:

$\mathcal{R}$-UTILITARIAN KNAPSACK ($\mathcal{R}$-UK)

Input: A set of items $\mathscr{P} = \{p_1, \ldots, p_m\}$, a cost function $c : \mathscr{P} \to \mathbb{N}$, a set of n voters $\mathscr{V}$, a utility function $\text{util}_v : \mathscr{P} \to \mathbb{N}$ for every voter $v \in \mathscr{V}$, a budget $b \in \mathbb{N}$, and a target $u_{tgt} \in \mathbb{N}$.

Question: Does there exist a set $\mathscr{X} \subseteq \mathscr{P}$ such that the cost of $\mathscr{X}$ is at most b and the total *satisfaction* (defined below) of voters is at least u_{tgt} under the voting rule $\mathcal{R}$?

Notably, COMMITTEE SELECTION problem is a special case of PB (or UK) where the *candidates* (items) are of unit cost, and we are looking for a *committee* (bundle) of size exactly b. An important and well-studied class of voting rules for COMMITTEE SELECTION is the class of *Committee Scoring Rules* (CSR) [15–17]. The *satisfaction* of a voter is given by a function that only depends on the utility of committee members, assigned by the voter. Towards this, for a committee S, we define the *utility vector* of a voter v as a vector of the utilities of the candidates in S, assigned by v. For example, let $\mathscr{P} = \{p_1, p_2, p_3\}$. Consider a voter v such that $\text{util}_v(p_1) = 3$, $\text{util}_v(p_2) = 2$, and $\text{util}_v(p_3) = 5$. Then, for $S = \{p_1, p_2\}$, the utility vector of v is $(3, 2)$. The scoring function, f, takes the utility vector and returns a natural number as the satisfaction of a voter. In these voting rules, the goal is to maximize the summation of satisfactions of all the voters.

Background. In PB, the city residents are asked for their opinion on the projects to be funded for the city, and then the preferences of all the voters are aggregated using some voting rule, which is used to decide the projects for the city. Initiated in Brazil in 1989 in the municipality of Porto Alegre [38], PB has become quite popular worldwide [19], including in the United States [33] and Europe [34]. In the last few years, PB has gained considerable attention from computer scientists [5–7,18,24–27,31,37]. We would like to note that Goel et al. [20] introduced the topic of *knapsack voting* that captures the process of aggregating the preferences of voters in the context of PB. The authors state that their motivation was to incorporate the classical knapsack problem by making the voter choose projects under the budget constraint but in a manner that aligns the constraints on the voters' decisions with those of the decision-makers. Their study is centered around strategic issues and extends knapsack voting further to more general settings with revenues, deficits, and surpluses.

Fluschnik et al. [18] propose three CSRs for $\mathcal{R}$-PB, and one of the rules uses a scoring function that takes the utility vector of a voter and returns the maximum utility in that vector as the satisfaction of the voter. This rule is a generalization of Chamberlin-Courant multiwinner (CC, in short, also known as the 1-median) voting rule, where the utilities are given by the Borda scores. This is called the DIVERSE KNAPSACK problem.

In this paper, we consider the *median scoring function* (and the *best scoring function*) [35], in which given a value $\lambda \in \mathbb{N}$, the scoring function takes a utility vector and returns the λth maximum value (the sum of the top λ values) as the satisfaction of the voter. If the size of the bundle is less than λ, then the utility of every voter under the median scoring function is 0. For a bundle $\mathcal{X} \subseteq \mathcal{P}$ and $\lambda \in \mathbb{N}$, let $\mathtt{sat}_v^\lambda(\mathcal{X})$ ($\mathtt{sat}_v^{[\lambda]}(\mathcal{X})$) denote the satisfaction of the voter v from the set $\mathcal{X}$ under median (best) scoring function. These rules are generalizations of the 1-median and k-best rules that are widely found in the literature. To the best of our knowledge, none of these are actually applied to real-life situations to determine the "winner". As with almost any mathematical model dealing with social choice, these are also proposals with provable guarantees and may be applied in practice. Let us consider a hypothetical PB situation where the proposed projects are building schools in different parts of the city, and the citizens are invited to vote for possible locations of choice. It is highly unlikely that after the facilities are built, all the students are admitted to their top choice. Suppose that they are guaranteed to be admitted to one of their top three choices. Then, aggregating preferences based on maximizing the utility of the λth most-preferred school for $\lambda = 3$ takes that uncertainty into account; and gives a lower bound on the utility derived by all who get admitted to one of their top λ choices.

Formally stated, we consider the problems MEDIAN-UTILITARIAN KNAPSACK (MEDIAN-UK, in short) and BEST-UTILITARIAN KNAPSACK (BEST-UK, in short) where the inputs are the same, and the goal is to decide if there exists a bundle $\mathcal{X} \subseteq \mathcal{P}$ such that $\sum_{p \in \mathcal{X}} \mathsf{c}(p) \leq \mathsf{b}$ and $\sum_{v \in \mathcal{V}} \mathtt{sat}_v^\lambda(\mathcal{X}) \geq \mathsf{u}_{\mathsf{tgt}}$ for the former; and $\sum_{v \in \mathcal{V}} \mathtt{sat}_v^{[\lambda]}(\mathcal{X}) \geq \mathsf{u}_{\mathsf{tgt}}$ for the latter. In the optimization version of this problem, we maximize the satisfaction of the voters. We use both variants (decision and optimization) in our algorithmic presentation, and the distinction will be clear from the context. Without loss of generality, we may assume that the cost of each item is at most the budget b. To be consistent with the literature, when $\lambda = 1$, we refer to both MEDIAN-UK and BEST-UK as DIVERSE KNAPSACK.[1]

---

[1] Missing details and proofs are present in a longer version of the paper at https://arxiv.org/abs/2208.02766.

A Generalized Model. It is worthwhile to point out that our work and that of Fluschnik et al. [18] are initial attempts at studying the vast array of problems in the context of multiagent knapsack with various preference elicitation schemes and various voting rules - this includes $\mathcal{R}$-UK as a subclass. For over a decade, researchers have studied scenarios where the votes/preferences are consistent with utility functions [1–4, 8, 9, 11, 32].

Algorithmic Concepts. A central notion in parameterized complexity is *fixed-parameter tractability*. A parameterized problem $L \subseteq \Sigma^* \times \mathbb{N}$ is *fixed-parameter tractable* (FPT) with respect to the parameter k (also denoted by FPT(k)), if for a given instance (x, k), its membership in L (i.e., $(x, k) \in L$) can be decided in time $f(k) \cdot \mathsf{poly}(|x|)$, where $f(\cdot)$ is an arbitrary computable function and $\mathsf{poly}(\cdot)$ is a polynomial function. However, all parameterized problems are not FPT. Contrastingly, W-hardness, captures the intractability in parameterized complexity. An XP algorithm for L with respect to k can decide $(x, k) \in L$ in time $|x|^{f(k)}$, where $f(\cdot)$ is an arbitrary computable function.

In some of our algorithms, we use the tool of a *reduction rule*, defined as a rule applied to the given instance of a problem to produce another instance of the same problem. A reduction rule is said to be *safe* if it is sound and complete, i.e., applying it to the given instance produces an equivalent instance. We refer the reader to books [12, 14, 30]. A central notion in the field of approximation algorithms is the *fully polynomial-time approximation scheme* (FPTAS). It is an algorithm that takes as input an instance of the problem and a parameter $\epsilon > 0$. It returns as output a solution whose value is at least $(1 - \epsilon)$ (resp. $(1 + \epsilon)$) times the optimal solution if it is a maximization (minimization) problem and runs in time polynomial in the input size and $1/\epsilon$.

Our Contributions. In this paper, we study the computational and parameterized complexity of MEDIAN-UK with respect to various natural input parameters and have put more effort towards identifying tractable special cases amid a sea of intractability. Moreover, we extend those results to BEST-UK.

For a start, we show that both the problems are NP-hard for every λ (Theorem 1). Since the COMMITTEE SELECTION problem with median scoring function is W[1]-hard with respect to n, the number of voters, for all $\lambda > 1$ [10], it follows that its generalization MEDIAN-UK must be as well (Corollary 1). Hence, even though an FPT algorithm with respect to n is unlikely for MEDIAN-UK, we are able to present an XP algorithm with respect to n (Theorem 2), that runs in time $\mathcal{O}((m(\lambda + 1))^n \mathsf{poly}(n, m))$. Additionally, it is known that MEDIAN-UK with $\lambda = 1$ (i.e. DIVERSE KNAPSACK), binary utilities, and unary costs is W[2]-hard with respect to budget b [18]. In the case of parameterization by both n and b, we obtain an XP algorithm for both the problems. While there is a trivial $\mathcal{O}^\star(2^m)$ algorithm, where m denotes the number of items, for both the problems, unless ETH fails there cannot be a $\mathcal{O}(2^{o(m+n)} \mathsf{poly}(n, m))$ algorithm [18]. In light of these dead ends, we turn our attention to identifying tractable special cases, and our search forked into three primary directions: the value of λ, preference profiles, and encoding of utilities and costs. When $\lambda = 1$, the problem is NP-hard [18] and remains so even when profile is single-crossing or single-peaked. We show that despite this intractability, when the profile is *unanimous*, that is, all the voters have the exact same top preference, DIVERSE KNAPSACK is polynomial-time solvable (Lemma 3); but both MEDIAN-UK and BEST-UK remain NP-hard for unanimous as well as single-crossing profiles (Theorem 1) when $\lambda > 1$. In the situation where there is a priority ordering over all the items, captured by a *strongly unanimous* profile, MEDIAN-UK is polynomial-time solvable. Furthermore, we observe that given an instance, the "closer" it is to a strongly unanimous profile, the faster the solution can be computed. Consequently, we consider d, the distance away from strong unanimity, as a parameter and show that DIVERSE KNAPSACK is FPT with respect to d. In the special case where $\lambda = 1$ and the utilities or costs are polynomially bounded or are encoded in unary, we have three different FPT algorithms, all of which are improvements on the one given by [18]. Additionally, when $\lambda = 1$ and the profile is single-peaked (or crossing) the problem admits an FPTAS (Theorem 6), an improvement over the $(1 - 1/e)$-factor approximation algorithm by [18]. Table 1 summarizes the main results with precise running times of our algorithms.

Table 1. Results on MEDIAN-UK and BEST-UK. Here n, m, b, and d denote the number of voters, the number of items, the budget, and the distance away from SU, respectively. The abbreviations SC, SP, SU, and U are preference restrictions which are defined in Sect. 1.

Restriction	Result	Ref.
SC $\lambda > 1$ & U	NP-hard	Theorem 1
$\lambda = 1$	$\mathcal{O}(n!\,\mathrm{poly}(\hat{u}, n, m))$	[18]
	$\mathcal{O}(4^n \mathrm{poly}(\bar{u}, n, m))$	Theorem 4
	$\mathcal{O}(4^n \mathrm{poly}(\mathbf{b}, n, m))$	Corollary 4
	$\mathcal{O}(2^n \mathrm{poly}(\bar{u}, \mathbf{b}, n, m))$	Theorem 5
$\lambda > 1$ & MEDIAN-UK	$\mathcal{O}((m(\lambda + 1))^n \mathrm{poly}(n, m))$	Theorem 2
	W [1]-hard w.r.t. n	[10]
No restriction	$\mathcal{O}^*(\mathbf{b}^2 (\mathbf{b}!)^n m^{\mathbf{b}})$	Theorem 3
SU & MEDIAN-UK	$\mathrm{poly}(n, m)$	Lemma 2
SU & BEST-UK	$\mathrm{poly}(\bar{u}, n, m), \mathrm{poly}(\mathbf{b}, n, m)$	Lemma 2
MEDIAN-UK	XP w.r.t. d, XP w.r.t. $\mathbf{b}, d$	Corollary 3
BEST-UK	XP w.r.t. $\mathbf{b}, d$	Corollary 3
$\lambda = 1$	FPT w.r.t. d for unary utilities	Corollary 2
$\lambda = 1$ & SP/SC	FPTAS	Theorem 6
$\lambda = 1$ & U	$\mathcal{O}(m)$	Lemma 3

Preliminaries. Let $\mathscr{V} = \{v_1, \ldots, v_n\}$ be a set of n voters and $\mathscr{P} = \{p_1, \ldots, p_m\}$ be a set of m items. The preference ordering of a voter $v \in \mathscr{V}$ over the set of items is given by the utility function $\mathtt{util}_v : \mathscr{P} \to \mathbb{N}$. That is, if $\mathtt{util}_v(p) \geq \mathtt{util}_v(p')$, then v prefers p more than p', and we use the notation $p \succ_v p'$. We drop the subscript when it is clear from the context. For a subset $Y \subseteq \mathscr{V}$, we use $\mathtt{util}_Y(p)$ to denote the utility of the item p to voters in Y: $\mathtt{util}_Y(p) = \sum_{v \in Y} \mathtt{util}_v(p)$. The set of utility functions form the *utility profile*, denoted by $\{\mathtt{util}_v\}_{v \in \mathscr{V}}$. For integers i, j, we use $[i, j]$ to denote the set $\{i, i+1, \ldots, j\}$. For an integer i, we use $[i]$ to denote $[1, i]$. In an instance of MEDIAN-UK, we say that an item $p \in \mathscr{P}$ is a *representative* of a voter $v \in \mathscr{V}$ in a bundle $\mathscr{Z} \subseteq \mathscr{P}$ if p is the λth most preferred item of v in $\mathscr{Z}$.

Preference profiles. A preference profile (P, in short) is said to be

- *unanimous* (U) if all voters have the same top preference;
- *strongly unanimous* (SU) if all voters have identical preference orderings;
- *single-crossing* (SC) if there exists an ordering σ on voters $\mathscr{V}$ such that for each pair of items $\{p, p'\} \subseteq \mathscr{P}$, the set of voters $\{v \in \mathscr{V} : \mathtt{util}_v(p) \geq \mathtt{util}_v(p')\}$ forms a consecutive block according to σ;
- *single-peaked* (SP) if the following holds for some ordering, denoted by $\lhd$, on the items $\mathscr{P}$: Let $\mathtt{top}_v$ denote voter v's most preferred item. Then, for each pair of items $\{p, p'\} \subseteq \mathscr{P}$ and each voter $v \in \mathscr{V}$, such that $p \lhd p' \lhd \mathtt{top}_v$ or $\mathtt{top}_v \lhd p' \lhd p$ we have that v weakly prefers p' over p, i.e. $\mathtt{util}_v(p') \geq \mathtt{util}_v(p)$.

2   Hardness of Median-UK

We dedicate this section to identifying intractable cases. Clearly, MEDIAN-UK is NP-hard as its unweighted version, COMMITTEE SELECTION, under the same rule, is NP-hard [35]. We begin with the following strong intractability result.

Theorem 1. MEDIAN-UK *and* BEST-UK *are* NP-hard *for every* $\lambda \in \mathbb{N}$ *even for SCPs. Furthermore, for* $\lambda > 1$, *both are* NP-hard *even for UPs.*

Proof. We give a polynomial-time reduction from the DIVERSE KNAPSACK problem, which is known to be NP-hard for SCPs [18]. Let $\mathcal{I} = (\mathscr{P}, \mathscr{V}, \mathtt{c}, \{\mathtt{util}_v\}_{v \in \mathscr{V}}, \mathtt{b}, \mathtt{u_{tgt}})$ be an instance of DIVERSE KNAPSACK with SCP. Let $\sigma = (v_1, \ldots, v_n)$ be an SC ordering of the voters in $\mathscr{V}$. Without loss of generality, let the preference order of v_1 be $p_1 \succ p_2 \succ \ldots \succ p_m$. Let u_{max} denote the maximum utility that a voter assigns to an item.

To construct an instance of MEDIAN-UK, the plan is to add $\lambda - 1$ new items, say $\mathscr{P}_{new} = \{p_{m+1}, \ldots, p_{m+\lambda-1}\}$, to $\mathscr{P}$ and ensure that they are in any feasible bundle by setting appropriate utilities and costs. Formally, we construct an instance $\mathcal{I}' = (\mathscr{P}', \mathscr{V}, \mathtt{c}', \{\mathtt{util}'_v\}_{v \in \mathscr{V}}, \mathtt{b}', \mathtt{u_{tgt}}, \lambda)$ of MEDIAN-UK as follows. Let $\mathscr{P}' = \mathscr{P} \cup \mathscr{P}_{new}$. For $p \in \mathscr{P}$, let $\mathtt{c}'(p) = \mathtt{c}(p)$, and for $p \in \mathscr{P}_{new}$, let $\mathtt{c}'(p) = 1$; and we set the budget $\mathtt{b}' = \mathtt{b} + \lambda - 1$. Note that the set of voters and the target utility in the instance $\mathcal{I}'$ are the same as in the instance $\mathcal{I}$. Next, we construct the utility function for a voter $v \in \mathcal{V}$ as

$$\text{util}'_v(p_j) = \begin{cases} u_{max} + j - m & \text{if } p_j \in \mathscr{P}_{new}, \\ \text{util}_v(p_j) & \text{otherwise.} \end{cases}$$

Clearly, the construction is doable in polynomial time. Without loss of generality, we assume that $u_{tgt} > 0$, otherwise we return an empty bundle.

In the case of BEST-UK, we can use the same reduction as above. From the construction, it follows that $\mathscr{P}_{new}$ belongs to any feasible bundle. We obtain the hardness result proceeding similarly.   □

Since COMMITTEE SELECTION problem is W[1]-hard with respect to n for $\lambda > 1$ [10], we have the following result.

Corollary 1. MEDIAN-UK *is* W[1]-hard *with respect to* n *for* $\lambda > 1$.

Hence, it is unlikely to be FPT. Now, we show that there exists an XP algorithm for MEDIAN-UK with respect to n. The intuition is as follows. For every voter $v \in \mathcal{V}$, we guess the representative in the solution. We know that for every voter v, we must pick $\lambda - 1$ other items whose utility is larger than the representative's. Thus, we reduce the problem to the WEIGHTED SET MULTICOVER (WSMC) problem, which is defined as follows:

WEIGHTED SET MULTICOVER (WSMC)
Input: A universe U, a family, $\mathcal{F}$, of subsets of U, a cost function $c \colon \mathcal{F} \to \mathbb{N}$, a budget $b \in \mathbb{N}$, and a positive integer $k \in \mathbb{N}$.
Question: Find a subset $\mathcal{F}' \subseteq \mathcal{F}$ such that every element of U is in at least k sets in $\mathcal{F}'$, and $\sum_{F \in \mathcal{F}'} c(F) \leq b$.

If a set F contains an element $u \in U$, then we say that F *covers* u. We first present an FPT algorithm for WSMC with respect to $k + n$, which is similar to the FPT algorithm for SET COVER with respect to n.

Lemma 1. WSMC *can be solved in* $\mathcal{O}((k+1)^n |\mathcal{F}|)$ *time.*

Proof. We give a dynamic-programming algorithm. Let $(U, \mathcal{F}, c, b, k)$ be an instance of WSMC. Let $U = \{u_1, \ldots, u_n\}$ and $\mathcal{F} = \{F_1, \ldots, F_m\}$. For a set $S \subseteq U$, let $\chi(S)$ denote the characteristic vector of S, i.e., it is an $|U|$-length vector such that $\chi(S)_i = 1$ if and only if $u_i \in S$. For any two $|U|$-length vectors $\vec{A}, \vec{B}$, we define the difference $\vec{C} = \vec{A} - \vec{B}$ as $\vec{C}_i = \max(0, \vec{A}_i - \vec{B}_i)$ for each $i \in [|U|]$. Let $\mathcal{R} = \{0, 1, \ldots, k\}^n$. We define the dynamic-programming table as follows. For every $\vec{X} \in \mathcal{R}$ and $j \in [0, m]$, let $T[\vec{X}, j]$ be the minimum cost of a subset of $\{F_1, \ldots, F_j\}$ that covers u_i at least $\vec{X}_i$ times for every $i \in [n]$. We compute the table entries as follows.

Base Case: For each $\overrightarrow{X} \in \mathcal{R}$, we set

$$T[\overrightarrow{X}, 0] = \begin{cases} 0, & \text{if } \overrightarrow{X} = \{0\}^n, \\ \infty, & \text{otherwise.} \end{cases} \tag{1}$$

Recursive Step: For each $\overrightarrow{X} \in \mathcal{R}$ and $j \in [m]$, we set

$$T[\overrightarrow{X}, j] = \min\{T[\overrightarrow{X}, j-1], T[\overrightarrow{X} - \chi(F_j), j-1] + \mathsf{c}(F_j)\}. \tag{2}$$

If there exists an $\overrightarrow{X} \in \mathcal{R}$ such that $\overrightarrow{X}_i \geq k$ for each $i \in [n]$, and the value in the entry $T[\overrightarrow{X}, m]$ is at most b, then we return "YES", else we return "NO". $\square$

Now, we are ready to give an XP algorithm for MEDIAN-UK.

Theorem 2. MEDIAN-UK *can be solved in* $\mathcal{O}((m(\lambda + 1))^n poly(n, m))$ *time.*

Proof. We begin by guessing the representative of every voter. Let r_v denote the guessed representative of a voter v, and R be the set of all guessed representatives. Next, we construct an instance of WSMc as follows. For every voter $v \in \mathcal{V}$, we add an element e_v to U. Corresponding to every item $p \in \mathcal{P}$, we have a set F_p in $\mathcal{F}$, where $F_p = \{e_v : \mathtt{util}_v(p) \geq \mathtt{util}_v(r_v)\}$. Note that F_p contains the elements corresponding to the set of voters who prefer p at least as much as their representative. For every $F_p \in \mathcal{F}$, $\mathsf{c}(F_p) = \mathsf{c}(p)$. We set $\mathfrak{b} = \mathsf{b}$ and $k = \lambda$. Let $\mathcal{F}' \subseteq \mathcal{F}$ be the solution of the WSMc instance $(U, \mathcal{F}, \mathsf{c}, \mathfrak{b}, k)$. We construct a set $P' \subseteq \mathcal{P}$ that contains items corresponding to the sets in $\mathcal{F}'$, i.e., $P' = \{p \in \mathcal{P} : F_p \in \mathcal{F}'\}$. Note that the cost of P' is at most b. Also, every element of U is covered at least λ times. Therefore, for every voter v, P' contains at least λ items whose utilities are at least that of r_v. Thus, the total utility is at least $\sum_{v \in \mathcal{V}} \mathtt{util}_v(r_v)$. Note that the cost and utility of a bundle can be computed in time polynomial in n and m. Since "guessing" the representatives takes m^n steps, the running time follows due to Lemma 1. $\square$

We cannot expect a similar result for BEST-UK due to the following reduction. Let $(A = \{a_1, \ldots, a_m\}, \mathsf{b}, w, \mathsf{c}', \mathsf{v}_{\mathsf{tgt}})$ be an instance of KNAPSACK problem where each item a_i costs $\mathsf{c}'(a_i)$ and has value $w(a_i)$. The goal is to determine if there exists a knapsack with value at least $\mathsf{v}_{\mathsf{tgt}}$ and cost at most b. Construct an instance of BEST-UK with exactly one voter v_1, set of items A, budget b, target utility $\mathsf{v}_{\mathsf{tgt}}$, and $\lambda = m$. An XP algorithm for BEST-UK will give us a polynomial time algorithm for KNAPSACK, implying P=NP.

Due to Bredereck et al. [10], BEST-UK is FPT(n, λ) when the item costs are equal to 1. We extend their algorithm to solve both MEDIAN-UK and BEST-UK.

Theorem 3. MEDIAN-UK *and* BEST-UK *can be solved in* $\mathcal{O}^*(\mathsf{b}^2(\mathsf{b}!)^n m^{\mathsf{b}})$ *time.*

Proof. Since the item costs are natural numbers, the maximum cardinality of a solution bundle is $\min(\mathsf{b}, m)$. Let $S = \{s_1, \ldots, s_k\}$ be a solution bundle. First, we guess the value of $k = |S|$. Next, for each voter v_i, we guess the permutation

ψ_i of $[k]$: $\psi_i(j) = l$ if and only if s_j is the lth most preferred item of v_i in S. Note that ψ_i gives us the projection of S on the preference order of v_i: $\succ_i$. Let $\mathbb{I}[\psi_i(j)]$ indicate whether $\psi_i(j) = \lambda$. We consider the following complete bipartite graph G with bipartitions $[k]$ and $\mathscr{P}$: the weight of an edge (l, p) is given by $w(l, p) = \sum_{v_i \in \mathscr{V}} \mathbb{I}[\psi_i(l)] \cdot \mathtt{util}_{v_i}(p)$. An edge (l, p) captures item p taking the role of s_l in S: $w(l, p)$ is the contribution of s_l to the utility of S. Our goal is to find a matching M of size k that agrees with the guesses such that the cost of the bundle corresponding to M is at most $\mathtt{b}$, and the weight of M is maximum. Each of the m items in $\mathscr{P}$ can be encoded using $ln = \lfloor \log_2 m \rfloor + 1$ bits. Thus, any matching of size k in G can be encoded using $k \cdot ln$ bits: the first ln bits represent the item that s_1 is matched to, and so on. To find M, we iterate through all $2^{k \cdot ln}$ binary strings of length $k \cdot ln$ and look for a desired matching.

Let $M \subseteq E(G)$ be a matching with the maximum weight over all possible guesses of k and $\{\psi_i : v_i \in \mathscr{V}\}$. We return "YES" if the weight of M is at least $\mathtt{u_{tgt}}$; otherwise we return "NO".

Complexity: There are $\mathtt{b}$ guesses for the cardinality of S, and $\sum_{k \in [\mathtt{b}]} (k!)^n$ guesses for the permutations $\{\psi_i : v_i \in \mathscr{V}\}$. For each guess k, $\{\psi_i : v_i \in \mathscr{V}\}$, we iterate through $2^{k \cdot ln}$ strings of length $k \cdot ln$, and for each string we spend polynomial time to process. Thus, the total running time is $\mathcal{O}^*(\mathtt{b} \cdot \mathtt{b}(\mathtt{b}!)^n \cdot m^\mathtt{b})$.

By redefining the indicator function $\mathbb{I}$ as $\mathbb{I}(\psi_i(j)) = 1$ if and only if $\psi_i(j) \leq \lambda$ and continuing as before, we obtain the result for BEST-UK. □

3   Algorithms for Special Cases

We now present algorithms for some restrictions in the input. First, we consider the case of a SUP in which all voters have identical preference orders. We first apply the following reduction rule that simplifies the input.

Reduction Rule 1 (†). *If there exist two voters $v_1, v_2 \in \mathscr{V}$ such that their preference orders are identical, then we can "merge" the two voters: Set $\mathscr{V}' = (\mathscr{V} \setminus \{v_1, v_2\}) \cup \{v_{12}\}$ with $\mathtt{util}'_{v_{12}}(p) = \mathtt{util}_{v_1}(p) + \mathtt{util}_{v_2}(p)$, for all $p \in \mathscr{P}$, and for all other voters $\mathtt{util}'_v(p) = \mathtt{util}_v(p)$. The new instance is $\mathcal{I}' = (\mathscr{P}, \mathscr{V}', \mathtt{c}, \{\mathtt{util}'_v\}_{v \in \mathscr{V}'}, \mathtt{b}, \mathtt{u_{tgt}}, \lambda)$.*

After the exhaustive application of Reduction Rule 1, the profile contains only one voter. In the case of BEST-UK with one voter, it is equivalent to a knapsack problem on the items. Thus, we have the following lemma.

Lemma 2 (†). MEDIAN-UK *can be solved in polynomial time for SUPs. Furthermore,* BEST-UK *can be solved in polynomial time for SUPs, when either the budget, or the utilities are encoded in unary.*

Now, we consider the distance from an easy instance as a parameter, which is a natural parameter in parameterized complexity. Let d be the minimum number of voters that need to change their utility function so that the profile is

SU. Given an instance of MEDIAN-UK, d can be computed in polynomial time as follows. First, we sort the utility profile. Let l be the length of the longest block of voters with the same utility function. Then, $d = n - l$. We consider the problem parameterized by d. This parameter has also been studied earlier for the CONNECTED CC problem [22].

Without loss of generality, we assume that voters $\{v_{d+1}, \ldots, v_n\}$ are SU. We first apply Reduction Rule 1 exhaustively. Note that after the exhaustive application of the rule, the instance has at most $d + 1$ voters. Since DIVERSE KNAPSACK is FPT with respect to n when the utilities are encoded in unary [18], we have the following corollary.

Corollary 2. DIVERSE KNAPSACK is FPT *with respect to* d *when utilities are encoded in unary.*

Furthermore, due to Theorems 2 and 3, we have the following.

Corollary 3

(1) MEDIAN-UK *can be solved in* $\mathcal{O}((m(\lambda + 1))^{d+1} poly(d, m))$ *time.*
(2) MEDIAN-UK *and* BEST-UK *can be solved in* $\mathcal{O}^*(\mathtt{b}^2(\mathtt{b}!)^{d+1} m^{\mathtt{b}})$ *time.*

3.1   When $\lambda = 1$: Diverse Knapsack

Unlike MEDIAN-UK for $\lambda > 1$, DIVERSE KNAPSACK can be solved in polynomial time for UPs. The optimal solution contains the most preferred item of all voters.

Lemma 3 (†). DIVERSE KNAPSACK *can be solved in polynomial time for UPs.*

Let $\mathcal{I} = (\mathscr{P}, \mathscr{V}, \mathtt{c}, \{\mathtt{util}_v\}_{v \in \mathscr{V}}, \mathtt{b}, \mathtt{u}_{\mathrm{tgt}})$ be an instance of DIVERSE KNAP-SACK. For our subsequent discussions, we define

$$\bar{u} = \sum_{v \in \mathscr{V}} \max_{p \in \mathscr{P}} \mathtt{util}_v(p), \text{ and } \hat{u} = \sum_{p \in \mathscr{P}} \sum_{v \in \mathscr{V}} \mathtt{util}_v(p).$$

Next, we design FPT algorithms with respect to n for DIVERSE KNAPSACK. Fluschnik et al. [18] gave an algorithm that runs in $\mathcal{O}(n! \, poly(\hat{u}, n, m))$, which is an FPT algorithm with respect to n when the total utility is either unary encoded or bounded by $poly(n, m)$.

We first give an algorithm that runs in $\mathcal{O}(4^n \, poly(\bar{u}, n, m))$ time. Further, we give an algorithm that runs in $\mathcal{O}(2^n \, poly(\bar{u}, \mathtt{b}, n, m))$ time. It is worth mentioning that the best known algorithm for CHAMBERLIN COURANT, a special case of DIVERSE KNAPSACK, also runs in $\mathcal{O}(2^n \, poly(n, m))$ time [23].

A $\mathcal{O}(4^n \, poly(\bar{u}, n, m))$ algorithm. To design the algorithm, we first reduce the problem to the following variant of the SET COVER problem, which we call KNAPSACK COVER due to its similarity with the KNAPSACK and SET COVER problems. Then, using the algorithm for KNAPSACK COVER (Theorem 5), we obtain the desired algorithm for DIVERSE KNAPSACK.

KNAPSACK COVER

Input: Two sets of universe $U_1 = \{u_1^1, \ldots, u_1^n\}$ and $U_2 = \{u_2^1, \ldots, u_2^m\}$, a family of sets $\mathscr{F} = \{\{F, u\} : F \subseteq U_1, u \in U_2\}$, a profit function $\texttt{profit} : \mathscr{F} \to \mathbb{N}$, a cost function $\texttt{cost} : \mathscr{F} \to \mathbb{N}$, budget $\texttt{b} \in \mathbb{N}$, and total profit $\texttt{p}$.

Question: Does there exist a set $\mathscr{Z} \subseteq \mathscr{F}$ such that (i) for every two sets $\{F, u\}$ and $\{F', u'\}$ (u can be equal to u') in $\mathscr{Z}$, $F \cap F' = \emptyset$, (ii) $\bigcup_{\{F,u\} \in \mathscr{Z}} F = U_1$, (iii) $\sum_{\{F,u\} \in \mathscr{Z}} \texttt{cost}(\{F, u\}) \leq \texttt{b}$, and (iv) $\sum_{\{F,u\} \in \mathscr{Z}} \texttt{profit}(\{F, u\}) \geq \texttt{p}$?

We first discuss the intuition behind reducing DIVERSE KNAPSACK to KNAPSACK COVER. Consider a non-empty bundle S. For each voter, there exists an item that represents the voter in the bundle. An item can represent more than one voter. For each subset of voters X and each item p, we create a set $\{X, p\}$. The goal is to find a family of sets such that the set of voters is disjointly covered by the voter subsets, and the set of items forms a bundle with utility at least $\texttt{p}$ and cost at most $\texttt{b}$. We first present the reduction from DIVERSE KNAPSACK to KNAPSACK COVER, which formalizes the intuition, in the following lemma.

Lemma 4. DIVERSE KNAPSACK *can be reduced to* KNAPSACK COVER *in* $\mathcal{O}(2^n poly(n, m))$ *time.*

Proof. Given an instance $\mathcal{I}$, we create an instance $\mathcal{J} = (U_1, U_2, \mathscr{F}, \texttt{cost}, \texttt{profit}, \texttt{b}', \texttt{p})$ of KNAPSACK COVER as follows. We first construct universe sets: $U_1 = \mathscr{V}$, and $U_2 = \mathscr{P}$. For each $X \subseteq U_1$ and each $r \in U_2$, we add a set $\{X, r\}$ in the set family $\mathscr{F}$. Note that the size of $\mathscr{F}$ is $2^n m$. Next, we define the cost and the profit functions. For every set $\{X, r\} \in \mathscr{F}$, $\texttt{cost}(\{X, r\}) = \texttt{c}(r)$ and $\texttt{profit}(\{X, r\}) = \texttt{util}_X(r)$. We set $\texttt{b}' = \texttt{b}$ and $\texttt{p} = \texttt{u}_{tgt}$. This completes the proof of Lemma 4. $\qquad\square$

Next, we design an algorithm for KNAPSACK COVER.

Lemma 5 (†). KNAPSACK COVER *can be solved in* $\mathcal{O}(2^{|U_1|}|\mathscr{F}|p_{max})$ *time, where* $p_{max} = \sum_{F \in \mathscr{F}} \texttt{profit}(F)$.

Lemmas 4 and 5 give us an algorithm for DIVERSE KNAPSACK. Analogous to Lemma 2, an alternative dynamic-programming approach to solve KNAPSACK COVER is that instead of finding the minimum cost of a subset with a particular utility, we find the maximum utility of a subset with a particular cost. This results in an algorithm running in time $\mathcal{O}(2^{|U_1|}|\mathscr{F}|\texttt{b})$. Thus, we have the following results.

Theorem 4. DIVERSE KNAPSACK *can be solved in* $\mathcal{O}(4^n poly(\bar{u}, n, m))$ *time.*

Corollary 4. DIVERSE KNAPSACK *can be solved in* $\mathcal{O}(4^n poly(\texttt{b}, n, m))$ *time.*

A $\mathcal{O}(2^n poly(\bar{u}, \texttt{b}, n, m))$ algorithm. Next, we give an algorithm that improves the exponential dependency on n, but it additionally has a dependency on the budget. In particular, we prove the following theorem.

Theorem 5. DIVERSE KNAPSACK *can be solved in* $\mathcal{O}(2^n \, poly(\bar{u}, \mathbf{b}, n, m))$ *time.*

To prove Theorem 5, we use the technique of polynomial multiplication, which has also been recently used to give a $\mathcal{O}(2^n \, poly(n, m))$-time algorithm for CC [23]. Our algorithm is similar to the one for CC. Here, we additionally keep track of the budget.

Note that there exists a solution S of $\mathcal{I}$ that induces an $|S|$-sized partition of the voter set, because if an item is not a representative of any voter, then we can delete this item from the solution, and it still remains a solution. We use the method of polynomial multiplication to find such a partition that is induced by some solution. We begin with defining some notations and terminologies, which are same as in [23].

Let $U = \{u_1, \ldots, u_n\}$. For a subset $X \subseteq U$, let $\chi(X)$ denote the characteristic vector of the set X: an $|U|$-length vector whose j^{th} bit is 1 if and only if $u_j \in X$. Two binary strings of length n are said to be *disjoint* if for each $i \in [n]$, the ith bit of both strings is not same. Let $\mathcal{H}(S)$ denote the *Hamming weight* of a binary string S: the number of 1s in S. The Hamming weight of a monomial x^i, where i is binary vector, is the Hamming weight of i. Let $\mathcal{H}_s(P(x))$ denote the *Hamming projection* of a polynomial $P(x)$ to a non-negative integer s: the sum of all monomials in $P(x)$ with Hamming weight s. Let $\mathcal{R}(P(x))$ denote the *representative polynomial* of $P(x)$: if the coefficient of a monomial is non-zero in $P(x)$, then the coefficient of the monomial is one in $\mathcal{R}(P(x))$. Next, we state a basic result following which we prove Theorem 5.

Lemma 6 [13,21]. *Subsets $S_1, S_2 \subseteq U$ are disjoint if and only if Hamming weight of the string $\chi(S_1) + \chi(S_2)$ is $|S_1| + |S_2|$.*

Proof (Proof of Theorem 5). We define the polynomials as follows. We first construct the Type-1 polynomial, in which for each $s \in [n]$, $\alpha \in [\bar{u}]$, and $\beta \in [\mathbf{b}]$, the non-zero polynomial $P^1_{s,\alpha,\beta}(x)$ denotes that there exists an s-sized subset of voters $Y \subseteq \mathcal{V}$ corresponding to which there is an item $p \in \mathscr{P}$ such that $\mathtt{c}(p) = \beta$ and $\mathtt{util}_Y(p) = \alpha$:

$$P^1_{s,\alpha,\beta}(x) = \sum_{\substack{Y \subseteq \mathcal{V}: \, |Y|=s \\ \exists p \in \mathscr{P}: \, \mathtt{util}_Y(p)=\alpha \\ \mathtt{c}(p)=\beta \leq \mathbf{b}}} x^{\chi(Y)}.$$

Next, for every $s \in [n]$, $\alpha \in [\bar{u}]$, $\beta \in [\mathbf{b}]$, and $j \in [2, n]$, we iteratively define the Type-j polynomial as follows:

$$P^j_{s,\alpha,\beta}(x) = \sum_{\substack{s_1,s_2 \in [n]: \, s_1+s_2=s \\ \alpha_1,\alpha_2 \in [\bar{u}]: \, \alpha_1+\alpha_2=\alpha \\ \beta_1,\beta_2 \in [\mathbf{b}]: \, \beta_1+\beta_2=\beta}} \mathcal{R}(\mathcal{H}_s(P^1_{s_1,\alpha_1,\beta_1} \times P^{j-1}_{s_2,\alpha_2,\beta_2})).$$

A non-zero polynomial $P^j_{s,\alpha,\beta}(x)$ denotes that there exists j disjoint voter subsets $Y_1, \ldots, Y_j$ such that $|Y_1| + \ldots + |Y_j| = s$, and there exists items $p_1, \ldots, p_j$

such that $\sum_{i\in[j]} \mathtt{util}_{Y_i}(p_i) = \alpha$ and $\sum_{i\in[j]} \mathsf{c}(p_i) = \beta \le \mathsf{b}$. Among all polynomials, we check if for some $\alpha \ge \mathsf{u_{tgt}}$ and $\beta \le \mathsf{b}$, the polynomial $P^k_{n,\alpha,\beta}$ is non-zero. If so, we return "YES", otherwise we return "NO".

Note that the total number of polynomials is at most $n\bar{u}\mathsf{b}$. A polynomial has at most 2^n monomials, each of which has degree at most 2^n. Given two polynomials of degree 2^n, we can compute their product in time $\mathcal{O}(2^n n)$ [29]. Hence, the running time is $\mathcal{O}(2^n \text{ poly}(m, n, \bar{u}, \mathsf{b}))$. $\qquad\square$

FPTAS for Diverse Knapsack. Due to the submodularity, DIVERSE KNAPSACK admits a factor $(1 - 1/e)$-approximation algorithm [18,36]. Here, we give an FPTAS when the profile is either SP or SC.

The idea is to first scale down the utilities, round-off, and then use the known polynomial time algorithms for SP and SC profile under unary utilities [18].

Proposition 1 *[18]. DIVERSE KNAPSACK can be solved in time poly$(n, m, \hat{u})$, when the utility profile is encoded in unary, and is either SP or SC.*

Let $u_{max} = \max_{v\in\mathcal{V}, p\in\mathcal{P}} \mathtt{util}_v(p)$, the maximum utility that a voter assigns to an item. Let $0 < \epsilon \le 1$ be the error parameter. We scale down the utility of every voter for every item by a factor of s, where $s = (\epsilon/2n)u_{max}$, as follows: $\overline{\mathtt{util}}_v(p) = \lceil \mathtt{util}_v(p)/s \rceil$. Next, we round the utilities as follows: $\widetilde{\mathtt{util}}_v(p) = s \cdot \overline{\mathtt{util}}_v(p)$. Clearly,

$$\mathtt{util}_v(p) \le \widetilde{\mathtt{util}}_v(p) \le \mathtt{util}_v(p) + s. \tag{3}$$

For simplicity of analysis, we assume that s is an integer. From now on, by $\mathtt{util}(S)$, we denote $\sum_{v\in\mathcal{V}} \max_{p\in S} \mathtt{util}_v(p)$. Next, we show that the total utility of the optimal bundle under the utilities $\widetilde{\mathtt{util}}$ is not very far from the total utility of the optimal bundle under the utilities $\mathtt{util}$. In particular, we prove the following lemma, where $\widetilde{\mathcal{I}} = (\mathcal{P}, \mathcal{V}, \mathsf{c}, \{\widetilde{\mathtt{util}}_v\}_{v\in\mathcal{V}}, \mathsf{b})$.

Lemma 7 (†). *Let $S^\star$ be an optimal solution to $\widetilde{\mathcal{I}}$. Let S be any subset of $\mathcal{P}$ such that $\sum_{p\in S} \mathsf{c}(p) \le \mathsf{b}$. Then, $\mathtt{util}(S) \le (1 + \epsilon)\, \mathtt{util}(S^\star)$.*

In light of Lemma 7, our goal is reduced to finding an optimal solution for $\widetilde{\mathcal{I}}$. Note that for any $v \in \mathcal{V}$ and $p \in \mathcal{P}$, $\widetilde{\mathtt{util}}_v(p)$ and $\overline{\mathtt{util}}_v(p)$ differ by a factor of s. Thus, the utility of optimal solutions under these functions differ by a factor of s. Thus, our goal is reduced to finding an optimal solution for $\overline{\mathcal{I}} = (\mathcal{P}, \mathcal{V}, \mathsf{c}, \{\overline{\mathtt{util}}_v\}_{v\in\mathcal{V}}, \mathsf{b})$. Note that by scaling the utilities, the profile still remains SP or SC. Thus, we can solve $\overline{\mathcal{I}}$ using Proposition 1 in polynomial time. The running time is due to the fact that for any voter, the utility for any item under $\overline{\mathtt{util}}$ is at most $2n/\epsilon$. Thus, we have the following theorem.

Theorem 6. *There exists an FPTAS for DIVERSE KNAPSACK when the utility profile is either SP or SC.*

4 Conclusion

In this paper, we studied the computational and parameterized complexity of three multiagent variants of the knapsack problem. For DIVERSE KNAPSACK, we presented improved FPT algorithms parameterized by n, an FPTAS under SP and SC restrictions, and algorithms for special cases. Additionally, we gave some hardness results and algorithms for MEDIAN-UK and BEST-UK.

DIVERSE KNAPSACK is NP-hard even for SPPs, as shown in [18]. Further, for unary encoded utilities, DIVERSE KNAPSACK can be solved in polynomial time for SPP and SCP. An open question is the complexity of MEDIAN-UK/BEST-UK with SPP for $\lambda \geq 2$. In this paper, we studied two aggregation rules. One can study multiagent knapsack with other combinations of preference elicitation schemes and aggregation rules.

Acknowledgements. Pallavi Jain received funding from Seed Grant (IITJ/R&D/2022–23/07). Sushmita Gupta received funding from SERB's Matrics Grant (MTR/2021/000869). Additionally, both were supported by SERB'S SUPRA Grant (SPR/2021/000860).

References

1. Anshelevich, E., Bhardwaj, O., Elkind, E., Postl, J., Skowron, P.: Approximating optimal social choice under metric preferences. Artif. Intell. **264**, 27–51 (2018)
2. Anshelevich, E., Bhardwaj, O., Postl, J.: Approximating optimal social choice under metric preferences. In: AAAI, pp. 777–783 (2015)
3. Anshelevich, E., Postl, J.: Randomized social choice functions under metric preferences. J. Artif. Intell. Res. **58**(1), 797–827 (2017)
4. Anshelevich, E., Sekar, S.: Blind, greedy, and random: algorithms for matching and clustering using only ordinal information. In: AAAI, pp. 383–389 (2016)
5. Aziz, H., Lee, B.E.: Proportionally representative participatory budgeting with ordinal preferences. In: AAAI, pp. 5110–5118 (2021)
6. Aziz, H., Lee, B.E., Talmon, N.: Proportionally representative participatory budgeting: axioms and algorithms. In: AAMAS, pp. 23–31 (2018)
7. Aziz, H., Shah, N.: Participatory budgeting: models and approaches. In: Rudas, T., Péli, G. (eds.) Pathways Between Social Science and Computational Social Science. CSS, pp. 215–236. Springer, Cham (2021). https://doi.org/10.1007/978-3-030-54936-7_10
8. Benadé, G., Procaccia, A.D., Nath, S., Shah, N.: Preference elicitation for participatory budgeting. Manage. Sci. **67**(5), 2813–2827 (2021)
9. Boutilier, C., Caragiannis, I., Haber, S., Lu, T., Procaccia, A.D., Sheffet, O.: Optimal social choice functions: a utilitarian view. Artif. Intell. **227**, 190–213 (2015)
10. Bredereck, R., Faliszewski, P., Kaczmarczyk, A., Knop, D., Niedermeier, R.: Parameterized algorithms for finding a collective set of items. In: AAAI, pp. 1838–1845 (2020)
11. Caragiannis, I., Procaccia, A.D.: Voting almost maximizes social welfare despite limited communication. Artif. Intell. **175**(9), 1655–1671 (2011)
12. Cygan, M., et al.: Parameterized Algorithms. Springer, Cham (2015). https://doi.org/10.1007/978-3-319-21275-3

13. Cygan, M., Pilipczuk, M.: Exact and approximate bandwidth. Theoret. Comput. Sci. **411**(40–42), 3701–3713 (2010)
14. Downey, R.G., Fellows, M.R.: Fundamentals of Parameterized Complexity, vol. 4. TCS, Springer, London (2013). https://doi.org/10.1007/978-1-4471-5559-1
15. Elkind, E., Faliszewski, P., Skowron, P., Slinko, A.: Properties of multiwinner voting rules. Soc. Choice Welfare **48**(3), 599–632 (2017). https://doi.org/10.1007/s00355-017-1026-z
16. Endriss, U.: Trends in computational social choice. Lulu. com (2017)
17. Faliszewski, P., Skowron, P., Slinko, A., Talmon, N.: Committee scoring rules: axiomatic characterization and hierarchy. TEAC **7**(1), 1–39 (2019)
18. Fluschnik, T., Skowron, P., Triphaus, M., Wilker, K.: Fair knapsack. In: AAAI, pp. 1941–1948 (2019)
19. Ganuza, E., Baiocchi, G.: The power of ambiguity: how participatory budgeting travels the globe. J. Public Deliberation **8**(2), 1–12 (2012)
20. Goel, A., Krishnaswamy, A.K., Sakshuwong, S., Aitamurto, T.: Knapsack voting for participatory budgeting. ACM Trans. Econ. Comput. **7**(2), 1–27 (2019)
21. Gupta, S., Jain, P., Panolan, F., Roy, S., Saurabh, S.: Gerrymandering on graphs: computational complexity and parameterized algorithms. In: SAGT, pp. 140–155 (2021)
22. Gupta, S., Jain, P., Saurabh, S.: Well-structured committees. In: IJCAI, pp. 189–195 (2020)
23. Gupta, S., Jain, P., Saurabh, S., Talmon, N.: Even more effort towards improved bounds and fixed-parameter tractability for multiwinner rules. In: IJCAI, pp. 217–223 (2021)
24. Hershkowitz, D.E., Kahng, A., Peters, D., Procaccia, A.D.: District-fair participatory budgeting. In: AAAI, pp. 5464–5471 (2021)
25. Jain, P., Sornat, K., Talmon, N.: Participatory budgeting with project interactions. In: IJCAI, pp. 386–392 (2020)
26. Jain, P., Sornat, K., Talmon, N., Zehavi, M.: Participatory budgeting with project groups. In: IJCAI, pp. 276–282 (2021)
27. Jain, P., Talmon, N., Bulteau, L.: Partition aggregation for participatory budgeting. In: AAMAS, pp. 665–673 (2021)
28. Kellerer, H., Pferschy, U., Pisinger, D.: Knapsack problems. Springer (2010). https://doi.org/10.1007/978-3-540-24777-7
29. Moenck, R.T.: Practical fast polynomial multiplication. In: SYMSAC86, pp. 136–148 (1976)
30. Niedermeier, R.: Invitation to Fixed-Parameter algorithms. Oxford University Press (2006)
31. Pierczyński, G., Skowron, P., Peters, D.: Proportional participatory budgeting with additive utilities. In: Advances in Neural Information Processing Systems 34 (2021)
32. Procaccia, A.D., Rosenschein, J.S.: The distortion of cardinal preferences in voting. In: Proceedings of the 10th International Conference on Cooperative Information Agents, pp. 317–331. CIA2006 (2006)
33. Gilman, H.R.: Transformative deliberations: participatory budgeting in the United States. J. Public Deliberation **8**(2), 1–20 (2012)
34. Sintomer, Y., Herzberg, C., Röcke, A.: Participatory budgeting in Europe: potentials and challenges. Int. J. Urban Reg. Res. **32**(1), 164–178 (2008)
35. Skowron, P., Faliszewski, P., Lang, J.: Finding a collective set of items: from proportional multirepresentation to group recommendation. Artif. Intell. **241**, 191–216 (2016)

36. Sviridenko, M.: A note on maximizing a submodular set function subject to a knapsack constraint. Oper. Res. Lett. **32**(1), 41–43 (2004)
37. Talmon, N., Faliszewski, P.: A framework for approval-based budgeting methods. In: AAAI, pp. 2181–2188 (2019)
38. Wampler, B.: Participatory budgeting in brazil: contestation, cooperation, and accountability. Penn State Press (2010)

Graph Drawing and Visualization

Dominance Drawings for DAGs
with Bounded Modular Width

Giacomo Ortali[1]([⊠]) [iD] and Ioannis G. Tollis[2] [iD]

[1] Università degli Studi di Perugia, Perugia, Italy
giacomo.ortali@studenti.unipg.it
[2] Computer Science Department, University of Crete, Heraklion, Crete, Greece
tollis@csd.uoc.gr

Abstract. A weak dominance drawing Γ of a DAG $G = (V, E)$ is a d-dimensional drawing such that $D(u) < D(v)$ for every dimension D of Γ if there is a directed path from a vertex u to a vertex v in G, where $D(w)$ is the coordinate of vertex $w \in V$ in dimension D of Γ. If $D(u) < D(v)$ for every dimension D of Γ, but there is no path from u to v, we have a *falsely implied path (fip)*. Minimizing the number of fips is an important theoretical and practical problem, which is NP-hard. We show that it is an FPT problem for graphs having bounded modular width mw and when d is bounded. This result in weak dominance, which is interesting by itself, lets us prove our main contributions. Computing the dominance dimension of G, that is, the minimum number of dimensions for which G has a dominance drawing (a weak dominance drawing with 0 fips), is a well-known NP-hard problem. We show that the dominance dimension of G is bounded by $\frac{mw}{2}$ (mw, if $mw < 4$) and that computing the dominance dimension of G is an FPT problem with parameter mw.

Keywords: Modular width · (Weak) dominance drawings · FPT-algorithms

1 Introduction

A *directed acyclic graph* (DAG) $G = (V, E)$, with n vertices and m edges, is a directed graph with no directed cycles. For any dimension D of a drawing Γ of G, we denote by $D(v)$ the coordinate of vertex $v \in V$ in dimension D. A d-dimensional dominance drawing Γ of G is a d-dimensional drawing of G where, given any pair of vertices $u, v \in V$, $D(u) < D(v)$ for every dimension D of Γ if and only if there exists a (directed) path connecting u to v in G. The efficient computation of dominance drawings of DAGs has many applications, including computational geometry [7], graph drawing [8], and databases [18]. The *dominance dimension* of G is the minimum d^* such that there exists a d^*-dimensional dominance drawing of G.

A *partially ordered set (poset)* is a mathematical formalization of the concept of ordering. Any poset P can be viewed as a transitive DAG G^*, i.e., as a DAG

© The Author(s), under exclusive license to Springer Nature Switzerland AG 2023
L. Gąsieniec (Ed.): SOFSEM 2023, LNCS 13878, pp. 65–79, 2023.
https://doi.org/10.1007/978-3-031-23101-8_5

that contains its transitive closure graph. See [19] for a formal definition of a poset and the dimension of a poset. Since the dimension of a DAG G is the same as the dimension of its transitive closure G^*, the results on the dimension of posets transfer directly to DAGs and their dominance dimension. The literature concerning dominance dimension of DAGs is vast, we report here some results.

Testing if a DAG has dominance dimension 2 can be done in linear time [6, 15], while it is NP-complete to decide if the dominance dimension is d for any $d \geq 3$ [19]. A linear-time algorithm that constructs straight-line 2-dimensional dominance drawings of upward planar graphs is described in [4] (see also [8]). Two vertices $u, v \in G$ are *incomparable* if there is no path from u to v or from v to u in G. The dominance dimension of a DAG G with n vertices ($n \geq 4$) is bounded by $\frac{n}{2}$ [2,10] and by the *width* of the graph [5], that is, the maximum cardinality of a set of incomparable vertices of G.

Most DAGs have dominance dimension greater than two and, in general, computing dominance drawings with a bounded number of dimensions is difficult. For this reason, a relaxed version of the concept of dominance drawings, the *weak dominance drawings*, was introduced in [12]. In weak dominance, the "if and only if" of the definition of dominance becomes an "if". More formally, in a weak dominance drawing Γ of a DAG $G = (V, E)$, for any two vertices $u, v \in V$, $D(u) < D(v)$ for every dimension D of Γ if there is a path from u to v in G. We have a *falsely implied path (fip)* when $D(u) < D(v)$ for every dimension D of Γ, but u and v are incomparable.

For any DAG G and any value d, G admits a d-dimensional weak dominance drawing. Recently, the concept of weak dominance drawing was adopted in order to construct compact representations of the reachability information of large graphs that are produced by large datasets in the database community [14,18]. In these works the focus lies on the computational time required by a reachability query. In a reachability query, we ask if there is a path connecting two vertices of the graph. Given a d-dimensional weak dominance drawing, it is possible to test in $O(d)$ time (i.e., very fast, since d is usually much smaller than n) if two vertices u and v of G are incomparable. Hence, by minimizing the number of fips of the drawings we maximize the number of reachability queries that we can perform in $O(d)$ time. The number of fips (or false positives in their terminology) plays a crucial role. However, the problem of computing a weak dominance drawing with the minimum number of fips is NP-hard [11,12].

We consider a parameter denoted by *modular width*, introduced in [9] and recently used to produce FPT-algorithms for several problems both in its undirected and directed version [3,13,17]. In particular, we are interested in the directed modular width. In the rest of the paper we omit the word "directed" and we denote our parameter simply by *modular width*. The modular width is a parameter based on the concept of *module*, which we define in Sect. 2. The concept of module is defined in different ways in the literature. Here, we consider the module as defined in [1]. In Sect. 2 and in the last paragraphs of Sect. 5 we observe that the results obtained by using such definition directly imply the same results when using other definitions.

Our Contributions: Let mw be the modular width of DAG G. We show that computing a d-dimensional weak dominance drawing of a DAG G with the minimum number of fips is an FPT problem when parameters mw and d are bounded. A key property that we use is described by Compaction Lemma, where we show an interesting property of any weak dominance drawing of G with the minimum number of fips. This result in weak dominance, which is interesting by itself, lets us prove our main contributions. We show that the dominance dimension of G is bounded by $\frac{mw}{2}$, if $mw \geq 4$, and by mw, otherwise. The above results imply that computing the dominance dimension of G is an FPT problem with parameter mw.

2 Preliminaries

In this section we introduce two concepts that we use in the rest of the paper. In order to prove Theorem 1, which is our main contribution, our strategy is to iteratively consider graphs obtained from the input DAG G by merging some of its vertices into "super-vertices". This way of merging and the parameter mw, that we use in our fixed-parameter algorithm, are introduced in Sect. 2. A key ingredient of our algorithm is the fact that every super-vertex has a *cost*, that is equal to the number of vertices of G it contains. In this section we define the concept of *cost-minimum weak dominance drawing*.

Modules and Directed Modular-Width. Let $G = (V, E)$ be a DAG. An *edge-based module* M of G is a subset of V so that every vertex of M is adjacent to the same set of vertices of $V \setminus M$. More formally, $|M| = 1$ (in this case the module is *trivial*) or, for any two vertices $v_1, v_2 \in M$ and any vertex $u \in V \setminus M$, $(v_1, u) \in E$ if and only if $(v_2, u) \in E$ and $(u, v_1) \in E$ if and only if $(u, v_2) \in E$. In the literature, edge-based modules are simply called "modules" [16].

An *edge-based congruence partition* C_P of V is a partition of V into edge-based modules. The *edge-based quotient graph* G/C_P is the graph obtained from G by merging into a vertex each edge-based module in C_P. The *edge-based modular decomposition* of G is a tree T describing a decomposition of G based into its edge-based modules. The root of T is the edge-based module V and any leaf of T is a trivial edge-based module $\{v\}$, where $v \in V$. For further details about the concepts defined so far see [16]. Figure 1(a) depicts a DAG G and its edge-based modules. The non-trivial modules are $M_1 = \{2, 3\}$ and $M_2 = \{6, 11\}$. Figure 1(b) depicts the edge-based modular decomposition tree of G.

In this paper we consider *path-based modules*. A path-based module M of G is a non-empty subset of V so that $|M| = 1$ or, for any two vertices $v_1, v_2 \in M$ and any vertex $u \in V \setminus M$: There is a path connecting v_1 to u if and only if there is a path connecting v_2 to u; there is a path connecting u to v_1 if and only if there is a path connecting u to v_2. The path-based modules are used in [1] as a generalization of the concept of edge-based modules.

Notice that a path-based module of G is an edge-based module for the transitive closure graph G^* of G. We define the concepts of *path-based congruence* partition, path-based quotient graph, and path-based modular decomposition tree also for path-based modules. Since the edge-based modular decomposition tree can be computed in $O(m)$ time [16], the path-based modular decomposition tree can be computed in $O(nm)$ time (which is needed to compute G^*).

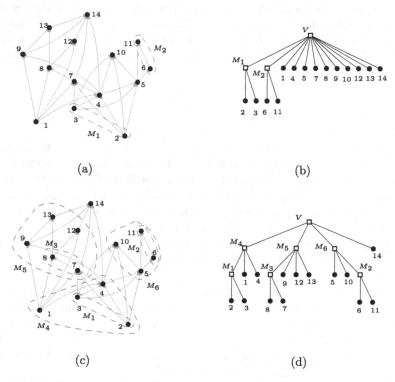

(a)   (b)

(c)   (d)

Fig. 1. (a–b) A DAG G and the edge-based modular decomposition tree of G. (c–d) G and the path-based modular decomposition tree of G.

Figure 1(c) depicts a DAG G and its path-based modules. Notice that M_1 and M_2 are also edge-based modules. Figure 1(d) depicts the path-based modular decomposition tree of G. Figure 2(a) depicts a DAG G and a path-based congruence partition $C/P = \{M_1, M_2, M_3, M_4, M_5, M_6\}$ of G. Figure 2(b) depicts the path-based quotient graph G/C_P, where every vertex v_i is the super-vertex associated to the path-based module $M_i \in C/P$ ($i \in [1, 6]$).

As already observed, the edge-based modules are simply denoted as modules in literature and the *modular width* mw is the maximum number of children of a node in the modular decomposition tree. So far we introduced two different definitions of modules; the *edge-based* and the *path-based*. Hence, we distinguish between *edge-based* and *path-based* modular width. Since the definition of an edge-based module is more restrictive than the definition of a path-based module (i.e., an edge-based module is always a path-based module, but not vice versa), the path-based modular width is less than or equal to the edge-based modular width.

See, for example, Fig. 1(a) and (c). If we consider its edge-based modular decomposition tree, depicted in Fig. 1(b), we have $mw = 12$. If we consider its path-based modular decomposition tree, depicted in Fig. 1(d), we have $mw = 4$.

In this paper we focus on path-based modules and path-based modular width. Every result that we obtain is also implied for edge-based modules and edge-based modular width. Since here we only consider path-based modules and modular width, from now on for simplicity we omit the word "path-based".

We denote by G_M the subgraph of G induced by the vertices in Module M. From now on, given the modular decomposition tree T of G and a module M, if $|M| > 1$ we associate to G_M the congruence partition induced by the children of the vertex associated to M in T. Consider Fig. 1. The root of the tree is the trivial module V and $G_V = G$. We associate to G the congruence partition $\{M_4, M_5, M_6, \{14\}\}$. We associate to G_{M_4} the congruence partition $\{M_1, \{1\}, \{4\}\}$.

Cost-Minimum Weak Dominance Drawings. Let H be a DAG such that every vertex v is assigned a *cost* $c(v)$. Let Γ be a weak dominance drawing of H. The cost of a fip (u, v) in Γ is $c(u, v) = c(u) \cdot c(v)$. The *cost* of Γ is the sum of the costs of its fips.

Let G be a DAG, $C_P = \{M_1, \ldots, M_h\}$ be a congruence partition of G, and v_i be the super-vertex representing $M_i \in C_P$ in the quotient graph G/C_P ($i \in [1, h]$). We assign the cost to the vertices of G and G/C_P such that: For any $v \in G$, $c(v) = 1$; For any $v_i \in G/C_P$, $c(v_i) = |M_i|$ ($i \in [1, h]$). Observe that, with such a cost assignment, the cost of a weak dominance drawing of G is equal to its number of fips.

Figure 2(a) depicts a DAG G and a congruence partition $C_P = \{M_1, \ldots, M_6\}$ of G. Figure 2(b) depicts the quotient graph G/C_P. Figure 2(c) depicts three 2-dimensional weak dominance drawings of G/C_P.

- Drawing Γ_1 has the following six fips: *#1* (v_1, v_2); *#2* (v_1, v_3); *#3* (v_1, v_6); *#4* (v_2, v_5); *#5* (v_3, v_4); *#6* (v_4, v_5); *#7* (v_6, v_5). The cost of Γ_1 is $c(v_1, v_2) + c(v_1, v_3) + c(v_1, v_6) + c(v_2, v_5) + c(v_3, v_4) + c(v_4, v_5) + c(v_6, v_5) = 6 + 12 + 6 + 1 + 18 + 9 + 1 = 53$.
- Drawing Γ_2 contains only the fip (v_3, v_4) and its cost is $c(v_3, v_4) = 18$.
- Drawing Γ_3 contains only the fip (v_2, v_5) and its cost is $c(v_2, v_5) = 1$.

Since G/C_P is a crown graph, the cost of a weak dominance drawing of G/C_P it is at least 1 [12] . Hence, Γ_3 is a cost-minimum 2-dimensional weak dominance drawing of G/C_P.

3   The Compaction Lemma

In this section we introduce and prove Lemma 1, that we denote by Compaction Lemma. This lemma shows an interesting property of a weak dominance drawing

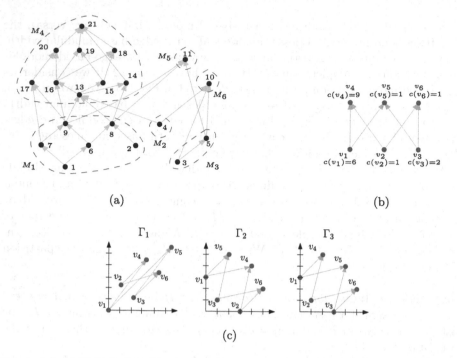

(a)   (b)

(c)

Fig. 2. (a) A DAG G and a congruence partition $C_P = \{M_1, \ldots, M_6\}$. (b) The correspondent quotient graph G/C_P and the costs of its vertices. (c) Three different 2-dimensional weak dominance drawings Γ_1, Γ_2, and Γ_3 of G/C_P. Drawing Γ_3 is cost-minimum.

with the minimum number of fips with respect to its modules. Before introducing the lemma, we define some notation.

A module M of a DAG G is *compact* in a dimension D of a weak dominance drawing Γ of G if the coordinates of the vertices of M in D are consecutive. Also, M is *compact* in Γ if it is compact in every dimension of Γ. A congruence partition C_P of G is compact in Γ if every module $M \in C_P$ is compact in Γ.

Figure 3 depicts three 2-dimensional weak dominance drawings of the same graph. Module $M = \{5, 7, 10, 12, 14\}$ is: Not compact in any dimension in Fig. 3(a); compact in Dimension Y and not in X, in Fig. 3(b); compact (in both dimensions) in Fig. 3(c). Figure 4(e) depicts a 2-dimensional weak dominance drawing where the congruence partition $C_P = \{M_1, \ldots, M_6\}$ is compact.

Recall that we assume $c(v) = 1$ for every vertex v of G. Hence, by definition of $c(\cdot)$, a cost-minimum weak dominance drawing of G is a weak dominance drawing of G with the minimum number of fips.

Lemma 1 (Compaction Lemma). *Let G be a DAG, C_P be any congruence partition of G and d be a constant. There exists a cost-minimum d-dimensional weak dominance drawing of G where C_P is compact.*

Before proving Lemma 1 we introduce some notation and prove some intermediate results. Let M be a module of G, Γ be a weak dominance drawing of G, and D be a dimension of Γ. The *separator* of M in D is the maximal set $S \subseteq V \backslash M$ so that, for any $v \in S$, there exist two vertices $u, w \in M$ so that $D(u) < D(v) < D(w)$. For example, consider Fig. 3(a) and the module $M = \{5, 7, 10, 12, 14\}$. The separator of M in dimension X and Y is $\{2, 3, 9, 13\}$ and $\{6, 8, 9, 11, 16\}$, respectively. Note that if $S = \emptyset$, M is compact in D.

Lemma 2. *Let Γ be a weak dominance drawing of G. Let M be a module of G, D be a dimension of Γ, and S be the separator of M in D. Any vertex $u \in S$ is incomparable with the vertices of M.*

Proof. Let $v \in S$ and $u, w \in M$ such that $D(u) < D(v) < D(w)$. There is no path from v to u, since $D(v) > D(u)$. Similarly, there is no path from w to v, since $D(w) > D(v)$. Hence, v is incomparable to the vertices of M.   □

Given a weak dominance drawing Γ of G and a congruence partition C_P of G, a fip (u, v) of Γ is an *inner-fip* if $u, v \in M$, where M is a module of C_P. Otherwise, (u, v) is an *outer-fip*. Consider the 2-dimensional weak dominance drawing in Fig. 4(e) and the congruence partition $C_P = \{M_1, \ldots, M_6\}$: Fip $(16, 19)$ is an inner-fip; Fip $(4, 11)$ is an outer-fip.

Let Γ be a weak dominance drawing of G. In the rest of the section, given a module M, we consider the congruence partition $C_P = \{M, V \backslash M\}$ of G. In this setting, a fip (u, v) of Γ is an inner-fip of Γ if $u, v \in M$ or $u, v \in V \backslash M$. Otherwise, it is an outer-fip of Γ. Switching u and v in dimension D is equivalent to setting $\alpha = D(u)$, $D(u) = D(v)$, and $D(v) = \alpha$. For any $v \in M$, let out_v be the number of outer-fips involving v. We now describe an operation that we denote by *compaction* of M. This operation modifies Γ such that the coordinates of all the vertices of M are consecutive in every dimension of Γ, without increasing the number of fips of Γ.

Compaction(Γ, M): Let p be a vertex of M with the minimum number of outer fips out_p. For every dimension D of Γ, having separator S, perform the following computation: While there are two vertices $u \in S$ and $v \in M$ such that $D(v) = D(u) - 1 < D(p)$ or $D(p) < D(u) = D(v) - 1$, switch u and v in D.

Illustration of the Operation of Compaction. Refer to Fig. 3(a–c) and the module $M = \{5, 7, 10, 12, 14\}$. Consider the 2-dimensional weak dominance drawing in Fig. 3(a). We have:

- $out_5 = 6$ (outer-fip $(3, 5)$ and $(5, w)$ $\forall$ $w \in \{9, 11, 16, 17, 18\}$).
- $out_7 = 5$ (outer-fip $(7, w)$ $\forall$ $w \in \{9, 13, 16, 17, 18\}$).
- $out_{10} = 5$ (outer-fip $(v, 10)$ $\forall$ $v \in \{2, 3, 6\}$) and $(5, w)$ $\forall$ $w \in \{16, 18\}$).
- $out_{12} = 7$ (outer-fip $(v, 12)$ $\forall$ $v \in \{3, 6, 8\}$) and $(12, w)$ $\forall$ $w \in \{13, 16, 17, 18\}$).
- $out_{14} = 6$ (outer-fip $(v, 14)$ $\forall$ $v \in \{3, 6, 8, 9\}$) and $(12, w)$ $\forall$ $w \in \{17, 18\}$).

The vertex p of M having minimum out_p can be either 7 or 10. We chose $p = 7$. Figure 3(b) shows the drawing after that the operation of compaction of M in Γ

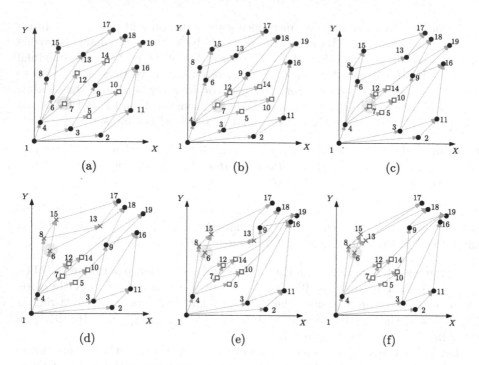

Fig. 3. Illustration of the operation of compaction.

performed its step on dimension Y (and still not on dimension X). Figure 3(c) shows the graph resulting after the operation of compaction of M in Γ.

By construction, after the compaction of M in Γ, M is compact in Γ ($S = \emptyset$). We have the following lemma.

Lemma 3. *Let M be a module of G and let M' be a module different from M that is compact in Γ. By performing the compaction of M, we have that: (1) Γ is (still) a weak dominance drawing; (2) the number of fips of Γ is not increased; (3) module M' is (still) compact in Γ.*

Proof. Recall that p is a vertex of M with the minimum number of outer fips out_p. Notice that, after the compaction of M, the relative positions of two vertices u and v in Γ changed if and only if $u \in M$ and $v \in S$ or vice versa. This fact has two implications. First, by Lemma 2, Γ remains a weak dominance drawing and we have that Property (1) is verified. Second, we have that:

(i) The number of inner-fips of Γ does not change (recall $S \subseteq V \backslash M$).

By construction, the relative position of any $u \in G$ and p do not change. Also, recall that M is compact in Γ. Hence:

(ii) The number of fips of Γ involving p does not change.

(iii) The relative position between any $v \in V \backslash M$ and p is the same as the relative position between v and any $u \in M$.

Consideration (iii) implies that every vertex $v \in M$ is involved in out_p outer-fips of M and, by (ii) and since $out_p \leq out_v$ before the compaction, we have that the number of outer-fips involving v is not augmented. Hence, by (i), the number of fips of Γ is not increased. It follows that Property (2) of the lemma is verified. It remains to show Property (3). Let D be any dimension of Γ and S be the separator of M in D. If $M' \cap S = \emptyset$ the compaction of M does not modify the position of the vertices of M' and M' remains compact. Suppose $M' \cap S \neq \emptyset$. Since M' is compact we have $M' \subseteq S$. Hence, if we switch $u \in M'$ and $v \in M$, then we switch v with all the vertices of M'. Hence, M' remains compact.   □

Illustration of Lemma 3 ($d = 2$). We now illustrate the three properties of the lemma with an example, when $d = 2$.
- `Property (1)`. The drawing in Fig. 3(c), obtained by performing the compaction on module $M = \{5, 7, 10, 12, 14\}$ in the weak dominance drawing in Fig. 3(a), is a weak dominance drawing.
- `Property (2)`. The outer-fips involving the any vertex $v \in M$ in the weak dominance drawing in Fig. 3(c) are $(3, v)$ and (v, w) $\forall w \in \{9, 11, 16, 17, 18\}$. Before the compaction, Fig. 3(a), any vertex v where involved in not less outer-fips. Notice that vertex $p = 7$ is involved in the same fips in both drawings. The inner-fips in Fig. 3(a) and (c) are the same. Hence, the drawing in Fig. 3(c) has no more fips than the one in Fig. 3(a).
- `Property (3)`. Refer to Fig. 3(d–f). Denote now $M = \{6, 8, 13, 15\}$ and $M' = \{5, 7, 10, 12, 14\}$. Consider the weak dominance drawing in Fig. 3(d). Figure 3(e) shows the drawing after that the operation of compaction of M in Γ performed its step on dimension Y (and still not on dimension X). Figure 3(f) shows the graph resulting after the operation of compaction of M in Γ. Notice that Module M', that is compact in Fig. 3(d), is still compact in Fig. 3(e) and (f).

Given Lemma 3, we have all the ingredients to prove the `Compaction Lemma`.

Proof of the `Compaction Lemma`. Let Γ be a cost-minimum weak dominance drawing of G. We show that, given Γ, it is possible to compute a cost-minimum weak dominance drawing $\overline{\Gamma}$ of G having the same number of fips of Γ and where C_P is compact. We initialize $\overline{\Gamma} = \Gamma$. For every $M \in C_P$ we perform the compaction of M in $\overline{\Gamma}$. After every compaction we have that: M is compact in $\overline{\Gamma}$ by construction; $\overline{\Gamma}$ is still a weak dominance drawing by Property (1) of Lemma 3; $\overline{\Gamma}$ has the minimum number of fips by Property (2) of Lemma 3; every module M' the that was compact before the compaction of M remains compact by Property (3) of Lemma 3. Hence, $\overline{\Gamma}$ is a weak dominance drawing of G with the minimum number of fips and where C_P is compact.   □

4 Minimizing the Number of Fips

In this section we present the main contribution of this paper, stated in Theorem 1. Given a DAG G and a congruence partition C_P of G, let $opt(G/C_P)$ be the cost of a cost-minimum weak dominance drawing of G/C_P. For example,

Γ_3 in Fig. 2(c) is a cost-minimum 2-dimensional weak dominance drawing of the graph G/C_P in Fig. 2(b) and its cost is 1. Hence, $opt(G/C_P) = 1$.

The next lemma, which is proved in the appendix, relates the number of outer-fips of a weak dominance drawing of G where C_P is compact to the value $opt(G/C_P)$.

Lemma 4. *Let G be a DAG and C_P be a congruence partition of G. For any weak dominance drawing Γ of G such that C_P is compact in Γ and having t outer-fips, we have $t \geq opt(G/C_P)$.*

Proof. Let $C_P = \{M_1, \ldots, M_h\}$. Let Γ be a weak dominance drawing of G where C_P is compact and with t outer-fips. It is possible to construct a weak dominance drawing Γ' of G/C_P by: Contracting every $M_i \in C_P$ ($i \in [1, h]$) to a vertex v_i in Γ; assigning $c(v_i) = |M_i|$. For example, Fig. 4(e) depicts a 2-dimensional weak dominance drawing Γ of the graph in Fig. 2(a) where the congruence partition $C_P = \{M_1, \ldots, M_6\}$ is compact. By contracting every module M_i to a vertex v_i and by assigning $c(v_i) = |M_i|$ ($i \in [1, 6]$) we obtain the weak dominance drawing of G/C_P in Fig. 4(d). Let t' be the cost of Γ'. Notice that $t' \geq opt(G/C_P)$ by definition. We now prove $t' = t$. Since C_P is compact in Γ and by definition of a module, we have that: For every fip (u, v) of Γ such that $u \in M_i$ and $v \in M_j$, where $i, j \in [1, h]$ and $i \neq j$, there is a fip (u', v') for any couple of vertices u' and v' such that $u' \in M_i$ and $v' \in M_j$. Hence, any fip (u, v) of Γ implies the existence of $|M_i| \cdot |M_j|$ outer-fips in Γ. Also, since Γ' is obtained by contracting the vertices of every module of C_P and since C_P is compact in Γ, fip (u, v) implies a fip (v_i, v_j) having a cost $c(v_i) \cdot c(v_j) = |M_i| \cdot |M_j|$ in Γ'. Hence, $t' \geq t$. The argument holds also in the other direction: Every fip (v_i, v_j) in Γ' implies the existence of $|M_i| \cdot |M_j|$ outer-fips in Γ. Hence, $t' \leq t$. Since $t' \geq t$ and $t' \leq t$ we have $t' = t$. Since $t' = t$ and $t' \geq opt(G/C_P)$, we have $t \geq opt(G/C_P)$. □

The following theorem is the main contribution of this section

Theorem 1. *Let G be a DAG with n vertices and m edges, let mw be the modular width of G and let k be any value such that $k \geq mw$. For any d, it is possible to compute a d-dimensional weak dominance drawing of G with the minimum number of fips in $O(nm + n2^{kd\log(k)})$ time.*

Proof. Recall that, by definition of $c(\cdot)$ and since we assign $c(v) = 1$ for every $v \in V$, for any module M of G a cost-minimum weak dominance drawing of G_M has the minimum number of fips. If $M = V$, $G_V = G$. It is possible to compute a cost-minimum d-dimensional weak dominance drawing of any graph H having k vertices in $O(dk^2(k!)^d)$ time by using the following brute force algorithm: (a) Compute all the possible d-dimensional weak dominance drawings of H in $O((k!)^d)$ time; (b) test in $O(dk^2)$ time, for each drawing, its cost; (c) select the drawing with the minimum cost. Notice that $O(dk^2(k!)^d) \in O(2^{kd\log(k)})$. In order to compute the cost-minimum d-dimensional weak dominance drawing of G drawing we do a bottom-up traversal of the modular decomposition tree T.

Base Step: Let M be a module of G such that the children of the corresponding vertex in T are leaves of T. Every module of the congruence partition C_P associated to G_M is a trivial module with cardinality 1. Hence, G_M is equal to the quotient graph G_M/C_P and it has less than k vertices. It is possible to compute a cost-minimum weak dominance drawing of G_M in $O(2^{kd \log(k)})$ time.

Recursive Step: Let M be a module of G such that the corresponding vertex in T has $k' \le k$ children that are not all leaves of T (i.e., one of them is an internal vertex of T). In order to simplify the notation, w.l.o.g., suppose $k' = k$ and $M = V$ (i.e., $G_M = G$). Let $C_P = \{M_1, \dots, M_h\}$ be the congruence partition that we associate to G given T. By inductive hypothesis, the cost-minimum weak dominance drawing Γ_{M_i} of G_{M_i} is given for every $M_i \in C_P$ ($i \in [1, h]$).

For example, consider the DAG G in Fig. 2(a) and the case $d = 2$. For any $i \in \{1, 2, 3, 5, 6\}$, G_{M_i} is planar and Γ_{M_i} is a dominance drawing. Drawing Γ_{M_1} is in Fig. 4(a), while Γ_{M_2}, Γ_{M_3}, Γ_{M_5}, and Γ_{M_6} are very simple and they are depicted in Fig. 4(b). Figure 4(c) depicts Γ_{M_4}. We have that G_{M_4} contains the crown graph and Γ_{M_4} is cost-minimum, since it has one fip, that is (16, 19). Drawing Γ' is Γ_3 of Fig. 2(c), which is cost-minimum.

We now compute a cost-minimum d-dimensional weak dominance drawing Γ of G. Since G/C_P has k vertices, we can compute a cost-minimum weak

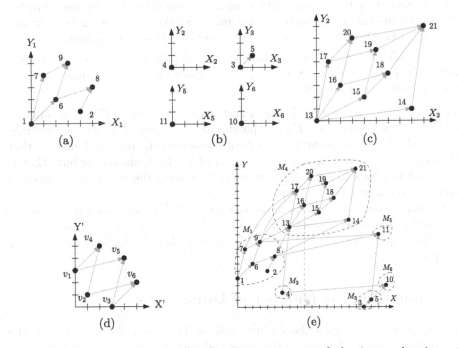

Fig. 4. Refer to G and C_P in Fig. 2. The cost-minimum weak dominance drawings of: (a) G_{M_1}; (b) G_{M_2}, G_{M_3}, G_{M_5}, G_{M_6}; (c) G_{M_4}; (d) G/C_P; (e) G.

dominance drawing Γ' of G/C_P in $O(2^{kd \log(k)})$ time. Recall that, for any $M_i \in C_P$ $(i \in [1, h])$, $c(v_i) = |M_i|$, where v_i is the vertex of G/C_P associated to M_i.

We compute a weak dominance drawing Γ of G by expanding the vertex v_i to the drawing Γ_{M_i} in Γ', for any $i \in [1, h]$. See Fig. 4(d-e). More formally, let D' a dimension of Γ' and let $V_{D'}^j$ be a set of vertices of G/C_P so that $D'(v) < D'(v_j)$ for any $v \in V_{D'}^j$. We perform the following coordinate assignment operation for every $v \in G$ and for every dimension D of Γ:

Coordinates Assignment Operation (v, D): Let M_i be the module of C_P containing v $(i \in [1, h])$. Let D' and D_i be the dimension corresponding to D in Γ' and Γ_i. I.e., if D is the gth dimension of Γ, D' and D_i are the gth dimension of Γ' and Γ_i, respectively. Set $D(v) = D'(v_i) + \sum_{u \in V_{D'}^i} (c(u) - 1) + D_i(v)$.

For example, refer to Fig. 4. Figure 4(e) depicts Γ after the Coordinates Assignment Operation is performed for every $v \in G$ and every dimension of Γ. The graphs G and G/C_P are depicted in Fig. 2(a) and (b). Refer to Vertex 16 and dimension X. We have $16 \in M_4$ and $V_X^4 = \{v_1, v_2\}$. Hence, $X(16) = X'(v_4) + (|c(v_1)| - 1) + (|c(v_2)| - 1) + X_4(16) = 2 + (6-1) + (1-1) + 2 = 9$.

Recall that, for $i \in [1, h]$, the vertices in Γ_{M_i} are contained in the same module M_i of G. Hence, since we obtain Γ by expanding the vertices of Γ' to the drawings $\Gamma_{M_1}, \ldots, \Gamma_{M_h}$ and since $\Gamma', \Gamma_{M_1}, \ldots, \Gamma_{M_h}$ are weak dominance drawing, we have that Γ is a weak dominance drawing and that C_P is compact in Γ. We now show that Γ is a cost-minimum d-dimensional dominance drawing of G having the minimum number of fips among all the d-dimensional dominance drawings of G where C_P is compact. By Lemma 1, Γ is cost-minimum.

The Inner-Fips: Notice that the drawing Γ restricted to the vertices of any module $M_i \in G/C_P$ is Γ_{M_i}. Since Γ_{M_i} is cost-minimum, we have that Γ has the minimum number of inner-fips.

The Outer-Fips: Notice that Γ' has a cost $opt(G/C_P)$. Since G/C_P is compact in Γ and since we obtained Γ by expanding the vertices of Γ' to the drawings $\Gamma_{M_1}, \ldots, \Gamma_{M_h}$, we can prove by argument similar to the ones of Lemma 4 that Γ has $opt(G/C_P)$ outer-fips, that is the cost of Γ'. By Lemma 4 we have that Γ is the d-dimensional dominance drawing of G having the minimum number of outer-fips and where C_P is compact.

For any vertex of T we perform the $O(2^{kd \log(k)})$ time operation that we described in the Base Step and in the Recursive Step. Hence, we have that the algorithm requires $O(n2^{kd \log(k)})$ time. As already observed in Sect. 2, computing T requires $O(nm)$. □

5   Minimizing the Number of Dimensions

In this section we discuss the implications of Theorem 1 with respect to the known results about dominance drawings and their dominance dimension.

Recall that testing if a DAG G has dominance dimension equal to any constant value $d \geq 3$ is NP-complete [19]. Observe that a constant d is the dominance dimension of a DAG $G = (V, E)$ if G admits a dominance drawing (i.e., a weak dominance drawing with 0 fips) with d dimensions, but not with $d - 1$. Hence, the following theorem follows by Theorem 1.

Theorem 2. *Let G be a DAG with n vertices and m edges, let mw be the modular width of G, and let k be any value such that $k \geq mw$. For any d, we can test if d is the dominance dimension of G in $O(nm + n2^{kd \log(k)})$ time. Also, if the test is positive, it is possible to construct a d-dimensional dominance drawing of G in the same time.*

We have the following simple, but interesting, corollary of Theorem 2.

Corollary 1. *Let G be a DAG with modular width k. It is possible to test if the dominance dimension of G is equal to 3 in $O(nm + n2^{k \log(k)})$ time.*

In the proof of Theorem 1 we compute dominance drawings of graphs having k or less vertices, where $k \geq mw$. Also, while merging dominance drawings of different DAGs $G_{M_1}, \ldots, G_{M_h}$ in order to compute the dominance drawing of their parent G_M in the modular decomposition tree, we do not add more dimensions with respect to the ones used to compute drawings $\Gamma_M, \Gamma_{M_1}, \ldots, \Gamma_{M_h}$. Since the dominance dimension of every DAG with n-vertices is at most $\frac{n}{2}$ if $n \geq 4$ [10] and it is lower than n in general, we have the following theorem.

Theorem 3. *Let G be a DAG, let mw be the modular width of G, and let d^* be the dominance dimension of G. Then, $d^* \leq \frac{mw}{2}$ if $mw \geq 4$; otherwise, $d^* \leq mw$.*

As already observed in the introduction, d^* is a parameter that is bounded not only by $\frac{n}{2}$ [10], but also by the width w of G [5]. Observe that the modular width mw is not bounded by w. For example, for cographs $mw = 2$ [17], but their width, w, is not bounded. Similarly, it is not difficult to construct a graph with bounded width and unbounded modular width. Consider a graph having width 2, i.e., that can be partitioned in two chains. Suppose that these two chains have the same number of edges and that, for any vertex v of the first chain in position i and any vertex w in position $i + 1$ there is an edge (v, w). In this case, no two vertices of the first chain can belong to a same module and, consequently, mw is unbounded. Theorems 2 and 3 imply the following result:

Theorem 4. *Let G be a DAG, mw be the modular width of G and k be a value such that $k \geq mw$. It is possible to compute the dominance dimension d^* of G and a d^*-dimensional dominance drawing of G in $O(nm + n2^{k^2 \log(k)})$ time.*

We conclude this section with a further discussion of the computational time of Theorems 1, 2, and 4. We have that the additive term "nm" in the time complexity of the theorems is required to compute G^* and consequently the path-based modular decomposition tree T of G. Our results hold if we use the concept of edge-based module instead of the concept of path-based module. In this case, since computing the edge-based modular decomposition requires $O(m)$ time, the time complexity of the theorems has an "m" instead of the "nm". However, as we already observed in Sect. 2, mw for edge-based modules is typically greater than mw for path-based modules.

Notice that in general the concept of module can be defined in many ways by restricting the internal structure of the modules, as observed in [3]. For example, in [1] the authors consider only path-based modules M where G_M is a simple path or a set of $|M|$ incomparable vertices. Given any definition of a module, there is a corresponding modular width and assume that the time complexity to compute the modular decomposition tree is y. Then Theorems 1 and 4 hold with "y" instead of "nm", in their the time complexity.

6   Concluding Remarks

In this paper we present a fixed parameter algorithm solving the fip-minimization problem, which is NP-hard. In particular, we show that if the modular width mw of DAG G is a constant, then the problem is polynomial-time solvable for any constant number of dimensions. In order to prove this result we use the `Compaction Lemma`, that shows that G always admits a weak dominance drawing with the minimum number of fips such that each module of G is compact in the drawing. Concerning dominance drawings (0 fips) we present an FPT algorithm with parameter mw to compute a dominance drawing of G in d dimensions, if it exists. Also, we show that $\frac{mw}{2}$ (mw, if $mw < 4$) is an upper bound for the dominance dimension of any DAG G. These two results imply an FPT algorithm with parameter mw that computes the dominance dimension of G.

Open Problems: From a practical point of view, it would be interesting to use the results of this paper to invent heuristics to compute weak dominance drawings with a smaller number of fips than is known in the literature [14,18]. Another interesting problem is to find algorithms computing drawings with a bounded number of fips. In this case, the problem of computing all the fips efficiently could become very interesting also in practice. Finally, we also believe that computing weak dominance drawings where the number of vertices involved in fips is minimized could be an important step forward in this line of research.

References

1. Anirban, S., Wang, J., Saiful Islam, Md.: Modular decomposition-based graph compression for fast reachability detection. Data Sci. Eng. **4**(3), 193–207 (2019)
2. Bogart, K.P.: Maximal dimensional partially ordered sets I. Hiraguchi's theorem. Discrete Math. **5**(1), 21–31 (1973)
3. Coudert, D., Ducoffe, G., Popa, A.: Fully polynomial FPT algorithms for some classes of bounded clique-width graphs. ACM Trans. Algorithms **15**(3), 33:1–33:57 (2019)
4. Battista, G.D., Tamassia, R., Tollis, I.G.: Area requirement and symmetry display of planar upward drawings. Discrete Comput. Geom. **7**(4), 381–401 (1992). https://doi.org/10.1007/BF02187850
5. Dilworth, R.P.: A decomposition theorem for partially ordered sets. Ann. Math. **52**, 161–166 (1950)
6. Dushnik, B., Miller, E.W.: Partially ordered sets. Am. J. Math. **63**, 600–610 (1941)

7. ElGindy, H.A., Houle, M.E., Lenhart, W., Miller, M., Rappaport, D., Whitesides, S.: Dominance drawings of bipartite graphs. In: Proceedings of the 5th Canadian Conference on Computational Geometry, Waterloo, Ontario, Canada, August 1993, pp. 187–191. University of Waterloo (1993)
8. Di Battista, G., Eades, P., Tamassia, R., Tollis, I.G.: Graph Drawing: Algorithms for the Visualization of Graphs, pp. 112–127. Prentice Hall, Upper Saddle River (1998)
9. Gajarský, J., Lampis, M., Ordyniak, S.: Parameterized algorithms for modular-width. In: Gutin, G., Szeider, S. (eds.) IPEC 2013. LNCS, vol. 8246, pp. 163–176. Springer, Cham (2013). https://doi.org/10.1007/978-3-319-03898-8_15
10. Hiraguchi, T.: On the dimension of partially ordered sets. Sci. Rep. Kanazawa Univ. 77–94 (1951)
11. Kornaropoulos, E.M., Tollis, I.G.: Weak dominance drawings and linear extension diameter. CoRR, abs/1108.1439 (2011)
12. Kornaropoulos, E.M., Tollis, I.G.: Weak dominance drawings for directed acyclic graphs. In: Graph Drawing - 20th International Symposium, GD 2012, Redmond, WA, USA, 19–21 September 2012, Revised Selected Papers, pp. 559–560 (2012)
13. Kratsch, S., Nelles, F.: Efficient and adaptive parameterized algorithms on modular decompositions. In: Azar, Y., Bast, H., Herman, G. (eds.) 26th Annual European Symposium on Algorithms, ESA 2018, 20–22 August 2018, Helsinki, Finland. LIPIcs, vol. 112, pp. 55:1–55:15. Schloss Dagstuhl - Leibniz-Zentrum für Informatik (2018)
14. Li, L., Hua, W., Zhou, X.: HD-GDD: high dimensional graph dominance drawing approach for reachability query. World Wide Web 20(4), 677–696 (2017)
15. McConnell, R.M., Spinrad, J.P.: Linear-time transitive orientation. In: Saks, M.E. (ed.) Proceedings of the Eighth Annual ACM-SIAM Symposium on Discrete Algorithms, 5–7 January 1997, New Orleans, Louisiana, USA, pp. 19–25. ACM/SIAM (1997)
16. McConnell, R.M., Spinrad, J.P.: Modular decomposition and transitive orientation. Discret. Math. 201(1–3), 189–241 (1999)
17. Steiner, R., Wiederrecht, S.: Parametrised algorithms for directed modular width. CoRR, abs/1905.13203 (2019)
18. Veloso, R.R., Cerf, L., Meira Jr., W., Zaki, M.J.: Reachability queries in very large graphs: a fast refined online search approach. In: Proceedings of the 17th International Conference on Extending Database Technology, EDBT 2014, Athens, Greece, 24–28 March 2014, pp. 511–522 (2014)
19. Yannakakis, M.: The complexity of the partial order dimension problem. SIAM J. Algebr. Discrete Methods 3, 303–322 (1982)

Morphing Planar Graph Drawings
Through 3D

Kevin Buchin[1], Will Evans[2], Fabrizio Frati[3(✉)], Irina Kostitsyna[4],
Maarten Löffler[5], Tim Ophelders[4,5], and Alexander Wolff[6]

[1] Technische Universität Dortmund, Dortmund, Germany
kevin.buchin@tu-dortmund.de
[2] University of British Columbia, Vancouver, Canada
will@cs.ubc.ca
[3] Roma Tre University, Rome, Italy
fabrizio.frati@uniroma3.it
[4] TU Eindhoven, Eindhoven, The Netherlands
i.kostitsyna@tue.nl
[5] Utrecht University, Utrecht, The Netherlands
{m.loffler,t.a.e.ophelders}@uu.nl
[6] Universität Würzburg, Würzburg, Germany

Abstract. In this paper, we investigate crossing-free 3D morphs
between planar straight-line drawings. We show that, for any two (not
necessarily topologically equivalent) planar straight-line drawings of an
n-vertex planar graph, there exists a piecewise-linear crossing-free 3D
morph with $O(n^2)$ steps that transforms one drawing into the other. We
also give some evidence why it is difficult to obtain a linear lower bound
(which exists in 2D) for the number of steps of a crossing-free 3D morph.

Keywords: Linear morph · 3D graph drawing · Morphing steps

1 Introduction

A *morph* is a continuous transformation between two given drawings of the
same graph. A morph is required to preserve specific topological and geometric
properties of the input drawings. For example, if the drawings are planar and
straight-line, the morph is required to preserve such properties throughout the
transformation. A morphing problem often assumes that the input drawings are
"topologically equivalent", that is, they have the same "topological structure".
For example, if the input drawings are planar, they are required to have the
same rotation system (i.e., the same clockwise order of the edges incident to
each vertex) and the same walk bounding the outer face; this condition is obviously necessary (and, if the graph is connected, also sufficient [6,10]) for a morph

This research was partially supported by MIUR Project "AHeAD" under PRIN
20174LF3T8.

to exist between the given drawings. A *linear* morph is a morph in which each vertex moves along a straight-line segment, all vertices leave their initial positions simultaneously, move at uniform speed, and arrive at their final positions simultaneously. A *piecewise-linear* morph consists of a sequence of linear morphs, called *steps*. A recent line of research culminated in an algorithm by Alamdari et al. [2] that constructs a piecewise-linear morph with $O(n)$ steps between any two topologically equivalent planar straight-line drawings of the same n-vertex planar graph; this bound is worst-case optimal.

What can one gain by allowing the morph to use a third dimension? That is, suppose that the input drawings still lie on the plane $z = 0$, does one get "better" morphs if the intermediate drawings are allowed to live in 3D? Arseneva et al. [4] proved that this is the case, as they showed that, for any two planar straight-line drawings of an n-vertex tree, there exists a crossing-free (i.e., no two edges cross in any intermediate drawing) piecewise-linear 3D morph between them with $O(\log n)$ steps. Later, Istomina et al. [9] gave a different algorithm for the same problem. Their algorithm uses $O(\sqrt{n} \log n)$ steps, however it guarantees that any intermediate drawing of the morph lies on a 3D grid of polynomial size.

Our Contribution. We prove that the use of a third dimension allows us to construct a morph between any two, possibly *topologically non-equivalent*, planar drawings. Indeed, we show that $O(n^2)$ steps always suffice for constructing a crossing-free 3D morph between any two planar straight-line drawings of the same n-vertex planar graph; see Sect. 2. Our algorithm defines some 3D morph "operations" and applies a suitable sequence of these operations in order to modify the embedding of the first drawing into that of the second drawing. The topological effect of our operations on the drawing is similar to, although not the same as, that of the operations defined by Angelini et al. in [3]. Both the operations defined by Angelini et al. and ours allow to transform an embedding of a biconnected planar graph into any other. However, while our operations are 3D crossing-free morphs, we see no easy way to directly implement the operations defined by Angelini et al. as 3D crossing-free morphs. We stress that the input of our algorithm consists of a pair of planar drawings in the plane $z = 0$; the algorithm cannot handle general 3D drawings as input.

We then discuss the difficulty of establishing non-trivial lower bounds for the number of steps needed to construct a crossing-free 3D morph between planar straight-line drawings; see Sect. 3. We show that, with the help of the third dimension, one can morph, in a constant number of steps, two topologically equivalent drawings of a nested-triangle graph (see Fig. 8) that are known to require a linear number of steps in any crossing-free 2D morph [2].

We conclude with some open problems in Sect. 4.

2   An Upper Bound

This section is devoted to a proof of the following theorem.

Theorem 1. *For any two planar straight-line drawings (not necessarily with the same embedding) of an n-vertex planar graph, there exists a crossing-free piecewise-linear 3D morph between them with $O(n^2)$ steps.*

We first assume that the given planar graph G is biconnected and describe four operations (Sect. 2.1) that allow us to morph a given 2D planar straight-line drawing of G into another one, while achieving some desired change in the embedding. We then show (Sect. 2.2) how these operations can be used to construct a 3D crossing-free morph between any two planar straight-line drawings of G. Finally, we remove our biconnectivity assumption (see Sect. 2.3 and the full version of this paper [5]).

We give some definitions. Throughout this paragraph, every considered graph is assumed to be connected. Two planar drawings of a graph are (*topologically*) *equivalent* if they have the same rotation system and the same clockwise order of the vertices along the boundary of the outer face. An *embedding* is an equivalence class of planar drawings of a graph. A *plane graph* is a graph with an embedding; when we talk about a planar drawing of a plane graph, we always assume that the embedding of the drawing is that of the plane graph. The *flip* of an embedding $\mathcal{E}$ produces an embedding in which the clockwise order of the edges incident to each vertex and the clockwise order of the vertices along the boundary of the outer face are the opposite of the ones in $\mathcal{E}$.

A pair of vertices of a biconnected graph G is a *separation pair* if its removal disconnects G. A *split pair* of G is a separation pair or a pair of adjacent vertices. A *split component* of G with respect to a split pair $\{u, v\}$ is the edge (u, v) or a maximal subgraph G_{uv} of G such that $\{u, v\}$ is not a split pair of G_{uv}. A plane graph is *internally-triconnected* if every split pair consists of two vertices both incident to the outer face.

2.1   3D Morph Operations

We begin by describing four operations that morph a given planar straight-line drawing into another with a different embedding; see Fig. 1.

Operation 1: Graph Flip. Let G be a biconnected plane graph, let u and v be two vertices of G, and let Γ be a planar straight-line drawing of G.

Lemma 1. *There exists a 2-step 3D crossing-free morph from Γ to a planar straight-line drawing Γ'' of G whose embedding is the flip of the embedding that G has in Γ; moreover, u and v do not move during the morph.*

We implement Operation 1, which proves Lemma 1, as follows. Let Π be the plane $z = 0$, which contains Γ. Let Π' be the plane that is orthogonal to Π and contains the line ℓ_{uv} through u and v. Let Γ' be the image of Γ under a clockwise rotation around ℓ_{uv} by 90°. Note that Γ' is contained in Π'. Now let Γ'' be the image of Γ' under another clockwise rotation around ℓ_{uv} by 90°. Note that Γ'' is a flipped copy of Γ and is contained in Π. Consider the linear morphs $\langle \Gamma, \Gamma' \rangle$ and $\langle \Gamma', \Gamma'' \rangle$. In each of them, every vertex travels on a line that makes a 45°-angle with both Π and Π', and all these lines are parallel. Due to the linearity of the morph and the fact that both pre-image and image are planar, all vertices stay coplanar during both linear morphs (although, unlike in a true rotation, the intermediate drawing size changes continuously). In particular, every intermediate drawing is crossing-free, and u and v (as well as all the points on ℓ_{uv}) are fixed points.

Operation 2: Outer Face Change. Let G be a biconnected plane graph, let Γ be a planar straight-line drawing of G, and let f be a face of Γ.

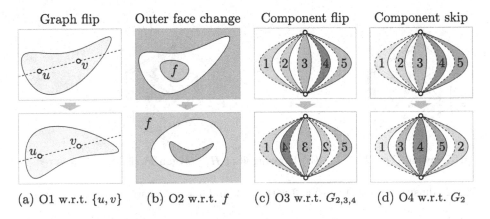

Graph flip	Outer face change	Component flip	Component skip
(a) O1 w.r.t. $\{u,v\}$	(b) O2 w.r.t. f	(c) O3 w.r.t. $G_{2,3,4}$	(d) O4 w.r.t. G_2

Fig. 1. The four operations that are the building blocks for our piecewise-linear morphs.

Lemma 2. *There exists a 4-step 3D crossing-free morph from Γ to a planar straight-line drawing Γ''' of G whose embedding is the same as the one of Γ, except that the outer face of Γ''' is f.*

We implement Operation 2, which proves Lemma 2, using the stereographic projection. Let Π be the plane $z = 0$, which contains Γ. Let S be a sphere that contains Γ in its interior and is centered on a point in the interior of f. Let Γ' be the 3D straight-line drawing obtained by projecting the vertices of G from their positions in Γ vertically to the Northern hemisphere of S. Let Γ'' be determined by projecting the vertices of Γ' centrally from the North Pole of S to Π. Both projections define linear morphs: $\langle \Gamma, \Gamma' \rangle$ and $\langle \Gamma', \Gamma'' \rangle$. Indeed, any intermediate drawing is crossing-free since the rays along which we project are parallel in $\langle \Gamma, \Gamma' \rangle$ and diverge in $\langle \Gamma', \Gamma'' \rangle$, and there is a one-to-one correspondence between the points in the pre-image and in the image. Since the morph also inverts the rotation system of Γ'' with respect to Γ, we apply Operation 1 to Γ'', which, within two morphing steps, flips Γ'' and yields our final drawing Γ'''.

Operation 3: Component Flip. Let G be a biconnected plane graph, and let $\{u,v\}$ be a split pair of G. Let $G_1,\dots,G_k$ be the split components of G with respect to $\{u,v\}$. Let Γ be a planar straight-line drawing of G in which u and v are incident to the outer face, as in Fig. 2a. Relabel $G_1,\dots,G_k$ so that they appear in clockwise order $G_1,\dots,G_k$ around u, where G_1 and G_k are incident to the outer face of Γ. Let i and j be two (not necessarily distinct) indices with $1 \le i \le j \le k$ and with the following property[1]: If G contains the edge (u,v), then this edge is one of the components $G_i,\dots,G_j$. Operation 3 allows us to flip the embedding of the components $G_i,\dots,G_j$ (and to incidentally reverse their order), while leaving the embedding of the other components of G unchanged. This is formalized in the following.

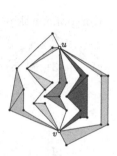
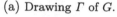

(a) Drawing Γ of G.

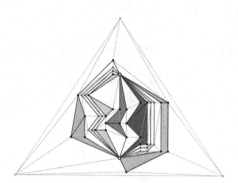

(b) Drawing Ψ of H; polygon P_{in} is blue, P_{out} is red.

Fig. 2. Illustration for Operation 3 with $i = 2$ and $j = 4$: Construction of Ψ from Γ.

Lemma 3. *There exists an $O(n)$-step 3D crossing-free morph from Γ to a planar straight-line drawing Γ' of G in which the embedding of G_ℓ is the flip of the embedding that G_ℓ has in Γ, for $\ell = i,\dots,j$, while the embedding of G_ℓ is the same as in Γ, for $\ell = 1,\dots,i-1,j+1,\dots,k$. The order of $G_1,\dots,G_k$ around u in Γ' is $G_1,\dots,G_{i-1},G_j,G_{j-1},\dots,G_i,G_{j+1},\dots,G_k$.*

In order to implement Operation 3, which proves Lemma 3, ideally we would like to apply Operation 1 to the drawing of the graph $G_i \cup G_{i+1} \cup \dots \cup G_j$ in Γ. However, this would result in a drawing which might contain crossings between edges of $G_i \cup G_{i+1} \cup \dots \cup G_j$ and edges of the rest of the graph. Thus, we first move $G_i \cup G_{i+1} \cup \dots \cup G_j$, via a 2D crossing-free morph, into a polygon that is symmetric with respect to the line through u and v and that does not contain any edges of the rest of the graph. Applying Operation 1 to $G_i \cup G_{i+1} \cup \dots \cup G_j$ now results in a drawing in which $G_i \cup G_{i+1} \cup \dots \cup G_j$ still lies inside the same symmetric polygon, which ensures that the edges of $G_i \cup G_{i+1} \cup \dots \cup G_j$ do not cross the edges of the rest of the graph.

---

[1] This is a point where our operations differ from the ones of Angelini et al. [3]. Indeed, their flip operation applies to any sequence of components of G, while ours does not.

We now describe the details of Operation 3; refer to Fig. 2b. We start by drawing a triangle (a, b, c) surrounding Γ. Then we insert in Γ two polygons P_{in} and P_{out} with $O(n)$ vertices, which intersect Γ only at u and v; the vertices of $G_1, \ldots, G_{i-1}, G_{j+1}, \ldots, G_k$ (except u and v) and a, b, and c lie outside P_{out}; the vertices of $G_i, \ldots, G_j$ (except u and v) lie inside P_{in}; P_{out} contains P_{in}; and the two paths of P_{in} connecting u and v have the same number of vertices. We let P_{in} and P_{out} "mimic" the boundary of the drawing of $G_i \cup G_{i+1} \cup \cdots \cup G_j$ in Γ.

We triangulate the exterior of P_{out}; that is, we triangulate each region inside (a, b, c) and outside P_{out} bounding a face of the current drawing. If this introduces a chord (x, y) with respect to P_{out}, let (x, y, w) and (x, y, z) be the two faces incident to (x, y); we subdivide (x, y) with a vertex and connect this vertex to w and z. We also triangulate the interior of P_{in}. Let Ψ be the resulting planar straight-line drawing of this plane graph H. Let C_{out} and C_{in} be the cycles of H represented by P_{out} and P_{in} in Ψ, let H_{out} be the subgraph of H induced by the vertices that lie outside or on P_{out}, and let H_{in} be the subgraph of H induced by the vertices that lie inside or on P_{in}. Note that H_{out} is a triconnected plane graph, as each of its faces is delimited by a 3-cycle, except for one face, which is delimited by a cycle C_{out} without chords. Further, H_{in} is an internally-triconnected plane graph, as each of its internal faces is delimited by a 3-cycle, while the outer face is delimited by a cycle C_{in} which may have chords.

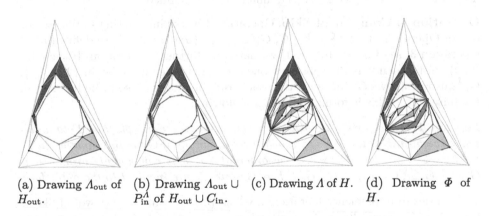

(a) Drawing Λ_{out} of H_{out}.

(b) Drawing $\Lambda_{out} \cup P_{in}^{\Lambda}$ of $H_{out} \cup C_{in}$.

(c) Drawing Λ of H.

(d) Drawing Φ of H.

Fig. 3. Illustration for Operation 3.

We now construct another planar straight-line drawing of H, as follows. Construct a strictly convex drawing Λ_{out} of H_{out}, e.g., by means of the algorithm by Hong and Nagamochi [8] or of the algorithm by Tutte [11], as in Fig. 3a. Let P_{out}^{Λ} be the strictly convex polygon representing C_{out} in Λ_{out}. As in Fig. 3b, plug a strictly convex drawing P_{in}^{Λ} of C_{in} in the interior of P_{out}^{Λ} (except at u and v) so that P_{in}^{Λ} is symmetric with respect to the line through u and v. This can be achieved because the two paths of C_{in} connecting u and v have the same number of vertices and because P_{out}^{Λ} is strictly convex, hence the segment $\overline{uv}$ lies in its interior, and thus also a polygon P_{in}^{Λ} sufficiently close to $\overline{uv}$ does. Finally, plug into $\Lambda_{\text{out}} \cup P_{\text{in}}^{\Lambda}$ a strictly convex drawing Λ_{in} of H_{in} in which C_{in} is represented by P_{in}^{Λ}, as in Fig. 3c; this drawing can be constructed again by means of [8,11]. This results in a planar straight-line drawing Λ of H.

We now describe the morph that occurs in Operation 3. We first define a morph $\langle \Psi, \ldots, \Phi \rangle$ from Ψ to another planar straight-line drawing Φ of H, as the concatenation of two morphs $\langle \Psi, \ldots, \Lambda \rangle$ and $\langle \Lambda, \ldots, \Phi \rangle$. The morph $\langle \Psi, \ldots, \Lambda \rangle$ is an $O(n)$-step crossing-free 2D morph obtained by applying the algorithm of Alamdari et al. [2]. The morph $\langle \Lambda, \ldots, \Phi \rangle$ is an $O(1)$-step 3D morph that is obtained by applying Operation 1 to Λ_{in} only, with u and v fixed; Fig. 3d shows the resulting drawing Φ. In order to prove that Operation 3 defines a crossing-free morph, it suffices to observe that, during $\langle \Lambda, \ldots, \Phi \rangle$, the intersection of H_{in} with the plane on which Λ_{out} lies is (a subset of) the segment $\overline{uv}$, which lies in the interior of a face of Λ_{out}; hence, H_{in} does not cross H_{out}. That no other crossings occur during $\langle \Psi, \ldots, \Phi \rangle$ is a consequence of the results of Alamdari et al. [2] (which ensure that $\langle \Psi, \ldots, \Lambda \rangle$ has no crossings) and of the properties of Operation 1 (which ensure that $\langle \Lambda, \ldots, \Phi \rangle$ has no crossings between edges of H_{out}). Finally, Operation 3 is the morph $\langle \Gamma, \ldots, \Gamma' \rangle$ obtained by restricting the morph $\langle \Psi, \ldots, \Phi \rangle$ to the vertices and edges of G. Note that the effect of Operation 1, applied only to Λ_{in}, is the one of flipping the embeddings of $G_i, \ldots, G_j$ (and also reversing their order around u), while leaving the embeddings of $G_1, \ldots, G_{i-1}, G_{j+1}, \ldots, G_k$ unaltered, as claimed.

Operation 4: Component Skip. Operation 4 works in a setting similar to the one of Operation 3. Specifically, G, $G_1, \ldots, G_k$, $\{u, v\}$, and Γ are defined as in Operation 3; see Fig. 4a. However, we have one further assumption: If the edge (u, v) exists, then it is the split component G_1. Let i be any index in $\{2, \ldots, k\}$. Operation 4 allows G_i to "skip" the other components of G, so to be incident to the outer face. This is formalized in the following.

Lemma 4. *There exists an $O(n)$-step 3D crossing-free morph from Γ to a planar straight-line drawing Γ' in which the embedding of G_ℓ is the same as in Γ, for $\ell = 1, \ldots, k$, and the clockwise order of the split components around u is $G_1, \ldots, G_{i-1}, G_{i+1}, \ldots, G_k, G_i$, where G_1 and G_i are incident to the outer face.*

In order to implement Operation 4, which proves Lemma 4, we would like to first move G_i vertically up from the plane $z = 0$ to the plane $z = 1$, to then send G_i "far away" by modifying the x- and y-coordinates of its vertices, and to finally project G_i vertically back to the plane $z = 0$. There are two complications to this plan, though. The first one is given by the vertices u and v, which belong both to G_i and to the rest of the graph. When moving u and v on the plane $z = 1$, the edges incident to them are dragged along, which might result in these edges crossings each other. The second one is that there might be no far away position that allows the drawing of G_i to be vertically projected back to the plane $z = 0$ without introducing any crossings. This is because the rest of the graph might be arbitrarily mingled with G_i in the initial drawing Γ. As in Operation 3, convexity comes to the rescue. Indeed, we first employ a 2D crossing-free morph which makes the boundary of the outer face of G convex and moves G_i into a convex polygon. After moving G_i vertically up to the plane $z = 1$, sending G_i far away can be simply implemented as a scaling operation, which ensures that the edges incident to u and v do not cross each other during the motion of G_i on the plane $z = 1$ and that projecting G_i vertically back to the plane $z = 0$ does not introduce crossings with the edges of the rest of the graph.

We now provide the details of Operation 4, which works slightly differently if the edge (u, v) exists and if it does not. We first describe the latter case. Refer to Fig. 4b. We insert two polygons P_{in} and P_{out} with $O(n)$ vertices in Γ. As in Operation 3, they intersect Γ only at u and v, with P_{in} inside P_{out} (except at u and v). All the vertices of G_i (except u and v) lie inside P_{in} and all the vertices of $G_1, \ldots, G_{i-1}, G_{i+1}, \ldots, G_k$ (except u and v) lie outside P_{out}. We also insert in Γ a polygon P_{ext}, with $O(n)$ vertices, that intersects Γ only at u and v, and that contains all the vertices of G and P_{out} (except u and v) in its interior.

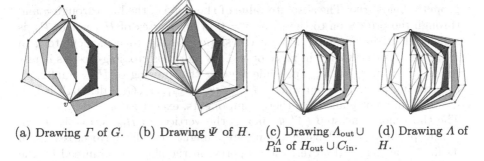

(a) Drawing Γ of G.   (b) Drawing Ψ of H.   (c) Drawing $\Lambda_{out} \cup P_{in}^\Lambda$ of $H_{out} \cup C_{in}$.   (d) Drawing Λ of H.

Fig. 4. Illustration for Operation 4: P_{in} is blue, P_{out} is red, and P_{ext} is purple. (Color figure online)

We now triangulate the region inside P_{ext} and outside P_{out}, without introducing chords for P_{out}. We also triangulate the interior of P_{in} without introducing chords for P_{in}. Let Ψ be the resulting planar straight-line drawing of a plane graph H. Let C_{out}, C_{in}, and C_{ext} be the cycles of H represented by P_{out}, P_{in}, and P_{ext} in Ψ, respectively, and let H_{out} (H_{in}) be the subgraph of H induced by the vertices that lie outside or on P_{out} (resp. inside or on P_{in}). Note that H_{out} is an internally-triconnected plane graph and H_{in} is a triconnected plane graph.

We now construct another planar straight-line drawing of H, as follows. First, construct a strictly convex drawing Q_{ext} of C_{ext} such that the angle of Q_{ext} at u (and the angle at v) is cut by the segment $\overline{uv}$ into two angles both smaller than $90°$. Next, construct a strictly convex drawing Λ_{out} of H_{out} in which C_{ext} is represented by Q_{ext}, by means of [8,11]. Let P_{out}^Λ be the strictly convex polygon representing C_{out} in Λ_{out}. As in Fig. 4c, plug a strictly convex drawing P_{in}^Λ of C_{in} in the interior of P_{out}^Λ, except at u and v, so that the path $\mathcal{P}_{in}$ that is traversed when walking in clockwise direction along C_{in} from u to v is represented by the straight-line segment $\overline{uv}$. Finally, plug into $\Lambda_{out} \cup P_{in}^\Lambda$ a convex drawing Λ_{in} of H_{in}

in which the outer face is delimited by P_{in}^Λ, by means of [8,11]. This results in a planar straight-line drawing Λ of H, see Fig. 4d.

We now describe the morph that occurs in Operation 4. We first define a morph $\langle \Psi, \ldots, \Phi \rangle$ from Ψ to an "almost" planar straight-line drawing Φ of H, as the concatenation of two morphs $\langle \Psi, \ldots, \Lambda \rangle$ and $\langle \Lambda, \ldots, \Phi \rangle$. The first morph $\langle \Psi, \ldots, \Lambda \rangle$ is an $O(n)$-step crossing-free 2D morph obtained by applying the algorithm in [2]. Translate and rotate the Cartesian axes so that, in Λ, the y-axis passes through u and v and u has a smaller y-coordinate than v. The second morph $\langle \Lambda, \ldots, \Phi \rangle$ is a 3-step 3D morph defined as follows.

- The first morphing step $\langle \Lambda, \Lambda' \rangle$ moves all the vertices of H_{in}, except for u and v, vertically up, to the plane $z = 1$. As the projection to the plane $z = 0$ of every drawing of H in $\langle \Lambda, \Lambda' \rangle$ coincides with Λ, the morph is crossing-free.
- The second morphing step $\langle \Lambda', \Lambda'' \rangle$ is such that Λ'' coincides with Λ', except for the x-coordinates of the vertices of H_{in}, which are all multiplied by the same real value $s > 0$. The value s is large enough so that, in Λ'', the following property holds true: The absolute value of the slope of the line through u and through the projection to the plane $z = 0$ of any vertex of H_{in} not in $\mathcal{P}_{in}$ is smaller than the absolute value of the slope of every edge incident to u in H_{out}; and likewise with v in place of u. This morph is crossing-free, as it just scales the drawing of H_{in} up, while leaving the drawing of H_{out} unaltered. Intuitively, this is the step where G_i "skips" $G_{i+1}, \ldots, G_k$ (although it still lies on a different plane than those components, except for u and v).
- The third morphing step $\langle \Lambda'', \Phi \rangle$ moves the vertices of H_{in} vertically down, to the plane $z = 0$. This morphing step might actually have crossings in its final drawing Φ. However, the property on the slopes guaranteed by the second morphing step ensures that the only crossings are those involving edges incident to vertices of $\mathcal{P}_{in}$ different from u and v, which do not belong to G. Hence, the restriction of $\langle \Lambda'', \Phi \rangle$ to G is a crossing-free morph.

(a) Drawing obtained by restricting Λ to G.

(b) Drawing Γ' obtained by restricting Φ to G.

Fig. 5. Illustration for Operation 4: construction of Γ' from the restriction of Λ to G.

As in Operation 3, the actual planar morph $\langle \Gamma, \ldots, \Gamma' \rangle$ is obtained by restricting the morph $\langle \Psi, \ldots, \Phi \rangle$ to G, see Fig. 5.

We now discuss the case that the edge (u, v) exists; then G_1 is such an edge. Now $\mathcal{P}_{in}$ and $\mathcal{P}_{out}$ surround all the components $G_1, \ldots, G_i$, and not just G_i;

consequently, H_{in} comprises $G_1, \ldots, G_i$. The description of Operation 4 remains the same, except for two differences. First, P_{in}^A is strictly convex; in particular, $\mathcal{P}_{\text{in}}$ is not represented by a straight-line segment, so that the edge (u, v) lies in the interior of P_{in}^A. Second, in the 3-step 3D morph $\langle A, A', A'', \Phi \rangle$, not all the vertices of H_{in} are lifted to the plane $z = 1$, then scaled, and then projected back to the plane $z = 0$, but only those of G_i. The arguments for the fact that the restriction of such a morph to G is crossing-free remain the same.

2.2   3D Morphs for Biconnected Planar Graphs

We now describe an algorithm that constructs an $O(n^2)$-step morph between any two planar straight-line drawings Γ and Φ of the same n-vertex biconnected planar graph G. It actually suffices to construct an $O(n^2)$-step morph from Γ to *any* planar straight-line drawing A of G with *the same embedding* as Φ, as then an $O(n)$-step morph from A to Φ can be constructed by means of [2]. And even more, it suffices to construct an $O(n^2)$-step morph from Γ to *any* planar straight-line drawing Ψ of G that has *the same rotation system* as A, as then an $O(1)$-step morph from Ψ to A can be constructed by means of Operation 2.

As proved by Di Battista and Tamassia [7], starting from a planar graph drawing (in our case, Γ), one can obtain the rotation system of any other planar drawing (in our case, Φ) of the same graph by: (i) suitably changing the permutation of the components in some *parallel compositions*; that is, for some split pairs $\{u, v\}$ that define three or more split components, changing the clockwise (circular) ordering of such components; and (ii) flipping the embedding for some *rigid compositions*; that is, for some split pairs that define a maximal split component that is biconnected, flipping the embedding of the component. Thus, it suffices to show how to implement these modifications by means of Operations 2–4 from Sect. 2.1. We first take care of the flips, not only in the description, but also algorithmically: All the flips are performed before all the permutation rearrangements since the flips might cause some permutation changes, which we then fix later.

Let $\{u, v\}$ be a split pair that defines a maximal biconnected split component K of G, and suppose that we want to flip the embedding of K in Γ (the drawing we deal with undergoes modifications, however for the sake of simplicity we always denote it by Γ). Note that K is not the edge (u, v), as otherwise we would not need to flip its embedding. Further, (u, v) does not belong to K, as otherwise K would not be a maximal split component. However, (u, v) might belong to $E(G) - E(K)$. Apply Operation 2 to morph Γ so that the outer face becomes any face incident to u and v. Let $G_1, \ldots, G_k$ be the split components of G with respect to $\{u, v\}$, in clockwise order around u, where G_1 and G_k are incident to the outer face. Let $\ell \in \{1, \ldots, k\}$ be such that $G_\ell = K$. We distinguish two cases, depending on whether the edge (u, v) belongs to G or not.

- If the edge (u, v) does not belong to G, then we simply apply Operation 3, with $i = j = \ell$, in order to morph Γ to flip the embedding of $G_\ell = K$.
- If (u, v) belongs to G, then let $m \in \{1, \ldots, k\}$ be such that G_m is (u, v). Assume that $\ell < m$, the other case is symmetric. Apply Operation 3 with $i = \ell$

and $j = m$, in order to morph Γ to flip the embeddings of $G_\ell, G_{\ell+1}, \ldots, G_m$. If we again denote by $G_1, \ldots, G_k$ the split components of G with respect to $\{u, v\}$, in clockwise order around u, where G_1 and G_k are incident to the outer face, G_ℓ is now the edge (u, v) and G_m is K. Apply Operation 3 a second time, with $i = \ell$ and $j = m - 1$, in order to morph Γ to flip the embeddings of $G_\ell, G_{\ell+1}, \ldots, G_{m-1}$ back to the embeddings they originally had. As desired, only the embedding of K is actually flipped.

Flipping the embedding of K is hence done in $O(n)$ morphing steps. Since there are $O(n)$ maximal biconnected split components whose embedding might need to be flipped, all such flips are performed in $O(n^2)$ morphing steps.

Let $\{u, v\}$ be a split pair of G that defines three or more split components and suppose that we want to change the clockwise (circular) ordering of such components around u to a different one. If the edge (u, v) exists, then apply Operation 2 to morph Γ so that the outer face becomes the one to the left of (u, v), when traversing (u, v) from u to v; otherwise, apply Operation 2 to morph Γ so that the outer face becomes any face incident to u and v. Let $G_1, \ldots, G_k$ be the split components of G with respect to $\{u, v\}$, in clockwise order around u, where G_1 and G_k are incident to the outer face; note that, if (u, v) exists, then it coincides with G_1. Let $G_1, G_{\sigma(2)}, G_{\sigma(3)}, \ldots, G_{\sigma(k)}$ be the desired clockwise order of the split components of G with respect to $\{u, v\}$ around u; since we are only required to fix a clockwise *circular* order of these components, then we can assume G_1 to be the first component in the desired clockwise *linear* order of such components around u that starts at the outer face.

We apply Operation 4 with index $\sigma(2)$, then again with index $\sigma(3)$, and so on until the index $\sigma(k)$. The first j applications make $G_{\sigma(2)}, G_{\sigma(3)}, \ldots, G_{\sigma(j+1)}$ the last j split components of G with respect to $\{u, v\}$ in the clockwise linear order of the components around u that starts at the outer face. Hence, after the last application we obtain the desired order. Each application of Operation 4 requires $O(n)$ morphing steps, hence changing the clockwise order around u of the split components of G with respect to a split pair $\{u, v\}$ takes $O(nk)$ morphing steps, where k is the number of split components with respect to $\{u, v\}$. Since the total number of split components with respect to every split pair of G that defines a parallel composition is in $O(n)$ [7], this sums up to $O(n^2)$ morphing steps. This concludes the proof of Theorem 1 for biconnected planar graphs.

2.3   3D Morphs for General Planar Graphs

We now sketch our algorithm for the case in which G is connected. A complete description of this case and of the case that G is not connected are in the full version [5] of this paper. Let Γ and Φ be the given planar straight-line drawings of G. Let (u, v) and (u, w) be edges of G that are consecutive in the circular order of the edges incident to u and that belong to different blocks of G, where the block B containing (u, v) is a leaf of the block-cut-vertex tree of G. We are going to augment G with a path (v, p, w). Repeating this augmentation makes G biconnected and then the algorithm from Sect. 2.2 can be applied. While v and

w can be chosen so that (v, p, w) can be planarly inserted in Φ, they are not necessarily incident to the same face of Γ, as in Fig. 6a. Thus, we let v and w share a face by suitably morphing Γ.

Triangulate Γ into a planar straight-line drawing Ψ of a maximal planar graph H, as in Fig. 6b, and then apply Operation 2 to change the outer face to a face (u, w, q), as in Fig. 6c. By means of the algorithm by Hong and Nagamochi [8] or of the algorithm by Tutte [11], construct a planar straight-line drawing Λ of H in which (u, w, q) is a triangle whose angle at u is smaller than $45°$. We obtain an $O(n)$-step crossing-free 2D morph from Ψ to Λ by means of [2]. Let $\langle \Gamma, \dots, \Gamma' \rangle$ be the restriction of this morph to G, see Fig. 6d.

Translate and rotate the Cartesian axes so that Γ' lies on the plane $z = 0$, the origin is at u, and the positive y-half-axis cuts the interior of the face to the right of (u, v). We now make u and v incident to the same face in three morphing steps. The first step moves all the vertices of B, except for u, vertically up, to the

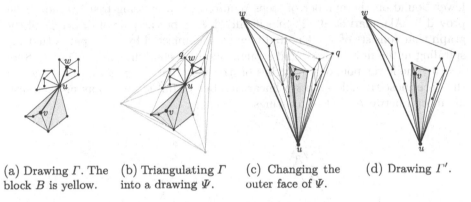

(a) Drawing Γ. The block B is yellow. (b) Triangulating Γ into a drawing Ψ. (c) Changing the outer face of Ψ. (d) Drawing Γ'.

Fig. 6. Illustration for the morph that allows path (v, p, w) to be inserted in Γ.

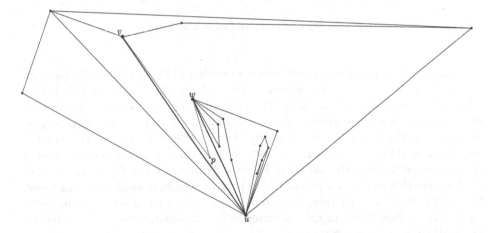

Fig. 7. Illustration for the morph that allows the path (v, p, w) to be inserted in Γ. Scaling B up so that it surrounds the rest of the graph.

plane $z = 1$. The second step scales the x and y-coordinates of all the vertices of B by a vector (α, β), where α and β are large enough so that: (i) every vertex in $V(B) - \{u\}$ has a y-coordinate larger than the one of every vertex in $V(G) - V(B)$; and (ii) the slope of every edge (u, r) of B is either between $0°$ and $45°$ (if r has positive x-coordinates) or between $135°$ and $180°$ (if r has negative x-coordinates). The third step moves all the vertices of B vertically down, to the plane $z = 0$. Now v and w are incident to the same face in Γ' and can be planarly connected via a path (v, p, w), as in Fig. 7.

3   Discussion: Lower Bounds

Though the algorithm of Sect. 2 uses a quadratic number of steps, we are not aware of any super-constant lower bound for crossing-free 3D morphs between planar straight-line graph drawings. The nested-triangles graph provides a linear lower bound on the number of steps required for a crossing-free *2D* morph, as proved by Alamdari et al. [2]. Specifically, let T_k be the pair of drawings of the graph that consists of $k + 1$ nested triangles, connected by three paths that are spiraling in the first drawing and straight in the second drawing, as in Fig. 8 for $k = 3$. The lower bound of [2] relies on the fact that the innermost triangle or the outermost triangle makes a linear number of full turns in any crossing-free 2D morph between the two drawings.

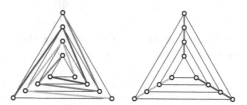

Fig. 8. The lower bound example of [2].

Even in 3D, it might seem that a linear number of linear morphs is required. However, the extra dimension allows us to perform the "turns" in parallel by "flipping" several triangles at once. The key operation is to morph T_6 in a constant number of steps without moving the innermost and outermost triangles, as shown in Fig. 9 and animated in [1]. Then for any k, we can construct a crossing-free 3D morph between the two drawings in T_{6k} in a constant number of steps by performing the morph of Fig. 9 in parallel for the k nested copies of T_6. Observe that in this morph the $(6i + 1)$-th outermost triangle does not move, for any $i = 0, \ldots, k$. Each morphing step of T_6 avoids a small tetrahedron above and below its innermost triangle, allowing different nested copies of T_6 to morph in parallel without intersecting.

The above example gives hope that the number of steps required to construct a crossing-free 3D morph between any two given planar straight-line graph drawings could be far smaller than quadratic – potentially even constant. However, it is unclear how to generalize our procedure.

The approach of Fig. 9 relies on the sequence of nested triangles to be *independent*, as we can untangle each one locally without affecting the others. This is not necessarily the case. For instance, the example in Fig. 10 shows a tree of nested triangles that are recursively twisted by 120° at every level. Here, each path in the tree has the same structure as a nested-triangles graph thus, in total, it requires $\Omega(\log n)$ morphing steps in 2D. It is unclear to us how to handle the dependencies between different tree branches.

4   Open Problems

Our research raises several other open problems. An immediate one is to reduce our quadratic upper bound for the number of steps that are needed to construct a crossing-free 3D morph between any two planar straight-line graph drawings. Extending the result of Arseneva at al. [4], we ask whether planar graph families richer than trees, e.g., outerplanar graphs and series-parallel graphs, admit crossing-free 3D morphs with a sub-linear number of steps.

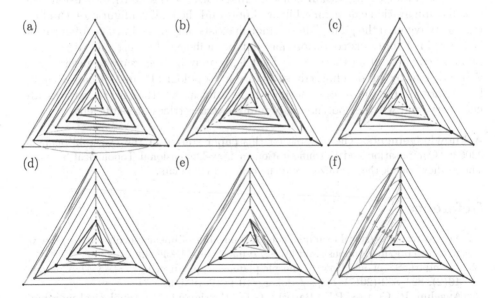

Fig. 9. Morphing T_6 in 3D without moving the innermost and outermost triangles. Orange arrows show the vertices that exchange position in the next step. Empty / large disks indicate that a vertex lies below / above the plane containing the initial drawing. The drawing obtained by the morph is of the type of the right drawing in Fig. 8. (Color figure online)

Fig. 10. A potential lower bound construction.

We have given an example of two topologically equivalent planar straight-line drawings of a triconnected graph that can be untangled in 3D using only $O(1)$ steps. Still we think that there are examples of planar graphs with topologically equivalent drawings where this is not the case. More specifically, we suspect that in 3D, as in 2D, a linear number of steps is sometimes necessary.

If the initial configuration can also make use of the third dimension, the initial configuration can be an arbitrary knot, and the final configuration can be a regular polygon in the plane. Then, a morph exists only if the initial configuration is the unknot, but this condition may not be sufficient because our edges must remain straight during the morph. That is, it may be necessary to insert extra vertices (e.g. by subdividing edges) before a (topological) unknot can actually be unknotted by one of our morphs. It is unclear whether extra vertices are necessary, and whether polynomially many extra vertices are sufficient.

Acknowledgements. The research for this paper started at the Dagstuhl Seminar 22062: "Computation and Reconfiguration in Low-Dimensional Topological Spaces". The authors thank the organizers and the other participants.

References

1. Nested triangles/spiral example: constant number of linear morphs. https://www.geogebra.org/m/djmqqhst and https://vimeo.com/718624499
2. Alamdari, S., et al.: How to morph planar graph drawings. SIAM J. Comput. **46**(2), 824–852 (2017). https://doi.org/10.1137/16M1069171
3. Angelini, P., Cortese, P.F., Battista, G.D., Patrignani, M.: Topological morphing of planar graphs. Theor. Comput. Sci. **514**, 2–20 (2013). https://doi.org/10.1016/j.tcs.2013.08.018
4. Arseneva, E., et al.: Pole dancing: 3D morphs for tree drawings. J. Graph Algorithms Appl. **23**(3), 579–602 (2019). https://doi.org/10.7155/jgaa.00503
5. Buchin, K., et al.: Morphing planar graph drawings through 3D. arXiv report. https://doi.org/10.48550/arXiv.2210.05384

6. Cairns, S.S.: Deformations of plane rectilinear complexes. Am. Math. Monthly **51**(5), 247–252 (1944). https://doi.org/10.1080/00029890.1944.11999082

7. Di Battista, G., Tamassia, R.: On-line planarity testing. SIAM J. Comput. **25**(5), 956–997 (1996). https://doi.org/10.1137/S0097539794280736

8. Hong, S.-H., Nagamochi, H.: Convex drawings of hierarchical planar graphs and clustered planar graphs. J. Discrete Algorithms **8**(3), 282–295 (2010). https://doi.org/10.1016/j.jda.2009.05.003

9. Istomina, A., Arseneva, E., Gangopadhyay, R.: Morphing tree drawings in a small 3D grid. In: Mutzel, P., Rahman, M.S., Slamin (eds.) WALCOM 2022. LNCS, vol. 13174, pp. 85–96. Springer, Cham (2022). https://doi.org/10.1007/978-3-030-96731-4_8

10. Thomassen, C.: Deformations of plane graphs. J. Comb. Theor. Ser. B **34**(3), 244–257 (1983). https://doi.org/10.1016/0095-8956(83)90038-2

11. Tutte, W.T.: How to draw a graph. Proc. Lond. Math. Soc. **3**(13), 743–767 (1963). https://doi.org/10.1112/plms/s3-13.1.743

Visualizing Multispecies Coalescent Trees: Drawing Gene Trees Inside Species Trees

Jonathan Klawitter[1,2](✉) , Felix Klesen[1] , Moritz Niederer[3],
and Alexander Wolff[1]

[1] Universität Würzburg, Würzburg, Germany
[2] University of Auckland, Auckland, New Zealand
jo.klawitter@gmail.com
[3] HTW Saar, Saarbrücken, Germany

Abstract. We consider the problem of drawing multiple gene trees inside a single species tree in order to visualize multispecies coalescent trees. Specifically, the drawing of the species tree fills a rectangle in which each of its edges is represented by a smaller rectangle, and the gene trees are drawn as rectangular cladograms (that is, orthogonally and downward, with one bend per edge) inside the drawing of the species tree. As an alternative, we also consider a style where the widths of the edges of the species tree are proportional to given effective population sizes.

In order to obtain readable visualizations, our aim is to minimize the number of crossings between edges of the gene trees in such drawings. We show that planar instances can be recognized in linear time and that the general problem is NP-hard. Therefore, we introduce two heuristics and give an integer linear programming (ILP) formulation that provides us with exact solutions in exponential time. We use the ILP to measure the quality of the heuristics on real-world instances. The heuristics yield surprisingly good solutions, and the ILP runs surprisingly fast.

1 Introduction

Visualizations of trees to present information have been used for centuries [25] and the study of producing readable, compact representations of trees has a long tradition [32, 34]. Trees and their drawings are also an ubiquitous and fundamental tool in the field of phylogenetics. In particular, a phylogenetic tree is used to model the evolutionary history and relationships of a set X of taxa such as species, genes, or languages [35]. There exist many different models, but most commonly a *phylogenetic tree* on X is a tree T whose leaves are bijectively labeled with X; see Fig. 1. In a *rooted* phylogenetic tree, each internal vertex represents a branching

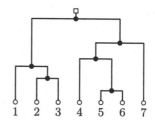

Fig. 1. A rectangular cladogram drawing of a rooted binary phylogenetic tree on seven taxa.

L. Gąsieniec (Ed.): SOFSEM 2023, LNCS 13878, pp. 96–110, 2023.
https://doi.org/10.1007/978-3-031-23101-8_7

event (such as species divergence); time (or genetic distance) is represented by the edge lengths from the root towards the leaves. In most applications, the tree is *binary*, that is, each internal vertex has indegree one, outdegree two, and thus represents a bifurcation event. An *unrooted* phylogenetic tree, on the other hand, models only the relatedness of the taxa. A phylogenetic tree where the taxa are species is called a *species tree*. If the taxa are biological sequences, such as particular genes or protein sequences, the tree is called a *gene tree*.

Multispecies Coalescent Models. One of the main tasks in phylogenetics is the inference of a phylogenetic tree for some given data and model. When inferring a species tree based on sequencing data, one might be inclined to set the species tree as that of an inferred gene tree. However, gene trees can differ from the species tree in the presence of so-called *incomplete lineage sorting*[1] or when divergence times are small[2], which can lead to inaccurate edge lengths or even to an incorrectly inferred species tree [1,26,29,33]. To address these issues, *multispecies coalescent (MSC) models* have been developed. An MSC model provides a framework for inferring species trees while accounting for conflicts between gene trees and species trees [15,18,31]. Roughly speaking, by using multiple samples (genes) per species, the model coestimates multiple gene trees that are constrained within their shared species tree. In doing so, the model can infer not only divergence times for inner vertices but also the *effective population size* for each edge (*branch*) in the species tree. There exist several models for population sizes [39], two of which we define here. In the *continuous linear model*, for each branch, the population size between the top and the bottom is linearly interpolated, and for a branch not incident to a leaf, the population size at the bottom equals the sum of the population sizes at the top of its two child branches; see Fig. 2a. In the *piecewise constant model*, the population size of each branch is constant from the top to the bottom of the branch and there are no restrictions between adjacent branches [12]; see Fig. 2b.

For a phylogenetic tree T, let $V(T)$ be the vertex set of T, let $E(T)$ be the edge set of T, and let $L(T)$ be the leaf set of T. We define an *MSC tree* as a triple $\langle S, T, \varphi \rangle$ consisting of a species tree S, a gene tree T, and a mapping $\varphi \colon L(T) \to L(S)$ with the following properties. Both S and T are rooted binary phylogenetic trees where all vertices have an associated height h that are strictly decreasing from root to leaf. We consider only the case where $h(\ell)$ is zero for each leaf ℓ in $L(S)$ and $L(T)$. For gene trees, we use the terms *leaf*, *vertex*, and *edge*; whereas we use the terms *species*, *node*, and *branch* if we want to stress that we talk about species trees. Each branch in $E(S)$ is associated with an upper and a lower population size. The mapping φ describes which leaves of T belong to which species in S. Next, consider two leaves ℓ and ℓ' of T with

---

[1] We speak of *incomplete lineage sorting* if (i) in a population of an ancestral species two (or more) variants of a gene were present, say red and blue, and (ii) when the species diverged, this did not result in one child species having the red variant and the other having the blue variant, but, e.g., one child species having both variants [33].

[2] A small divergence time corresponds to a short edge in the phylogenetic tree, which can be hard to infer correctly.

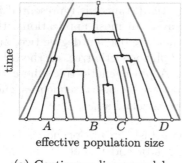

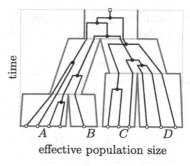

(a) Continuous linear model.   (b) Piecewise constant model.

Fig. 2. Multispecies coalescent of a species tree on four species A, B, C, D and a gene tree on eleven taxa under two different models.

$\varphi(\ell) \neq \varphi(\ell')$. Let v be the lowest common ancestor of ℓ and ℓ'. In the MSC model we have that a divergence event of v occurred before the divergence event at a node s of S that ultimately split $\varphi(\ell)$ and $\varphi(\ell')$. Hence, $h(v) > h(s)$ and we can extrapolate φ to a mapping of each inner vertex v of T to a branch of S. Lastly, we assume that the input consists of a single gene tree; otherwise we merge multiple given gene trees by connecting all their roots to a super root.

Visualizing MSC trees. Visualizations of an MSC tree usually show the species and gene tree together. This allows the user to detect any discordance between them such as whether they have different topologies and where incomplete lineage sorting occurs. It is also interesting to see where these events occur with respect to the inferred population sizes. Furthermore, these drawings are used to diagnose whether the parameters of the model are set up well. E.g., if all inner vertices of the gene tree occur directly above nodes of the species tree or if all occur near the root of the species tree, parameters may have been chosen poorly.

Wilson et al. [39] suggested a *tree-in-tree* style for an MSC tree $\langle S, T, \varphi \rangle$ under continuous models similar to the one shown in Fig. 2a. There, the species tree S is drawn in a space-filling fashion such that the branch widths of S reflect the associated population sizes and the gene tree T is then drawn into S as a classic node-link diagram. Without these constraints on the branch widths, T could be drawn as a classic rectangular cladogram as in Fig. 1. As noted above, the MSC model ensures that T can be drawn inside S without edges of T crossing edges of S since, for each edge uv of T, we have that, if u and v lie inside the branches e_u and e_v of S, respectively, then either e_u precedes e_v or $e_u = e_v$.

Douglas [12] developed the tool `UglyTrees` that generates tree-in-tree drawings for MSC trees under the piecewise constant model; Fig. 2b resembles such a drawing. He points out that the results are in many cases visually unpleasing (as reflected in his choice for the tool's name), in particular if the difference in width between parent and child is large. This is amplified in practice by the inverse relationship between the number of gene tree vertices and population sizes, which results in clusters of vertices in the narrowest branches [12].

Fig. 3. Representation of co-phylogenetic trees with the host tree as background shape and the parasite tree drawn with a node-link diagram (after Calamoneri et al. [9]).

Related Work. There exist several applications where multiple phylogenetic trees are displayed together. In a *tanglegram*, two phylogenetic trees on the same set of taxa are drawn opposite each other and the corresponding leaves are connected by line segments for easy comparison [8,14]. The tool `DensiTree` [7] allows the user to compare many trees simultaneously by drawing them on top of each other. A *co-phylogenetic tree* consists of two rooted phylogenetic trees, namely, a *host tree H* and a *parasite tree P*, together with a mapping (*reconciliation*) of the vertices of P to vertices of H. Other than in an MSC tree, the vertices of P commonly do not have heights but are mapped to nodes of H, the host branches do not have associated population sizes, and the edges of P can go from one subtree of H to another, representing so-called *host switches*. Several tools visualize the reconciliation of co-phylogenetic trees [10,11,27,36]. Commonly, the branches of H are drawn with thick lines such that P can be embedded into H; see Fig. 3. Recently, Calamoneri et al. [9] suggested a tree-in-tree style for reconciliation similar to the one for MSC trees above. They draw H in a space-filling way and embed P into H as an orthogonal node-link diagram.

More generally, visualizations have been studied for various models in phylogenetics, such as rooted phylogenetic trees [2,6,32], in conjunction with a geographic map [28,30], unrooted phylogenetic trees, and split networks [13,24,37]. In recent years, research has been extended from drawings of trees to drawings of *phylogenetic networks* [19,20,22,23,38], which are more general.

All these applications share the main combinatorial objective of finding drawings where the number of crossings between edges is minimized. To this end, good embeddings of the trees (or networks) have to be found, which are mostly fully defined by the order of the leaves. For example, Calamoneri et al. [9] investigated the problem of minimizing the number of crossings of the parasite tree in their drawings. They showed that this problem is in general NP-hard, though planar instances can be identified efficiently, and they suggested two heuristics.

Contribution. Motivated by the drawing styles of Wilson et al. [39] and by the recently proposed space-filling drawing style for reconciliation [9] and phylogenetic networks [38], we formally define tree-in-tree drawing styles for MSC trees (Sect. 2). In our base, rectangular style, we draw the species tree such that it completely fills a rectangle; the branch widths are based on the number of leaves in the respective gene subtree. Additionally, population sizes can be represented, e.g., by a background color gradient. This avoids visual overload and can be used for any population size model. Nonetheless, based on this, we also define a style where the branch widths are proportional to the associated population sizes.

We then study the problem of minimizing the number of crossings between edges of the gene tree both for the case when the embedding of the species tree is already *fixed* and when it is left *variable*. We show that the crossing minimization problem is NP-hard in both cases (Sect. 3). On the positive side, we show that crossing-free instances can be identified in linear time (Sect. 4) and we introduce two heuristics and an integer linear program (ILP) formulation for the non-planar cases (Sect. 5). We measure the performance of the heuristics on real-world instances by comparing them to optimal solutions obtained via the ILP, which we have tuned to solve medium-size instances in reasonable time.

Complete proofs to some of our claims and detailed descriptions of our algorithms can be found in the full version [21]. Implementations of our algorithm are shared upon request.

2   Drawing Style

In this section, we define styles for tree-in-tree drawings of an MSC tree $\langle S, T, \varphi \rangle$. A drawing is defined for particular leaf orders $\pi(S)$ and $\pi(T)$ of S and T, respectively, and we assume that they satisfy the following requirements. (i) At least one leaf is mapped to each species. (ii) The leaf order $\pi(T)$ is consistent with φ and $\pi(S)$, that is, the sets of leaves of T mapped by φ to a species s are consecutive in $\pi(T)$ and succeed all leaves mapped to the species that precede s in $\pi(S)$. (iii) If all the leaves of a subtree T' of T are mapped to the same species s, then these leaves must be consecutive in $\pi(T)$, and T' must admit a plane drawing above the leaves. We first describe the *rectangular style* where branch widths are proportional to the number of leaves in subtrees, and then the *proportional style* where branch widths are proportional to the population sizes. Finally, we define the crossing minimization problem for tree-in-tree drawings.

Rectangular Style. Our drawing area is an axis-aligned rectangle R. The width of R is twice the number of leaves of T. We assume that the roots of S and T have out-degree 1. We scale h such that the heights of the roots equal the height of R. The given heights of vertices and nodes thus correspond to heights in R.

The species tree S is drawn as follows; see Fig. 4. For a species $s \in L(S)$, we define $n(s) = |\varphi^{-1}(s)|$, that is, the number of leaves of T mapped to s by φ. The branches of S are represented by internally disjoint rectangles whose union covers R. Of each such rectangle we only draw the left and the right border – the *delimiters*. Their y-coordinates are defined by the heights of their start and target nodes. The x-coordinates are defined recursively: A branch incident to a species s has width $2n(s)$ and an internal branch has width equal to the width of its two child branches; see Fig. 4b. Note that the branch incident to the root has a width equal to the width of R.

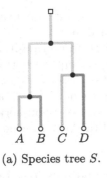

(a) Species tree S.

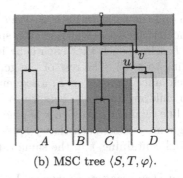

(b) MSC tree $\langle S, T, \varphi \rangle$.

Fig. 4. In the rectangular drawing style for an MSC tree, the branch widths are proportional to the number of leaves in the contained subtree.

The gene tree T is drawn in a classical orthogonal cladogram style: The leaves of T are evenly distributed at the base of R by placing them on odd coordinates and ordered by $\pi(T)$. Since $\pi(T)$ is consistent with φ and $\pi(S)$, for each species s, the leaves $\varphi^{-1}(s)$ are thus placed at the baseline of the branch incident to s. Each inner vertex of T is centered horizontally between its two children and placed at its respective height.

Note that two or more vertical line segments can end up with the same x-coordinate. Suppose that this is the case for two vertical line segments e_u and e_v that also overlap vertically and that have end vertices u and v, respectively; see Fig. 4b for an example. Further suppose that e_u ends below e_v. We then shift u slightly in the direction of its parent and v into the opposite direction. Overlaps of horizontal line segments could be handled analogously, though one would have to point out that the given heights are then misrepresented.

In this style, the population sizes are not represented by the branch widths. Instead, one could set the background color of each branch to a corresponding intensity. We advocate the rectangular tree-in-tree style (with or without coloring) since it yields a clear representation for MSC trees. This is helpful for model diagnosis and for finding incomplete lineage sorting events.

Proportional Style. The proportional style conceptually follows the rectangular style, though here the population sizes of each branch are represented by its width in the drawing; see Fig. 2a for an example. We require that the shape of S is symmetric with respect to the central vertical axis. Therefore, each branch e is represented by a sequence of trapezoids whose widths are derived from the population sizes associated with e. Embedding T into S may force the non-horizontal line segments of T to take various different slopes, "following" the trapezoids. The combinatorial properties of a drawing in the rectangular style and a drawing in the proportional style may thus differ.

A proportional-style drawing can be computed as follows. First, we draw S bottom to top by adding one row of trapezoids for each inner node u of S encountered. More precisely, at the height of u, we calculate the width of each "active" branch and the total width of S. We can then extend the delimiters between the branches. Second, the width of each species s is subdivided into $2|\varphi^{-1}(s)|$ pieces at the baseline, such that each gene leaf can be placed at an odd position and according to $\pi(T)$. The rest of the gene tree can then be computed bottom-up. For each inner vertex v of T, a sequence of line segments is drawn from each of its two children up to the height of v. The two ends are connected with a horizontal line segment on which v is placed centrally. The slope of a non-horizontal line segment f is set such that f splits the top edge and the bottom edge of the trapezoid containing f in the same ratio.

Crossing Minimization Problems. In the drawing styles above, by our assumptions for S, T, and φ, no edge of T crosses a segment that represents S. However, edges of T may cross each other. Such crossings are determined by $\pi(S)$ and $\pi(T)$. We do not know $\pi(T)$, and we consider two subproblems: in one $\pi(S)$ is given, in the other $\pi(S)$ is not given. Our objective is to find a leaf order of $\pi(T)$ and possibly of $\pi(S)$ such that the number of crossings among edges of T is minimized.

We define this problem formally for both drawing styles. In the VARIABLE TREE-IN-TREE DRAWING CROSSING MINIMIZATION (VTT) problem, we are given an MSC tree $\langle S, T, \varphi \rangle$ and an integer k, and the task is to find a tree-in-tree drawing (in rectangular or proportional style) such that T has at most k crossings; a solution is specified by leaf orders $\pi(S)$ and $\pi(T)$. In the FIXED TREE-IN-TREE DRAWING CROSSING MINIMIZATION (FTT) problem, we have the same task, but we are additionally given a leaf order $\pi(S)$; a solution is specified by a leaf order $\pi(T)$.

3   NP-Hardness

In this section, we show that VTT and FTT are NP-complete. For showing hardness, we reduce from MAX-CUT, which is NP-hard [16]. Recall that in an instance of MAX-CUT, we are given a graph G and a positive integer c. The task is to decide whether there exists a bipartition $\{A, B\}$ of the vertex set $V(G)$ of G such that at least c edges have one endpoint in A and one endpoint in B.

In the proofs below, we use the rectangular style. Since the branch widths of the rectangles can also be seen as population sizes, the proofs also hold for the proportional style. We make use of the following construction where we replace a single leaf with a particular subtree. Let ℓ be a leaf of T with its parent p at height $h(p)$. Suppose that we replace ℓ with a full binary subtree T_ℓ that has a specific number of leaves, say n_ℓ many. (Recall that a binary tree is *full* if every vertex has either 0 or 2 children.) Now we have two options to influence the shape of T_ℓ in the solution drawing. In option 1, we set the height of the lowest inner vertex of T_ℓ to at least $h(p) - \varepsilon$ for some appropriately small $\varepsilon > 0$. Now if the vertical line segment incident to ℓ is initially crossed by at least one horizontal segment, then any drawing of T_ℓ will contain at least n_ℓ many crossings. In this case, we call T_ℓ a *thick expanded leaf*. In option 2, we set the height of the root

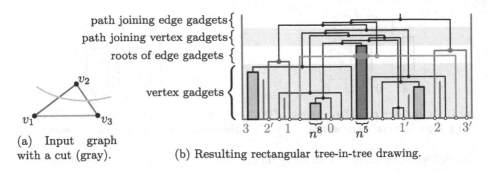

path joining edge gadgets
path joining vertex gadgets
roots of edge gadgets

vertex gadgets

v_2

v_1 v_3

3   2'   1 $\overset{\frown}{n^8}$ 0 $\overset{\frown}{n^5}$ 1'   2   3'

(a) Input graph
with a cut (gray).

(b) Resulting rectangular tree-in-tree drawing.

Fig. 5. Example for the reduction of a given graph to a rectangular tree-in-tree drawing with variable species tree embedding. Each edge gadget is drawn in the respective color.

of T_ℓ to ε', for some appropriately small $\varepsilon' > 0$. Then a drawing of T_ℓ will require n_ℓ horizontal space. In this case, we call T_ℓ a *wide expanded leaf*.

Theorem 1. *The* VTT *problem is NP-complete.*

Proof. The problem is in NP since, given an MCS tree $\langle S, T, \varphi \rangle$, an integer k, and leaf orders $\pi(S)$ and $\pi(T)$, we can check in polynomial time whether the resulting drawing has at most k crossings. To prove NP-hardness, we reduce from MAX-CUT as follows.

For a MAX-CUT instance (G, c), we construct an instance $(\langle S, T, \varphi \rangle, k)$ of the VTT problem with a species tree S, a gene tree T, a leaf mapping φ, and a positive integer k; see Fig. 5. Let $V(G) = \{v_1, \ldots, v_n\}$ ($n \geq 3$), let $m = |E(G)|$, and let $\{A, B\}$ be some partition of $V(G)$. Let S be a caterpillar tree on $2n + 1$ species labeled $0, 1, 1', \ldots, n, n'$ with decreasing depth, that is, S contains phylogenetic subtrees on species sets $\{0, 1\}, \{0, 1, 1'\}, \ldots, \{0, 1, 1', \ldots, n, n'\}$. For each $i \in \{1, \ldots, n\}$, we add to T a *vertex gadget* (described below) to enforce that species i and i' are on opposite sites of 0. Then species i being to the left of 0 corresponds to v_i being in A, whereas i being to the right of 0 corresponds to v_i being in B. Furthermore, for each edge $\{v_i, v_j\} \in E(G)$ with $i < j$, we add to T an *edge gadget* that consists of a *cherry* (i.e., a subtree on two leaves) from i to j' and that induces n^5 crossings if and only if i and j are both to the left or both to the right of 0. By construction, all pairs of vertex gadgets will induce at most n^2 crossings, all pairs of edge gadgets will induce at most n^4 crossings, and all pairs of vertex and edge gadgets will induce at most $2n^3$ crossings. In total, these gadgets induce at most $2n^4$ crossings (using $n \geq 3$). Hence, by setting $k = (m - c)n^5 + 2n^4$, we get that a tree-in-tree drawing of $\langle S, T, \varphi \rangle$ with less than k crossings exists if and only if G admits a cut containing at least c edges.

A *vertex gadget* consists of two cherries; see Fig. 6. The first cherry has one leaf each in species 0 and i' and their parent p gets some height $h(p)$. The second cherry has one leaf each in species 0 and i and their parent gets height $h(p) + 1$. We replace the leaf in i with a thick expanded leaf on n^8 many leaves. Note that if i and i' are on the same side of 0, then i lies between i' and 0. Hence, in this

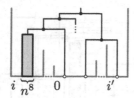
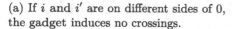

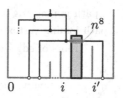

(a) If i and i' are on different sides of 0, the gadget induces no crossings.

(b) If i and i' are on the same side of 0, the gadget induces n^8 many crossings.

Fig. 6. The vertex gadget for v_i forces the species i and i' on opposite sites of species 0.

case, the horizontal line segment through p crosses the thick expanded leaf and causes n^8 crossings. Since $n^8 > k$, the vertex gadgets work as intended.

We set the heights of the roots of the edge gadget cherries above those of all vertex gadgets. Furthermore, we add a thick expanded leaf on n^5 leaves in species 0 with the lowest inner vertex higher than any edge gadget. Hence, the horizontal line segment of an edge gadget crosses n^5 vertical segments if and only if i and j are both in A or both in B.

To tie everything together in T, we introduce a path from the thick expanded leaf in 0 to the root. To this path, going upwards, we first connect the cherries of the vertex gadgets, whose leaves are in $1, 1', \ldots, n, n'$, in this order. Above those, we then connect the cherries of the edge gadgets to the path. □

The complete proof of the following statement can be found in the full version [21] of this paper.

Theorem 2. *The* FTT *problem is NP-complete.*

Proof (sketch). To prove NP-hardness, we again reduce from MAX-CUT. For a MAX-CUT instance (G, c), we construct an instance $(\langle S, T, \varphi \rangle, \pi(S), k)$ of FTT. Let $V(G) = \{v_1, \ldots, v_n\}$, and let $\{A, B\}$ be some partition of $V(G)$. Our construction consists of three parts and uses several different gadgets; see Fig. 7. On the left side, we have a *vertex gadget* for each vertex v_i. For each edge $v_i v_j$, we have an *edge gadget* that connects the vertex gadgets of v_i and v_j. The gadget has a further leaf at the far right. We simulate v_i being in either partition by having a thick expanded leaf always being either left or right of all attached edge gadgets; otherwise it would cause too many crossings. Using *spacer gadgets*, the leaves of edge gadgets to the far right are horizontally placed such that the root of each edge gadget lies exactly where we place a *cut gadget*. The cut gadget will induce n^4 crossings with the incoming edge of the root of each edge gadget only if the respective vertices are in the same partition. While some parts of our construction induce a fixed number of crossings, others cause in total at most n^3 crossings. Hence, as in the proof of Theorem 1, we can set k with respect to c such that the instance admits a tree-in-tree drawing with at most k crossings if and only if G admits a cut with at least c edges. □

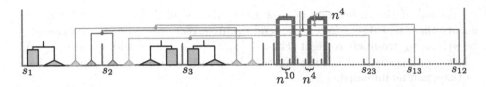

Fig. 7. Sketch of the reduction of the graph from Fig. 5a to a rectangular tree-in-tree drawing with fixed species tree embedding. Each edge gadget is drawn in the color of the respective edge in Fig. 5a. The gadget for the edge v_1v_3 (orange) has n^4 crossings more than the other edge gadgets; namely with the cut gadget (purple). (Color figure online)

4 Planar Instances

In this section, we show that we can decide in linear time whether an FTT or VTT instance admits a planar drawing.

Theorem 3. *Both when the embedding of S is fixed or variable, we can decide, in linear time, whether an MSC tree $\langle S, T, \varphi \rangle$ admits a planar rectangular tree-in-tree drawing. If yes, such a drawing can be constructed within the same time bound.*

Proof. Bertolazzi et al. [5] devised a constructive linear-time algorithm for upward planarity testing of a single-source (or single-sink) digraph, that is, whether the given digraph can be drawn with each edge uv drawn as a monotonic upward curve from u to v. For both the VTT and FTT problem, we can extend T to a single-source digraph $\bar{T}$ that admits an upward planar embedding if and only if $\langle S, T, \varphi \rangle$ admits a planar tree-in-tree drawing (respecting any given leaf order for S). We can thus apply Bertolazzi et al.'s algorithm to $\bar{T}$.

First, suppose that the embedding of S is variable. Let $L_1, L_2, \ldots, L_m$ be the subsets of $L(T)$ corresponding to the m species of S. For $i \in \{1, \ldots, m\}$, we merge all vertices in L_i into a single vertex v_i. We then connect the vertices $v_1, \ldots, v_m$ to a new vertex t; see Fig. 8b. We use the resulting single-source digraph as $\bar{T}$, which clearly has the desired properties.

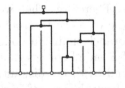

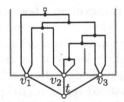

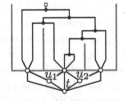

(a) A planar tree-in-tree instance $\langle S, T, \varphi \rangle$.

(b) Extending T if the embedding of S is variable.

(c) Extending T further if the embedding of S is fixed.

Fig. 8. We can test efficiently whether a tree-in-tree instance $\langle S, T, \varphi \rangle$ admits a planar drawing with a single-source upward planarity test on an extended gene tree.

Second, if the embedding of S is fixed, we extend $\bar{T}$ from above further to ensure that the subsets $L_1, \ldots, L_m$ end up in correct order. Let the species of S be $s_1, \ldots, s_m$ from left to right. For $i \in \{1, \ldots, m-1\}$, we add a vertex u_i and edges $v_i u_i$, $v_{i+1} u_i$, and $u_i t$; see Fig. 8c. The resulting graph is our new $\bar{T}$, which works again as intended.

In both cases, $\bar{T}$ has linear size and can be constructed in linear time.   □

5   Algorithms

For non-planar instances of the FTT and the VTT problem, we propose a heuristic as well as an ILP. We describe the main ideas of the algorithms here; more details can be found in the full version [21]. The ILP, which models a drawing in a straightforward fashion, is also described in the full version [21]. Overlaps of vertical segments in an ILP solution are resolved in a post-processing step. We focus here on the rectangular tree-in-tree style, though the heuristics can also be set up analogously for the proportional style. However, since the computation for the proportional style is more involved, as alternative, one can simply use leaf orders computed for the rectangular style.

Heuristic for FTT. Let $\langle S, T, \varphi \rangle$ be an MSC tree and $\pi(S)$ a leaf order for S. The idea of the heuristic is to greedily sort the leaves in each species from the left and from the right towards the centre. To this end, the algorithm (i) goes through the inner vertices in order of increasing height and (ii) when the subtree $T(v)$ of an inner vertex v has leaves in more than one species, then any unplaced leaves of $T(v)$ are put on a *left stack* or a *right stack* of their respective species; see Fig. 9. In doing so, we aim at a placement that minimizes the horizontal dimension of a drawing of $T(v)$. In particular, $T(v)$ initially has unplaced leaves in at most two species. Therefore, we place the leaves in the left species s on the right stack of s and the leaves in the right species s' on the left stack of s'; see Fig. 9b. When leaves are pushed on a stack, it is ensured that any subtree with all leaves in one species admits a planar drawing. This can be done in linear time.

Heuristic for VTT. We extend the heuristic for FTT to also compute a leaf order for S as follows. The main idea is to set the rotation of inner nodes of S such that subtrees of T horizontally span over few species. Therefore, when we handle an inner vertex v with children x and y and we try to move the roots of $T(x)$ and $T(y)$ close together. Suppose that x lies in the branch ending at node x' of S. Let $S(x')$ be the minimal phylogenetic subtree of S on all species that contain a leaf of $T(x)$; define $S(y')$ analogously. If $S(x')$ and $S(y')$ are disjoint, then we set the rotation of each unfixed vertex on the path from the root of $S(x)$ to the root of $S(y)$ such that the species of $S(x)$ and $S(y)$ get as close together as possible; see Fig. 10. Only then is v processed with the FTT heuristic. There are a few other cases to consider, which can be handled along the same line (see the full version [21] for details). Overall, handling an inner vertex of T can be done in linear time and so the overall running time is quadratic.

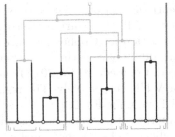
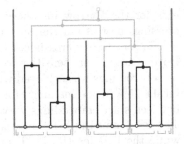
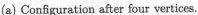

(a) Configuration after four vertices.   (b) Configuration after six vertices.

Fig. 9. The heuristic sorts the leaves in each species from the sides towards the centre by using a left stack and a right stack for each species (plus a central bucket of unplaced (orange) leaves), here on the example from Figs. 2 and 4. (Color figure online)

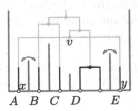

$A\ \ B\ \ C\ \ D\ \ \ \ E$

Fig. 10. The heuristic for the VTT problem rotates inner nodes of S to bring the leaves of the currently handled gene subtree closer together. Here, for the second inner vertex v of T, two nodes would be rotated to bring the species A and E together.

Note if an instance admits a planar solution, then the heuristics find one. That is, because any rotation of a node of S or an assignment to stacks keeps the leaves of a subtree of T consecutively whenever possible.

Experimental Evaluation. We tested the heuristic and the ILP on three different real world data sets: Gopher (S on 8 species, T on 26 gene taxa, 1083 instances, i.e., different topologies and heights for pairs of S and T) [4], Barrow (21 species, 88 gene taxa, 312 instances) [3], and Hamilton (36 species, 83 gene taxa, 99 instances) [17]. On a laptop with 4 cores, 8 GB of RAM, Ubuntu 20.04, and CPLEX 12.10 we tested each heuristic and the ILP on each instance once with the default (start) embedding of S from the input file and 10 times with a random (start) embedding for S. A proper experimental evaluation is out of scope for this paper, but we observed the following:

- The VTT heuristic got a better result than the FTT heuristic for 60–75% of the instances, the same result for 6–27%, and a worse result for 13–20%. For the Barrow instances, they improved the average number of 24.5 crossings of the default embeddings to 10.3 (FTT) and 7.2 (VTT) or even to 6.6 and 5.7 using random starting embeddings of S.
- Concerning FTT, the optimal solutions found by the ILP show that the FTT heuristic also found the optimal solution for about 50–55% of the instances;

e.g., for the Barrow instances, the FTT heuristic had on average only 1.3 crossings more than the optimal. Concerning VTT, the heuristics also got within zero to few crossings to the best ILP solution for Gopher instances.
– Both heuristics are sensitive to the initial embedding of S as the lowest number of crossings was achieved with a random start embedding for 70–90% and for 44–75% of the instances for FTT and for VTT, respectively. The results between start embeddings vary more for VTT than for FTT.
– The FTT and the VTT heuristic run in a fraction of a second per instance, while the ILP for the FTT problem takes about 1–4 s for most instances. The ILP for the VTT problem only found solutions for the Gopher instance within reasonable time for some instances.

Since the heuristics are so fast, our recommendation is to run both heuristics for several different start embeddings of S and then take the best found solution.

To the best of our knowledge, this is the first software to visualize MSC trees for the continuous linear model and so we hope that this will help researchers in the emerging field of MSC to visualize their results.

Acknowledgments. We thank the reviewers for their comments and J. Douglas for providing us with the test data and his helpful explanations concerning MSC.

References

1. Arbogast, B.S., Edwards, S.V., Wakeley, J., Beerli, P., Slowinski, J.B.: Estimating divergence times from molecular data on phylogenetic and population genetic timescales. Ann. Rev. Ecol. Syst. **33**, 707–740 (2002). https://doi.org/10.2307/3069277
2. Bachmaier, C., Brandes, U., Schlieper, B.: Drawing phylogenetic trees. In: Deng, X., Du, D.-Z. (eds.) ISAAC 2005. LNCS, vol. 3827, pp. 1110–1121. Springer, Heidelberg (2005). https://doi.org/10.1007/11602613_110
3. Barrow, L.N., Ralicki, H.F., Emme, S.A., Lemmon, E.M.: Species tree estimation of North American chorus frogs (Hylidae: Pseudacris) with parallel tagged amplicon sequencing. Mol. Phyl. Evol. **75**, 78–90 (2014). https://doi.org/10.1016/j.ympev.2014.02.007
4. Belfiore, N.M., Liu, L., Moritz, C.: Multilocus phylogenetics of a rapid radiation in the genus Thomomys (Rodentia: Geomyidae). Syst. Biol. **57**(2), 294–310 (2008). https://doi.org/10.1080/10635150802044011
5. Bertolazzi, P., Di Battista, G., Mannino, C., Tamassia, R.: Optimal upward planarity testing of single-source digraphs. SIAM J. Comput. **27**(1), 132–169 (1998). https://doi.org/10.1137/S0097539794279626
6. Besa, J.J., Goodrich, M.T., Johnson, T., Osegueda, M.C.: Minimum-width drawings of phylogenetic trees. In: Li, Y., Cardei, M., Huang, Y. (eds.) COCOA 2019. LNCS, vol. 11949, pp. 39–55. Springer, Cham (2019). https://doi.org/10.1007/978-3-030-36412-0_4
7. Bouckaert, R.R.: DensiTree: making sense of sets of phylogenetic trees. Bioinformatics **26**(10), 1372–1373 (2010). https://doi.org/10.1093/bioinformatics/btq110
8. Buchin, K., et al.: Drawing (complete) binary tanglegrams – hardness, approximation, fixed-parameter tractability. Algorithmica **62**(1–2), 309–332 (2012). https://doi.org/10.1007/s00453-010-9456-3

9. Calamoneri, T., Di Donato, V., Mariottini, D., Patrignani, M.: Visualizing co-phylogenetic reconciliations. Theoret. Comput. Sci. **815**, 228–245 (2020). https://doi.org/10.1016/j.tcs.2019.12.024

10. Chevenet, F., Doyon, J., Scornavacca, C., Jacox, E., Jousselin, E., Berry, V.: SylvX: a viewer for phylogenetic tree reconciliations. Bioinformatics **32**(4), 608–610 (2016). https://doi.org/10.1093/bioinformatics/btv625

11. Conow, C., Fielder, D., Ovadia, Y., Libeskind-Hadas, R.: Jane: a new tool for the cophylogeny reconstruction problem. Algorithms Molecul. Biol. **5**(1), 1–10 (2010). https://doi.org/10.1186/1748-7188-5-16

12. Douglas, J.: UglyTrees: a browser-based multispecies coalescent tree visualizer. Bioinformatics **37**(2), 268–269 (2020). https://doi.org/10.1093/bioinformatics/btaa679

13. Dress, A.W.M., Huson, D.H.: Constructing splits graphs. Trans. Comput. Biol. Bioinf. **1**(3), 109–115 (2004). https://doi.org/10.1145/1041503.1041506

14. Fernau, H., Kaufmann, M., Poths, M.: Comparing trees via crossing minimization. J. Comput. Syst. Sci. **76**(7), 593–608 (2010). https://doi.org/10.1016/j.jcss.2009.10.014

15. Flouri, T., Jiao, X., Rannala, B., Yang, Z.: Species tree inference with BPP using genomic sequences and the multispecies coalescent. Molecul. Biol. Evol. **35**(10), 2585–2593 (2018). https://doi.org/10.1093/molbev/msy147

16. Garey, M.R., Johnson, D.S.: Computers and Intractability: A Guide to the Theory of NP-Completeness. W. H. Freeman & Co., San Francisco (1979)

17. Hamilton, C.A., Lemmon, A.R., Lemmon, E.M., Bond, J.E.: Expanding anchored hybrid enrichment to resolve both deep and shallow relationships within the spider tree of life. BMC Evol. Biol. **16**(1), 1–20 (2016). https://doi.org/10.1186/s12862-016-0769-y

18. Heled, J., Drummond, A.J.: Bayesian inference of species trees from multilocus data. Mol. Biol. Evol. **27**(3), 570–580 (2009). https://doi.org/10.1093/molbev/msp274

19. Huson, D.II.: Drawing rooted phylogenetic networks. IEEE/ACM Trans. Comput. Biol. Bioinf. **6**(1), 103–109 (2009). https://doi.org/10.1109/TCBB.2008.58

20. Huson, D.H., Rupp, R., Scornavacca, C.: Phylogenetic Networks: Concepts, Algorithms and Applications. Cambridge University Press, New York (2010)

21. Klawitter, J., Klesen, F., Niederer, M., Wolff, A.: Visualizing multispecies coalescent trees: drawing gene trees inside species trees. arXiv report (2022). https://doi.org/10.48550/arXiv.2210.06744

22. Klawitter, J., Mchedlidze, T.: Upward planar drawings with two slopes. J. Graph Algorithms Appl. **26**(1), 171–198 (2022). https://doi.org/10.7155/jgaa.00587

23. Klawitter, J., Stumpf, P.: Drawing tree-based phylogenetic networks with minimum number of crossings. In: Auber, D., Valtr, P. (eds.) GD 2020. LNCS, vol. 12590, pp. 173–180. Springer, Cham (2020). https://doi.org/10.1007/978-3-030-68766-3_14

24. Kloepper, T.H., Huson, D.H.: Drawing explicit phylogenetic networks and their integration into splitstree. BMC Evol. Biol. **8**(1), 22 (2008). https://doi.org/10.1186/1471-2148-8-22

25. Lima, M.: The Book of Trees: Visualizing Branches of Knowledge. Princeton Architectural Press, New York (2014)

26. Mendes, F.K., Hahn, M.W.: Gene tree discordance causes apparent substitution rate variation. Syst. Biol. **65**(4), 711–721 (2016). https://doi.org/10.1093/sysbio/syw018

27. Merkle, D., Middendorf, M.: Reconstruction of the cophylogenetic history of related phylogenetic trees with divergence timing information. Theory Biosci. **123**(4), 277–299 (2005). https://doi.org/10.1016/j.thbio.2005.01.003

28. Page, R.: Visualising geophylogenies in web maps using geojson. PLoS Currents **7**, (2015). https://www.ncbi.nlm.nih.gov/pmc/articles/PMC4481111

29. Pamilo, P., Nei, M.: Relationships between gene trees and species trees. Mol. Biol. Evol. **5**(5), 568–583 (1988). https://doi.org/10.1093/oxfordjournals.molbev.a040517

30. Parks, D.H., et al.: Gengis 2: geospatial analysis of traditional and genetic biodiversity, with new gradient algorithms and an extensible plugin framework. PLoS ONE **8**(7), 1–10 (2013). https://doi.org/10.1371/journal.pone.0069885

31. Rannala, B., Edwards, S.V., Leaché, A., Yang, Z..: The multi-species coalescent model and species tree inference. In: Scornavacca, C., Delsuc, F., Galtier, N. (eds.) Phylogenetics in the Genomic Era, chapter 3.3, pp. 3.3:1–3.3:21. HAL (2020). https://hal.archives-ouvertes.fr/hal-02535070v3

32. Rusu, A.: Tree drawing algorithms. In: Tamassia, R. (ed.) Handbook on Graph Drawing and Visualization, chapter 3, pp. 155–192. Chapman and Hall/CRC (2013)

33. Schrempf, D., Szöllősi, G.: The sources of phylogenetic conflicts. In: Scornavacca, C., Delsuc, F., Galtier, N. (eds.) Phylogenetics in the Genomic Era, chapter 3.1, pages 3.1:1–3.1:23. HAL (2020). https://hal.archives-ouvertes.fr/hal-02535070v3

34. Schulz, H.: Treevis.net: a tree visualization reference. IEEE Comput. Graphics Appl. **31**(6), 11–15 (2011). https://doi.org/10.1109/MCG.2011.103

35. Semple, C., Steel, M.A.: Phylogenetics. vol. 24 of Oxford Lect. Ser. Math. & Its Appl. Oxford University Press (2003)

36. Sennblad, B., Schreil, E., Berglund Sonnhammer, A.-C., Lagergren, J., Arvestad, L.: Primetv: a viewer for reconciled trees. BMC Bioinf. **8**(148), (2007). https://doi.org/10.1186/1471-2105-8-148

37. Spillner, A., Nguyen, B.T., Moulton, V.: Constructing and drawing regular planar split networks. IEEE/ACM Trans. Comput. Biol. Bioinf. **9**(2), 395–407 (2012). https://doi.org/10.1109/TCBB.2011.115

38. Tollis, I.G., Kakoulis, K.G.: Algorithms for visualizing phylogenetic networks. Theoret. Comput. Sci. **835**, 31–43 (2020). https://doi.org/10.1016/j.tcs.2020.05.047

39. Wilson, I.J., Weale, M.E., Balding, D.J.: Inferences from DNA data: Population histories, evolutionary processes and forensic match probabilities. J. Royal Stat. Soc. Ser. A **166**(2), 155–188 (2003). https://doi.org/10.1111/1467-985X.00264

Parameterized Approaches to Orthogonal Compaction

Walter Didimo[1], Siddharth Gupta[2]($\boxtimes$), Philipp Kindermann[3] (iD),
Giuseppe Liotta[1], Alexander Wolff[4] (iD), and Meirav Zehavi[5]

[1] Universitá degli Studi di Perugia, Perugia, Italy
{walter.didimo,giuseppe.liotta}@unipg.it
[2] University of Warwick, Coventry, UK
siddharth.gupta.1@warwick.ac.uk
[3] Universität Trier, Trier, Germany
kindermann@uni-trier.de
[4] Universität Würzburg, Würzburg, Germany
[5] Ben-Gurion University of the Negev, Beersheba, Israel

Abstract. Orthogonal graph drawings are used in applications such as
UML diagrams, VLSI layout, cable plans, and metro maps. We focus
on drawing planar graphs and assume that we are given an *orthogonal
representation* that describes the desired shape, but not the exact coordinates of a drawing. Our aim is to compute an orthogonal drawing on
the grid that has minimum area among all grid drawings that adhere to
the given orthogonal representation.

This problem is called orthogonal compaction (OC) and is known
to be NP-hard, even for orthogonal representations of cycles [Evans et
al. 2022]. We investigate the complexity of OC with respect to several
parameters. Among others, we show that OC is fixed-parameter tractable
with respect to the most natural of these parameters, namely, the number of *kitty corners* of the orthogonal representation: the presence of pairs
of kitty corners in an orthogonal representation makes the OC problem
hard. Informally speaking, a pair of kitty corners is a pair of reflex corners of a face that point at each other. Accordingly, the number of kitty
corners is the number of corners that are involved in some pair of kitty
corners.

Keywords: Orthogonal graph drawing · Orthogonal representation ·
Compaction · Parameterized complexity

This research was initiated at Dagstuhl Seminar 21293: Parameterized Complexity
in Graph Drawing. Work partially supported by: (i) Dep. of Engineering, Perugia
University, grant RICBA21LG: Algoritmi, modelli e sistemi per la rappresentazione
visuale di reti, (ii) Engineering and Physical Sciences Research Council (EPSRC) grant
EP/V007793/1, (vi) European Research Council (ERC) grant termed PARAPATH.

L. Gąsieniec (Ed.): SOFSEM 2023, LNCS 13878, pp. 111–125, 2023.
https://doi.org/10.1007/978-3-031-23101-8_8

1 Introduction

In a *planar orthogonal drawing* of a planar graph G each vertex is mapped to a distinct point of the plane and each edge is represented as a sequence of horizontal and vertical segments. A planar graph G admits a planar orthogonal drawing if and only if it has vertex-degree at most four. A *planar orthogonal representation* H of G is an equivalence class of planar orthogonal drawings of G that have the same "shape", i.e., the same planar embedding, the same ordered sequence of bends along the edges, and the same vertex angles. A planar orthogonal drawing belonging to the equivalence class H is simply called a *drawing of H*. For example, Figs. 1a and b are drawings of the same orthogonal representation, while Fig. 1c is a drawing of the same graph with a different shape.

Given a planar orthogonal representation H of a connected planar graph G, the *orthogonal compaction* problem (OC for short) for H asks to compute a minimum-area drawing of H. More formally, it asks to assign integer coordinates to the vertices and to the bends of H such that the area of the resulting planar orthogonal drawing is minimum over all drawings of H. The area of a drawing is the area of the minimum bounding box that contains the drawing. For example, the drawing in Fig. 1a has area $7 \times 5 = 35$, whereas the drawing in Fig. 1b has area $7 \times 4 = 28$, which is the minimum for that orthogonal representation.

The area of a graph layout is considered one of the most relevant readability metrics in orthogonal graph drawing (see, e.g., [16,27]). Compact grid drawings are desirable as they yield a good overview without neglecting details. For this reason, the OC problem is widely investigated in the literature. Bridgeman et al. [11] showed that OC can be solved in linear time for a subclass of planar orthogonal representations called *turn-regular*. Informally speaking, a face of a planar orthogonal representation H is turn-regular if it does not contain any pair of so-called *kitty corners*, i.e., a pair of reflex corners (turns of 270°) that point to each other; a representation is turn-regular if all its faces are turn-regular. See Fig. 1 and refer to Sect. 2 for a formal definition. On the other hand, Patrignani [31] proved that, unfortunately, OC is NP-hard in general. Evans et al. [23] showed that OC remains NP-hard even for orthogonal representations of simple cycles. Since cycles have constant pathwidth (namely 2), this immediately shows that we cannot expect an FPT (or even an XP) algorithm parameterized by pathwidth alone unless P = NP. The same holds for parametrizations with respect to treewidth since the treewidth of a graph is upper bounded by its pathwidth.

In related work, Bannister et al. [3] showed that several problems of compacting *not necessarily planar* orthogonal graph drawings to use the minimum number of rows, area, length of longest edge, or total edge length cannot be approximated better than within a polynomial factor of optimal (if P$\neq$NP). They also provided an FPT algorithm for testing whether a drawing can be compacted to a small number of rows. Note that their algorithm does not solve the planar case because the algorithm is allowed to change the embedding.

The research in this paper is motivated by the relevance of the OC problem and by the growing interest in parameterized approaches for NP-hard prob-

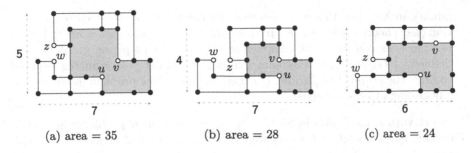

Fig. 1. (a) Drawing of a non-turn-regular orthogonal representation H; vertices u and v point to each other in the filled internal face, thus they represent a pair of kitty corners. Vertices w and z are a pair of kitty corners in the external face. (b) Another drawing of H with minimum area. (c) Minimum-area drawing of a turn-regular orthogonal representation of the same graph.

lems in graph drawing [25]. Recent works on the subject include parameterized algorithms for book embeddings and queue layouts [2,7,8,10,28], upward planar drawings [10,12], orthogonal planarity testing and grid recognition [18,26], clustered planarity and hybrid planarity [15,29,30], 1-planar drawings [1], and crossing minimization [4,21,22].

Contribution. Extending this line of research, we initiate the study of the parameterized complexity of OC and investigate several parameters:

- *Number of kitty corners.* Given that OC can be solved efficiently for orthogonal representations without kitty corners, the number of kitty corners (that is, the number of corners involved in some pair of kitty corners) is a very natural parameter for OC. We show that OC is fixed-parameter tractable (FPT) with respect to the number of kitty corners (Theorem 1 in Sect. 3).
- *Number of faces.* Since OC remains NP-hard for orthogonal representations of simple cycles [23], OC is para-NP-hard when parameterized by the number of faces. Hence, we cannot expect an FPT (or even an XP) algorithm in this parameter alone, unless P = NP. However, for orthogonal representations of simple cycles we show the existence of a polynomial kernel for OC when parameterized by the number of kitty corners (Theorem 2 in Sect. 4).
- *Maximum face-degree.* The maximum face-degree is the maximum number of vertices on the boundary of a face. Since both the NP-hardness reductions by Patrignani [31] and Evans et al. [23] require faces of linear size, it is interesting to know whether faces of constant size make the problem tractable. We prove that this is not the case, i.e., OC remains NP-hard when parameterized by the maximum face degree (Theorem 3 in Sect. 5).
- *Height.* The *height* of an orthogonal representation is the minimum number of distinct y-coordinates required to draw the representation. Since a $w \times h$ grid has pathwidth at most h, graphs with bounded height have bounded pathwidth, but the converse is generally not true [9]. In fact, we show that OC

admits an XP algorithm parameterized by the height of the given representation (see Theorem 4 in Sect. 6). In this context, we remark that a related problem has been considered by Chaplick et al. [13]. Given a planar graph G, they defined $\bar{\pi}_2^1(G)$ to be the minimum number of distinct y-coordinates required to draw the graph straight-line. (In their version of the problem, however, the embedding of G is not fixed.)

We start with some basics in Sect. 2 and close with open problems in Sect. 7. Theorems marked with "$\star$" are proven in detail in the full version [19] of this paper.

2   Basic Definitions

Let $G = (V, E)$ be a connected planar graph of vertex-degree at most four and let Γ be a planar orthogonal drawing of G. We assume that in Γ all the vertices and bends have integer coordinates, i.e., we assume that Γ is an integer-coordinate grid drawing. Two planar orthogonal drawings Γ_1 and Γ_2 of G are *shape-equivalent* if: (*i*) Γ_1 and Γ_2 have the same planar embedding; (*ii*) for each vertex $v \in V$, the geometric angles at v (formed by any two consecutive edges incident on v) are the same in Γ_1 and Γ_2; (*iii*) for each edge $e = (u, v) \in E$ the sequence of left and right bends along e while moving from u to v is the same in Γ_1 and Γ_2. An *orthogonal representation* H of G is a class of shape-equivalent planar orthogonal drawings of G. It follows that an orthogonal representation H is completely described by a planar embedding of G, by the values of the angles around each vertex (each angle being a value in the set $\{90°, 180°, 270°, 360°\}$), and by the ordered sequence of left and right bends along each edge (u, v), moving from u to v; if we move from v to u, then this sequence and the direction (left/right) of each bend are reversed. If Γ is a planar orthogonal drawing in the class H, then we also say that Γ is a drawing of H. Without loss of generality, we also assume that an orthogonal representation H comes with a given "orientation", i.e., for each edge segment $\overline{pq}$ of H (where p and q correspond to vertices or bends), we fix whether p lies to the left, to the right, above, or below q.

Turn-Regular Orthogonal Representations. Let H be a planar orthogonal representation. For the purpose of the OC problem, and without loss of generality, we always assume that each bend in H is replaced by a degree-2 vertex. Let f be a face of a planar orthogonal representation H and assume that the boundary of f is traversed counterclockwise (resp. clockwise) if f is internal (resp. external). Let u and v be two reflex vertices of f. Let $\mathrm{rot}(u, v)$ be the number of convex corners minus the number of reflex corners encountered while traversing the boundary of f from u (included) to v (excluded); a reflex vertex of degree one is counted like two reflex vertices. We say that u and v is a pair of *kitty corners* of f if $\mathrm{rot}(u, v) = 2$ or $\mathrm{rot}(v, u) = 2$. A vertex is a kitty corner if it is part of a pair of kitty corners. A face f of H is *turn-regular* if it does not contain a pair of kitty corners. The representation H is *turn-regular* if all faces are turn-regular.

Parameterized Complexity. Let Π be an NP-hard problem. In the framework of parameterized complexity, each instance of Π is associated with a *parameter k.* Here, the goal is to confine the combinatorial explosion in the running time of an algorithm for Π to depend only on k. Formally, we say that Π is *fixed-parameter tractable (FPT)* if any instance (I, k) of Π is solvable in time $f(k) \cdot |I|^{O(1)}$, where f is an arbitrary computable function of k. A weaker request is that for every fixed k, the problem Π would be solvable in polynomial time. Formally, we say that Π is *slice-wise polynomial (XP)* if any instance (I, k) of Π is solvable in time $f(k) \cdot |I|^{g(k)}$, where f and g are arbitrary computable functions of k.

A companion notion of fixed-parameter tractability is that of *kernelization.* A *kernelization algorithm* is a polynomial-time algorithm that transforms an arbitrary instance of the problem to an equivalent instance of the same problem whose size is bounded by some computable function g of the parameter of the original instance. The resulting instance is called a *kernel*, and we say that the problem admits a kernel of size $g(k)$ where k is the parameter. If g is a polynomial function, then it is called a *polynomial kernel*, and we say that the problem admits a polynomial kernel. For more information on parameterized complexity, we refer to books such as [14, 20, 24].

3   Number of Kitty Corners: An FPT Algorithm

Turn-regular orthogonal representation can be compacted optimally in linear time [11]. We recall this result and then describe our FPT algorithm.

Upward Planar Embeddings and Saturators. Let $D = (V, E)$ be a plane DAG, i.e., an acyclic digraph with a given planar embedding. An *upward planar drawing* Γ of D is an embedding-preserving drawing of D where each vertex v is mapped to a distinct point of the plane and each edge is drawn as a simple Jordan arc monotonically increasing in the upward direction. Such a drawing exists if and only if D is the spanning subgraph of a plane *st-graph*, i.e., a plane digraph with a unique source s and a unique sink t, which are both on the external face [17]. Let S be the set of sources of D, T be the set of sinks, and $I = V \setminus (S \cup T)$. D is *bimodal* if, for every vertex $v \in I$, the outgoing edges (and hence the incoming edges) of v are consecutive in the clockwise order around v. If an upward planar drawing Γ of D exists, then D is necessarily bimodal and Γ uniquely defines the left-to-right orderings of the outgoing and incoming edges of each vertex. This set of orderings (for all vertices of D) is an *upward planar embedding* of D, and is regarded an equivalence class of upward planar drawings of D. A plane DAG with a given upward planar embedding is an *upward plane DAG.*

Let e_1 and e_2 be two consecutive edges on the boundary of a face f of a bimodal plane digraph D, and let v be their common vertex. Vertex v is a *source switch of f* (resp. a *sink switch of f*) if both e_1 and e_2 are outgoing edges (resp. incoming edges) of v. Note that, for each face f, the number n_f of source switches of f equals the number of sink switches of f. The *capacity of f* is the

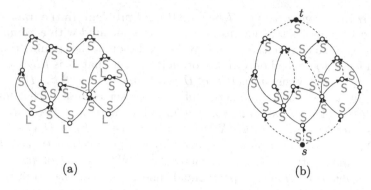

(a)   (b)

Fig. 2. (a) An upward plane DAG D and the corresponding upward labeling. (b) A plane st-graph obtained by augmenting D with a complete saturator (dotted edges).

function $\mathsf{cap}(f) = n_f - 1$ if f is an internal face and $\mathsf{cap}(f) = n_f + 1$ if f is the external face. If Γ is an upward planar drawing of D, then each vertex $v \in S \cup T$ (i.e., a source or a sink) has exactly one angle larger than 180°, called a *large angle*, in one of its incident faces, and $\deg(v) - 1$ angles smaller than 180°, called *small angles*, in its other incident faces. For a source or sink switch of f, assign either a label L or a label S to its angle in f, depending on whether this angle is large or small. For each face f of D, the number of L-labels determined by Γ equals $\mathsf{cap}(f)$ [6]. Conversely, given an assignment of L- and S-labels to the angles at the source and sink switches of D; for each vertex v, $\mathsf{L}(v)$ (resp. $\mathsf{S}(v)$) denotes the number of L- (resp. of S-) labels at the angles of v. For each face f, $\mathsf{L}(f)$ (resp. $\mathsf{S}(f)$) denotes the number of L- (resp. of S-) labels at the angles in f. Such an assignment corresponds to the labels induced by an upward planar drawing of D if and only if the following properties hold [6]: (a) $\mathsf{L}(v) = 0$ for each $v \in I$ and $\mathsf{L}(v) = 1$ for each $v \in S \cup T$; (b) $\mathsf{L}(f) = \mathsf{cap}(f)$ for each face $f \in F$. We call such an assignment an *upward labeling of D*, as it uniquely corresponds to (and hence describes) an upward planar embedding of D; see Fig. 2a. We will implicitly assume that a given upward plane DAG is described by an upward labeling.

Given an upward plane DAG D, a *complete saturator* of D is a set of vertices and edges, not belonging to D, used to augment D to a plane st-graph D'. More precisely, a complete saturator consists of a source s and a sink t, which will belong to the external face of D', and of a set of edges where each edge (u, v) is called a *saturating edge* and fulfills one of the following conditions (see, e.g., Fig. 2b): (i) $u, v \notin \{s, t\}$ and u, v are both source switches of the same face f such that u has label S in f and v has label L in f; in this case u *saturates* v. (ii) $u, v \notin \{s, t\}$ and u, v are both sink switches of the same face f such that u has label L in f and v has label S in f; in this case v *saturates* u. (iii) $u = s$ and v is a source switch of the external face with an L angle. (iv) $v = t$ and u is a sink switch of the external face with an L angle.

We now recall how to compact in linear time a turn-regular orthogonal representation. Let H be an orthogonal representation that is not necessarily turn-regular. Let D_x be the plane DAG whose vertices correspond to the maximal vertical chains of H and such that two vertices of D_x are connected by an edge oriented rightward, if the corresponding vertical chains are connected by a horizontal segment in H. Define the upward plane DAG D_y symmetrically, where the vertices correspond to the maximal horizontal chains of H and where the edges are oriented upward. Refer to Fig. 3. D_x and D_y are both upward plane DAGs (for D_x rotate it by $90°$ to see all edges flowing in the upward direction). For a vertex v of H, $c_x(v)$ (resp. $c_y(v)$) denotes the vertex of D_x (resp. of D_y) corresponding to the maximal vertical (resp. horizontal) chain of H that contains v. For any two vertices u and v of H such that $c_x(u) \neq c_x(v)$, we write $u \rightsquigarrow_x v$ if there exists a directed path from $c_x(u)$ to $c_x(v)$ in D_x. We also write $u \leftrightsquigarrow_x v$ if either $u \rightsquigarrow_x v$ or $v \rightsquigarrow_x u$, while $u \nrightsquigarrow_x v$ means that neither $u \rightsquigarrow_x v$ nor $v \rightsquigarrow_x u$. The notations $u \rightsquigarrow_y v$, $v \rightsquigarrow_y u$, $u \leftrightsquigarrow_y v$, and $u \nrightsquigarrow_y v$ are used symmetrically referring to D_y when $c_y(u) \neq c_y(v)$.

Bridgeman et al. [11] showed that H is turn-regular if and only if, for every two vertices u and v in H, we have $u \leftrightsquigarrow_x v$, or $u \leftrightsquigarrow_y v$, or both. This is equivalent to saying that the relative position along the x-axis or the relative position along the y-axis (or both) between u and v is fixed over all drawings of H. Under this condition, the OC problem for H can be solved by independently solving in $O(n)$ time a pair of 1D compaction problems for H, one in the x-direction and the other in the y-direction. The 1D compaction in the x-direction consists of: (i) augmenting D_x to become a plane st-graph by means of a complete saturator; (ii) computing an optimal topological numbering X of D_x (see [16], p. 89); each vertex v of H receives an x-coordinate $x(v)$ such that $x(v) = X(c_x(v))$. We recall that a topological numbering of a DAG D is an assignment of integer numbers to the vertices of D such that if there is a path from u to v then u is assigned a number smaller than the number of v. A topological numbering is optimal if the range of numbers that is used is the minimum possible. Regarding step (i) of the 1D compaction, note that D_x admits a unique complete saturator when H is turn-regular [11]. This is due to the absence of kitty corners in each face of H. The 1D compaction in the y-direction is solved symmetrically, so that each vertex v receives a y-coordinate $y(v) = Y(c_y(v))$. Figure 3 illustrates this process.

Unfortunately, if H is not turn-regular, the aforementioned approach fails. This is because there are in general many potential complete saturators for augmenting the two upward plane DAGs D_x and D_y to plane st-graphs. Also, even when an st-graph for each DAG is obtained from a complete saturator, computing independently an optimal topological numbering for each of the two st-graphs may lead to non-planar drawings if no additional relationships are established for the coordinates of kitty corner pairs, because for a pair $\{u, v\}$ of kitty corners we have $u \nrightsquigarrow_x v$ and $u \nrightsquigarrow_y v$. We now prove that OC is fixed-parameter-tractable when parameterized by the number of kitty corners.

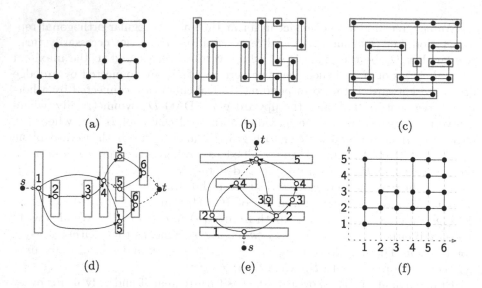

Fig. 3. (a) A turn-regular orthogonal representation H. (b)–(c) The maximal horizontal and vertical chains of H are highlighted. (d) The upward plane DAG D_x with its complete saturator (dashed edges) and an optimal topological numbering. (e) The same for D_y. (f) A minimum-area drawing of H where the x- and y-coordinates correspond to the two optimal topological numberings.

Theorem 1. *Let H be a planar orthogonal representation with n vertices and $k > 0$ kitty corners. There exists an $O(2^{11k} n \log n)$-time algorithm that computes a minimum-area drawing of H.*

Proof. Let H be an orthogonal representation and let k be the number of kitty corners of H. For each pair $\{u, v\}$ of kitty corners, we guess the relative positions of u and v in a drawing of H, i.e., $x(u) \lesseqgtr x(v)$ and $y(u) \lesseqgtr y(v)$.

Namely, we generate all maximal plane DAGs (together with an upward planar embedding) that can be incrementally obtained from D_x by repeatedly applying the following sequence of steps: Guess a pair $\{u, v\}$ of kitty corners of H such that $c_x(u)$ and $c_x(v)$ belong to the same face; for such a pair either add a directed edge $(c_x(u), c_x(v))$ (which establishes that $x(u) < x(v)$), or add a directed edge $(c_x(v), c_x(u))$ (which establishes that $x(u) > x(v)$), or identify $c_x(u)$ and $c_x(v)$ (which establishes that $x(u) = x(v)$); this last operation corresponds to adding in H a vertical segment between u and v, thus merging the vertical chain of u with the vertical chain of v. Analogously, we generate from D_y a set of maximal plane DAGs (together with an upward planar embedding). Let $\overline{D_x}$ and $\overline{D_y}$ be two upward plane DAGs generated as above. We augment $\overline{D_x}$ (resp. $\overline{D_y}$) with a complete saturator that makes it a plane st-graph. Observe that, by construction, neither $\overline{D_x}$ nor $\overline{D_y}$ contain two non-adjacent vertices in the same face whose corresponding chains of H have a pair of kitty corners. Hence their complete saturators are uniquely defined. We finally compute a pair of optimal topological numberings to determine the x- and the y-coordinates of each vertex of H as in [11]. Note that, for some pairs of $\overline{D_x}$ and $\overline{D_y}$, the procedure described above may assign x- and y-coordinates to the vertices of H that do not lead to a planar orthogonal drawing. If so, we discard the solution. Conversely, for those solutions that correspond to (planar) drawings of H, we compute the area and, at the end, we keep one of the drawings having minimum area.

Figure 4 shows a non-turn-regular orthogonal representation; Fig. 4d depicts four drawings resulting from different pairs of upward plane DAGs, each establishing different x- and y-relationships between pairs of kitty corners. One of the drawings has minimum area; another one is not planar and therefore discarded.

We now analyze the runtime. Let $\{f_1, f_2, \ldots, f_h\}$ be the set of faces of H, and let k_i be the number of kitty corners in f_i. Denote by a_i the number of distinct maximal planar augmentations of f_i with edges that connect pairs of kitty corners. An upper bound to the value of a_i is the number c_{k_i} of distinct maximal outerplanar graphs with k_i vertices, which corresponds to the number of distinct triangulations of a convex polygon with k_i vertices. It is known that c_{k_i} equals the (k_i-2)-nd Catalan number (see, e.g., [32]), whose standard estimate is $c_{k_i-2} \sim \frac{4^{k_i-2}}{(k_i-2)^{3/2}\sqrt{\pi}}$. Therefore, $a_i \in O(4^{k_i})$. Note that all distinct triangulations of a convex polygon can easily be generated with a recursive approach.

Now, for each edge (u, v) of a maximal planar augmentation of f_i such that $\{u, v\}$ is a pair of kitty corners in H, we have to consider three alternative possibilities: $\overline{D_x}$ has a directed edge $(c_x(u), c_x(v))$, or $\overline{D_x}$ has a directed edge $(c_x(v), c_x(u))$, or $c_x(u)$ and $c_x(v)$ are identified in $\overline{D_x}$. The same happens for $\overline{D_y}$. Therefore, since the number of edges of a maximal outerplanar graph on k_i vertices is $2k_i - 3$, the number of different configurations to be considered for each face f_i both in $\overline{D_x}$ and in $\overline{D_y}$ is $O(3^{2k_i} 4^{k_i}) \cdot O(3^{2k_i} 4^{k_i}) = O(36^{2k_i}) = O(2^{11k_i})$. By combining these possible configurations over all faces of H, we obtain $O(2^{11k})$ pairs of possible configurations for $\overline{D_x}$ and in $\overline{D_y}$ (clearly, $\sum_{i=0}^{h} k_i = k$). For each such pair, we augment each of the two upward plane DAGs to a plane st-graph and compute an optimal topological numbering in $O(n)$ time. Then we test whether the drawing resulting from the two topological numberings is planar, which can be done in $O(n \log n)$ time by a sweep line algorithm (see, e.g., [5,33]). It follows that the whole testing algorithm takes $O(2^{11k} n \log n)$ time. □

4 A Polynomial Kernel for Cycle Graphs

In this section, we sketch our proof of the following theorem.

Theorem 2. (⋆) *Parameterized by the number of kitty corners, OC admits a compression with linear number of vertices (and a polynomial kernel) on cycle graphs.*

Let G be a cycle graph with an orthogonal representation H. We traverse the (single) internal face of H in counterclockwise direction to define a labeled digraph $G^{\rightarrow}$: label each edge E, W, N, or S based on its direction, and label each vertex F, C and R based on whether it is flat, convex or reflex from the internal

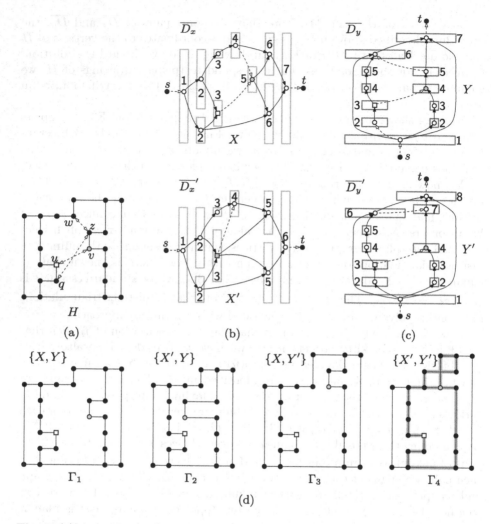

Fig. 4. (a) An orthogonal representation H with three pairs of kitty corners, $\{u,v\}$, $\{w,z\}$, and $\{q,z\}$. (b) Two distinct (saturated) upward plane DAGs $\overline{D_x}$ and $\overline{D_x}'$, and their optimal topological numberings X and X'; in $\overline{D_x}'$ the nodes $c_x(u)$ and $c_x(v)$ are identified (filled square). (c) Two distinct (saturated) upward plane DAGs $\overline{D_y}$ and $\overline{D_y}'$ and their optimal topological numberings Y and Y'. (d) Drawings derived from the four different combinations of the topological numberings: Γ_1 and Γ_3 have sub-optimal areas, Γ_2 has minimum area, and Γ_4 is non-planar (the bold red face is self-crossing). (Color figure online)

face. Given two vertices u and v, let $P_{u,v}$ be the directed path from u to v in $G^{\rightarrow}$. For an edge e, let weight(e) be the weight of e. (The addition of edge weights will yield a compression, which can be turned into a kernel.)

Let $\langle c_1, \ldots, c_k, c_{k+1} = c_1 \rangle$ be the cyclic order of kitty corners of H in $G^{\rightarrow}$. For each $P_{c_i, c_{i+1}}$, we bound the number of internal vertices. As $G^{\rightarrow}$ is the union of these paths, this bounds the size of the reduced instance. We now present reduction rules to reduce the number of vertices on these paths. We always apply them in the given order. We first reduce a path of F-vertices to a weighted edge:

Reduction Rule 1. We reduce every path $P_{u,v}$, whose internal vertices are all labeled F, to a directed edge (u,v) with $\mathsf{weight}((u,v)) = \sum_{e \in E(P_{u,v})} \mathsf{weight}(e)$.

Thus, next assume that $G^\rightarrow$ does not have any F-vertex. Observe that if $P_{c_i, c_{i+1}}$ has at least 7 internal vertices, then either all the internal vertices are labeled R or $P_{c_i, c_{i+1}}$ has an internal subpath with a labeling from $\{\mathsf{RCR}, \mathsf{RCCC}, \mathsf{RCCR}, \mathsf{RRRC}, \mathsf{CRC}, \mathsf{CRRC}, \mathsf{CRRR}, \mathsf{CCCR}\}$. So, in the former case, we give a counting rule to count all the R vertices against the kitty corners. Moreover, in the later case, we give reduction rules to reduce those paths. This, in turn, will bound the size of the reduced instance. Due to lack of space, we refer the readers to the full version [19] of this paper for the details.

5 Maximum Face Degree: Parameterized Hardness

We show that the problem remains NP-hard even if all faces have constant degree. Our proof elaborates on ideas of Patrignani's NP-hardness proof for OC [31].

Theorem 3. (⋆) *OC is para-NP-hard when parameterized by the maximum face degree.*

Proof (sketch). Patrignani [31] reduces from SAT to OC. For a SAT instance ϕ with n variables and m clauses, he creates an orthogonal representation H_ϕ that admits an orthogonal grid drawing of size $w_\phi \cdot h_\phi$ if and only if ϕ is satisfiable.

Every variable is represented by a *variable rectangle* inside a *frame*; see Fig. 5. Between the frame and the rectangles, there is a *belt*: a long path of 4 reflex vertices alternating with 4 convex vertices that ensures that every variable rectangle is either shifted to the top (true) or to the bottom (false).

Every clause is represented by a *chamber* through the variable rectangles with one or two *blocker rectangles* depending on the occurrence of the variable in the clause; see Fig. 6. Into the chamber, a *pathway* is inserted that can only be drawn if there is a gap between blocker rectangles is vertically aligned with a gap between two variable-clause rectangles, which represents a fulfilled literal.

We now briefly sketch how to adjust the reduction.

For the clause gadgets, there are two large faces of size $O(m)$; above and below the pathway. To avoid these, we connect the pathway to the top and the bottom boundary of each of the variable-clause rectangles as in see Fig. 7.

For the face around the variable rectangles, we refine the left and the right side (that both have $O(m)$ vertices) by adding $O(m)$ rectangles of constant degree in a tree-like shape; see Fig. 8a. Instead of a single long belt, we use a small belt of constant length around every variable rectangle that lies inside its own frame and extend the variable rectangles vertically; see Fig. 8b.

After these adjustments, all faces have constant degree as desired.   □

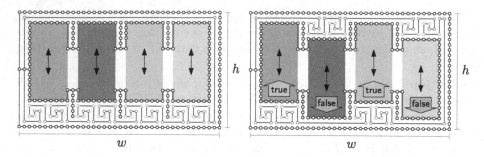

Fig. 5. The shifting variable rectangles (shaded) and the belt (the path with hexagonal vertices) in the NP-hardness proof by Patrignani [31].

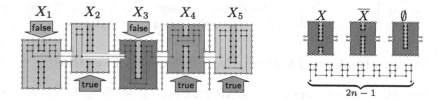

Fig. 6. A clause gadget in the NP-hardness proof by Patrignani [31] for the clause $\overline{X_1} \vee X_2 \vee \overline{X_4}$: the variable-clause rectangles (color shaded), the blocking rectangles (gray shaded), and the pathway (the path with diamond vertices). The segment that corresponds to a fulfilled variable assignment for this clause (X_2) is highlighted.

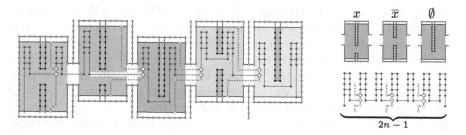

Fig. 7. The clause gadget of Fig. 6 adjusted to constant face degree.

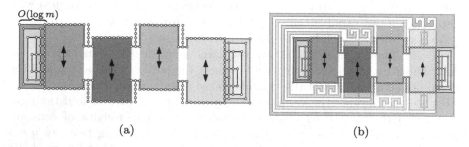

Fig. 8. The frame in our adjusted NP-hardness reduction. (a) The left and right extensions of the variable rectangles; (b) the belts around the variable rectangles.

6   Height of the Representation: An XP Algorithm

By "guessing" for every column of the drawing what lies on each grid point, we obtain an XP algorithm for OC parameterized by the height of the representation.

Theorem 4. (⋆) OC *is XP when parameterized by the height of a given orthogonal representation of a connected planar graph of maximum degree 4.*

Proof (sketch). Let H be the given orthogonal representation, let n be the number of vertices of H, let b the number of bends in H, and let $h \geq 1$. We want to decide, in $(O(n + b))^{O(h)}$ time, whether H admits an orthogonal drawing on a grid with h horizontal lines. Given a solution, that is, a drawing of H, we can remove any grid column that does not contain any vertex or bend point. Hence it suffices to check if there exists a drawing of H on a grid of width $w \leq n + b$.

To this end, we use dynamic programming (DP) with a table B. Each entry of $B[c, t]$ corresponds to a column c of the grid and an h-tuple t. (The full version [19] contains a figure with an example.) Each component of t represents an object (if any) that lies on the corresponding grid point in column c. In a drawing of H, a grid point g can either be empty or it is occupied by a vertex, bend, or edge. Let $\mathcal{T}$ be the set of h-tuples constructed in this way. Note that $|\mathcal{T}| \in (O(n + b))^h$.

The table entry $B[c, t]$ stores a Boolean value that is true if an orthogonal drawing of $left_H(t)$ on a grid of size $c \times h$ exists, false otherwise. For a given column $c \in \{2, \ldots, w\}$, we check for each $t \in \mathcal{T}$, whether t can be extended to the left by one unit. We do this by going through all $t' \in \mathcal{T}$ and checking whether $B[c - 1, t'] = $ true and whether t' and t "match". The DP returns true if and only if, for any $c \in \{1, \ldots, w\}$ and $t \in \mathcal{T}$, it holds that $B[c, t] = $ true and t is such that all elements of H lie on t or to the left of t. The desired runtime is easy to see.   □

7   Open Problems

The following interesting questions remain open. (1) Can we find a polynomial kernel for OC with respect to the number of kitty corners, or at least with respect to the number of kitty corners plus the number of faces, for general graphs? (2) Does OC admit an FPT algorithm parameterized by the height of the orthogonal representation? (3) Is OC solvable in $2^{O(\sqrt{n})}$ time? This bound would be tight assuming that the Exponential Time Hypothesis is true. (4) If we parameterize by the number of *pairs* of kitty corners, can we achieve substantially better running times?

References

1. Bannister, M.J., Cabello, S., Eppstein, D.: Parameterized complexity of 1-planarity. J. Graph Algorithms Appl. **22**(1), 23–49 (2018). https://doi.org/10.7155/jgaa. 00457

2. Bannister, M.J., Eppstein, D.: Crossing minimization for 1-page and 2-page drawings of graphs with bounded treewidth. J. Graph Algorithms Appl. **22**(4), 577–606 (2018). https://doi.org/10.7155/jgaa.00479

3. Bannister, M.J., Eppstein, D., Simons, J.A.: Inapproximability of orthogonal compaction. J. Graph Algorithms Appl. **16**(3), 651–673 (2012). https://doi.org/10.7155/jgaa.00263

4. Bannister, M.J., Eppstein, D., Simons, J.A.: Fixed parameter tractability of crossing minimization of almost-trees. In: Wismath, S., Wolff, A. (eds.) GD 2013. LNCS, vol. 8242, pp. 340–351. Springer, Cham (2013). https://doi.org/10.1007/978-3-319-03841-4_30

5. Bentley, J.L., Ottmann, T.: Algorithms for reporting and counting geometric intersections. IEEE Trans. Comput. **28**(9), 643–647 (1979). https://doi.org/10.1109/TC.1979.1675432

6. Bertolazzi, P., Di Battista, G., Liotta, G., Mannino, C.: Upward drawings of triconnected digraphs. Algorithmica **12**(6), 476–497 (1994). https://doi.org/10.1007/BF01188716

7. Bhore, S., Ganian, R., Montecchiani, F., Nöllenburg, M.: Parameterized algorithms for book embedding problems. J. Graph Algorithms Appl. **24**(4), 603–620 (2020). https://doi.org/10.7155/jgaa.00526

8. Bhore, S., Ganian, R., Montecchiani, F., Nöllenburg, M.: Parameterized algorithms for queue layouts. In: GD 2020. LNCS, vol. 12590, pp. 40–54. Springer, Cham (2020). https://doi.org/10.1007/978-3-030-68766-3_4

9. Biedl, T.: Small drawings of outerplanar graphs, series-parallel graphs, and other planar graphs. Discrete Comput. Geom. **45**(1), 141–160 (2010). https://doi.org/10.1007/s00454-010-9310-z

10. Binucci, C., Da Lozzo, G., Di Giacomo, E., Didimo, W., Mchedlidze, T., Patrignani, M.: Upward book embeddings of st-graphs. In: Symposium on Computational Geometry (SoCG), vol. 129 of LIPIcs, pp. 1–22. Schloss Dagstuhl - Leibniz-Zentrum für Informatik, 2019. https://doi.org/10.4230/LIPIcs.SoCG.2019.13

11. Bridgeman, S.S., Di Battista, G., Didimo, W., Liotta, G., Tamassia, R., Vismara, L.: Turn-regularity and optimal area drawings of orthogonal representations. Comput. Geom. **16**(1), 53–93 (2000). https://doi.org/10.1016/S0925-7721(99)00054-1

12. Chaplick, S., Di Giacomo, E., Frati, F., Ganian, R., Raftopoulou, C.N., Simonov, K.: Parameterized algorithms for upward planarity. arXiv (2022). https://doi.org/10.48550/arXiv.2203.05364

13. Chaplick, S., Fleszar, K., Lipp, F., Ravsky, A., Verbitsky, O., Wolff, A.: Drawing graphs on few lines and few planes. J. Comput. Geom. **11**(1), 433–475 (2020). https://doi.org/10.20382/jocg.v11i1a17

14. Cygan, M., et al.: Parameterized Algorithms. Springer, Cham (2015). https://doi.org/10.1007/978-3-319-21275-3

15. Da Lozzo, G., Eppstein, D., Goodrich, M.T., Gupta, S.: Subexponential-time and FPT algorithms for embedded flat clustered planarity. In: International Workshop on Graph-Theoretic Concepts in Computer Science (WG), vol. 11159 of LNCS, pp. 111–124. Springer (2018). https://doi.org/10.1007/978-3-030-00256-5_10

16. Di Battista, G., Eades, P., Tamassia, R., Tollis, I.G.: Graph Drawing: algorithms for the Visualization of Graphs. Prentice-Hall (1999)

17. Di Battista, G., Tamassia, R.: Algorithms for plane representations of acyclic digraphs. Theor. Comput. Sci. **61**, 175–198 (1988). https://doi.org/10.1016/0304-3975(88)90123-5

18. Di Giacomo, E., Liotta, G., Montecchiani, F.: Orthogonal planarity testing of bounded treewidth graphs. J. Comput. Syst. Sci. **125**, 129–148 (2022). https://doi.org/10.1016/j.jcss.2021.11.004

19. Didimo, W., Gupta, S., Kindermann, P., Liotta, G., Wolff, A., Zehavi, M.: Parameterized approaches to orthogonal compaction. arXiv (2022). https://doi.org/10.48550/arXiv.2210.05019

20. Downey, R.G., Fellows, M.R.: Fundamentals of parameterized complexity, vol. 4 of TCS. Springer (2013).https://doi.org/10.1007/978-1-4471-5559-1

21. Dujmović, V., et al.: On the parameterized complexity of layered graph drawing. Algorithmica **52**(2), 267–292 (2008). https://doi.org/10.1007/s00453-007-9151-1

22. Dujmović, V., Fernau, H., Kaufmann, M.: Fixed parameter algorithms for one-sided crossing minimization revisited. J. Discrete Algorithms **6**(2), 313–323 (2008). https://doi.org/10.1016/j.jda.2006.12.008

23. Evans, W.S., Fleszar, K., Kindermann, P., Saeedi, N., Shin, C.-S., Wolff, A.: Minimum rectilinear polygons for given angle sequences. Comput. Geom. **100**(101820), 1–39 (2022). https://doi.org/10.1016/j.comgeo.2021.101820

24. Fomin, F.V., Lokshtanov, D., Saurabh, S., Zehavi, M.: Kernelization: Theory of Parameterized Preprocessing. Cambridge University Press (2019)

25. Ganian, R., Montecchiani, F., Nöllenburg, M., Zehavi, M.: Parameterized complexity in graph drawing (Dagstuhl Seminar 21293). Dagstuhl Rep. **11**(6), 82–123 (2021). https://doi.org/10.4230/DagRep.11.6.82

26. Gupta, S., Sa'ar, G., Zehavi, M.: Grid recognition: classical and parameterized computational perspectives. In: International Symposium on Algorithms and Computation (ISAAC), vol. 212 of LIPIcs, pp. 1–15. Schloss Dagstuhl - Leibniz-Zentrum für Informatik (2021). https://doi.org/10.4230/LIPIcs.ISAAC.2021.37

27. Kaufmann, M., Wagner, D. (eds.): Drawing Graphs. LNCS, vol. 2025. Springer, Heidelberg (2001). https://doi.org/10.1007/3-540-44969-8

28. Kobayashi, Y., Ohtsuka, H., Tamaki, H.: An improved fixed-parameter algorithm for one-page crossing minimization. In: Lokshtanov, D., Nishimura, N. (eds.) 12th International Symposium on Parameterized and Exact Computation (IPEC), vol. 89 of LIPIcs, pp. 1–12. Schloss Dagstuhl - Leibniz-Zentrum für Informatik (2018). https://doi.org/10.4230/LIPIcs.IPEC.2017.25

29. Liotta, G., Rutter, I., Tappini, A.: Parameterized complexity of graph planarity with restricted cyclic orders. In: International Workshop on Graph-Theoretic Concepts in Computer Science (WG), vol. 13453 of LNCS, pp. 383–397. Springer (2022). https://doi.org/10.1007/978-3-031-15914-5_28

30. Da Lozzo, G., Eppstein, D., Goodrich, M.T., Gupta, S.: C-planarity testing of embedded clustered graphs with bounded dual carving-width. Algorithmica **83**(8), 2471–2502 (2021). https://doi.org/10.1007/s00453-021-00839-2

31. Patrignani, M.: On the complexity of orthogonal compaction. Comput. Geom. **19**(1), 47–67 (2001). https://doi.org/10.1016/S0925-7721(01)00010-4

32. Pickover, C.A.: The Math Book. Sterling (2009)

33. Shamos, M.I., Hoey, D.: Geometric intersection problems. In: 17th Annual Symposium on Foundations of Computer Science (FOCS), pp. 208–215 (1976). https://doi.org/10.1109/SFCS.1976.16

NP-Hardness and Fixed Parameter Tractability

Hardness of Bounding Influence via Graph Modification

Robert D. Barish$^{(\boxtimes)}$ ⓘ and Tetsuo Shibuya ⓘ

Division of Medical Data Informatics, Human Genome Center,
Institute of Medical Science, University of Tokyo,
4-6-1 Shirokanedai, Minato-ku, Tokyo 108-8639, Japan
rbarish@ims.u-tokyo.ac.jp, tshibuya@hgc.jp

Abstract. We consider the problem of minimally modifying graphs and
digraphs by way of exclusively deleting vertices, exclusively deleting
edges, or exclusively adding new edges, with or without connectivity
constraints for the resulting graph or digraph, to ensure that centrality-
based influence scores of all vertices satisfy either a specified lowerbound
$\mathcal{A}$ or upperbound $\mathcal{B}$. Here, we classify the hardness of exactly or approximately solving this problem for: (1) all vertex- and edge-deletion cases
for betweenness, harmonic, degree, and in-degree centralities; (2) all
vertex-deletion cases for eigenvector, Katz, and PageRank centralities;
(3) all edge-deletion cases for eigenvector, Katz, and PageRank centralities under a connectivity or weak-connectivity constraint; and (4) a set
of edge-addition cases for harmonic, degree, and in-degree centralities.
We show that some of our results, in particular multiple results concerning betweenness, eigenvector, Katz, and PageRank centralities, hold for
planar graphs and digraphs. Finally, under a variety of constraints, we
establish that no polynomial time constant factor approximation algorithm can exist for computing the cardinality of a minimum set of vertices
or minimum set of edges whose deletion ensures a lowerbound betweenness centrality score, or a lower- or upperbound eigenvector, Katz, or
PageRank centrality score (unless $P = NP$).

Keywords: NP hardness · Approximation hardness · PageRank
centrality · Katz centrality · Eigenvector centrality · Betweenness
centrality · Closeness centrality · Harmonic centrality · Degree
centrality · Vertex deletion · Edge deletion · Edge augmentation

1 Introduction

A fundamental measure of the robustness of a network – be it one of individuals, organisms, objects, or system states – is the uniformity of the distribution
of influence among its components, where we can quantify a component's influence by the magnitude of the perturbation resulting from its failure or removal.

Supported by JSPS Kakenhi grants {20K21827, 21H05052, 20H05967}.

L. Gąsieniec (Ed.): SOFSEM 2023, LNCS 13878, pp. 129–143, 2023.
https://doi.org/10.1007/978-3-031-23101-8_9

Equidistribution of influence is accordingly a core design principal in the field of network engineering. As evidence of this, we need look no further than the telecommunications infrastructure underlying the modern internet. In particular, this infrastructure traces its origins to the fault-tolerant adaptable packet switching models of Baran and Davies [3–5,16], developed (originally in the form of the late 1960's ARPANET project [5,38]) with the explicit motivation of allowing communication systems to survive natural and manmade disasters.

In the other direction, understanding how to optimally minimize influence is important in the context of allocating sparse resources to mitigate or deconstruct "pathological" or "dark" networks. Botnets serve as a case in point, as their topologies typically consist of more important "command and control" and "master" servers directing a large number of lower influence "zombie" nodes [22,44,45]. This is often also the case for networks modeling the spread of pathogens, such as contact tracing networks or compartmental models in epidemiology (e.g., the well-known Susceptible, Infectious, or Recovered (SIR) model [41]).

As one might expect, the problem of finding the most influential nodes in a network, and the problem of optimizing the influence of specified nodes, have both received significant attention in the literature (see Sect. 2 "Related work"). Surprisingly however, very little appears to be known concerning the hardness or tractability of modifying directed or undirected networks to ensure the global equidistribution of influence among its components, or to otherwise eliminate components with too much or too little influence. In this work, we attempt to address this gap by analyzing the problem of minimally modifying the topology of a graph or digraph by exclusively deleting vertices, exclusively deleting edges, or exclusively adding edges to ensure a lowerbound influence score of $\mathcal{A}$ or upperbound influence score of $\mathcal{B}$ for all nodes. We also consider the constraint that the graph or digraph remains connected or weakly-connected, respectively, after vertex- or edge-deletion operations.

As detailed in Table 1, in Sect. 4 we determine the complexity of this problem for: (1) all vertex- and edge-deletion cases for betweenness, harmonic, degree, and in-degree centralities; (2) all vertex-deletion cases for eigenvector, Katz, and PageRank centralities; (3) all edge-deletion cases for eigenvector, Katz, and PageRank centralities under a connectivity or weak-connectivity constraint; and (4) a set of edge-addition cases for harmonic, degree, and in-degree centralities. In many instances we are also able to prove that our hardness results hold on planar or bipartite classes of graphs and digraphs.

We further establish that no constant factor approximation algorithm can exist for determining minimum cardinality sets of: (Corollary 1) vertices whose deletion ensures a lowerbound betweenness centrality $\geq \mathcal{A}$ for graphs and digraphs with and without a connectivity or weak-connectivity constraint; (Corollary 2) edges whose deletion ensures a lowerbound betweenness centrality $\geq \mathcal{A}$ for graphs with and without a connectivity constraint; and (Corollary 3) exclusively vertices or exclusively edges whose deletion, under a connectivity constraint, ensures a lower- or upperbound eigenvector, Katz, and PageRank centrality $\geq \mathcal{A}$ or $\leq \mathcal{B}$, respectively.

2   Related Work

Since at least the early 21st century, significant effort has been expended on the problem of understanding and quantifying the perturbations of vertex centrality scores caused by adding, removing, or rearranging edges in both directed and undirected networks [2,9,10,12–15,18,23,25,26,29,36].

Concerning research focused on PageRank centrality [10,11,30,43], following work by Bianchini et al. [10] on the influence of graph topological features on PageRank centralities, Avrachenkov and Litvak [2] conducted an asymptotic analysis on the effect of adding multiple links to a single vertex. In particular, they established that, while the addition of a single directed edge will necessarily increase the PageRank of the terminal vertex [10], the addition of multiple edges may not necessarily improve the PageRank scores of the recipient nodes. The authors then examined optimal strategies for improving the score of a fixed node provided control over only its outlinks. Following this work, Kerchove et al. [29] conducted a study to determine which local neighborhood link topologies maximize a node's PageRank centrality, and Ishii and Tempo [26] examined the impact of removing broken, inconsistent, or otherwise noisy edges (what the authors refer to as *fragile links*) on the PageRank centralities for all nodes in a network. Later, Csaji et al. [13,14] explicitly considered the problem of adding or removing the aforementioned fragile links to optimize (e.g., maximize or minimize) the PageRank centrality of a fixed node, detailed a polynomial time linear programming algorithm for the problem, and moreover proved that the same problem becomes NP-hard if mutually exclusive fragile links are permitted.

Aside from PageRank, there has also been substantial work on optimizing the centrality score for a given node in a network according to measures such as betweenness centrality [19,30,46] and closeness centrality [6,8,24,30,40]. In particular, Crescenzi et al. [12] defined the Maximum Betweenness Improvement (MBI) and Maximum Closeness Improvement (MCI) problems (for both directed and undirected graphs) of adding or removing a set of at most k edges incident to a vertex to maximize its betweenness centrality score and closeness centrality score, respectively. The authors then proved that no Polynomial-Time Approximation Scheme (PTAS) can exist for either problem (unless $P = NP$), and detailed a greedy algorithm for both cases that achieves an almost tight approximation ratio. Subsequently, Dangelo et al. [15] showed that the MBI problem for undirected graphs likewise cannot admit a PTAS (unless $P = NP$), and furthermore showed that a greedy algorithm for the MBI problem can have an unbounded approximation ratio in the worst case.

3   Preliminaries

3.1   Graph Theoretic Terminology

We will generally follow definitions that are more-or-less standard (see, e.g., Diestel [17]). However, for some brief clarifications, when we use the term *graph* we are everywhere referring to simple undirected and unweighted graphs, and when we use the term *digraph* we are everywhere referring to simple (i.e., loop and multi-edge-free) and unweighted directed graphs that allow for edges (equiv. arcs) of opposite orientation between the same pair of vertices (often referred to as antiparallel arcs). Here, a graph is called *cubic* if and only if all of its vertex degrees are uniformly equal to 3, and *subcubic* if and only if it has maximum vertex degree 3. When we refer to the degree of a node in a digraph, we are referring to the sum of its in-degree and out-degree. As a final clarification, when we refer to the length of a path or cycle, we are referring to its edge count.

3.2   Centrality Measures

Letting G be a graph or digraph with vertex set V_G, edge set E_G, and $n = |V_G|$ total vertices, we consider the following vertex centrality measures:

Betweenness centrality, $\mathcal{C}_{Betweenness}$ – Letting $f_{(SP,all)}\,(G, v_a, v_b)$ and $f_{(SP,v_i)}\,(G, v_a, v_b)$ be functions which return the number of shortest paths from a vertex $v_a \in V_G$ to a vertex $v_b \in V_G$ and the number of such paths traversing the vertex $v_i \notin \{v_a, v_b\}$, respectively, the *betweenness centrality* [19,30,46] for a vertex $v_i \in V_G$ is given by: $\mathcal{C}_{Betweenness}\,(v_i) =$

$$\sum_{(a,b\in[1,n]\wedge a<b\wedge a\neq i\wedge b\neq i)} \left\{ \begin{array}{ll} \left(\dfrac{f_{(SP,v_i)}(G,v_a,v_b)}{f_{(SP,all)}(G,v_a,v_b)} \right), & f_{(SP,all)}\,(G, v_a, v_b) \neq 0 \\ 0, & f_{(SP,all)}\,(G, v_a, v_b) = 0 \end{array} \right\}$$

in the case of graphs. For digraphs we change the constraint $a < b$ to $a \neq b$ in the sum.

Eigenvector centrality, $\mathcal{C}_{Eigenvector}$ – Letting $\mathcal{M}$ be the adjacency matrix of a connected graph or weakly-connected digraph with primary eigenvector x_1 corresponding to the eigenvalue λ_1 (such that $\mathcal{M}x_1 = \lambda_1 x_1$), and letting x'_1 be a normalization of x_1 such that all of its entries sum to unity, the *eigenvector centrality* [21,30,43] or *Gould index* of the ith vertex in a graph (where vertex indexing must be the same as for $\mathcal{M}$) corresponds to the ith entry in x'_1. Recall here that the applicability of the Perron-Frobenius theorem to non-negative irreducible square matrices (see e.g., [33] and references therein) guarantees a unique primary eigenvector with only real positive entries.

Katz centrality, $\mathcal{C}_{Katz}$ – Adopting the prior definitions from the description of $\mathcal{C}_{Eigenvector}$, *Katz centrality* [28,30,43] is a weighted and adjusted variant of eigenvector centrality where, for some *attenuation factor* $0 \le \alpha \le \frac{1}{\lambda_1}$ and vector of correction factors (or a scalar) β, we have that $x_1 = \alpha\mathcal{M}^\mathsf{T}.x_1 + \beta$. We then generate the normalized vector x'_1 from x_1 as before.

PageRank centrality, $\mathcal{C}_{PageRank}$ – Adopting the prior definitions from the description of $\mathcal{C}_{Eigenvector}$ and $\mathcal{C}_{Katz}$, with the exception that we now assume an attenuation factor with the bounds $0 \le \alpha \le 1$ (typically we set $\alpha \approx 0.85$), *PageRank centrality* [11,30,43] is a variation on Katz centrality where, letting $\mathcal{D}$ be a diagonal matrix in which the ith entry encodes the reciprocal of the degree (or 1 in the case of a degree 0 vertex) for graphs and reciprocal of the out-degree (or 1 in the case of an out-degree 0 vertex) for digraphs of the ith vertex (note that vertex indexing must be the same as for $\mathcal{M}$), we have that

$x_1 = \alpha \mathcal{M}^{\mathsf{T}}.\mathcal{D}.x_1 + \beta$. We then generate the normalized vector x_1' from x_1 as before.

Harmonic centrality, $\mathcal{C}_{Harmonic}$ – The *harmonic centrality* [34,39] is a variation on closeness centrality [6,8,24,30,40] where, in unnormalized form,

$$\mathcal{C}_{Harmonic}(v_i) = \sum_{(j \in [1,n] \wedge j \neq i)} \left\{ \begin{array}{ll} \left(\frac{1}{d(G,v_i,v_j)}\right), & \text{for } d(G,v_i,v_j) \neq \infty \\ 0, & \text{for } d(G,v_i,v_j) = \infty \end{array} \right\}.$$

Degree centrality, $\mathcal{C}_{Degree}$ – The *degree centrality* [30] of a vertex $v_i \in V_G$ is defined as its degree (i.e., the number of vertices it is adjacent to) for graphs, and the sum of its in- and out-degree in the case of digraphs.

In-degree centrality, $\mathcal{C}_{Degree-In}$ – The *in-degree centrality* of a vertex $v_i \in V_G$ in a digraph is simply defined as its in-degree (i.e., the number of inward-oriented adjacent arcs).

4   Bounding the Influence of Vertex Centrality Scores

In this section, excluding results with straightforward or trivial proof arguments (i.e., this is an extended abstract), we will establish the claims stated in Table 1. Here, letting G be a graph, we write (Constraint Set 1) to refer to the constraint that G is initially connected, and write (Constraint Set 2) to refer to the constraint that G is initially connected and remains so after modification. Likewise, letting G be a digraph, we write (Constraint Set 3) to refer to the constraint that G is initially weakly-connected, and write (Constraint Set 4) to refer to the constraint that G is initially weakly-connected and remains so after modification. When we write that a result holds true under all constraint sets, we are referring to (Constraint Set 1) through (Constraint Set 4), or (Constraint Set 3) and (Constraint Set 4) in the special case of in-degree centrality defined only for digraphs. Unless otherwise specified, we everywhere let $\mathcal{A} \in \mathbb{R}$ and $\mathcal{B} \in \mathbb{R}$ be a lowerbound and upperbound, respectively, for vertex centrality scores.

Definition 1. *Triangle-replaced cubic graph. A triangle-replaced cubic graph G' is the graph generated from a cubic graph G with vertex set V_G by replacing each vertex $v_i \in V_G$ with the 3-cycle $\{v_{(i,1)} \leftrightarrow v_{(i,2)}, v_{(i,1)} \leftrightarrow v_{(i,3)}, v_{(i,2)} \leftrightarrow v_{(i,3)}\}$, such that vertices $v_{(i,1)}, v_{(i,2)},$ and $v_{(i,3)}$ are each the endpoints of a distinct edge formerly adjacent to $v_i \in V_G$.*

Lemma 1. *For any of the centrality metrics discussed in the Sect. 3 "Preliminaries" of this work applicable to graphs, if it is NP-hard to find a minimum set of vertices to delete to ensure a minimum vertex centrality score $\geq \mathcal{A}$ or maximum vertex centrality score $\leq \mathcal{B}$, then the problem is NP-hard for digraphs.*

Proof. It suffices to observe that we can generate a digraph from a graph G by replacing all undirected edges with pairs of antiparallel arcs, and that this will have no additional consequences beyond doubling betweenness centralities.

Table 1. Complexity of modifying a graph by way of exclusively $\leq k$ vertex deletions, exclusively $\leq k$ edge deletions, or exclusively $\leq k$ edge additions, to ensure that the minimum vertex centrality is $\geq \mathcal{A}$ for some $\mathcal{A} \in \mathbb{R}$, or the maximum vertex centrality is $\leq \mathcal{B}$ for some $\mathcal{B} \in \mathbb{R}$; the label NPH implies that a problem is NP-hard; the label $\widehat{NPH}$ implies that a problem is NP-hard to approximate within any constant factor under at least one constraint set; subscript labels "a", "b", "c", and "d" imply the stated result holds for connected graphs (Constraint Set 1), connected graphs that must remain connected post-modification (Constraint Set 2), weakly-connected digraphs (Constraint Set 3), and weakly-connected digraphs that must remain weakly-connected post-modification (Constraint Set 4), respectively; superscript labels $T*$, $P*$, or $C*$ refer to the theorem, proposition, or corollary establishing the stated result (or a sub-result of the stated result), respectively; the superscript symbol $\ddagger$ (which occurs twice) implies that the stated result is originally due to Yannakakis and Lewis [31, 47]; the superscript symbol $\triangle$ implies that the proof of the stated result has been omitted in this extended abstract.

Measure	Constraint	Vertex deletion	Edge deletion	Edge addition
$\mathcal{C}_{Betweenness}$	$\geq \mathcal{A}$	$\widehat{NPH}^{T1,C1}_{[a,b,c,d]}$	$\widehat{NPH}^{T2,C2}_{[a,b,c,d]}$	—
	$\leq \mathcal{B}$	$NPH^{P1}_{[a,b,c,d]}$	$NPH^{P2}_{[a,b,c,d]}$	—
$\mathcal{C}_{Eigenvector}$	$\geq \mathcal{A}$	$\widehat{NPH}^{T3,P3,C3}_{[a,b,c,d]}$	$\widehat{NPH}^{T3,C3}_{[b,d]}$	—
	$\leq \mathcal{B}$	$\widehat{NPH}^{T3,P3,C3}_{[a,b,c,d]}$	$\widehat{NPH}^{T3,C3}_{[b,d]}$	—
$\mathcal{C}_{Katz}$	$\geq \mathcal{A}$	$\widehat{NPH}^{T3,P3,C3}_{[a,b,c,d]}$	$\widehat{NPH}^{T3,C3}_{[b,d]}$	—
	$\leq \mathcal{B}$	$\widehat{NPH}^{T3,P3,C3}_{[a,b,c,d]}$	$\widehat{NPH}^{T3,C3}_{[b,d]}$	—
$\mathcal{C}_{PageRank}$	$\geq \mathcal{A}$	$\widehat{NPH}^{T3,P3,C3}_{[a,b,c,d]}$	$\widehat{NPH}^{T3,C3}_{[b,d]}$	—
	$\leq \mathcal{B}$	$\widehat{NPH}^{T3,P3,C3}_{[a,b,c,d]}$	$\widehat{NPH}^{T3,C3}_{[b,d]}$	—
$\mathcal{C}_{Harmonic}$	$\geq \mathcal{A}$	$\mathcal{O}\left(|V|^2 \cdot |E|\right)^{\triangle}_{[a,b,c,d]}$	$\mathcal{O}\left(|V| \cdot |E|\right)^{\triangle}_{[a,b,c,d]}$	—
	$\leq \mathcal{B}$	$NPH^{\triangle}_{[a,b,c,d]}$	$NPH^{\triangle}_{[a,b,c,d]}$	$\mathcal{O}\left(|V| \cdot |E|\right)^{\triangle}_{[a,b,c,d]}$
$\mathcal{C}_{Degree}$	$\geq \mathcal{A}$	$\mathcal{O}\left(|E|\right)^{\triangle}_{[a,b,c,d]}$	$\mathcal{O}\left(|E|\right)^{\triangle}_{[a,b,c,d]}$	$\mathcal{O}\left(|V|^{\frac{5}{2}} \cdot |E|\right)^{\triangle}_{[a,b,c,d]}$
	$\leq \mathcal{B}$	$NPH^{\ddagger,\triangle}_{[a,b,c,d]}$ [31, 47]	$\mathcal{O}\left(|V|^2\right)^{\triangle}_{[a,c]}$ $NPH^{\triangle}_{[b,d]}$	$\mathcal{O}\left(|E|\right)^{\triangle}_{[a,b,c,d]}$
$\mathcal{C}_{Degree-In}$	$\geq \mathcal{A}$	$\mathcal{O}\left(|E|\right)^{\triangle}_{[c,d]}$	$\mathcal{O}\left(|E|\right)^{\triangle}_{[c,d]}$	$\mathcal{O}\left(|E|\right)^{\triangle}_{[c,d]}$
	$\leq \mathcal{B}$	$NPH^{\ddagger,\triangle}_{[c,d]}$ [31, 47]	$\mathcal{O}\left(|E|\right)^{\triangle}_{[c]}$ $NPH^{\triangle}_{[d]}$	$\mathcal{O}\left(|E|\right)^{\triangle}_{[c,d]}$

Theorem 1. *It is NP-complete under all constraint sets to determine if $\leq k$ vertices can be deleted to ensure a minimum betweenness centrality in a planar graph or digraph of $\geq \mathcal{A}$.*

Proof. Concerning first (Constraint Set 1) and (Constraint Set 2), for a cubic planar graph G with vertex set V_G and edge set E_G, we proceed via reduction from the problem of deciding the existence of an induced st-path between a pair of adjacent vertices, $v_s, v_t \in V_G$, of length $r = \frac{1}{3}(2 \cdot |V_G| - 1)$, where we are furthermore guaranteed that r is the longest possible induced path length. It is straightforward to observe that this problem is NP-complete as a consequence of the fact that the Hamiltonian cycle problem is NP-complete for cubic planar graphs under the constraint that the Hamiltonian cycle traverses a specified edge [20], and the fact that any Hamiltonian cycle in a cubic planar graph M with vertex set V_M will correspond to a set of longest possible induced paths of length $\frac{1}{3}(2 \cdot |V_M| - 1)$ in the triangle-replaced cubic graph M' generated from M.

To begin, we construct a planar graph H from G via the following steps: (step 0) we delete the edge between v_s and v_t; (step 1) we create two copies of a cycle graph of length Υ; (step 2) letting v_x be any vertex in one copy of the cycle graph and v_y be any vertex in the other, we add an edge between v_x and v_s as well as an edge between v_y and v_t; and (step 3) generating 7 copies of a path graph of length r, we add an edge between one degree 1 vertex in each path graph and v_x, and add an edge between the other degree 1 vertex in each path graph and v_y. Accordingly, letting V_H and E_H be the vertex and edge sets for the graph H, respectively, we have that $|V_H| = |V_G| + 2\Upsilon + 7r + 7$ and $|E_H| = |E_G| + 2\Upsilon + 7r + 16$.

We next note Freeman's observation [19] that the betweenness centrality of a vertex in a graph on n vertices can be at most $\frac{1}{2}(n-1)(n-2)$, which is achieved by the central vertex of a star graph. This allows us to write a naïve upperbound for the betweenness centrality of any vertex $v_i \in V_H$ not falling along a shortest path between v_x and v_y, or falling along one of ≥ 9 shortest paths between v_x and v_y, as a sum of the terms (everywhere letting references to cycles of length Υ refer to the cycle graphs constructed in (step 1) of creating H): (term 1) $\frac{1}{2}(|V_G| + 7r + 6)(|V_G| + 7r + 5)$, corresponding to the maximum possible contribution to the betweenness centrality of v_i from the subgraph in H disjoint from the two cycles of length Υ; (term 2) 2Υ, to account for shortest paths between vertices in either of the two cycles of length Υ and vertices disjoint from these two cycles; and (term 3) $\frac{\Upsilon^2}{9}$, to account for shortest paths between pairs of vertices in the two cycles of length Υ. We can also write a naïve lowerbound for the betweenness centrality of any vertex $v_i \in V_H$ along one of exactly 8 shortest paths between v_x and v_y as (term 4) $\frac{\Upsilon^2}{8}$.

We now note a result of Unnithan et al. [42] that the betweenness centrality of a vertex in a cycle graph with n vertices is equal to $\frac{1}{8}(n-2)^2$ for n even and $\frac{1}{8}(n-1)(n-3)$ for n odd, implying that the minimum betweenness centrality for any vertex in a cycle graph will be at least $\frac{1}{8}(\Upsilon - 1)(\Upsilon - 3)$ for $\Upsilon \geq 3$. Finally, we can observe that setting $\mathcal{A} = \frac{1}{8}(\Upsilon - 1)(\Upsilon - 3)$ and $\Upsilon = \lceil (\sqrt{2}\sqrt{18 \cdot |V_G|^2 + 252 \cdot r \cdot |V_G| + 198 \cdot |V_G| + 882 \cdot r^2 + 1386r + 4577}) \rceil +$ 90 will guarantee that the sum of (term 1), (term 2), and (term 3) will be $< \mathcal{A}$ and that $\frac{\Upsilon^2}{8}$ (term 4) will be $> \mathcal{A}$ for all $|V_G| \geq 1$.

Putting everything together, assuming $k < |V_G|$ vertices are deleted, we have that every $v_i \in V_H$ disjoint from the two cycles of length Υ created in (step 1) will fall along one of exactly 8 shortest paths between v_x and v_y if and only if $k \leq |V_G| - (r+1)$ vertices can be deleted from G to yield a path graph of length r corresponding to a longest possible induced path in G ($\implies k = |V_G| - (r+1)$).

Accordingly, as a simple case analysis yields that deleting any vertex $v_i \in V_H \backslash V_G$ will ensure the minimum betweenness centrality of H will be equal to 0 unless $> |V_G|$ additional vertices are deleted, we have that $k \leq |V_G| - (r + 1)$ vertices can be deleted from H to ensure the minimum betweenness centrality of every vertex $v_i \in V_H$ is $\geq \mathcal{A}$ (as earlier specified) if and only if an induced st-path of length r exists in G. As the problem of deciding if such a vertex cut exists is clearly in NP, this yields the theorem in the case of (Constraint Set 1) and (Constraint Set 2). Finally, appealing to Lemma 1, we can straightforwardly extend this result to (Constraint Set 3) and (Constraint Set 4).

Corollary 1. *Under all constraint sets, unless $P = NP$, no polynomial time algorithm exists for approximating within a constant factor the minimum number of vertices that must be deleted in a graph or digraph to ensure a minimum betweenness centrality of $\geq \mathcal{A}$.*

Proof. The problem of finding a longest induced st-path in a cubic graph, and hence longest induced st-path in a triangle-replaced cubic graph (see Definition 1), does not admit a polynomial time constant factor approximation algorithm unless $P = NP$ [7]. This directly implies that, unless $P = NP$, there can be no polynomial time constant factor approximation algorithm for the number of vertices that must be deleted to yield an induced st-path in a cubic graph. Accordingly, dropping the planarity constraint, we can simply follow the proof argument given in Theorem 1 to establish that no polynomial time constant factor approximation algorithm exists for the minimum number of vertices that must be deleted in a graph or digraph to ensure a minimum betweenness centrality of $\geq \mathcal{A}$.

Theorem 2. *Determining if $\leq k$ edges can be deleted to ensure the minimum betweenness centrality in a planar graph or digraph is $\geq \mathcal{A}$ is NP-complete under all constraint sets.*

Proof. Concerning first (Constraint Set 1) and (Constraint Set 2), observe that the st-Hamiltonian path problem for cubic planar graphs is NP-complete [20]. Accordingly, letting G be a cubic planar graph with vertex set V_G and edge set E_G, we can follow almost exactly the proof argument in Theorem 1 to show that $\leq k = \frac{2}{3}|V_G|$ edges can be deleted from the graph H (constructed from G in the same manner as before) to ensure the minimum betweenness centrality of all vertices is $\geq \mathcal{A}$ if and only if $k = \frac{2}{3}|V_G|$ edges can be deleted from G to yield a path graph of length $r = |V_G| - 1$ with v_s and v_t as its endpoints (corresponding to an st-Hamiltonian path for G). In particular, we can again perform a simple case analysis to show that deleting any edge $e_i \in E_H \setminus E_G$ will ensure the minimum betweenness centrality of H will be equal to 0 unless $> |V_G|$ additional edges are deleted.

Concerning (Constraint Set 3) and (Constraint Set 4), we proceed along the same lines via reduction from the NP-complete st-Hamiltonian path problem for cubic planar digraphs [37]. Here, letting G be a cubic planar digraph where we wish to find an st-Hamiltonian path between a pair of vertices v_s and v_t, we construct a graph H via the following steps: (step 1) we create two copies of an undirected cycle graph of length Υ, then replace all edges with pairs of antiparallel arcs; (step 2) letting v_x be any vertex in one copy of the cycle graph and v_y be any vertex in the other, we add the edges $v_x \rightarrow v_s$ and $v_t \rightarrow v_y$; and (step 3) we generate 7 copies of a directed path graph of length $r = |V_G| - 1$, then add an edge between the vertex of in-degree 0 in each path graph and v_x oriented away v_x, and add an edge between the vertex of out-degree 0 in each path graph and v_y oriented towards v_y. We can now observe that the same arguments as in the undirected case can be followed to complete the reduction. Putting everything together yields the theorem.

Corollary 2. *Under (Constraint Set 1) and (Constraint Set 2), unless $P = NP$, no polynomial time algorithm exists for approximating within a constant factor the minimum number of edges that must be deleted in a graph to ensure a minimum betweenness centrality of $\geq \mathcal{A}$.*

Proof. It is known that the problem of finding a longest st-path in a cubic graph does not admit a polynomial time constant factor approximation algorithm unless $P = NP$ [7]. As this result specifically concerns bounded degree cubic graphs, unless $P = NP$, there can be no polynomial time constant factor approximation algorithm for the minimum number of edges that must be deleted to yield an st-path. Accordingly, dropping the planarity constraint, we can follow the proof argument given in Theorem 2 to establish that no polynomial time constant factor approximation algorithm exists for the minimum number of edges that must be deleted in a graph to ensure a minimum betweenness centrality of $\geq \mathcal{A}$.

Proposition 1. *Determining if $\leq k$ vertices can be deleted to ensure the maximum betweenness centrality is $\leq \mathcal{B}$ is NP-complete under all constraint sets.*

Proof. Concerning first (Constraint Set 1), set $\mathcal{B} = 0$ and observe that we now have the problem of finding $\leq k$ vertices to delete to satisfy the property that the resulting graph is a disjoint union of cliques. By a result of Yannakakis & Lewis [31,47], this problem is NP-complete under (Constraint Set 1) and (Constraint Set 2). Finally, the instances of this problem under (Constraint Set 3) and (Constraint Set 4) are addressed by Lemma 1.

Proposition 2. *Determining if $\leq k$ edges can be deleted to ensure the maximum betweenness centrality is $\leq \mathcal{B}$ is NP-complete under all constraint sets.*

Proof. To briefly treat (Constraint Set 1), observe that by setting $\mathcal{B} = 0$ we ensure any witness set of $\leq k$ edges will decompose G into a disjoint union of cliques. It now suffices to observe that such a minimum cardinality set of edges is also a witness for the NP-complete cluster deletion problem [35]. For (Constraint Set 2), we proceed via reduction from the NP-hard problem of finding a Hamiltonian path between a specified pair of vertices v_s and v_t [27] in a graph G with vertex set V_G and edge set E_G. Here, this can be done by generating a path graph with $\geq |V_G|^2$ vertices, adding an edge between one path end and v_s, and adding an edge between the other path end and v_y, to generate a graph H with vertex set V_H. Recalling the result of Unnithan et al. [42] that the betweenness centrality of a vertex in a cycle graph with n vertices is equal to $\frac{1}{8}(n-2)^2$ for n even and $\frac{1}{8}(n-1)(n-3)$ for n odd, we can then observe that the bound $\mathcal{B} = \frac{1}{8}(|V_H|-2)^2$ (for $|V_H|$ even) or $\mathcal{B} = \frac{1}{8}(|V_H|-1)\cdot(|V_H|-3)$ (for $|V_H|$ odd) will be satisfied if and only if, assuming $|V_G| \geq 3$, $k = |E_G| - |V_G| + 1$ edges can be removed from H to yield a cycle graph with $|V_H|$ total vertices. Finally, we can observe that this will only be possible if there exists a Hamiltonian path in G with endpoints at v_s and v_t.

Concerning (Constraint Set 3) and (Constraint Set 4), we proceed via reduction from the problem of deciding if $\leq r$ edges can be deleted in a triangle-free 2-subdivision of a cubic graph G (i.e., where we replace each edge with a path of length 3) to make the graph bipartite, which can equivalently be formulated as a maximum cut problem on the same graph. Here, we appeal to a result of Yannakakis [47] that finding a minimum set of edges to delete to make a graph bipartite is NP-complete for cubic graphs, even if we require the resulting bipartite graph to be connected, and appeal to a Karp reduction given by P. Irzhavsky (see ref. "1648" from the Information System on Graph Classes and their Inclusions (ISGCI) website [1]) from the maximum cut problem on graphs to the maximum cut problem on 2-subdivisions of graphs.

For the reduction, we begin by transforming G into a digraph H with edge set E_H by replacing all undirected edges in G with pairs of antiparallel arcs. We next specify $\mathcal{B} = 0$, correspondingly tasking us with deleting $\leq k$ edges in H to decompose the digraph into a collection of zero or more vertex disjoint 2-cycles and zero or more 2-cycle-free bipartite digraphs, in the latter case where all vertices in one partite set are sources and all vertices in the other partite set are sinks (i.e., so that no directed paths of length ≥ 2 exist).

Here, a simple case analysis shows that, as a consequence of G being a 2-subdivision of a connected cubic graph, if such a decomposition can be obtained by deleting k edges, then there will also exist a decomposition into only the aforementioned 2-cycle-free bipartite digraphs where all vertices are sources or sinks that can likewise be obtained by deleting k edges. In particular, observe that we can always delete one edge in a 2-cycle, then add an edge to connect the resulting out-degree 0 vertex to a sink vertex in a 2-cycle-free bipartite digraph without creating any directed paths of length ≥ 2. Putting everything together, we have that $k \leq r + \frac{1}{2}|E_H|$ edges can be deleted in H to ensure that maximum betweenness centrality is $\leq \mathcal{B} = 0$ if and only if $\leq r$ edges can be deleted in G to yield a bipartite graph. As determining if $\leq r$ edges can be deleted to make a cubic graph bipartite is again NP-complete even if we require the bipartite graph to be connected [47], this establishes the theorem under (Constraint Set 3) and (Constraint Set 4).

Theorem 3. *Determining if exclusively $\leq k$ vertices or exclusively $\leq k$ edges can be deleted in a planar graph or digraph to ensure the minimum eigenvector, Katz, or PageRank centrality is $\geq \mathcal{A}$, or to ensure the maximum values are $\leq \mathcal{B}$, is NP-complete in all cases under (Constraint Set 2) and (Constraint Set 4).*

Proof. Recall the definitions for eigenvector and Katz centrality in the Sect. 3 "Preliminaries" of the current work, and note that Katz centrality is equivalent to eigenvector centrality under the constraint that $\beta = 0$ and $\alpha = \frac{1}{\lambda_1}$, where λ_1 is the eigenvalue corresponding to the primary eigenvector x_1. Here, let $\mathcal{M}$ be the adjacency matrix for an arbitrary graph or digraph G with vertex set V_G, and specify $\alpha = \frac{1}{\lambda_1}$, $\beta = 0$, and $\mathcal{A} = \mathcal{B} = \frac{1}{|V_G|}$. Observe this will require $\mathcal{M}^\mathsf{T}.x_1$ to yield a vector where all entries are equivalent, and furthermore, that this will be possible if and only if G is a regular undirected graph, or alternatively, a digraph where all vertices have uniform in-degree.

Now consider the problem of deciding if $\leq k$ vertices can be deleted to ensure that the minimum and maximum eigenvector or Katz centralities in a graph or digraph G are $\geq \mathcal{A}$ or $\leq \mathcal{B}$ under the constraint that G must be connected or weakly-connected post-modification. Here, in the undirected case we have a straightforward reduction from the NP-complete problem of deciding the existence of a longest possible induced cycle in a subcubic planar graph of length $r \in \mathbb{N}$ having at least one vertex of degree 2 (see Garey et al. [20] and the proof argument for Theorem 1). Specifically, we can set $\mathcal{A} = \mathcal{B} = \left(\frac{1}{|V_G|-k}\right)$ and $r = |V_G| - k$, then observe that k vertex deletions must yield a cycle of length exactly r to simultaneously satisfy the lowerbound and upperbound constraint, and that this will necessarily be an induced cycle. The result can then be extended to the directed case via Lemma 1. We can proceed similarly in the case of deciding if $\leq k$ edges can be deleted to ensure that the minimum and maximum eigenvector or Katz centralities in a graph or digraph G are $\geq \mathcal{A}$ or $\leq \mathcal{B}$ by specifying $\mathcal{A} = \mathcal{B} = \left(\frac{1}{|V_G|-k}\right)$. In particular, we can reduce from the Hamiltonian cycle problem on subcubic planar graphs, and subcubic planar digraphs with exactly one vertex of degree 2 having in-degree and out-degree 1, and with all other vertices of degree 3 having in-degree and out-degree at most 2. The former problem is NP-complete by a trivial extension of the NP-completeness proof for the Hamiltonian cycle decision problem on cubic planar graphs [20], and that the latter problem is NP-complete by a straightforward extension of the NP-completeness proof for the Hamiltonian cycle decision problem on cubic planar digraphs where all vertices have in-degree and out-degree at most 2 [37].

Finally, for the classes of subcubic planar graphs and digraphs, we can observe that the aforementioned arguments also hold for PageRank centrality where we require the expression $\mathcal{M}^\mathsf{T}.\mathcal{D}.x_1$ to yield a vector where all entries are equivalent. Here, recall that $\mathcal{D}$ is a diagonal matrix with entrees corresponding to inverse vertex degrees or out-degrees (or an entry of 1 in the case where the degree or out-degree is 0) as detailed in the Sect. 3 "Preliminaries" of the current work. Accordingly, the dot product $W = \mathcal{M}^\mathsf{T}.\mathcal{D}$ will be a matrix where the entries in the ith row correspond to the inverse degrees (or 1 if the degree is 0) of the neighbors of the ith vertex $v_i \in V_G$ in the undirected case, or to the inverse out-degrees (or 1 if the out-degree is 0) of the vertices with edges directed towards the ith vertex $v_i \in V_G$ in the directed case, and where to satisfy the bounds

$\mathcal{A} = \mathcal{B} = \left(\frac{1}{|V_G|-k}\right)$, the sum of the entries in each row must be equal to a constant.

Now observe in the undirected subcubic case that any row in W corresponding to a vertex of degree 1, 2 and 3 will have exactly one, two, and three non-zero entries drawn from the set $\{\frac{1}{3}, \frac{1}{2}, 1\}$. Here, a simple case analysis – or, alternatively, checking the PageRank centralities for all possible root vertices and their neighbors in the 466008 trees on ≤ 22 vertices having maximum degree ≤ 3 – shows that, in order for $\mathcal{M}^{\intercal}.\mathcal{D}.x_1$ to yield a vector where all entries are equivalent, any degree 1 vertex will necessarily be adjacent to another degree 1 vertex (otherwise there will be two rows in W with an unequal sum), and no vertex of degree 2 can be adjacent to a vertex of degree 3. This correspondingly implies that the graph must be regular for the row sum constraint for W to be satisfied. In the case of subcubic digraphs having in-degree and out-degree at most 2, we can use a similar analysis to establish that no two vertices with distinct out-degrees can be adjacent. Putting everything together, we have that the same proof arguments in the case of eigenvector and Katz centrality can be applied to PageRank under the stated topological constraints, yielding the theorem.

Corollary 3. *Under (Constraint Set 2), unless $P = NP$, no polynomial time algorithm exists for approximating within a constant factor the minimum number of vertices or minimum number of edges that must be deleted to ensure that the minimum eigenvector, Katz, or PageRank centrality is $\geq \mathcal{A}$, or to ensure that the maximum eigenvector, Katz, or PageRank centrality is $\leq \mathcal{B}$.*

Proof. Concerning the edge deletion cases, by a result of Bazgan et al. [7] the problem of finding a longest cycle in a cubic graph does not admit a polynomial time constant factor approximation algorithm unless $P = NP$. As this result specifically concerns bounded degree cubic graphs, unless $P = NP$, there can be no polynomial time constant factor approximation algorithm for the minimum number of edges that must be deleted to yield a cycle. Accordingly, dropping the planarity constraint, and observing that the construction of Bazgan et al. [7] allows us to preserve the stated inapproximability result for a subcubic graph with a single degree 2 vertex, we can follow the proof argument given in Theorem 3 to establish that no polynomial time constant factor approximation algorithm exists for the minimum number of edges that must be deleted in a graph to ensure the stated lower- and upperbounds for eigenvector, Katz, and PageRank centralities. For the remaining vertex deletion cases, the transformation of a cubic graph into a triangle-replaced cubic graph (see Definition 1) yields a simple reduction from the aforementioned edge deletion cases.

Proposition 3. *Determining if exclusively $\leq k$ vertices can be deleted in a bipartite graph or digraph to ensure the minimum eigenvector, Katz, or PageRank vertex centrality is $\geq \mathcal{A}$, or determining if the same values are $\leq \mathcal{B}$, is NP-complete in all cases under (Constraint Set 1) and (Constraint Set 3).*

Proof. From the proof argument for Theorem 3, we have that the eigenvector, Katz, or PageRank vertex centralities for a subcubic graph G, with vertex set

V_G, will be uniform if and only if all vertices have uniform degree. Using this observation, we can proceed in the current context by reducing from the NP-complete problem of deciding the existence of an induced matching of size $\geq r$ in a bipartite graph with maximum degree 3 [32]. Here, generate a graph H from G by creating a star graph with central vertex v_a and k pendant (i.e., degree 1) vertices, subdivide each edge in the star graph, and finally add an edge between v_a and an arbitrary vertex $v_i \in V_G$. Observe now that if $k - 1$ vertices must be deleted in G to yield an 1-regular graph (corresponding to an induced matching for G), then k vertices must be deleted in H to yield a 1-regular graph. Putting everything together, by setting $\mathcal{A} = \mathcal{B} = \left(\frac{1}{|V_G|-k}\right)$ we will ensure that any witness for the problem of deleting $\leq k$ vertices in H to satisfy a lowerbound $\mathcal{A}$ or upperbound $\mathcal{B}$ for the eigenvector, Katz, or PageRank centrality will correspondingly serve as a witness for the problem of deciding if an induced perfect matching of size $\left(\frac{|V_G|-k+1}{2}\right)$ exists in G. Finally, observe that we can invoke Lemma 1 to nail the case of (Constraint Set 3).

References

1. de Ridder et al. H.N.: Information System on Graph Classes and their Inclusions (ISGCI). https://www.graphclasses.org. Accessed Sept 2021
2. Avrachenkov, K., Litvak, N.: The effect of new links on google PageRank. Stoch. Model. **22**(2), 319–331 (2006). https://doi.org/10.1080/15326340600649052
3. Baran, P.: Reliable digital communications systems using unreliable network repeater nodes. Document P-1995, pp. 1–30. the RAND Corporation, Santa Monica, CA (1960)
4. Baran, P.: On distributed communications: I. Introduction to distributed communications networks. Memorandum RM-3420-PR, pp. 1–51. the RAND Corporation, Santa Monica, CA (1964)
5. Baran, P.: The beginnings of packet switching: some underlying concepts. IEEE Commun. Mag. **40**(7), 42–48 (2002). https://doi.org/10.1109/MCOM.2002. 1018006
6. Bavelas, A.: Communication patterns in task-oriented groups. J. Acoust. Soc. Am. **22**(6), 725–730 (1950). https://doi.org/10.1121/1.1906679
7. Bazgan, C., Santha, M., Tuza, Z.: On the approximation of finding a(nother) Hamiltonian cycle in cubic Hamiltonian graphs. J. Algorithms **31**(1), 249–268 (1999). https://doi.org/10.1006/jagm.1998.0998
8. Beauchamp, M.A.: An improved index of centrality. Behav. Sci. **10**(2), 161–163 (1965). https://doi.org/10.1002/bs.3830100205
9. Bergamini, E., Crescenzi, P., D'Angelo, G., Meyerhenke, H., Severini, L., Velaj, Y.: Improving the betweenness centrality of a node by adding links. J. Exp. Algorithmics **23**(1), 1.5:1–1.5:32 (2018). https://doi.org/10.1145/3166071
10. Bianchini, M., Gori, M., Scarselli, F.: Inside PageRank. ACM Trans. Int. Technol. **5**(1), 92–128 (2005). https://doi.org/10.1145/1052934.1052938
11. Brin, S., Page, L.: The anatomy of a large-scale hypertextual web search engine. Comput. Netw. ISDN Syst. **30**(1–7), 107–117 (1998). https://doi.org/10.1016/S0169-7552(98)00110-X

12. Crescenzi, P., D'Angelo, G., Severini, L., Velaj, Y.: Greedily improving our own closeness centrality in a network. ACM Trans. Knowl. Discov. Data **11**(1), 9:1–9:32 (2016). https://doi.org/10.1145/2953882
13. Csáji, B.C., Jungers, R.M., Blondel, V.D.: PageRank optimization in polynomial time by stochastic shortest path reformulation. In: Hutter, M., Stephan, F., Vovk, V., Zeugmann, T. (eds.) ALT 2010. LNCS (LNAI), vol. 6331, pp. 89–103. Springer, Heidelberg (2010). https://doi.org/10.1007/978-3-642-16108-7_11
14. Csáji, B.C., Jungers, R.M., Blondel, V.D.: PageRank optimization by edge selection. Discret. Appl. Math. **169**, 73–87 (2014). https://doi.org/10.1016/j.dam.2014.01.007
15. D'Angelo, G., Severini, L., Velaj, Y.: On the maximum betweenness improvement problem. Electron. Notes Theor. Comput. Sci. **322**, 153–168 (2016). https://doi.org/10.1016/j.entcs.2016.03.011
16. Davies, D.W.: Proposal for a digital communication network. Unpublished memorandum, pp. 1–28. National Physical Laboratory, London (1966)
17. Diestel, R.: Graph Theory, 5th edn. Springer, Heidelberg (2017). https://doi.org/10.1007/978-3-662-53622-3
18. Fercoq, O., Akian, M., Bouhtou, M., Gaubert, S.: Ergodic control and polyhedral approaches to PageRank optimization. IEEE Trans. Autom. Contr. **58**(1), 134–148 (2013). https://doi.org/10.1109/TAC.2012.2226103
19. Freeman, L.C.: A set of measures of centrality based on betweenness. Sociometry **40**(1), 35–41 (1977). https://doi.org/10.2307/3033543
20. Garey, M.R., Johnson, D.S., Tarjan, R.E.: The planar Hamiltonian circuit problem is NP-complete. SIAM J. Comput. **5**(4), 704–714 (1976). https://doi.org/10.1137/0205049
21. Gould, P.R.: On the geographical interpretation of eigenvalues. Trans. Inst. Br. Geogr. **42**, 53–86 (1967). https://doi.org/10.2307/621372
22. Grizzard, J.B., Sharma, V., Nunnery, C., Kang, B.B., Dagon, D.: Peer-to-peer botnets: overview and case study. In: Proceedings of 1st Workshop on Hot Topics in Understanding Botnets (HotBots), pp. 1–8 (2007)
23. Han, C.G., Lee, S.H.: Analysis of effect of an additional edge on eigenvector centrality of graph. J. Korea Soc. Comput. Inf. **21**(1), 25–31 (2016). https://doi.org/10.9708/jksci.2016.21.1.025
24. Harary, F.: Status and contrastatus. Sociometry **22**(1), 23–43 (1959). https://doi.org/10.2307/2785610
25. Ishakian, V., Erdös, D., Terzi, E., Bestavros, A.: A framework for the evaluation and management of network centrality. In: Proceedings of 12th SIAM International Conference on Data Mining (SDM), pp. 427–438 (2012). https://doi.org/10.1137/1.9781611972825.37
26. Ishii, H., Tempo, R.: Computing the PageRank variation for fragile web data. SICE J. Control Meas. Syst. Integr. **2**(1), 1–9 (2009). https://doi.org/10.9746/jcmsi.2.1
27. Karp, R.M.: Reducibility among combinatorial problems. In: Miller, R.E., Thatcher, J.W., Bohlinger, J.D. (eds.) Complexity of Computer Computations. The IBM Research Symposia Series, pp. 85–103. Springer, Boston (1972). https://doi.org/10.1007/978-1-4684-2001-2_9
28. Katz, L.: A new status index derived from sociometric analysis. Psychometrika **18**(1), 39–43 (1953). https://doi.org/10.1007/BF02289026
29. de Kerchove, C., Ninove, L., van Dooren, P.: Maximizing PageRank via outlinks. Linear Algebra Appl. **429**(5–6), 1254–1276 (2008). https://doi.org/10.1016/j.laa.2008.01.023

30. Landherr, A., Friedl, B., Heidemann, J.: A critical review of centrality measures in social networks. Bus. Inf. Syst. Eng. **2**, 371–385 (2010). https://doi.org/10.1007/s12599-010-0127-3

31. Lewis, J.M., Yannakakis, M.: The node-deletion problem for hereditary properties is NP-complete. J. Comput. Syst. Sci. **20**(2), 219–230 (1980). https://doi.org/10.1016/0022-0000(80)90060-4

32. Lozin, V.V.: On maximum induced matchings in bipartite graphs. Inf. Process. Lett. **81**(1), 7–11 (2002). https://doi.org/10.1016/S0020-0190(01)00185-5

33. MacCluer, C.R.: The many proofs and applications of Perron's theorem. SIAM Rev. **42**(3), 487–498 (2000). https://doi.org/10.1137/S0036144599359449

34. Marchiori, M., Latora, V.: Harmony in the small-world. Phys. A **285**(3–4), 539–546 (2000). https://doi.org/10.1016/S0378-4371(00)00311-3

35. Natanzon, A.: Complexity and approximation of some graph modification problems. Masters thesis, pp. 1–60. Tel Aviv University, Department of Computer Science (1999)

36. Olsen, M., Viglas, A.: On the approximability of the link building problem. Theoret. Comput. Sci. **518**, 96–116 (2014). https://doi.org/10.1016/j.tcs.2013.08.003

37. Plesník, J.: The NP-completeness of the Hamiltonian cycle problem in planar digraphs with degree bound two. Inf. Process. Lett. **8**(4), 199–201 (1979). https://doi.org/10.1016/0020-0190(79)90023-1

38. Roberts, L.G.: Multiple computer networks and intercomputer communication. In: Proceedings of 1st ACM Symposium on Operating System Principles (SOSP), pp. 3.1–3.6 (1967). https://doi.org/10.1145/800001.811680

39. Rochat, Y.: Closeness centrality extended to unconnected graphs: the harmonic centrality index. In: Proceedings of 6th Conference on Applications of Social Network Analysis (ASNA) (2009)

40. Sabidussi, G.: The centrality index of a graph. Psychometrika **31**(4), 581–603 (1966). https://doi.org/10.1007/BF02289527

41. Tolles, J., Luong, T.B.: Modeling epidemics with compartmental models. J. Am. Med. Assoc. **323**(24), 2515–2516 (2020). https://doi.org/10.1001/jama.2020.8420

42. Unnithan, S.K.R., Kannan, B., Jathavedan, M.: Betweenness centrality in some classes of graphs. Int. J. Comb. **2014**(Article ID 241723), 1–12 (2014). https://doi.org/10.1155/2014/241723

43. Vigna, S.: Spectral ranking. Netw. Sci. **4**(4), 433–445 (2016). https://doi.org/10.1017/nws.2016.21

44. Vormayr, G., Zseby, T., Fabini, J.: Botnet communication patterns. IEEE Commun. Surv. Tut. **19**(4), 2768–2796 (2017). https://doi.org/10.1109/COMST.2017.2749442

45. Wang, P., Sparks, S., Zou, C.C.: An advanced hybrid peer-to-peer botnet. IEEE Trans. Depend. Secure Comput. **7**(2), 113–127 (2010). https://doi.org/10.1109/TDSC.2008.35

46. White, D.R., Borgatti, S.P.: Betweenness centrality measures for directed graphs. Soc. Netw. **16**(4), 335–346 (1994). https://doi.org/10.1016/0378-8733(94)90015-9

47. Yannakakis, M.: Node- and edge-deletion NP-complete problems. In: Proceedings of 10th Annual ACM Symposium on Theory of Computing (STOC), pp. 253–264 (1978). https://doi.org/10.1145/800133.804355

Heuristics for Opinion Diffusion via Local Elections

Rica Gonen[1] , Martin Koutecký[2] , Roei Menashof[1(✉)] ,
and Nimrod Talmon[3]

[1] The Open University of Israel, Raanana, Israel
`RoeiMena@gmail.com`
[2] Charles University, Prague, Czech Republic
`koutecky@iuuk.mff.cuni.cz`
[3] Ben-Gurion University, Be'er Sheva, Israel
`talmonn@bgu.ac.il`

Abstract. Most research on influence maximization considers asimple diffusion model, in which binary information is being diffused (i.e., vertices – corresponding to agents – are either *active* or *passive*). Here we consider a more involved model of opinion diffusion: In our model, each vertex in the network has either approval-based or ordinal-based preferences and we consider diffusion processes in which each vertex is influenced by its neighborhood following a local election, according to certain "local" voting rules. We are interested in externally changing the preferences of certain vertices (i.e., campaigning) in order to influence the resulting election, whose winner is decided according to some "global" voting rule, operating after the diffusion converges. As the corresponding combinatorial problem is computationally intractable in general, and as we wish to incorporate probabilistic diffusion processes, we consider classic heuristics adapted to our setting: A greedy heuristic and a local search heuristic. We study their properties for plurality elections, approval elections, and ordinal elections, and evaluate their quality experimentally. The bottom line of our experiments is that the heuristics we propose perform reasonably well on both the real world and synthetic instances. Moreover, examining our results in detail also shows how the different parameters (ballot type, bribery type, graph structure, number of voters and candidates, etc.) influence the run time and quality of solutions. This knowledge can guide further research and applications.

Keywords: Social choice · Influence maximization · Bribery in elections

---

Partially supported by Ministry of Science, Technology and Space Binational Israel-Taiwan grant, number 3-16542.

Partially supported by Charles University project UNCE/SCI/004 and by the project 22-22997S of GA ČR. Computational resources were supplied by the project "e-Infrastruktura CZ" (e-INFRA CZ LM2018140) supported by the Ministry of Education, Youth and Sports of the Czech Republic, and by the ELIXIR-CZ project (LM2018131), part of the international ELIXIR infrastructure.

L. Gąsieniec (Ed.): SOFSEM 2023, LNCS 13878, pp. 144–158, 2023.
https://doi.org/10.1007/978-3-031-23101-8_10

1   Introduction

Social networks are ubiquitous in our lives and, as such, they have extensive influence on the public opinion in our society (see, e.g., [9]). In this paper we model a scenario in which an external agent wishes to change the public opinion; say, to have its preferred candidate win in an upcoming election (one such classical example is the 2016 US presidential elections [1]).

The situation we set out to study is complex as it consists of an interplay between several factors – a social network, opinions, an external agent, and the public opinion. As a result, our high level modeling contains the following ingredients:

A Social Network. There is a social network where each node initially possesses their own opinion. We model this naturally as a labeled graph, in which each node corresponds to a voter and is labeled by her opinion, and edges correspond to mutual influence of voters. Of course, there are many ways to formalize human opinions; as we are interested in a setting in which there is an upcoming election to be held, we model opinions as ballots. Importantly, we consider several ballot types, in particular, plurality ballots, approval ballots, and ordinal ballots.

A Bribing Agent. There is an external agent that can influence some voters and cause them to change their opinions. We model this through the well-established line of work considering campaigning or bribery in elections (see, e.g., [11,12,15]).

A Diffusion Process. There is a process by which information propagates through the network, so that some voters may further change their opinions as they are influenced by their neighbors. We model this through a probabilistic diffusion process in which, repeatedly, voters look at the opinions of their neighbors and may change their own opinion as a result (intuitively, if the opinions of their neighbors are significantly different than their own opinion). Such processes are studied quite extensively (see, e.g., [18,22]), however our modeling is different from some existing work and generalizes others: Technically, we introduce the concept of a *local voting rule* that builds upon the neighboring opinions of a voter and returns a score for each alternative, and use it to define a probability distribution for the altered opinion of the voter.

A Voting Rule. There is a mechanism that takes the eventual opinions of the voters and declares a winner of the election. Such mechanisms are usually referred to as voting rules, and are a fundamental structure studied in computational social choice [2,10].

1.1   Our Contributions

Our first, conceptual contribution is our general model, that is able to capture the diffusion of complex opinions; we then realize our model with plurality ballots, approval ballots, and ordinal ballots. Moreover, as we use local voting rules in a stochastic way, our modeling is inherently stochastic; indeed, introducing further probabilistic diffusion processes to more complex kinds of opinions is a main motivation of our work.

We then take the point of view of the bribing agent and ask whether such an agent can efficiently find a bribery scheme that would maximize the chances of its preferred alternative winning the eventual election and maximize its winning gap. Not surprisingly, the corresponding combinatorial problem is computationally intractable in general. Thus, we describe several heuristic methods and evaluate their effectiveness experimentally. Technically, we consider both standard heuristic approaches (in particular, Simulated Annealing) and heuristics that proved to be effective for related tasks of opinion diffusion (in particular, greedy heuristics for Influence Maximization [20]).

1.2 Related Work

Our work fits naturally within the growing literature on opinion diffusion in social choice [17]. In particular, a recent paper [13] considers a similar setting that, while allowing some islands of computational tractability, differs in that the diffusion process is deterministic. Another paper [6] considers bribery and opinion diffusion for ordinal ballots, however their diffusion process is significantly different than ours and, in our opinion, somewhat problematic. This is because in their diffusion process, only the preferred candidate potentially moves up, but this means that the process depends on the point of view of the briber, while we hold that any definition of diffusion in the context of campaigning or bribery needs to be oblivious to the bribing agent(s). Other related works are the paper of Bredereck and Elkind [3] who consider a particular setting and approach it from a theoretical point of view. Wilder and Vorobeychik [24] consider a diffusion process related to the Linear Cascade model while we take an approach that is more in line with the Threshold Voter model.

More generally, our work relates to the study on Influence Maximization [20] (and hence, also to Target Set Selection [4]. These works usually deal with similar situations as we do, albeit in which the opinions are rather simple, usually binary. For this setting, there is a greedy heuristic [20] that was later improved [5,21], which our greedy heuristics builds upon.

2   Formal Model

We describe the ingredients of our setting.

2.1 Opinion Graphs

We have a simple undirected graph $G = (V, E)$, where each vertex corresponds to a voter. The vertices of G are labeled by the votes, such that the label of $v \in V$ corresponds to the vote of the agent v.

In the elections we consider there is always an underlying set of alternatives A. We consider several types of elections corresponding to the ballots a voter casts: plurality, approval, and ordinal. We identify a voter with their ballot, thus: In plurality elections, each voter $v \in V$ is some $v \in A$; in approval elections, each voter $v \in V$ is some $v \subseteq A$; and in ordinal elections, each voter $v \in V$ is some $v \in L(A)$, where $L(A)$ is the set of linear orders over A. (I.e., formally, V is a multisubset of A, 2^A and $L(A)$, for plurality, approval, and ordinal ballots, respectively.) Then, v is the label of the vertex corresponding to voter v, and such labeled graph is referred to as an *opinion graph*.

2.2 Campaigning and Bribery

We are interested in the problem of campaigning (also studied under the name bribery [12]) in our setting. Thus, we assume an external agent (i.e., the briber), who has a given budget, and can perform certain *bribery operations* on the voters, where a bribery operation operating on a certain voter $v \in V$ causes v to change her vote.[1] We consider several bribery settings, differing by the cost of each possible bribery operation:

- In *simple bribery* [14], the briber pays one coin to change the vote of a voter v to any vote the briber wishes.
- In *approval bribery* [23], which is relevant only for approval elections, the briber pays one coin for adding an alternative to the approval set of a vote v or removing an alternative from the approval set of a vote v.
- In *swap bribery* [8], which is relevant for ordinal elections, the briber pays one coin for a single swap of two consecutive alternatives in the vote of a voter v. We focus on a restricted variant, *shift bribery* [7], in which the briber is only allowed to move the preferred candidate, and to only move them up.

In our setting we have an *initial* society graph, modeling the society **before** the briber bribes; then, the society might change following the bribery operations of the briber, into a society **after** the bribery.

Remark 1. Technically we speak of bribery but conceptually our work relates to campaigning. One difference between the two is that in bribery one expects the bribed voter to stay loyal, but in campaigning, one attempts to influence the voter more indirectly and thus does not expect loyalty. Our model still allows for this definition of bribery by setting a high stubbornness parameter, defined below. That way, a bribed voter remains loyal, but also influences their peers.

2.3 Diffusion Processes via Local Elections

We are interested in the propagation of opinions after the bribery happens. Specifically, we focus on synchronous diffusion, where in each step of the diffusion all voters might change their labels simultaneously. An asynchronous diffusion process, where in each step of the diffusion only one voter might change her label,

---

[1] For simplicity, we assume bribery operations always succeed. A relaxation of this assumption is left for future work.

can be defined analogously. The specific way by which voters might change their labels is governed by two parameters: A *local voting rule* $\mathcal{R}_L$, and a *stubbornness parameter* $\alpha \geq 0$.

The local elections voting rule $\mathcal{R}_L$ is a function that takes a certain collection of votes and returns a score for each alternative $a \in A$. It is used as follows: In each step of the diffusion, each vertex $v \in V$ applies $\mathcal{R}_L$ on a collection of votes obtained by taking the votes of all their neighbors, plus α-times their own vote v. We refer to the set of votes obtained by taking the open neighborhood of v, plus $\alpha \cdot d(v)$ copies of v's vote, as the *local election*. In particular, if $\alpha = 0$, then the local election of v consists of her open neighborhood (i.e., her opinion is disregarded); if $\alpha = 1$, then the local election consists precisely of the closed neighborhood; and if $\alpha = 5$, then the local election consists of all neighbors of v plus 5 copies of v.

The scores reported by the local rule $\mathcal{R}_L$ are used to define the probability distribution according to which v changes her opinion. The definitions are specific to the different voting rules, and are given below.

Remark 2. The fact that the diffusion process is defined by the local election is a major extension of the existing models. Indeed, a main motivation for our work was to enrich existing models of opinion diffusion in social networks and push them closer to reality by considering various probabilistic processes.

A Diffusion Process for Plurality Elections. Here we use $\mathcal{R}_L$ that returns the plurality score of each alternative. Then, we swap the voter's label to an alternative a with probability which is the score of a in the local election, divided by the number of votes in the local election.

Example 1. Consider a voter v with open neighborhood $\{u_1, u_2\}$. Assume that v votes for (i.e., is labeled with) alternative c while u_1 and u_2 vote for alternative d. If $\alpha = 0$, then in the next timestep of a synchoronous diffusion process, v would surely change her vote to d. In contrast, if $\alpha = 1$ then v would change her vote to d with probability $1/3$, and with probability $2/3$ would still vote for c.

A Diffusion Process for Approval Elections. Here $\mathcal{R}_L$ is a function that returns the approval score of each alternative. Then, for each alternative $a \notin v$ (i.e., not currently approved by v), a is added to v with probability that equals the relative approval score of a (the *relative approval* score of an alternative is the fraction of voters approving the alternative); similarly, for each alternative $a \in v$ (i.e., currently approved by v), we remove a from v with probability that is one minus the relative approval score of a.

Example 2. Consider again a voter v with open neighborhood $\{u_1, u_2\}$. Assume v votes for $\{a, b\}$ while u_1 and u_2 each votes for $\{b, c\}$. If $\alpha = 1$, then v would: Definitely keep on approving b; with probability $2/3$ would also approve c; and with probability $2/3$ would cease approving a.

A Diffusion Process for Ordinal Elections. For ease of presentation, we describe our diffusion process for the case where $\mathcal{R}_L$ is the Borda rule; the description can be generalized to any ordinal voting rule that assigns scores to candidates (importantly, this includes also rules such as Copeland and STV, which can be defined as such).

We proceed as follows: Denote by $c_1, \ldots, c_m$ the candidates ordered by decreasing Borda scores in the local election centered at v; refer to this ordering as the *Borda-order* (in particular, the first candidate in the Borda-order is the Borda winner). The process is iterative, where in iteration i we consider c_i and do as follows: We look at position j of c_i in the ranking of v. Denote the ranking of v as $a_1, \ldots, a_m$; so, in particular, $a_j = c_i$ as c_i is the jth candidate in v's ranking. If $j = 1$ (i.e., if c_i is ranked first by v), then the iteration is complete. Otherwise, look at the Borda scores of a_j and of a_{j-1} (i.e., the candidate ranked, by v, just above c_i), and denote by $B(c)$ the Borda score of a candidate c. Now, define $x = \frac{B(a_j)}{B(a_{j-1})}$, with probability $\frac{x}{x+\frac{1}{x}}$, swap a_j and a_{j-1} (i.e., shift c_i one position up in v's ranking); otherwise (i.e., with the complement probability), the iteration is complete.

So, intuitively, we go over the candidates in decreasing Borda scores and we bubble-up each candidate with probability related to the Borda score of the candidate and the Borda score of the candidates in front of it in v's ranking.

Example 3. Consider voter v with open neighborhood $\{u_1, u_2\}$. Assume v has stubbornness $\alpha = 1$ and she votes for (x, y, z) while u_1 votes for (y, z, x) and u_2 votes for (z, y, x). Then the Borda score[2] would give $x = 5$, $y = 7$ and $z = 6$ and v's local Borda election will result with $c = (y, z, x)$. The first iteration over c would then be: $i = 1$, $c_i = y$, $j = 2$. With probability $\frac{\frac{7}{5}}{\frac{7}{5}+\frac{5}{7}} = 0.66$ candidate y will be bubbled up resulting with v voting for (y, x, z). If the first iteration resulted in v voting for (y, x, z), then the second iteration over c would be: $i = 2$, $c_i = z$, $j = 3$. With probability $\frac{\frac{6}{5}}{\frac{6}{5}+\frac{5}{6}} = 0.59$ candidate z will be bubbled up resulting with v voting for (y, z, x). If the second iteration resulted in v voting for (y, z, x), then the third iteration over c would be: $i = 3$, $c_i = x$, $j = 3$. With probability $\frac{\frac{5}{6}}{\frac{6}{5}+\frac{5}{6}} = 0.4$ candidate x will be bubbled up, and with probability 0.6 v's vote will not change from the last iteration and will remain (y, z, x). With the highest probability after all three iterations v's vote is identical to c's result, i.e., the Borda local election.

Now consider Copeland as $\mathcal{R}_L$ and v, u_1, u_2 vote in the same way above. According to Copeland tournament y beats x, z beats x, and y beats z, and so y has two outgoing arcs and z has one outgoing arc. In order to avoid division by zero, we normalize the scores by adding 1. Then the Copeland score would give $x = 1$, $y = 3$, and $z = 2$, and the Copeland v's local election will result with

---

[2] To avoid division by zero, we define the Borda score of a candidate ranked as jth to be $|A| - j + 1$ instead of $|A| - j$, although the latter is more common. These definitions are mathematically equivalent.

$c = (y, z, x)$. The first iteration over c would be then: $i = 1$, $c_i = y$, $j = 2$. With probability $\frac{\frac{3}{4}}{\frac{1}{4} + \frac{3}{4}} = 0.9$ candidate y will be bubbled up resulting with v voting for (y, x, z). Then the second iteration over c would be: $i = 2$, $c_i = z$, $j = 3$. With probability $\frac{\frac{2}{4}}{\frac{1}{4} + \frac{2}{4}} = 0.8$ candidate z will be bubbled up resulting with v voting for (y, z, x). Then, finally, the third iteration over c would be: $i = 3$, $c_i = x$, $j = 3$. With probability $\frac{\frac{1}{4}}{\frac{1}{4} + \frac{1}{4}} = 0.2$ candidate x will be bubbled up, and with probability 0.8 v's vote will not change from the last iteration and will remain (y, z, x).

2.4   Election Results via Global Voting Rules

Intuitively, we wish to study the society after the bribery and after the diffusion process halts. However, as the diffusion process is probabilistic and is not guaranteed to halt, let us consider the expected society at infinity. Let the Markov chain of our process be a directed graph in which the starting node is the society after bribery, and each node corresponds to a possible society reached during the diffusion process; we have an arc from a node to another node with probability p if p is the probability of transitioning from one node to the other. Then, imagining an infinite random walk in this network, we wish to study the distribution of probabilities of where we end up, over all nodes. In particular, the *resulting election* is a probability distribution over the set of votes (i.e., labels) at infinity (wrt. the diffusion steps). In the simple case in which there is one absorbing node (i.e., a node with no outgoing edges), it means that the diffusion would halt on a specific society. Finally, a *global voting rule* $\mathcal{R}_G$ takes the society and returns a single alternative as the winner.

We consider a society stable from the perspective of our problem if the winner of the election is unlikely to change. We ran a sample of simulations for a large number of diffusion steps to determine a number k of steps after which the likelihood of a change of winner becomes reasonably small. Our finding is that after 20 diffusion steps, the proportion of instances in which the winner changes is at most 0.2%, and the trend is clearly decreasing. For figure illustrating see full version [16]. Because modeling a diffusion step is computationally expensive, we will assume from now on that, with respect to who wins the election, the society is close enough to the state of the Markov chain at infinity after 20 steps.

2.5   Optimization Goals

In general, we would like to understand the effect of different bribery actions on the resulting winner. Since the process is stochastic, we define two measures of success:

Definition 1 (PoW).   *Given a society after bribery and diffusion, the PoW (Probability of Winning) is the probability mass on the Markov chain nodes in which p wins.*

Definition 2 (MoV). *Given a society after bribery and diffusion, the MoV (Margin of Victory) is the expected MoV of p, defined for a specific society as follows: If p wins, then the MoV is the difference between the score of p and the score of the runner-up (so, in particular, positive); if p loses, then the MoV is the difference between the score of p and the score of the winner (so, in particular, negative).*

To conclude, a specific model is characterized by:

1. A ballot type – Plurality, Approval, or Ordinal;
2. A bribery type – Simple bribery, Approval bribery, or Swap bribery;
3. A local voting rule $\mathcal{R}_L$ – Plurality, Approval, or Borda/Copeland;
4. Stubbornness parameter α;
5. A global voting rule $\mathcal{R}_G$ – Plurality, Approval, or Borda/Copeland;

For such models, we consider two computational problems, corresponding to optimizing either the PoW or the MoV. The input for both problems – referred to as Optimal-PoW and Optimal-MoV, respectively – contains an opinion graph G and a budget of b coins; Optimal-PoW or Optimal-MoV asks for finding a bribery scheme costing at most b that maximizes the PoW or MoV, respectively.

3   Computing Optimal Bribery Schemes

Not surprisingly, the problems we set out to solve are NP-hard. In fact, even if there is no graph at all, our problems are intractable in general, since they reduce to bribery in elections [12] when there is no graph, and it is known, e.g., that bribery is hard for approval elections [11, Theorem 4.2]. Our setting is drastically more involved as we also consider a graph and a stochastic diffusion process operating on it.

3.1   Heuristic Methods

As our problems are generally intractable, our aim is to evaluate the possibility of efficiently solving them by heuristic methods. We report on computational simulations performed on their implementations. In particular, we use two heuristic algorithms. While there are indeed many other possibilities of heuristic algorithms one might consider, here we concentrate on two classic methods that proved to be useful in the setting of influence maximization (see below). Note that, interestingly, the heuristics we consider are, in a sense, oblivious to the specifics of the diffusion process considered; that is, their specific operation does not depend on the specifics of the problem we consider (e.g., the ballot type and other problem parameters).

Greedy. Our first heuristic approach is an adaptation of an algorithm considered for Influence Maximization [20] that works as follows: We iterate for b times where in each iteration we bribe the vertex that, if bribed, would increase the probability of p winning after the diffusion. Notice that computing the probability of p winning after the diffusion is a non-trivial sub-problem. In our simulations we handle this issue as follows: We perform 50 independent runs of the diffusion process using Monte Carlo, where in each run we perform 20 diffusion steps. Then we use the average over the 50 runs as an estimation of this probability.

Simulated Annealing. Our second heuristic approach is a local search algorithm that is an adaptation of an algorithm considered for Influence Maximization [19] that works as follows: With budget b, we start by selecting b vertices and bribe them, each by one coin (e.g., for plurality elections this corresponds to selecting an initial solution uniformly at random). Then, we estimate the probability of p winning after the diffusion, again using 50 iterations of Monte Carlo. Then, in each iteration of local improvement, we select one of the currently-bribed voters and one of her neighbors, and instead of bribing her, we bribe the neighbor. If this small change to the current solution increases the probability of p winning (as estimated by our Monte Carlo repetitions), then we keep this local improvement; otherwise, with increasing probability, we reject it and consider a different local improvement.

4   Simulations

We implemented the heuristics described above and evaluated them in various settings. The main goal of the simulations was to understand the possibility of computing optimal briberies in practice and to better identify the problem parameters that make finding optimal bribery schemes hard. Below we describe our experimental design for plurality elections, approval elections, and ordinal elections.

4.1   Experimental Design

First, let us describe our input graphs. We use both synthetic and real world data.

Synthetic Graphs. We use the following models:

- $G(n,p)$ – We generate input graphs from $G(n,p)$, as follows: For a given number n of vertices and a value $0 \le p \le 1$, we first create n independent vertices. Then, for each pair of vertices, independently and uniformly at random, we flip a coin and with probability p put an edge between them. After we generate the $G(n,p)$ graph as just described, we assign labels to its vertices; we do so uniformly at random (thus, effectively, the resulting election behaves according to the Impartial Culture model).
- k-k-clusters – We create k subgraphs, each a $G(n/k, p_1)$ graph with some p_1. Then, for each pair of vertices u, v which are from different subgraphs, we put an edge with probability $p_2 < p_1$, independently and uniformly at random. This model creates the graph together with the labels as follows:

- For plurality elections – The label of the vertices in the jth subgraph ($j \in [k]$) is j.
- For approval elections – Make k random ballots and label them with $b_1, \ldots, b_k$. Denote b_j the cluster V_j's "base ballot." Set $0 < \alpha < 1$. Label each vertex V_j with an α-pertubation of b_j, where α-pertubation of a ballot b is defined as: take each candidate approved in ballot b and make it a not-approval with probability α, and take each candidate not approved in ballot b and make it an approval with probability α.
- For ordinal elections – Make k random ballots and label them with $b_1, \ldots, b_k$. Donate b_j the cluster V_j's "base ballot." Set $0 < \alpha < 1$. Label each vertex V_j with an α-pertubation of b_j. α-pertubation of a ballot $b = b_1, b_2, \ldots$ defined as: Begin with a blank ranking r (this is the ballot we are creating). For i from 1 to m: insert b_i into r at position $j \le i$ with probability $\alpha^{i-j}/(1 + \alpha + \cdots + \alpha^{i-1})$.

Intuitively, while the $G(n, p)$ model creates uniform graphs, the k-k-clusters model creates random graphs that aim at mimicking communities.

Real-World Graphs. We use graphs from the *email-Eu-core* network[3], referred to below as *real-world network graph*. Vertices in this graph correspond to real people, and there is a directed edge from one vertex to another if the person corresponding to the head of the directed edge sent at least one email to the person corresponding to the tail of the directed edge.

Metrics. We evaluate the heuristic algorithms described above by estimating the PoW and the MoV (see Definitions 1 and 2). To estimate PoW (MoV) we perform 50 Monte Carlo iterations to estimate the probability that p wins (the expected margin of victory of p) after the bribery operations performed by the heuristic algorithm and after 20 iterations of the diffusion process (recall Sect. 2.4 and the discussion above of why 20 iterations are a reasonable proxy to the stable state). Recall that the higher the PoW (MoV) the better. Furthermore, we report on the running times of our heuristics.

Model Settings summary of the various inputs for the model, as mentioned above, with the parameter settings:

- Voting rule – Plurality, Approval, Borda or Copeland.
- Graph type – $G(n, p)$, 5-k-clusters or Real-world Graphs.
- Budget – Supposedly, each coin can bribe a single node, there are b coins to use for bribery. We experimented with amounts between 5 and 50.
- Number of candidates – The number of candidates that are running for election is also expressed as m. We experimented with leaps of 5 between range of 5 to 50.

---

[3] http://snap.stanford.edu/data/email-Eu-core.html.

– Number of voters – The number of voters, also expressed as n. For $G(n, p)$ and 5-k-clusters we experimented with leaps of 100 between range of 100 to 1500 and also 1005, the values are fixed on 1005 nodes for Real-world Graphs.

4.2   Results

Our main results are threefold:

– When **plurality election** is used in the diffusion process and MoV is optimized at the cost of a slower run time, then Simulated Annealing performs the best.
– When **approval election** is used in the diffusion process and the available budget is relatively high then again Simulated Annealing performs the best with regards to all studied objectives, namely, it achieves higher PoW, higher MoV, and shorter run time than Greedy.
– When **ordinal election** is used in the diffusion process then Greedy and Simulated Annealing perform about the same though Greedy takes longer time to run. A significant factor in the run time is the choice of an ordinal rule, since computations for Borda are easier than computations for Copeland and indeed runs with Copeland as the ordinal rule tend to last ten times longer than runs with Borda as the ordinal rule. However, for Copeland, significantly higher MoV in the real world network graph and slightly higher PoW in both synthetic and real world network graphs is achieved.

More specifically, we compare the Greedy heuristic method performance with that of Simulated Annealing on the different optimization goal and the different diffusion processes (see full version [16] for additional details).

– In plurality and approval elections, the Simulated Annealing heuristic achieve better MoV and PoW than the Greedy heuristic as the **budget** increases, in particular with real world graph (See Fig. 2). For ordinal elections, however, we were mostly unable to find a correlation.
– The impact of the number of **candidates and voters** on the Greedy and Simulated Annealing heuristics is not significant.
– With high budgets, the Simulated Annealing heuristic performs better in terms of the **PoW run time**, because it finds the maximum point quickly.
– The greedy heuristic performs and scales better in terms of the **MoV run time**. The Simulated Annealing heuristic took longer to execute in most cases, and there was not always a clear correlation (see Fig. 2).

Several more in-depth observations of the impact of the various parameters in our analysis are:

1. We found that the PoW and the MoV performance of both the Greedy and the Simulated Annealing heuristics are positively correlated with the budget increase. This makes intuitive sense: the more money, the better the outcome. This observation is true for the diffusion processes of plurality, approval, and

Borda in the real world graph (slightly) and Copeland in $G(n,p)$ and the real world graph. However we could not find correlation for Borda in $G(n,p)$ and the 5-clusters graphs and Copeland in 5-clusters graph. We can see that even with a small number of candidates, there is a correlation to the budget; however, as the number of candidates grows, the budget becomes irrelevant. The intuitive reason is the implementations of the ordinal bribe and diffusion operations; the bribe increases the candidate's rank by one place only, while the diffusion favors high-ranking candidates, so the diffusion process may often change the ranking back down (see full version [16] for additional details). While the Greedy heuristic run time is also positively correlated with the budget increase with respect to the diffusion processes, the Simulated Annealing heuristic's run time is not correlated with the budget increase with respect to the plurality, approval, and ordinal diffusion processes. The intuitive reason is that, while the Greedy heuristic performs more iterations as the budget increases, Simulated Annealing operates by local improvement, irrespective of the budget.

2. We find that the Simulated Annealing heuristic takes less time to run than the Greedy heuristic for PoW, especially with high budgets with a large number of candidates and voters in any type of election. However, the MoV Simulated Annealing heuristic performs slower than the Greedy heuristic when used with a plurality or approval diffusion process. This is true for execution with the ordinal diffusion process only when budgets are relatively small.

3. The MoV performance of the Simulated Annealing heuristic is higher than that of the Greedy heuristic when measured as a function of the number of voters in terms of plurality and approval diffusion processes. In terms of the ordinal diffusion process, however, the Simulated Annealing heuristic's MoV performance is similar to that of Greedy.

4. The MoV performance of Simulated Annealing is similar to that of Greedy as a function of the number of candidates in terms of plurality, approval, and ordinal diffusion processes.

5. We found negative correlation between the number of voters and PoW performance for both heuristics with respect to all three diffusion processes. One exception to the above is no correlation between the number of voters and PoW performance using the Greedy heuristic for Copeland $G(n,p)$ and 5-clusters graphs.

The tables in Fig. 1 provide a thorough overview of our findings.

Results comparing the various graph types are shown in Fig. 2. The main insight is that, for plurality elections, the Simulated Annealing heuristic outperforms the Greedy heuristic in terms of MoV and PoW in the different types of graphs, but at a significantly higher time cost.

For graph comparisons with our observations see full version [16].

		Plurality			Approval		
		Candidates	Budget	Voters	Candidates	Budget	Voters
PoW	Comparison	SA ≅ greedy	SA > greedy (slightly)	SA > greedy	SA > greedy (slightly)	SA > greedy	SA ≥ greedy
	Greedy Correlation	not correlated negatively correlated (Gnp)	positively correlated	negatively correlated (slightly)	negatively correlated (slightly)	positively correlated	negatively correlated
	SA Correlation	not correlated	positively correlated	negatively correlated (slightly)	not correlated	positively correlated	negatively correlated (slightly)
MoV	Comparison	SA > greedy	SA > greedy (slightly)	SA > greedy	SA ≥ greedy	SA ≥ greedy	SA > greedy (slightly)
	Greedy Correlation	not correlated	positively correlated	not correlated	not correlated	positively correlated	negatively correlated (slightly)
	SA Correlation	not correlated	positively correlated	negatively correlated (slightly)	not correlated	positively correlated	negatively correlated (slightly)
Runtime PoW	Comparison	greedy > SA (low budgets) SA > greedy (high budgets)	greedy > SA (high budgets) SA > greedy (low budgets)	greedy > SA (high budgets) SA > greedy (low budgets)	greedy > SA (high budgets) SA > greedy (low budgets)	greedy > SA (high budgets) SA > greedy (low budgets)	SA ≅ greedy (with exception in high budgets)
	Greedy Correlation	not correlated	positively correlated	positively correlated	positively correlated	positively correlated	positively correlated
	SA Correlation	not correlated	negatively correlated	positively correlated	positively correlated	negatively correlated (slightly)	positively correlated
Runtime MoV	Comparison	SA > greedy	SA > greedy	SA > greedy	SA > greedy (With exception in GNP)	SA > greedy greedy > SA (GNP high budgets)	SA > greedy (except for high budgets with high number of voters)
	Greedy Correlation	negatively correlated (slightly)	positively correlated (slightly)	positively correlated (slightly)	positively correlated (slightly)	positively correlated	positively correlated
	SA Correlation	positively correlated (slightly)	not correlated	positively correlated	positively correlated	not correlated	positively correlated

		Ordinal Borda			Ordinal Copeland		
		Candidates	Budget	Voters	Candidates	Budget	Voters
PoW	Comparison	greedy ≅ SA	greedy ≅ SA	greedy ≅ SA	greedy ≅ SA	greedy ≅ SA	greedy ≅ SA
	Greedy Correlation	negatively correlated	not correlated	negatively correlated	not correlated (gnp & 5-clusters) negatively correlated (real-world graph)	not correlated (gnp & 5-clusters) positively correlated (real-world graph)	not correlated (gnp & 5-clusters) positively correlated (real-world graph)
	SA Correlation	negatively correlated	not correlated	negatively correlated	not correlated (gnp & 5-clusters) negatively correlated (real-world graph)	not correlated (gnp & 5-clusters) positively correlated (real-world graph)	not correlated
MoV	Comparison	greedy ≅ SA	greedy ≅ SA	greedy ≅ SA	greedy ≅ SA	greedy ≅ SA	greedy ≅ SA
	Greedy Correlation	negatively correlated	not correlated (Gnp & 5-clusters) positively correlated (real world graph, slightly)	negatively correlated	negatively correlated	positively correlated	positively correlated (slightly)
	SA Correlation	negatively correlated	not correlated	negatively correlated	negatively correlated	not correlated	not correlated
Runtime PoW	Comparison	greedy > SA (when budget > 7)	greedy > SA (high budgets) SA > greedy (low budgets)	greedy > SA (when budget > 7)	greedy > SA (high budgets) SA > greedy (low budgets)	greedy > SA (high budgets) SA > greedy (low budgets)	greedy > SA (high budgets with high number of voters)
	Greedy Correlation	positively correlated	positively correlated	positively correlated	positively correlated	positively correlated (gnp & 5-clusters) not correlated (real-world graph)	positively correlated (gnp & 5-clusters) not correlated (real-world graph)
	SA Correlation	positively correlated	not correlated	positively correlated	positively correlated	not correlated	not correlated
Runtime MoV	Comparison	greedy > SA (when budget > 15)	greedy > SA (high budgets) SA > greedy (low budgets)	greedy > SA (high budgets with high number of voters)	greedy > SA (high budgets) SA > greedy (low budgets)	greedy > SA (high budgets) SA > greedy (low budgets) (With exceptions in greedy gnp)	greedy > SA (high budgets with high number of voters)
	Greedy Correlation	positively correlated	positively correlated	positively correlated	positively correlated (slightly)	positively correlated	positively correlated
	SA Correlation	positively correlated	not correlated	positively correlated	positively correlated	not correlated	not correlated

Fig. 1. A summary of the performance of the Greedy and Simulated Annealing heuristics with respect to each other in the different scenarios of plurality and approval (top) ordinal Borda and Copeland (bottom) diffusion processes, optimization goal, and input parameters. Correlation calculated based on Pearson correlation coefficient.

Fig. 2. Results for comparing the Greedy and SA heuristics in plurality elections on 3 different graphs and labels. The leftmost (middle, rightmost) shows the results for PoW (respectively, MoV, running time) as a function of the budget b; The green (blue, purple) represents the real-world network graph, i.e., email-Eu-core network (respectively, $G(n, p)$, and k-clusters graph where $k = 5$ (named in the plot 5-connected). (Color figure online)

5   Outlook

We proposed a general model for diffusion of opinions in social networks and considered heuristics that optimize over bribery schemes for several realizations of the model. Our model is general enough to incorporate various diffusion processes, including different ballot types and quite complex stochastic elements. We performed simulations to evaluate the performance of heuristic solutions that solve the task at hand; while our task is theoretically computationally intractable, our simulations are quite encouraging, in the sense that they reach good results in reasonable time. We also highlighted several parameter factors affecting the quality of the heuristics. We briefly discuss some avenues for future research.

Improved Heuristics. A natural future direction is to design better heuristic solutions, in particular such that are oblivious to the specific type of diffusion; E.g., local search heuristics with better initial solutions, other greedy approaches, as well as methods based on general solvers would be natural to try.

Other Settings. While we considered plurality, approval, and ordinal elections, it is natural to also consider utility-based elections and elections with cumulative ballots. Furthermore, there are other natural ways to treat the diffusion of complex opinions, most notably ordinal opinions. Fitting real-world data to various modeling choices would be an interesting future direction.

References

1. Allcott, H., Gentzkow, M.: Social media and fake news in the 2016 election. J. Econ. Perspectives **31**(2), 211–36 (2017)
2. Brandt, F., Conitzer, V., Endriss, U., Lang, J.: P rocaccia. Handbook of computational social choice. Cambridge University Press, A.D. (2016)
3. Bredereck, R., Elkind, E.: Manipulating opinion diffusion in social networks. In: Proceedings of IJCAI 2017 (2017)

4. Chen, N.: On the approximability of influence in social networks. SIAM J. Discret. Math. **23**(3), 1400–1415 (2009). https://doi.org/10.1137/08073617X, https://doi.org/10.1137/08073617X

5. Chen, W., Wang, Y., Yang, S.: Efficient influence maximization in social networks. In: Proceedings of the 15th ACM SIGKDD International Conference on Knowledge Discovery and Data Mining, pp. 199–208 (2009)

6. Coro, F., Cruciani, E., D'Angelo, G., Ponziani, S.: Exploiting social influence to control elections based on scoring rules. In: Proceedings of IJCAI 2019 (2019)

7. Elkind, E., Faliszewski, P.: Approximation algorithms for campaign management. In: Proceedings of WINE 2010, pp. 473–482 (2010)

8. Elkind, E., Faliszewski, P., Slinko, A.: Swap bribery. In: Mavronicolas, M., Papadopoulou, V.G. (eds.) SAGT 2009. LNCS, vol. 5814, pp. 299–310. Springer, Heidelberg (2009). https://doi.org/10.1007/978-3-642-04645-2_27

9. Ellison, N.B., Lampe, C., Steinfield, C.: Social network sites and society: current trends and future possibilities. Interactions **16**(1), 6 (2009)

10. Endriss, U.: Trends in Computational Social Choice. AI Access (2017)

11. Faliszewski, P., Hemaspaandra, E., Hemaspaandra, L.: How hard is bribery in elections? J. Artif. Intell. Res. **35**, 485–532 (2009)

12. Faliszewski, P., Rothe, J.: Control and bribery in voting. In: Brandt, F., Conitzer, V., Endriss, U., Lang, J., Procaccia, A.D. (eds.) Handbook of Computational Social Choice, chap. 7. Cambridge University Press (2015)

13. Faliszewski, P., Gonen, R., Koutecký, M., Talmon, N.: Opinion diffusion and campaigning on society graphs. In: Proceedings of IJCAI 2018, pp. 219–225 (2018)

14. Faliszewski, P., Hemaspaandra, E., Hemaspaandra, L.A.: How hard is bribery in elections? J. Artif. Intell. Res. **35**, 485–532 (2009)

15. Faliszewski, P., Skowron, P., Talmon, N.: Bribery as a measure of candidate success: complexity results for approval-based multiwinner rules. In: Proceedings of AAMAS 2017, pp. 6–14 (2017)

16. Gonen, R., Koutecky, M., Menashof, R., Talmon, N.: Heuristics for opinion diffusion via local elections full version (2022). https://www.openu.ac.il/personal_sites/rica-gonen/

17. Grandi, U., Lorini, E., Perrussel, L.: Propositional opinion diffusion. In: Proceedings of AAMAS 2015, pp. 989–997 (2015)

18. Guille, A., Hacid, H., Favre, C., Zighed, D.A.: Information diffusion in online social networks: a survey. ACM SIGMOD Rec. **42**(2), 17–28 (2013)

19. Jiang, Q., Song, G., Cong, G., Wang, Y., Si, W., Xie, K.: Simulated annealing based influence maximization in social networks. In: Proceedings of AAAI 2011. vol. 11, pp. 127–132 (2011)

20. Kempe, D., Kleinberg, J., Tardos, É.: Maximizing the spread of influence through a social network. In: Proceedings of KDD 2003, pp. 137–146 (2003)

21. Leskovec, J., Krause, A., Guestrin, C., Faloutsos, C., VanBriesen, J., Glance, N.: Cost-effective outbreak detection in networks. In: Proceedings of the 13th ACM SIGKDD International Conference on Knowledge Discovery and Data Mining, pp. 420–429 (2007)

22. Shakarian, P., Bhatnagar, A., Aleali, A., Shaabani, E., Guo, R.: Diffusion in Social Networks. Springer (2015)

23. Wang, J., Su, W., Yang, M., Guo, J., Feng, Q., Shi, F., Chen, J.: Parameterized complexity of control and bribery for d-approval elections. Theoret. Comput. Sci. **595**, 82–91 (2015)

24. Wilder, B., Vorobeychik, Y.: Controlling elections through social influence. In: Proceedings of AAMAS 2018, pp. 265–273 (2018)

On the Parameterized Complexity
of s-club Cluster Deletion Problems

Fabrizio Montecchiani[ID], Giacomo Ortali[ID], Tommaso Piselli[ID],
and Alessandra Tappini[✉][ID]

Dipartimento di Ingegneria, Università degli Studi di Perugia, Perugia, Italy
{fabrizio.montecchiani,giacomo.ortali,alessandra.tappini}@unipg.it,
tommaso.piselli@studenti.unipg.it

Abstract. We study the parameterized complexity of the s-CLUB
CLUSTER EDGE DELETION problem: Given a graph G and two inte-
gers $s \geq 2$ and $k \geq 1$, is it possible to remove at most k edges from G
such that each connected component of the resulting graph has diameter
at most s? This problem is known to be NP-hard already when $s = 2$. We
prove that it admits a fixed-parameter tractable algorithm when parameterized by s and the treewidth of the input graph. The same result easily
transfers to the case in which we can remove at most k vertices, rather
than k edges, from G such that each connected component of the resulting graph has diameter at most s, namely to s-CLUB CLUSTER VERTEX
DELETION.

1   Introduction

Graph clustering [25] is a classical task in data mining, with important applications in numerous fields including computational biology [6], image processing [28], and machine learning [5]. A prominent formalization is CLUSTER EDIT-
ING (also known as CORRELATION CLUSTERING): Given a graph G and an integer k as input, the goal is to find a sequence of k operations, each of which
can be an edge or vertex insertion or removal, such that the resulting graph is
a so-called *cluster graph*, i.e., each of its connected components is a clique. If
we restrict the editing operations to be edge removals only, then the problem is
known as CLUSTER EDGE DELETION. Namely, CLUSTER EDGE DELETION takes
a graph G and an integer k as input, and asks for a set of k edges whose removal
yields a cluster graph. Equivalently, we seek for a partition of the vertices of
G into cliques, such that the inter-cluster edges (whose end-vertices belong to
different cliques) are at most k.

Unfortunately, both CLUSTER EDITING and CLUSTER EDGE DELETION are
well-known to be NP-complete in general [27], even when the input instances have

This work was partially supported by: (*i*) MIUR, grant 20174LF3T8 "AHeAD: efficient Algorithms for HArnessing networked Data"; (*ii*) University of Perugia, Fondi di
Ricerca di Ateneo, edizione 2021, project "AIDMIX - Artificial Intelligence for Decision
making: Methods for Interpretability and eXplainability".

L. Gąsieniec (Ed.): SOFSEM 2023, LNCS 13878, pp. 159–173, 2023.
https://doi.org/10.1007/978-3-031-23101-8_11

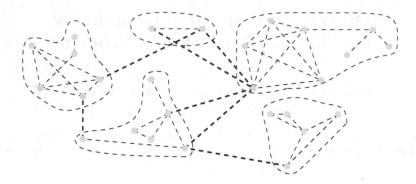

Fig. 1. Removing the seven thicker edges yields a partition of the vertices of the graph into five 3-clubs.

bounded vertex degree [17]. Indeed, their parameterized complexity with respect to the natural parameter k has been intensively investigated; in particular, both problems are in FPT [7,11], but do not allow subexponential-time parameterized algorithms unless ETH fails [14,17].

In many applications, modelling clusters with cliques might be a severe limitation, for instance, in presence of noise in the data collection process. Consequently, several notions of relaxed cliques have been introduced and investigated [4,16]. We focus on the concept of s-club, in which each pair of vertices is at distance at most $s \geq 2$ in the cluster. (Note that a 1-club is in fact a clique.) We remark that defining clusters as s-clubs proved to be effective in several application scenarios such as social network analysis and bioinformatics [2,3,18,21,22]. The s-CLUB CLUSTER EDGE DELETION problem can be stated analogously as CLUSTER EDGE DELETION by replacing cliques with s-clubs (refer to Sect. 2 for formal definitions); see also Fig. 1. Unfortunately, s-CLUB CLUSTER EDGE DELETION is NP-complete already for $s = 2$ [19]. Also, 2-CLUB CLUSTER EDGE DELETION belongs to FPT parameterized by k [1,19]. Indeed, by using bounded search trees (see [26, Theorem 6.3]), one easily verifies that s-CLUB CLUSTER EDGE DELETION is FPT parameterized by $k + s$. On the other hand, for any $s \geq 2$, the problem cannot be solved in time $2^{o(k)}n^{O(1)}$ unless ETH fails [20].

Contribution. Based on the above discussion, we know that it is unlikely that s-CLUB CLUSTER EDGE DELETION lies in FPT when parameterized by s, while it lies in FPT parameterized by $s + k$. The main focus of this paper is instead on those scenarios in which the solution size (measured by k) is large, and we still aim for tractable problems based on alternative parameterizations. In this respect, treewidth is a central parameter in the parameterized complexity analysis (see [13,24]). We prove that s-CLUB CLUSTER EDGE DELETION lies in FPT when parameterized by $s + \omega$, where ω is an upper bound for the treewidth of the input graph. More precisely, our main contribution can be summarized as follows.

Theorem 1. *There is an algorithm that, for any n-vertex graph G of treewidth at most ω, solves the s-CLUB CLUSTER EDGE DELETION problem on G in $O(2^{2^{O(\omega^2 \log s)}} \cdot n)$ time.*

The algorithm used to prove Theorem 1 can be easily adapted to solve a similar problem, namely s-CLUB CLUSTER VERTEX DELETION. The task is to find at most k vertices (rather than edges) whose removal yields a set of disjoint s-clubs. Again, it turns out that 2-CLUB CLUSTER VERTEX DELETION is NP-complete [10,19] but FPT parameterized by k [19]. We prove the following.

Theorem 2. *There is an algorithm that, for any n-vertex graph G of treewidth at most ω, solves the s-CLUB CLUSTER VERTEX DELETION problem on G in $O(2^{2^{O(\omega^2 \log s)}} \cdot n)$ time.*

Overview. We first observe that using Courcelle's theorem to prove that s-CLUB CLUSTER EDGE DELETION is in FPT using only s as parameter is not obvious. Instead, we provide an explicit algorithm. Namely, the main crux of our approach lies in the definition of sufficiently small records that allow to keep track of the distances between pairs of vertices in a (partial) s-club. In particular, we cannot afford to explicitly store the distance between all possible pairs of vertices, but we rather store sets of "requests for paths" placed by forgotten vertices. With such records at hand, we then apply a standard dynamic programming algorithm over a tree decomposition of the input graph, which still requires a nontrivial amount of technicalities in order to maintain such data structures.

Our records have some similarities, but also several key differences and extensions, with those used in a technique presented by Dondi and Lafond in [12, Thm. 14] to solve a different problem. Namely, they describe an FPT algorithm to decide whether the vertices of a graph can be covered with at most d (possibly overlapping) 2-clubs, parameterized by treewidth. The main novelties of the presented approach with respect to [12] will be suitably highlighted throughout the paper.

For reasons of space, some proofs have been omitted or sketched. The corresponding statements are marked with a $(\diamond)$ and can be found in [23].

2   Preliminaries

Problem Formulation. For any $d \in \mathbb{Z}^+$, we use $[d]$ as shorthand for the set $\{1, 2, \ldots, d\}$. Let $G = (V, E)$ be a graph. For a subset $W \subset V$, we denote by $G[W]$ the subgraph of G induced by the vertices of W. The *neighborhood* of a vertex v of G is defined as $N_G(v) = \{u : uv \in E\}$. Given two vertices $u, v \in V$, the *distance in G* between u and v, denoted by $d_G(u, v)$, is the number of edges in any shortest path between u and v in G. The *diameter* of G is the maximum distance in G between any two of its vertices. An *s-club* of G, with $s \geq 1$, is a subset $W \subseteq V$ such that the diameter of $G[W]$ is at most s. A *partition* of G is a collection of subsets $\mathcal{C} = \{C_i\}_{i \in [d]}$ such that: (a) $\bigcup_{i=1}^{d} C_i = V$, and (b) $C_i \cap C_j = \emptyset$ for each $i, j \in [d]$ with $i \neq j$. We denote by $E_\mathcal{C}$ the set of all edges uv of G such that $u, v \in C_i$, for some $i \in [d]$, and by $E_D = E \setminus E_\mathcal{C}$.

We study the following problem.

---

s-CLUB CLUSTER EDGE DELETION
Input: $G = (V, E)$, $k \geq 1$, $s \geq 2$.
Output: A partition $\mathcal{C} = \{C_i\}_{i \in [d]}$ of G such that C_i is an s-club for each $i \in [d]$, and $|E_D| \leq k$.

A *partial partition* of G is a collection of subsets $\mathcal{C} = \{C_i\}_{i \in [d]}$ such that: (a) $\bigcup_{i=1}^{d} C_i \subseteq V$, and (b) $C_i \cap C_j = \emptyset$ for each $i, j \in [d]$ with $i \neq j$. Denote by $V_\mathcal{C}$ the set of vertices v of G such that $v \in C_i$ for some $i \in [d]$, and by $V_D = V \setminus V_\mathcal{C}$.

We study the following problem.

---

s-CLUB CLUSTER VERTEX DELETION
Input: $G = (V, E)$, $k \geq 1$, $s \geq 2$.
Output: A partial partition $\mathcal{C} = \{C_i\}_{i \in [d]}$ of G such that C_i is an s-club for each $i \in [d]$, for each $uv \in E$ either there exists $C \in \mathcal{C}$ such that $u, v \in C$ or at least one of u, v is in V_D, and $|V_D| \leq k$.

Tree-Decompositions. Let $(\mathcal{X}, T)$ be a pair such that $\mathcal{X} = \{X_i\}_{i \in [\ell]}$ is a collection of subsets of vertices of a graph $G = (V, E)$, called *bags*, and T is a tree whose nodes are in one-to-one correspondence with the elements of $\mathcal{X}$. When this creates no ambiguity, X_i will denote both a bag of $\mathcal{X}$ and the node of T whose corresponding bag is X_i. The pair $(\mathcal{X}, T)$ is a *tree-decomposition* of G if: (i) $\bigcup_{i \in [\ell]} X_i = V$, (ii) for every edge uv of G, there exists a bag X_i that contains both u and v, and (iiii) for every vertex v of G, the set of nodes of T whose bags contain v induces a non-empty (connected) subtree of T. The *width* of $(\mathcal{X}, T)$ is $\max_{i=1}^{\ell} |X_i| - 1$, while the *treewidth* of G, denoted by $\mathrm{tw}(G)$, is the minimum width over all tree-decompositions of G. The problem of computing a tree-decomposition of width $\mathrm{tw}(G)$ is fixed-parameter tractable in $\mathrm{tw}(G)$ [8]. A tree-decomposition $(\mathcal{X}, T)$ of a graph G is *nice* if T is a rooted binary tree with the following additional properties [9].

1. If a node X_i of T has two children whose bags are X_j and $X_{j'}$, then $X_i = X_j = X_{j'}$. In this case, X_i is a *join bag*.
2. If a node X_i of T has only one child X_j, then $X_i \neq X_j$ and there exists a vertex $v \in G$ such that either $X_i = X_j \cup \{v\}$ or $X_i \cup \{v\} = X_j$. In the former case X_i is an *introduce bag*, while in the latter case X_i is a *forget bag*.
3. If a node X_i is the root or a leaf of T, then $X_i = \emptyset$.

Given a tree-decomposition of width ω of G, a nice tree-decomposition of G with the same width can be computed in $O(\omega \cdot n)$ time [15].

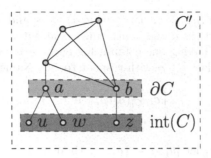

Fig. 2. A 4-club C' and a potential 4-club $C \subset C'$.

3 Algorithm for s-CLUB CLUSTER EDGE DELETION

The proof of Theorem 1 is based on a dynamic programming algorithm over a nice tree-decomposition. We first describe the records to be stored at each bag, and we then present the algorithm.

3.1 Definition of the Records

Let $G = (V, E)$ be an n-vertex graph and let $(\mathcal{X}, T)$ be a nice tree-decomposition of G of width ω. For each $i \in [\ell]$, let T_i be the subtree of T rooted at the bag $X_i \in \mathcal{X}$ and let $G_i = (V_i, E_i)$ be the subgraph of G induced by the vertices that belong to at least one bag of T_i. Moreover, given a subset of vertices $C \subseteq V_i$, we call C a *potential s-club*, and let $\partial C = C \cap X_i$ and $\text{int}(C) = C \setminus X_i$. We are now ready to describe the items of the record to be stored at each bag. Some of these items are similar to those described in [12], although with some important differences that allow us to deal with any fixed value of s.

Distances Between Vertices in ∂C. The first item of the record is a table that stores the pairwise distances of the vertices in ∂C. Namely, let $D(\partial C)$ be a table having one row and one column for each vertex in ∂C, and such that:

$$D(\partial C)[a, b] = \begin{cases} 0, & \text{if } a = b \\ d_{G[C]}(a, b), & \text{if } 1 \leq d_{G[C]}(a, b) \leq s \\ \infty, & \text{otherwise.} \end{cases}$$

For instance, in Fig. 2, $D(\partial C)[a, b] = \infty$. Observe that $D(\partial C)$ contains at most $(\omega + 1)^2 \in O(\omega^2)$ entries.

Distances Between a Vertex in $\text{int}(C)$ and a Vertex in ∂C. The second item of the record is a table that stores the distances between pairs of vertices such that one is in $\text{int}(C)$ and the other is in ∂C. Two vertices u, u' in $\text{int}(C)$ are *equivalent* with respect to ∂C, if for each vertex $a \in \partial C$, then either $1 \leq$

$d_{G[C]}(u,a) = d_{G[C]}(u',a) \leq s$, or $d_{G[C]}(u,a) > s$ and $d_{G[C]}(u',a) > s$. For example, in Fig. 2, vertices u and w are equivalent with respect to ∂C. Namely, let $H(\partial C)$ be a table having one column for each vertex $a \in \partial C$, and one row for each equivalence class $[u]_{\partial C}$ with respect to ∂C. Namely:

$$H(\partial C)[u,a] = \begin{cases} d_{G[C]}(u,a), & \text{if } 1 \leq d_{G[C]}(u,a) \leq s \\ \infty, & \text{otherwise.} \end{cases}$$

For instance, in Fig. 2, let $[u]_{\partial C} = \{u,w\}$, then we have $H(\partial C)[u,a] = 1$ and $H(\partial C)[u,b] = \infty$. Observe that $H(\partial C)$ contains at most $(s+1)^{(\omega+1)}$ rows, and hence $O(\omega \cdot s^{O(\omega)}) = O(2^{O(\omega \log s)})$ entries. It is worth observing that, if $H(\partial C)$ contains two rows r and r' such that cell-wise the values of r are smaller than or equal to those of r', then we can avoid storing r and keep only r' in $H(\partial C)$. However, this would not reduce the asymptotic size of $H(\partial C)$.

Distances Between Vertices in int(C). The third (and last) item of the record represents the key insight to extend the result in [12]. Roughly speaking, in the case $s = 2$, pairs of vertices in int(C) must have distance at most two in $G[C]$, otherwise there is no $C' \supseteq C$ such that C' is a 2-club of G. Unfortunately, this is not true in general when $s > 2$. On the other hand, suppose that C is a subset of an s-club C' of G and that there exist two vertices $u, u' \in \text{int}(C)$ whose distance in $G[C]$ is larger than s. Then, since each bag if T is a separator of G, the shortest path in G between u and u' goes through some pair of vertices in ∂C. We formalize this observation. Let $w, z \in \text{int}(C)$ be two vertices such that $d_{G[C]}(w,z) > s$. A *request for* ∂C, denoted by R_{wz}, is a table having one row and one column for each vertex in ∂C. Namely, for each $a, b \in \partial C$, let $2 \leq \delta \leq s - 2$ be the largest value (if any) such that connecting a and b with a path π of length δ makes the distance between w and z to be at most s, then $R_{wz}[a,b] = \delta$. If instead such a value δ does not exist, then $R_{wz}[a,b] = \star$. For instance, in Fig. 2, $d_{G[C]}(w,z) > 4$, while $\delta = 2$, hence we would set $R_{wz}[a,b] = 2$.

Observe that if there exist two requests R_{wz} and $R_{w'z'}$ such that $R_{wz}[a,b] = R_{w'z'}[a,b]$ for each pair $a, b \in \partial C$, then w and w' are equivalent with respect to ∂C (i.e., $w, w' \in [w]_{\partial C}$), and the same holds for z and z'. For example, in Fig. 2, this would be the case for R_{uz} and R_{wz}, and indeed we have already observed that u and w are equivalent with respect to ∂C. Therefore we can avoid storing duplicated requests, and we denote by $Q(\partial C)$ the set containing all distinct requests for ∂C. Also, Q contains at most $(s-2)^{(\omega+1)^2} \in O(s^{O(\omega^2)}) = O(2^{O(\omega^2 \log s)})$ distinct requests.

Solutions. Before describing our algorithm, we need to slightly extend our notation. If a potential s-club C is such that $\partial C = \emptyset$ (recall that $C \subseteq V_i$), then we call it *complete*. Consider a partitioning $\mathcal{P}_i^l$ of G_i into potential s-clubs and let $\mathcal{C}_i^l = \{C_{j,i}^l \mid j \in [d_l]\}$ be the potential s-clubs in $\mathcal{P}_i^l$ that are not complete, i.e., any $C \in \mathcal{C}_i^l$ is such that $\partial C \neq \emptyset$. In particular, let $\partial \mathcal{C}_i^l = \{\partial C_{j,i}^l \mid j \in [d_l]\}$.

Moreover, we let $\mathcal{D}_i^l = \{D(\partial C_{j,i}^l) \mid j \in [d_l]\}$, $\mathcal{H}_i^l = \{H(\partial C_{j,i}^l) \mid j \in [d_l]\}$, and $\mathcal{Q}_i^l = \{Q(\partial C_{j,i}^l) \mid j \in [d_l]\}$. A *solution* of X_i is a tuple $S_i^l = \langle \partial \mathcal{C}_i^l, \mathcal{D}_i^l, \mathcal{H}_i^l, \mathcal{Q}_i^l, k_i^l \rangle$. Here k_i^l is an integer, called *edge-counter*, equal to $|E_i \setminus \mathcal{P}_i^l(E_i)|$, i.e., k_i^l counts the number of edges having their endpoints in different potential s-clubs of $\mathcal{P}_i^l$, hence $k_i^l \leq k$. Two solutions $S_i^l = \langle \partial \mathcal{C}_i^l, \mathcal{D}_i^l, \mathcal{H}_i^l, \mathcal{Q}_i^l, k_i^l \rangle$ and $S_i^g = \langle \partial \mathcal{C}_i^g, \mathcal{D}_i^g, \mathcal{H}_i^g, \mathcal{Q}_i^g, k_i^g \rangle$ are *distinct* if $\partial \mathcal{C}_i^l \neq \partial \mathcal{C}_i^g$, or $\mathcal{D}_i^l \neq \mathcal{D}_i^g$, or $\mathcal{H}_i^l \neq \mathcal{H}_i^g$, or $\mathcal{Q}_i^l \neq \mathcal{Q}_i^g$. Observe that if S_i^l and S_i^g are *not* distinct but $k_i^l < k_i^g$, then it suffices to consider only S_i^l.

Lemma 1 ($\diamond$). *For a bag X_i, there exist $O(2^{2^{O(\omega^2 \log s)}})$ distinct solutions.*

3.2 Description of the Algorithm

We are now ready to describe our dynamic-programming algorithm over a nice tree-decomposition $(\mathcal{X}, T)$ of the input graph G. The main differences with respect to the algorithm in [12] lie in the management of the sets $Q(\cdot)$ (which do not exist in [12]), and on a more sophisticated updating of the tables $D(\cdot)$ and $H(\cdot)$, as a consequence of the non applicability of some simplifying assumptions that hold only when $s = 2$. Moreover, we also keep track of the number of edges having their end-vertices in different potential s-clubs.

Let X_i be the current bag visited by the algorithm. We compute the set of solutions for X_i based on the solutions computed for its child or children (if any). In case the resulting set of solutions is empty, the algorithm halts and returns a negative answer. We distinguish four cases based on the type of X_i.

X_i **is a Leaf Bag.** In this case X_i is empty and there exists only one trivial solution S_i^1, in which all tables are empty and $k_i^l = 0$.

X_i **is an Introduce Bag.** Let $X_j = X_i \setminus \{v\}$ be the child of X_i. The algorithm exhaustively extends each solution S_j^l of X_j as follows. It first generates at most d_l new partitions by placing v in each $\partial C' \in \partial \mathcal{C}_j^l$. Also, it generates a partition in which v forms a new potential s-club $C = \partial C = \{v\}$. Consider one of the new partitions generated by the algorithm. In order to build the corresponding new solution for X_i, we distinguish the following two cases.

Case A ($\partial C = \{v\}$). $D(\partial C)$ is trivially defined, $H(\partial C)$ and $Q(\partial C)$ are empty.

Case B ($\partial C = \partial C' \cup \{v\}$). The next observation immediately follows from the fact that $\partial C = \partial C' \cup \{v\}$ and $\text{int}(C) = \text{int}(C')$.

Observation 1. *Suppose that there exist $a, b \in \partial C'$ such that $d_{G[C']}(a, b) > d_{G[C]}(a, b)$, then any shortest path between a and b in $G[C]$ contains vertex v.*

We proceed as follows.

– Computing $D(\partial C)$ from $D(\partial C')$[1].

---

[1] Since the matrix is symmetric, when we update a cell $D(\partial C)[a, b]$ we assume that also $D(\partial C)[b, a]$ is updated with the same value.

1. We add a new row and a new column for vertex v.
2. For each vertex $a \in \partial C'$, let $\delta_{av} = \min_{b \in N_{G[X_i]}(v)} D(\partial C')[a, b]$, and note that $\delta_{av} = 0$ if edge av belongs to $G[C]$. Clearly, it holds that

$$D(\partial C)[a, v] = \begin{cases} \infty, & \text{if } \delta_{av} \in \{s, \infty\} \\ 1 + \delta_{av}, & \text{otherwise.} \end{cases}$$

3. By Observation 1, for each pair $a, b \in \partial C'$, the corresponding value of $D(\partial C)$ can be updated as follows:

$$D(\partial C)[a, b] = \min\{D(\partial C')[a, b], D(\partial C)[a, v] + D(\partial C)[b, v]\}.$$

– Computing $H(\partial C)$ from $H(\partial C')$.
   1. We add a new column for vertex v.
   2. For each equivalence class $[u]_{\partial C'}$, let $\delta_{uv} = \min_{a \in N_{G[X_i]}(v)} H(\partial C')[u, a]$. Since there is no edge uv such that $u \in \text{int}(C)$, it follows that

$$H(\partial C)[u, v] = \begin{cases} \infty, & \text{if } \delta_{uv} \in \{s, \infty\} \\ 1 + \delta_{uv}, & \text{otherwise.} \end{cases}$$

   3. By Observation 1, for each pair of vertices $u \in \text{int}(C')$ and $a \in \partial C'$, the corresponding value of $H(\partial C)$ can be updated as follows:

$$H(\partial C)[u, a] = \min\{H(\partial C')[u, a], H(\partial C)[u, v] + D(\partial C)[v, a]\}.$$

– Computing $Q(\partial C)$ from $Q(\partial C')$. Note that the addition of v cannot lead to new requests but it may actually yield the update of some request in $Q(\partial C')$.
   1. For each request R_{wz} in $Q(\partial C')$, we verify whether, as a consequence of the introduction of v, there exists a cell $R_{wz}[a, b]$ such that $D(\partial C)[a, b] \leq R_{wz}[a, b]$. If such a cell exists, we say that R_{wz} is *fulfilled*. We add R_{wz} to $Q(\partial C)$ if and only if R_{wz} is not fulfilled.
   2. If R_{wz} is not fulfilled, before adding it to $Q(\partial C)$, we update it as follows:
      (a) We add a row and a column for v.
      (b) For each pair $a, b \in \partial C'$, we compute

$$\delta_{ab} = \min\{(H(\partial C)[w, a] + H(\partial C)[z, b], H(\partial C)[z, a] + H(\partial C)[w, b]\}.$$

      Observe that $\delta_{ab} + D(\partial C)[a, b] > s$, otherwise the request would have been fulfilled before.
      (c) By definition of request, we have $R_{wz}[a, b] = s - \delta_{ab}$, if $\delta_{ab} < s - 1$, and $R_{wz}[a, b] = \star$, otherwise.

Finally, in both **Case A** and **Case B**, in order to obtain the edge-counter of the new solution, k_j^l needs to be increased by the number of edges incident to v whose other end-vertex is in X_i but not in C. If the resulting edge-counter is greater than k, the solution is discarded.

X_i **is a Forget Bag.** Let $X_j = X_i \cup \{v\}$ be the child of X_i. The algorithm updates each solution S_j^l of X_j as follows. It first identifies the potential s-club $C' \in \partial C_j^l$ that vertex v belongs to. Then, it verifies whether $\partial C' = \{v\}$, i.e., whether removing v from C' makes it complete.

Case A ($\partial C' = \{v\}$). The algorithm checks the following *completion conditions*:

1. The value of each cell of $D(\partial C')$ is at most s;
2. The value of each cell of $H(\partial C')$ is at most s;
3. The set $Q(\partial C')$ is empty.

Observation 2. *If C' is complete, then it is an s-club of G if and only if the completion conditions are satisfied.*

If any of the completion conditions is not satisfied, the solution S_j^l is discarded. Otherwise, we generate a new solution for X_i such that $\partial C_i^l = \partial C_j^l \backslash \{\partial C'\}$.

Case B ($\partial C \supset \{v\}$). First, we set $\partial C = \partial C' \setminus \{v\}$ and $\partial C_i^l = (\partial C_j^l \setminus \{\partial C'\}) \cup \{\partial C\}$. Note that the distance in $G[C]$ between any two vertices is the same as in $G[C']$.

- Computing $D(\partial C)$ from $D(\partial C')$. We simply remove the row and column of v.
- Computing $H(\partial C)$ from $H(\partial C')$.
  1. We remove the column corresponding to v.
  2. We check whether $H(\partial C')$ already contains a row that represents the distances of v with respect to the vertices in ∂C. Namely, we check if there is an equivalence class $[u]_{\partial C}$ such that for each vertex $a \in \partial C$, it holds $H(\partial C)[u, a] = D(\partial C')[a, v]$. If such a row does not exist, we add it to $H(\partial C)$.
- Computing $Q(\partial C)$ from $Q(\partial C')$. Forgetting vertex v causes the update of existing requests in $Q(\partial C')$, as well as the introduction of new requests.
  1. To update the existing requests, for each request R_{wz} in $Q(\partial C')$:
     (a) We remove the row and column corresponding to v.
     (b) We verify that there exist at least two vertices $a, b \in \partial C$ such that $R_{wz}[a, b] \neq \star$. If this is the case, we add the updated request to $Q(\partial C)$, otherwise we discard the solution S_j^l.
  2. To introduce new requests, we verify whether there exists a cell $H(\partial C')[u, v] = \infty$, which represents the existence of an equivalence class $[u]_{\partial C'}$ whose vertices have distance in C' from v larger than s. If such a cell exists, then:
     (a) We create a new request R_{uv}.
     (b) For each pair $a, b \in \partial C$, we compute

$$\delta_u = \min\{H(\partial C)[u, a], H(\partial C)[u, b]\}$$

and

$$\delta_v = \min\{H(\partial C)[v, a], H(\partial C)[v, b]\}.$$

     (c) By definition of request, we have $R_{uv}[a, b] = s - (\delta_u + \delta_v)$, if $\delta_u + \delta_v < s - 1$, and $R_{uv}[a, b] = \star$, otherwise.

(d) If the value of at least one cell of R_{uv} is different from $\star$, then we add R_{uv} to $Q_i^l(\partial C)$. Otherwise, we discard the solution S_j^l.

In both **Case A** and **Case B**, we observe that the edge-counter of the new solution can be set to be equal to the original k_j^l.

Finally, we observe that two solutions S_i^l and S_i^g, stemming from two distinct solutions of X_j, may now be the same as a consequence of the removal of v, up to the values of k_i^l and k_i^g. For each such a pair, it suffices to keep the solution with lower edge-counter.

X_i **is a Join Bag.** Let $X_j = X_{j'}$ be the two children of X_i. The algorithm exhaustively merges each pair of solutions S_j^l of X_j and $S_{j'}^{l'}$ of $X_{j'}$, if possible. A successful merge corresponds to a solution of X_i. Without loss of generality, we can avoid merging S_j^l and $S_{j'}^{l'}$ when $\partial C_j^l \neq \partial C_{j'}^{l'}$, because a resulting solution (if any), can be obtained by merging a different pair of solutions S_j^h of X_j and $S_{j'}^{h'}$ such that $\partial C_j^h = \partial C_{j'}^{h'}$. Therefore we assume $\partial C_j^l = \partial C_{j'}^{l'}$. In other words, for each ∂C in ∂C_j^l there exists $\partial C'$ in $\partial C_{j'}^{l'}$ such that $\partial C = \partial C'$. Also, in the following we denote by C^* the potential s-club such that $\partial C^* = \partial C = \partial C'$ and $\mathrm{int}(C^*) = \mathrm{int}(C) \cup \mathrm{int}(C')$. It remains to verify that, for each such C^*, each pair of vertices u, u' such that $u \in \mathrm{int}(C)$ and $u' \in \mathrm{int}(C')$ is either at distance at most s or we can generate a new request for u, u'. We remark that, when $s > 2$ (and hence differently from [12]), a new shortest path between u and u' may be formed by both vertices in $\mathrm{int}(C)$ and vertices in $\mathrm{int}(C')$. We proceed as follows.

- For each pair $a, b \in \partial C$, let $\omega_{ab} = \min\{D(\partial C)[a, b], D(\partial C')[a, b]\}$. We construct the weighted complete graph W^* on the vertex set ∂C^* and such that the weight of any edge ab is ω_{ab}.
- Computing $D(\partial C^*)$ from $D(\partial C)$ and $D(\partial C')$: By construction of W^*, it follows that for each pair $a, b \in \partial C^*$, $D(\partial C^*)[a, b]$ corresponds to the weighted shortest path between a and b in W^*.
- Computing $H(\partial C^*)$ from $H(\partial C)$ and $H(\partial C')$.
  1. We first merge $H(\partial C)$ and $H(\partial C')$ avoiding duplicated rows. Let $H(\partial C'')$ be the resulting table.
  2. Similarly as in the previous step, for each equivalence class $[u]_{\partial C''}$ of $H(\partial C'')$, we add a vertex u to W^* and, for each vertex a of W^* we add the edge ua with weight $\omega_{ua} = H(\partial C'')[u, a]$. Then $H(\partial C^*)[u, a]$ corresponds to the weighted shortest path between u and a in the resulting graph.
  3. As some rows of $H(\partial C^*)$ may be the same, we remove possible duplicates.
- Computing $Q(\partial C^*)$ from $Q(\partial C)$ and $Q(\partial C')$.
  1. We first merge the two sets $Q(\partial C)$ and $Q(\partial C')$ avoiding duplicated requests. Let $Q(\partial C'')$ be the resulting set of requests.
  2. We verify whether some of these requests have been fulfilled.
     (a) For each request R_{wz} in $Q(\partial C'')$, we add R_{wz} to $Q(\partial C^*)$ if and only if R_{wz} is not fulfilled (i.e., there is no cell $R_{wz}[a, b]$ such that $D(\partial C^*)[a, b] \leq R_{wz}[a, b]$).

(b) If R_{wz} is not fulfilled, before adding it to $Q(\partial C^*)$, we update it as follows:

   i For each pair of vertices $a, b \in \partial C^*$, we compute

$$\delta_{ab} = \min\{(H(\partial C^*)[w, a] + H(\partial C^*)[z, b],$$
$$H(\partial C^*)[z, a] + H(\partial C^*)[w, b]\}.$$

   Observe that $\delta_{ab} + D(\partial C^*)[a, b] > s$, otherwise the request would have been fulfilled before.

   ii Similarly as in the previous cases, by definition of requests it follows that $R_{wz}[a, b] = s - \delta_{ab}$, if $\delta_{ab} < s - 1$, and $R_{wz}[a, b] = \star$, otherwise.

   iii If the value of at least one cell of R_{wz} is different from $\star$, we add R_{wz} to $Q_i^l(\partial C^*)$. Otherwise, we discard the pair of solutions S_j^l, $S_{j'}^{l'}$.

3. We now generate new requests, if needed.

   (a) For each pair of rows $[w]_{\partial C_j} \in H(\partial C_j)$ and $[z]_{\partial C_{j'}} \in H(\partial C_{j'})$, we add two representative vertices w and z to W^* and, for each vertex v of W^* we add the edges wv and zv with weights $H(\partial C^*)[w, v]$ and $H(\partial C^*)[z, v]$, respectively. Let σ_{uv} be the shortest path between any two vertices u, v in this graph.

   (b) If $\sigma_{wz} > s$, we generate a new request R_{wz}.

      i For each pair $u, u' \in \partial C^*$, we compute

$$\delta_{uu'} = \min\{\sigma_{wu} + \sigma_{zu'}, \sigma_{wu'} + \sigma_{zu}\}.$$

      ii Again it follows that $R_{wz}[u, u'] = s - \delta_{uu'}$, if $\delta_{uu'} < s - 1$, and $R_{wz}[u, u'] = \star$, otherwise.

      iii If the value of at least one cell of R_{wz} is different from $\star$, we add R_{wz} to $Q_i^l(\partial C^*)$. Otherwise, we discard the solutions S_j^l, $S_{j'}^{l'}$.

Finally, let M the set of edges in the new solution between vertices of X_i that belong to different partial s-clubs. The edge-counter of the new solution is $k_i^l + k_i^{l'} - |M|$ (to avoid double-counting such edges). If the resulting edge-counter is larger than k, then the solution is discarded.

The next lemma establishes the correctness of the algorithm.

Lemma 2 ($\diamond$). *Graph G admits a solution for s-CLUB CLUSTER EDGE DELE-TION if and only if the algorithm terminates after visiting the root of T.*

Proof (Sketch). We sketch one direction, namely, to derive a contradiction, suppose that G admits a partition $\mathcal{P}$ into s-clubs that represents a valid solution but the algorithm terminates prematurely. If this is the case, there must be a bag X_i in which a solution S_i of X_i has been discarded but the partition $\mathcal{P}_i$ corresponding to S_i could be extended to $\mathcal{P}$. We can distinguish cases based on the type of bag X_i, we only describe the case in which X_i is an introduce bag. Then S_i is discarded only if the edge-counter becomes larger than k, which means that there exist at least $k + 1$ edges whose end-vertices are in distinct potential s-clubs of $\mathcal{P}_i$. Since any two distinct potential s-clubs of $\mathcal{P}_i$ will be subsets of two distinct s-clubs of $\mathcal{P}$, this is a contradiction with the fact that partition $\mathcal{P}$ represents a valid solution. □

Proof (of Theorem 1). The correctness of the algorithm derives from Lemma 2. It remains to argue about its time complexity. A tree decomposition of G of width ω can be computed in $O(2^{O(\omega^3)} \cdot n)$ [8] time, and from it a nice tree-decomposition of width ω can be derived in $O(\omega \cdot n)$ time [15].

For each bag X, by Lemma 1 and by the fact that we avoid storing duplicated solutions, we have $O(2^{2^{O(\omega^2 \log s)}})$ solutions. Hence, when building the solution set of X from its child or children, we process at most these many elements. The size of a solution S of X is $O(2^{O(\omega^2 \log s)})$, and each extension takes polynomial time in the size of the extended solution. Let $f(\omega, s) = 2^{2^{O(\omega^2 \log s)}}$ and $g(\omega, s) = 2^{O(\omega^2 \log s)}$, and observe that $g(\omega, s)^{O(1)} = o(f(\omega, s))$. It follows that constructing the solution set of X takes $O(f(\omega, s) \cdot g(\omega, s)^{O(1)}) = O(f(\omega, s)) = O(2^{2^{O(\omega^2 \log s)}})$. In addition, if X is a forget bag, we also need to remove possible duplicated solutions. This can be done by sorting the solutions, which takes $O(g(\omega, s)) \times O(f(\omega, s) \log f(\omega, s))$. Also, since $O(g(\omega, s)) = O(\log f(\omega, s))$, we have $O(f(\omega, s) \log f(\omega, s)) = O(f(\omega, s) \cdot g(\omega, s)) = O(f(\omega, s))$. The statement then follows because there are $O(n)$ bags. □

4 Algorithm for s-CLUB CLUSTER VERTEX DELETION

The algorithm presented in Sect. 3 can be adapted to solve s-CLUB CLUSTER VERTEX DELETION, and hence to prove Theorem 2. At high level, the modified algorithm does not allow inter-cluster edges but may ignore up to k vertices that will not be part of the partial partition. Since the modifications are rather minor, we simply describe such differences without repeating common parts.

Let X_i be the current bag visited by the algorithm. A solution S_i^l for X_i is modeled in the same way as for s-CLUB CLUSTER EDGE DELETION, with one minor difference. Namely, k_i^l is interpreted as a *vertex-counter* rather than an edge-counter. Clearly, Lemma 1 still holds.

Concerning the algorithm, if X_i is a leaf bag, again there exists only one trivial solution S_i^1, in which all tables are empty and $k_i^1 = 0$.

If X_i is an introduce bag, the only modifications that we need are as follows. Let $X_j = X_i \setminus \{v\}$. The algorithm still exhaustively extends each solution S_j^l of X_j. It first generates at most d_l new partitions by placing v in each $\partial C' \in \partial C_j^l$. However, we can place v in $\partial C'$ only if all v's neighbors in G_i (if any) belong to $\partial C'$. Similarly, the algorithm generates a partition in which v forms a new potential s-club $C = \partial C = \{v\}$, only if v does not have any neighbor in G_i. Observe that this is equivalent as running the algorithm described in Sect. 3 imposing that the edge counter of any solution be zero. Additionally, the algorithm can extend any solution S_j^l of X_j by ignoring v (i.e., assuming v to be part of V_D) and increasing the vertex-counter of the resulting solution by one. If the resulting vertex-counter is larger than k, the solution is discarded.

If X_i is a forget bag, we only have an additional case when the forgotten vertex v does not belong to any potential s-club. In this case, the algorithm simply keeps the solution as is, since v is not part of the partial partition in output. Similarly as in Sect. 3, two solutions S_i^l and S_i^g, stemming from two distinct solutions of X_j, may now be the same up to the values of k_i^l and k_i^g. For each such a pair, it suffices to keep the solution with lower vertex-counter.

If X_i is a join bag whose children are X_j and $X_{j'}$, again there is no substantial modification to be applied. Recall that we merge two solutions, say S_j^l of X_j and $S_{j'}^{l'}$ of $X_{j'}$, only if $\partial C_j^l = \partial C_{j'}^{l'}$. Let M be the set of vertices of X_i that are not part of any partial s-club of $\partial C_j^l = \partial C_{j'}^{l'}$. Then the vertex-counter of a new solution is $k_i^l + k_i^{l'} - |M|$ (to avoid double-counting such vertices). If the resulting vertex-counter is larger than k, then the solution is discarded.

5   Discussion and Open Problems

We have proved that the s-CLUB CLUSTER EDGE DELETION and s-CLUB CLUS-TER VERTEX DELETION problems parameterized by $s + \omega$ (where ω bounds the treewidth of the input graph) belong to FPT. On the other hand, we know that both problems parameterized by s alone are paraNP-hard. It remains open their complexity parameterized by ω alone.

We also observe that it is possible to adjust a reduction in [12] to prove that s-CLUB CLUSTER EDGE DELETION remains paraNP-hard even when parameterized by $s + d$, where d is the number of clusters in the sought solution. With respect to parameter d, we also know that the problem has no subexponential-time parameterized algorithm in $k + d$ [20]. Yet, whether $k + d$ is a tractable parameterization is an interesting question.

References

1. Abu-Khzam, F.N., Makarem, N., Shehab, M.: An improved fixed-parameter algorithm for 2-club cluster edge deletion. CoRR. arXiv:2107.01133 (2021)
2. Alba, R.: A graph-theoretic definition of a sociometric clique. J. Math. Soc. **3**, 113–126 (1973). https://doi.org/10.1080/0022250X.1973.9989826
3. Balasundaram, B., Butenko, S., Trukhanov, S.: Novel approaches for analyzing biological networks. J. Comb. Optim. **10**(1), 23–39 (2005). https://doi.org/10.1007/s10878-005-1857-x
4. Balasundaram, B., Pajouh, F.M.: Graph theoretic clique relaxations and applications. In: Pardalos, P.M., Du, D.-Z., Graham, R.L. (eds.) Handbook of Combinatorial Optimization, pp. 1559–1598. Springer, New York (2013). https://doi.org/10.1007/978-1-4419-7997-1_9
5. Bansal, N., Blum, A., Chawla, S.: Correlation clustering. Mach. Learn. **56**(1–3), 89–113 (2004). https://doi.org/10.1023/B:MACH.0000033116.57574.95

6. Ben-Dor, A., Shamir, R., Yakhini, Z.: Clustering gene expression patterns. J. Comput. Biol. **6**(3/4), 281–297 (1999). https://doi.org/10.1089/106652799318274
7. Böcker, S.: A golden ratio parameterized algorithm for cluster editing. J. Discrete Algorithms **16**, 79–89 (2012). https://doi.org/10.1016/j.jda.2012.04.005
8. Bodlaender, H.L.: A linear-time algorithm for finding tree-decompositions of small treewidth. SIAM J. Comput. **25**(6), 1305–1317 (1996). https://doi.org/10.1137/S0097539793251219
9. Bodlaender, H.L., Kloks, T.: Efficient and constructive algorithms for the pathwidth and treewidth of graphs. J. Algorithms **21**(2), 358–402 (1996). https://doi.org/10.1006/jagm.1996.0049
10. Chakraborty, D., Chandran, L.S., Padinhatteeri, S., Pillai, R.R.: Algorithms and complexity of s-club cluster vertex deletion. In: Flocchini, P., Moura, L. (eds.) IWOCA 2021. LNCS, vol. 12757, pp. 152–164. Springer, Cham (2021). https://doi.org/10.1007/978-3-030-79987-8_11
11. Chen, J., Meng, J.: A 2k kernel for the cluster editing problem. J. Comput. Syst. Sci. **78**(1), 211–220 (2012). https://doi.org/10.1016/j.jcss.2011.04.001
12. Dondi, R., Lafond, M.: On the tractability of covering a graph with 2-clubs. In: Gąsieniec, L.A., Jansson, J., Levcopoulos, C. (eds.) FCT 2019. LNCS, vol. 11651, pp. 243–257. Springer, Cham (2019). https://doi.org/10.1007/978-3-030-25027-0_17
13. Downey, R.G., Fellows, M.R.: Parameterized complexity. Monographs in Computer Science, 1st Edition. Springer (1999). https://doi.org/10.1007/978-1-4612-0515-9
14. Fomin, F.V., Kratsch, S., Pilipczuk, M., Pilipczuk, M., Villanger, Y.: Tight bounds for parameterized complexity of cluster editing with a small number of clusters. J. Comput. Syst. Sci. **80**(7), 1430–1447 (2014). https://doi.org/10.1016/j.jcss.2014.04.015
15. Kloks, T. (ed.): Treewidth. LNCS, vol. 842. Springer, Heidelberg (1994). https://doi.org/10.1007/BFb0045375
16. Komusiewicz, C.: Multivariate algorithmics for finding cohesive subnetworks. Algorithms **9**(1), 21 (2016). https://doi.org/10.3390/a9010021
17. Komusiewicz, C., Uhlmann, J.: Cluster editing with locally bounded modifications. Discret. Appl. Math. **160**(15), 2259–2270 (2012). https://doi.org/10.1016/j.dam.2012.05.019
18. Laan, S., Marx, M., Mokken, R.J.: Close communities in social networks: boroughs and 2-clubs. Soc. Netw. Anal. Min. **6**(1), 1–16 (2016). https://doi.org/10.1007/s13278-016-0326-0
19. Liu, H., Zhang, P., Zhu, D.: On editing graphs into 2-club clusters. In: Snoeyink, J., Lu, P., Su, K., Wang, L. (eds.) AAIM/FAW -2012. LNCS, vol. 7285, pp. 235–246. Springer, Heidelberg (2012). https://doi.org/10.1007/978-3-642-29700-7_22
20. Misra, N., Panolan, F., Saurabh, S.: Subexponential algorithm for d-cluster edge deletion: exception or rule? J. Comput. Syst. Sci. **113**, 150–162 (2020). https://doi.org/10.1016/j.jcss.2020.05.008
21. Mokken, R.: Cliques, clubs and clans. Qual. Quan. Int. J. Methodol. **13**(2), 161–173 (1979). https://doi.org/10.1007/BF00139635
22. Mokken, R.J., Heemskerk, E.M., Laan, S.: Close communication and 2-clubs in corporate networks: europe 2010. Soc. Netw. Anal. Min. **6**(1), 1–19 (2016). https://doi.org/10.1007/s13278-016-0345-x
23. Montecchiani, F., Ortali, G., Piselli, T., Tappini, A.: On the parameterized complexity of the s-club cluster edge deletion problem. CoRR. arXiv:2205.10834 (2022)
24. Robertson, N., Seymour, P.D.: Graph minors. II. algorithmic aspects of tree-width. J. Algorithms **7**(3), 309–322 (1986)

25. Schaeffer, S.E.: Graph clustering. Comput. Sci. Rev. **1**(1), 27–64 (2007). https://doi.org/10.1016/j.cosrev.2007.05.001
26. Schäfer, A.: Exact algorithms for s-club finding and related problems. Diploma thesis, Friedrich-Schiller-Universität Jena (2009)
27. Shamir, R., Sharan, R., Tsur, D.: Cluster graph modification problems. Discret. Appl. Math. **144**(1–2), 173–182 (2004). https://doi.org/10.1016/j.dam.2004.01.007
28. Wu, Z., Leahy, R.M.: An optimal graph theoretic approach to data clustering: theory and its application to image segmentation. IEEE Trans. Pattern Anal. Mach. Intell. **15**(11), 1101–1113 (1993). https://doi.org/10.1109/34.244673

SOFSEM 2023 Best Papers

Balanced Substructures in Bicolored Graphs

P. S. Ardra[1]($\boxtimes$), R. Krithika[1], Saket Saurabh[2,3], and Roohani Sharma[4]

[1] Indian Institute of Technology Palakkad, Palakkad, India
111914001@smail.iitpkd.ac.in, krithika@iitpkd.ac.in
[2] The Institute of Mathematical Sciences, HBNI, Chennai, India
saket@imsc.res.in
[3] University of Bergen, Bergen, Norway
[4] Max Planck Institute for Informatics, Saarland Informatics Campus, Saarbrücken, Germany
rsharma@mpi-inf.mpg.de

Abstract. An edge-colored graph is said to be *balanced* if it has an equal number of edges of each color. Given a graph G whose edges are colored using two colors and a positive integer k, the objective in the EDGE BALANCED CONNECTED SUBGRAPH problem is to determine if G has a balanced connected subgraph containing at least k edges. We first show that this problem is NP-complete and remains so even if the solution is required to be a tree or a path. Then, we focus on the parameterized complexity of EDGE BALANCED CONNECTED SUBGRAPH and its variants (where the balanced subgraph is required to be a path/tree) with respect to k as the parameter. Towards this, we show that if a graph has a balanced connected subgraph/tree/path of size at least k, then it has one of size at least k and at most $f(k)$ where f is a linear function. We use this result combined with dynamic programming algorithms based on *color coding* and *representative sets* to show that EDGE BALANCED CONNECTED SUBGRAPH and its variants are FPT. Further, using polynomial-time reductions to the MULTILINEAR MONOMIAL DETECTION problem, we give faster randomized FPT algorithms for the problems. In order to describe these reductions, we define a combinatorial object called *relaxed-subgraph*. We define this object in such a way that balanced connected subgraphs, trees and paths are relaxed-subgraphs with certain properties. This object is defined in the spirit of branching walks known for the STEINER TREE problem and may be of independent interest.

Keywords: Edge-colored graphs · Balanced subgraphs · Parameterized complexity

1 Introduction

Ramsey Theory is a branch of Combinatorics that deals with patterns in large arbitrary structures. In the context of edge-colored graphs where each edge is colored with one color from a finite set of colors, a fundamental problem in the area

L. Gąsieniec (Ed.): SOFSEM 2023, LNCS 13878, pp. 177–191, 2023.
https://doi.org/10.1007/978-3-031-23101-8_12

is concerned with the existence of *monochromatic* subgraphs of a specific type. Here, monochromatic means that all edges of the subgraph have the same color. For simplicity, we discuss only undirected graphs where each edge is colored either red or blue. Such a coloring is called a *red-blue coloring* and a graph associated with a red-blue coloring is referred to as a *red-blue graph*. In this work, we study questions related to the existence of and finding *balanced* subgraphs instead of monochromatic subgraphs, where by a balanced subgraph we mean one which has an equal number of edges of each color. These problems come under a subarea of Ramsey Theory known as Zero-Sum Ramsey Theory. Here, given a graph whose vertices/edges are assigned weights from a set of integers, one looks for conditions that guarantee the existence of a certain subgraph having total weight zero. For example, one may ask when is a graph whose all the edges are given weight -1 or 1 guaranteed to have a spanning tree with total weight of its edges 0. This is equivalent to asking when a red-blue graph is guaranteed to have a balanced spanning tree. Necessary and sufficient conditions have been established for complete graphs, triangle-free graphs and maximal planar graphs [7]. In the same spirit, one may ask a more general question like when is a red-blue graph G guaranteed to have a balanced connected subgraph of size (number of edges) k. An easy necessary condition is that there are at least $k/2$ red edges and at least $k/2$ blue edges in G. This condition is also sufficient (as we show in the proof of Theorem 3) if G is a complete graph (or more generally a split graph). However, we do not think that such a simple characterization will exist for all graphs. This brings us to the following natural algorithmic question concerning balanced connected subgraphs.

EDGE BALANCED CONNECTED SUBGRAPH   **Parameter:** k
Input: A red-blue graph G and a positive integer k
Question: Does G have a balanced connected subgraph of size at least k?

When the subgraph is required to be a tree or a path, the corresponding variants of EDGE BALANCED CONNECTED SUBGRAPH are called EDGE BALANCED TREE and EDGE BALANCED PATH, respectively. We show that these problems are NP-complete.

– (Theorems 1, 2, 4) EDGE BALANCED CONNECTED SUBGRAPH, EDGE BALANCED TREE and EDGE BALANCED PATH are NP-complete.

In fact, EDGE BALANCED CONNECTED SUBGRAPH and EDGE BALANCED TREE remain NP-complete on bipartite graphs, planar graphs and chordal graphs. However, EDGE BALANCED CONNECTED SUBGRAPH is polynomial-time solvable on split graphs (Theorem 3). Yet, EDGE BALANCED PATH is NP-complete even on split graphs.

Note that if a graph has a balanced connected subgraph/tree/path of size at least k, then it is not guaranteed that it has one of size equal to k. This brings us to the following combinatorial question: if a graph has a balanced connected subgraph/tree/path of size at least k, then can we show that it has a balanced connected subgraph/tree/path of size equal to $f(k)$ for some function

f? We answer these questions in the affirmative and show the existence of such a function which is linear in k.

- (Theorems 5, 6, 7) If a graph has a balanced connected subgraph/tree of size at least k, then it has one of size at least k and at most $3k + 3$. Further, if a graph has a balanced path of size at least k, then it has a balanced path of size at least k and at most $2k$.

Therefore, in order to find a balanced connected subgraph/tree/path of size at least k, it suffices to focus on the problem of finding a balanced connected subgraph/tree/path of size exactly k. This leads us to the following problem.

EXACT EDGE BALANCED CONNECTED SUBGRAPH   **Parameter:** k
Input: A red-blue graph G and a positive integer k
Question: Does G have a balanced connected subgraph of size k?

As before, when the connected subgraph is required to be a tree or a path, the corresponding variants of EXACT EDGE BALANCED CONNECTED SUBGRAPH are called EXACT EDGE BALANCED TREE and EXACT EDGE BALANCED PATH, respectively. These problems are also NP-complete and so we study them from the perspective of parameterized complexity. In this framework, each instance is associated with a non-negative integer ℓ called *parameter*, and a problem is said to be *fixed-parameter tractable* (FPT) with respect to ℓ if it can be solved in $\mathcal{O}^*(f(\ell))^2$ time for some computable function f. Algorithms with such running times are called FPT algorithms or parameterized algorithms. Focusing on solution size k as the parameter, we give randomized FPT algorithms for solving the three problems using reductions to the MULTILINEAR MONOMIAL DETECTION problem.

- (Theorems 8, 9, 10) EXACT EDGE BALANCED CONNECTED SUBGRAPH/ TREE/PATH can be solved by a randomized algorithm in $\mathcal{O}^*(2^k)$ time.

Many problems reduce to MULTILINEAR MONOMIAL DETECTION [17] and the current fastest algorithm solving it is a randomized algorithm that runs in $\mathcal{O}^*(2^k)$ time [16,17,23]. The reductions that we give to MULTILINEAR MONO-MIAL DETECTION use a combinatorial object called *relaxed-subgraph*. This object is defined in the spirit of *branching walks* known for the STEINER TREE problem [20]. We define this object in such a way that balanced connected subgraphs, trees and paths are relaxed-subgraphs with certain properties.

Then, using the *color-coding* technique [1,8] and *representative sets* [8,12,21], we give deterministic dynamic programming algorithms for the problems.

- (Theorems 11, 12, 13) EXACT EDGE BALANCED CONNECTED SUBGRAPH/ TREE can be solved in $\mathcal{O}^*((4e)^k)$ time and EXACT EDGE BALANCED PATH can be solved in $\mathcal{O}^*(2.619^k)$ time.

The method of representative sets is a generic approach for designing efficient dynamic programming based parameterized algorithms that may be viewed as

---

^{2} $\mathcal{O}^*$ notation supresses polynomial (in the input size) terms.

a deterministic-analogue to the color-coding technique. Representative sets have been used to obtain algorithms for several parameterized problems [12] and our algorithm adds to this list.

Road Map. The NP-completeness of the problems are given in Sect. 2. In Sect. 3, the combinatorial results related to the existence of small balanced connected subgraphs, trees and paths are proven. Section 4 discusses the deterministic and randomized algorithms for the problems. Section 5 concludes the work by listing some future directions.

Related Work. A variant of EXACT EDGE BALANCED CONNECTED SUB-GRAPH has recently been studied [2,3,9,15,18]. In order to state these results using our terminology, we define the notion of *vertex-balanced subgraphs* of *vertex-colored graphs*. A coloring of the vertices of a graph using red and blue colors is called a *red-blue vertex coloring*. A subgraph of a vertex-colored graph is said to be *vertex-balanced* if it has an equal number of vertices of each color. [2] and [3] study the EXACT VERTEX BALANCED CONNECTED SUBGRAPH problem where the interest is in finding a vertex-balanced connected subgraph on k vertices in the given graph associated with a red-blue vertex coloring. This problem is NP-complete and remains so on restricted graph classes like bipartite graphs, planar graphs, chordal graphs, unit disk graphs, outer-string graphs, complete grid graphs, and unit square graphs [2,3]. However, polynomial-time algorithms are known for trees, interval graphs, split graphs, circular-arc graphs and permutation graphs [2,3]. Further, the problem is NP-complete even when the subgraph required is a path [2]. FPT algorithms, exact exponential algorithms and approximation results for the problem are known from [3], [15] and [18]. Observe that finding vertex-balanced connected subgraphs in vertex-colored graphs reduces to finding vertex-balanced trees while the analogous solution in edge-colored graphs may have more complex structures.

Preliminaries. For $k \in \mathbb{N}_+$, $[k]$ denotes the set $\{1, 2, ..., k\}$. In this work, we only consider simple undirected graphs. For standard graph-theoretic terminology not stated here, we refer the reader to the book by Diestel [10]. For the necessary parameterized complexity background, we refer to the book by Cygan et al. [8]. For a graph G, its sets of vertices and edges, are denoted by $V(G)$ and $E(G)$, respectively. The *size* of a graph is the number of its edges and the *order* of a graph is the number of its vertices. An edge between vertices u and v is denoted as $\{u, v\}$ and u and v are the *endpoints* of the edge $\{u, v\}$. Two vertices u, v in $V(G)$ are *adjacent* if there is an edge $\{u, v\}$ in G. The *neighborhood* of a vertex v, denoted by $N_G(v)$, is the set of vertices adjacent to v. Similarly, two edges e, e' in $E(G)$ are *adjacent* if they have exactly one common endpoint and the *neighborhood* of an edge e, denoted by $N_G(e)$, is the set of edges adjacent to e. The *degree* of a vertex v is the size of $N_G(v)$. A *tree* is an undirected connected acyclic graph. A *clique* is a set of pairwise adjacent vertices and a *complete graph* is a graph whose vertex set is a clique. A *split graph* is a graph whose vertex set can be partitioned into a clique and an independent set. Given a graph G, its *line graph* $L(G)$ is defined as $V(L(G)) = \{e \mid e \in E(G)\}$ and $E(L(G)) = \{\{e, e'\} \mid e$ and e' are adjacent$\}$. It is well-known that a graph G without isolated vertices is connected if and only if $L(G)$ is connected.

Due to space constraints, for results labelled with a [⋆], proofs are omitted or only a proof sketch is given.

2 NP-hardness Results

We show the NP-hardness of EDGE BALANCED CONNECTED SUBGRAPH using a polynomial-time reduction from the STEINER TREE problem [14, ND12]. In this problem, given a connected graph G, a subset $T \subseteq V(G)$ (called *terminals*) and a positive integer k, the task is to determine if G has a subtree H (called a *Steiner tree*) with $T \subseteq V(H)$ and $|E(T)| \leq k$. The idea behind the reduction is to color all edges of G of the STEINER TREE instance blue and add exactly k red edges incident to the terminals such that each terminal has at least one red edge incident on it. Any balanced connected subgraph of size (at least) k of the resulting graph is required to include all the red edges and hence includes all the terminals which in turn corresponds to a Steiner tree of G.

Theorem 1. [⋆] EDGE BALANCED CONNECTED SUBGRAPH *is* NP-*complete.*

As the variant of the STEINER TREE problem where a tree on exactly k edges is required is also NP-complete, we have the following result.

Theorem 2. EDGE BALANCED TREE *is* NP-*complete.*

The reduction described in Theorem 1 is a polynomial parameter transformation and hence the infeasibility of the existence of polynomial kernels for STEINER TREE parameterized by the solution size (i.e., the size k of the tree) [8, 11] extends to EDGE BALANCED CONNECTED SUBGRAPH and EDGE BALANCED TREE as well. Further, since STEINER TREE has no $2^{o(k)}$ time algorithm assuming the Exponential Time Hypothesis, it follows that EDGE BALANCED CONNECTED SUBGRAPH and EDGE BALANCED TREE also do not admit subexponential FPT algorithms. Moreover, the reduction in Theorem 1 preserves planarity, bipartiteness and chordality. This property along with the NP-completeness of STEINER TREE (and its variant) on bipartite graphs [14], planar graphs [13] and chordal graphs [22] imply that EDGE BALANCED CONNECTED SUBGRAPH and EDGE BALANCED TREE are NP-complete on planar graphs, chordal graphs and bipartite graphs as well.

2.1 Complexity in Split Graphs

Next, we consider EDGE BALANCED CONNECTED SUBGRAPH on split graphs. Let (G, k) be an instance. An easy necessary condition for G to have a balanced connected subgraph of size (at least) k is that there are at least $k/2$ red edges and at least $k/2$ blue edges in G. We show that this condition is also sufficient if G is a split graph leading to the following result.

Theorem 3. [⋆] EDGE BALANCED CONNECTED SUBGRAPH *is polynomial-time solvable on split graphs.*

Now, we move on to EDGE BALANCED PATH which we show is NP-hard on split graphs by giving a polynomial-time reduction from LONGEST PATH. In the LONGEST PATH problem, given a graph G and a positive integer k, the task is to find a path P in G of length k. It is known that LONGEST PATH is NP-hard [14, ND29] and remains so on split graphs even when the starting vertex u_0 of the path is given as part of the input [14, GT39]. The reduction may be viewed as attaching a red path of length k (consisting of new internal vertices) starting from u_0 to the split graph G (whose edges are colored blue) of the LONGEST PATH instance and adding certain additional edges (colored blue) to ensure that the graph remains a split graph.

Theorem 4. [⋆] EDGE BALANCED PATH *is* NP-*complete on split graphs.*

As LONGEST PATH parameterized by the solution size (i.e., the size k of the path) in general graphs does not admit a polynomial kernel [4,8] and the reduction described (which is adaptable for general graphs) is a polynomial parameter transformation, it follows that EDGE BALANCED PATH does not admit polynomial kernels. Further, it is known that, assuming the Exponential Time Hypothesis, LONGEST PATH has no $2^{o(k)}$ time algorithm in general graphs. Hence, EDGE BALANCED PATH also does not admit subexponential FPT algorithms.

3   Small Balanced Paths, Trees and Connected Subgraphs

In this section, we prove the combinatorial result that if a graph has a balanced connected subgraph/tree/path of size at least k, then it has one of size at least k and at most $f(k)$ where f is a linear function. We begin with balanced paths.

Theorem 5. [⋆] *Let G be a red-blue graph and $k \geq 2$ be a positive integer. Then, if G has a balanced path of length at least $2k$, then G has a smaller balanced path of length at least k.*

Proof. **(sketch)** Let E_B be the set of blue edges and E_R be the set of red edges in G. Consider a balanced path P in G with at least $2k$ edges. If the terminal edges e and e' are of different colors, then delete e and e' to get a smaller path of length at least k. Otherwise, let $P = (v_1, v_2, \ldots, v_\ell)$ where e_i denotes the edge $\{v_i, v_{i+1}\}$ for each $i \in [\ell-1]$. Without loss of generality, let $e_1, e_{\ell-1} \in E_R$. Define the function $h : E(P) \to \mathbb{N}$ as follows.

$$h(e_i) = \begin{cases} 1, & \text{if } i = 1 \\ h(e_{i-1}) + 1, & \text{if } i > 1 \text{ and } e_i \in E_R \\ h(e_{i-1}) - 1, & \text{if } i > 1 \text{ and } e_i \in E_B \end{cases}$$

We show that there is an edge e_i with $i < \ell - 1$ and $h(e_i) = 0$. Then, the subpaths P_1 and P_2 with $E(P_1) = \{e_1, \ldots, e_i\}$ and $E(P_2) = \{e_{i+1}, \ldots, e_{\ell-1}\}$ are two balanced paths strictly smaller than P. Further, as $|E(P)| \geq 2k$, at least one of them has at least k edges.   □

Now, we move to the analogous result for balanced trees. An edge with an endpoint that has degree 1 is called a *pendant edge*.

Theorem 6. [⋆] *Let G be a red-blue graph and $k \geq 2$ be a positive integer. Then, if G has a balanced tree with at least $3k + 2$ edges, then G has a smaller balanced tree with at least k edges.*

Proof. **(sketch)** Let E_B be the set of blue edges and E_R be the set of red edges in G. Consider a balanced tree T in G with at least $3k+2$ edges. If T is a path, then by Theorem 5, we obtain the desired smaller tree (path). If T has pendant edges e and e' of different colors, then delete e and e' to get a smaller tree on at least k edges. Otherwise, without loss of generality, let all pendant edges of T be in E_R. Let n denote $|V(T)|$. Root T at an arbitrary vertex of degree at least 3. For a vertex $v \in V(T)$, let T_v denote the subtree of T rooted at v. Let u be a vertex with maximum distance from the root such that $|V(T_u)| > \frac{n}{3}$. Let $u_1, \ldots, u_\ell$ be the children of u. Observe that for each $i \in [\ell]$, $|V(T_{u_i})| \leq \frac{n}{3}$. Let i be the least integer in $[\ell]$ such that $\frac{n}{3} \leq | \bigcup_{1 \leq j \leq i} V(T_{u_j})| \leq \frac{2n}{3}$. Let S denote $\bigcup_{1 \leq j \leq i} V(T_{u_j})$ and R denote $V(T) \setminus S$. As $\frac{n}{3} \leq |S| \leq \frac{2n}{3}$, we have $\frac{n}{3} \leq |R| \leq \frac{2n}{3}$. Further, since $n \geq 3k+3$, we have $\frac{n}{3} \geq k+1$ and $\frac{2n}{3} \leq 2k+2$. Hence, $k+1 \leq |S|, |R| \leq 2k+2$. If $T[S \cup \{u\}]$ or $T[R]$ is balanced, then we get the desired result. Otherwise, consider the case when $T[S \cup \{u\}]$ and $T[R]$ have lesser edges from E_R than from E_B. As $E(T[S \cup \{u\}])$ and $E(T[R])$ partition $E(T)$, this case implies that T has more edges from E_B than from E_R contradicting that T is balanced. The remaining case is when $T[S \cup \{u\}]$ or $T[R]$ has more edges from E_R than from E_B. Suppose $T[R]$ has more edges from E_R than from E_B. Initialize T^* to be $T[R]$. As $T[R]$ has at least $k+1$ vertices (and therefore at least k edges), it has at least $k/2$ edges from E_R. Add the edges of $T[S \cup \{u\}]$ to T^* in the breadth-first order until T^* becomes balanced. We show that $T \neq T^*$ leading to the desired result.   □

Finally, we prove the result for balanced connected subgraphs using line graphs, vertex-balanced subgraphs and vertex-balanced trees. For a red-blue graph G, we define a red-blue coloring on $V(L(G))$ as follows: for each vertex x in $L(G)$ corresponding to a red (blue) edge $\{u, v\}$ in G, color x using red (blue). Then, G has a balanced connected subgraph of size ℓ if and only if $L(G)$ has a vertex-balanced connected subgraph of order ℓ. Now, it remains to show that if $L(G)$ has a vertex-balanced connected subgraph (equivalently, a vertex-balanced tree T) with at least $3k + 3$ vertices, then $L(G)$ has a smaller vertex-balanced connected subgraph (equivalently, a vertex-balanced tree T^*) with at least k vertices. The proof of this claim is similar to the proof of Theorem 6.

Theorem 7. [⋆] *Let G be a red-blue graph and $k \geq 2$ be a positive integer. Then, if G has a balanced connected subgraph with at least $3k + 3$ edges, then G has a smaller balanced connected subgraph with at least k edges.*

Due to Theorems 5, 6 and 7, it suffices to give FPT algorithms for EXACT EDGE BALANCED CONNECTED SUBGRAPH/TREE/PATH in order to obtain FPT algorithms for EDGE BALANCED CONNECTED SUBGRAPH/TREE/PATH.

4   FPT Algorithms

We now describe parameterized algorithms for EXACT EDGE BALANCED CON-
NECTED SUBGRAPH/TREE/PATH.

4.1   Randomized Algorithms

In this section, we show that EXACT EDGE BALANCED CONNECTED SUBGRAPH,
EXACT EDGE BALANCED PATH and EXACT EDGE BALANCED TREE admit ran-
domized algorithms that runs in $\mathcal{O}^*(2^k)$ time. We do so by reducing the problems
to MULTILINEAR MONOMIAL DETECTION. In order to define this problem, we
state some terminology related to polynomials from [17]. Let X denote a set of
variables. A *monomial* of degree d is a product of d variables from X, with multiplication assumed to be commutative. A monomial is called *multilinear* if no
variable appears twice or more in the product. A *polynomial* $P(X)$ over a ring is
a linear combination of monomials with coefficients from the ring. A polynomial
contains a certain monomial if the monomial appears with a non-zero coefficient
in the linear combination that constitutes the polynomial. Polynomials can be
represented as *arithmetic circuits* which in turn represented as *directed acyclic
graphs*. In the MULTILINEAR MONOMIAL DETECTION problem, given an arithmetic circuit (represented as a directed acyclic graph) representing a polynomial
$P(X)$ over $\mathbb{Z}_+$ and a positive integer k, the task is to decide whether $P(X)$
contains a multilinear monomial of degree at most k.

Proposition 1. [16,17,23] *Let $P(X)$ be a polynomial over $\mathbb{Z}_+$ represented by
a circuit. The MULTILINEAR MONOMIAL DETECTION problem for $P(X)$ can be
decided in randomized $\mathcal{O}^*(2^k)$ time and polynomial space.*

The reductions from EXACT EDGE BALANCED CONNECTED SUBGRAPH/
TREE/PATH to MULTILINEAR MONOMIAL DETECTION crucially use the notions
of a *color-preserving homomorphism* (also known as an edge-colored homomorphism in the literature [6]) and *relaxed-subgraphs*.

Definition 1. *Given graphs G and H with red-blue edge colorings col_G :
$E(G) \to \{red, blue\}$ and $col_H : E(H) \to \{red, blue\}$, a color-preserving homomorphism from H to G is a function $h : V(H) \to V(G)$ satisfying the following
properties: (1) For each pair $u, v \in V(H)$, if $\{u, v\} \in E(H)$, then $\{h(u), h(v)\} \in
E(G)$. (2) For each edge $\{u, v\}$ in H, $col_H(\{u, v\}) = col_G(\{h(u), h(v)\})$.*

Definition 2. *Given a red-blue graph G, a relaxed-subgraph is a pair $S = (H, h)$
where H is a red-blue graph and h is a color-preserving homomorphism from H
to G.*

The vertex set of a relaxed-subgraph $S = (H, h)$ is $V(S) = \{h(a) \in V(G) \mid
a \in V(H)\}$ and the edge set of S is $\{\{h(a), h(b)\} \in E(G) \mid \{a, b\} \in E(H)\}$.
We treat the vertex and edge sets of a relaxed-subgraph as multi-sets. The size
of S is the number of edges in H (equivalently, the size of $E(S)$). S is said to

be connected if H is connected and S is said to be balanced if H has an equal number of red edges and blue edges. S is said to be a *relaxed-path* if H is a path and a *relaxed-tree* if H is a tree. Next, we have the following observation that states that relaxed-subgraphs with certain specific properties correspond to balanced connected subgraphs, trees and paths.

Observation 1. [⋆] *The following hold for a red-blue graph G.*

- *G has a balanced connected subgraph of size k if and only if there is a balanced connected relaxed-subgraph S of size k such that $E(S)$ consists of distinct elements.*
- *G has a balanced path of size k if and only if there is a balanced relaxed-path (P, h) of size k where h is injective.*
- *G has a balanced tree of size k if and only if there is a balanced relaxed-tree (T, h) of size k where h is injective.*

Now, we are ready to describe the randomized algorithms for EXACT EDGE BALANCED CONNECTED SUBGRAPH/TREE/PATH based on Observation 1. First, we consider EXACT EDGE BALANCED CONNECTED SUBGRAPH.

Theorem 8. [⋆] EXACT EDGE BALANCED CONNECTED SUBGRAPH *admits a randomized $\mathcal{O}^*(2^k)$-time algorithm.*

Proof. **(sketch)** Consider an instance (G, k). Let E_R denote the set of red edges and E_B denote the set of blue edges in G. In order to obtain an instance of MULTILINEAR MONOMIAL DETECTION that is equivalent to (G, k), we will define a polynomial P over the variable set $\{x_e : e \in E(G)\}$ satisfying the following properties: (1) For each balanced connected relaxed-subgraph $S = (H, h)$ of size k there exists a monomial in P that corresponds to S. We say that a monomial M corresponds to S, if $M = \prod_{e \in E(S)} x_e$. (2) Each multilinear monomial in P corresponds to some balanced connected relaxed-subgraph S of size k where $E(S)$ has distinct elements.

If P is such a polynomial, then from Observation 1, G has a balanced connected subgraph of size k if and only if P has a multilinear monomial of degree k. This way, after the construction of P, we reduce the problem to MULTILINEAR MONOMIAL DETECTION and use Proposition 1. In order to construct P, we first construct polynomials $P_j(e, r, b)$ for each $e \in E(G)$, $j \in [k]$ and $0 \le r, b \le \frac{k}{2}$ with $r + b \ge 1$. Monomials of $P_j(e, r, b)$ will correspond to connected relaxed-subgraphs $S = (H, h)$ of size j such that H has r red edges, b blue edges and $e \in E(S)$. The construction of $P_j(e, r, b)$ is as follows. For an edge $e \in E(G)$,

$$P_1(e, 1, 0) = \begin{cases} x_e, & \text{if } e \in E_R \\ 0 & \text{otherwise.} \end{cases} \text{ and } P_1(e, 0, 1) = \begin{cases} x_e, & \text{if } e \in E_B \\ 0 & \text{otherwise.} \end{cases}$$

Also, $P_j(e, r, b) = 0$ if $j \neq r + b$. Now, if $e \in E_R$ and $r + b > 1$, then we have

$$P_j(e, r, b) = \sum_{\substack{e' \in N_G(e), \ell < j \\ r' + r'' = r, b' + b'' = b}} P_\ell(e', r', b') P_{j-\ell}(e, r'', b'') + \sum_{e' \in N_G(e)} x_e P_{j-1}(e', r-1, b).$$

If $e \in E_B$ and $r+b > 1$, then $P_j(e,r,b) = \displaystyle\sum_{\substack{e' \in N_G(e), \ell < j \\ r'+r''=r, b'+b''=b}} P_\ell(e', r', b') P_{j-\ell}(e, r'', b'')$

$+ \displaystyle\sum_{e' \in N_G(e)} x_e P_{j-1}(e', r, b-1).$

We now show that every multilinear monomial of $P_j(e,r,b)$ corresponds to a connected relaxed-subgraph $S = (H, h)$ of size j such that H has r red edges and b blue edges while $E(S)$ consists of distinct elements with $e \in E(S)$. We prove this claim by induction on j. The base case is easy to verify. Consider the induction step. Suppose $e = \{u, v\} \in E_R$ (the other case is symmetric). Let M be a multilinear monomial of $P_j(e,r,b)$ where $j > 1$.

Case 1: $M = x_e M'$ where M' is a multilinear monomial of $P_{j-1}(e', r-1, b)$ such that $e' \in N_G(e)$. Let $e' = \{v, w\}$. By induction, M' corresponds to a connected relaxed-subgraph $S' = (H', h')$ of size $j-1$ such that $e' \in E(S')$ and H' has $r-1$ red edges, b blue edges with $v \in V(h'(H'))$. Let $z = h'^{-1}(v)$ (well-defined due to the multilinearity of M). Note that $E(S')$ consists of distinct elements and $e \notin E(S')$. Let H denote the graph obtained from H' by adding a new vertex z' adjacent to z with the edge $\{z, z'\}$ colored red. Let $h : V(H) \to V(G)$ denote the homomorphism obtained from h' by extending its domain to include z' and setting $h(z') = u$. Then, $S = (H, h)$ is a connected relaxed-subgraph that M corresponds to.

Case 2: $M = M_1 M_2$ where M_1 is a multilinear monomial of $P_{j_1}(e', r', b')$ and M_2 is a multilinear monomial of $P_{j_2}(e, r'', b'')$ such that $e' \in N_G(e)$, $j_1, j_2 < j$, $r'' \leq r$ and $b'' \leq b$. Let $e' = \{v, w\}$. By induction, M_1 corresponds to a connected relaxed-subgraph $S_1 = (H_1, h_1)$ of size j_1 such that $e' \in E(S_1)$. Similarly, M_2 corresponds to a connected relaxed-subgraph $S_2 = (H_2, h_2)$ of size j_2 such that $e \in E(S_2)$. Further, H_1 has r' red edges, b' blue edges and H_2 has r'' red edges and b'' blue edges. Also, $v \in V(h_1(H_1)) \cap V(h_2(H_2))$ and $E(S_1) \cap E(S_2) = \emptyset$. Without loss of generality, assume that $V(H_1) \cap V(H_2) = \emptyset$ as this can be achieved by a renaming procedure. Let $z_1 = h_1^{-1}(v)$ and $z_2 = h_2^{-1}(v)$. Observe that z_1 and z_2 are well-defined due to the multilinearity of M_1 and M_2. Now, rename z_1 in S_1 and z_2 in S_2 as z. Let H denote the graph with vertex set $V(H_1) \cup V(H_2)$ and edge set $E(H_1) \cup E(H_2)$. Observe that H is a connected graph. Let $h : V(H) \to V(G)$ denote identity map. Then, $S = (H, h)$ is a connected relaxed-subgraph that M corresponds to.

Similarly, we show (by induction on j) that if there is a connected relaxed-subgraph $S = (H, h)$ of size j with r red edges, b blue edges and such that $e = \{u, v\} \in E(S)$, then there is a monomial of $P_j(e, r, b)$ that corresponds to it. Suppose $e \in E_R$ (the other case is symmetric). The base case is trivial. Consider the induction step ($j \geq 2$). Let $a = h^{-1}(u)$, $b = h^{-1}(v)$ and z denote the edge $\{a, b\}$ of H.

Case 1: $H - z$ is connected. Then, $S' = (H - z, h)$ is a connected relaxed-subgraph of size $j-1$ with $r-1$ red edges and b blue edges and contains an edge $e' \in N_G(e)$. By induction, there is a monomial M' corresponding to S' in $P_{j-1}(e', r-1, b)$. Then, the monomial $M = x_e M'$ which is in $P_j(e, r, b)$ corresponds to S.

Case 2: $H-z$ is disconnected. Then H has two components H_a (containing a) and H_b (containing b). Without loss of generality let H_a have at least one edge. Let H'_b denote the subgraph of H obtained from H_b by adding the vertex a and edge $\{a,b\}$. Let j_1 and j_2 denote the number of edges in H_a and H'_b, respectively. Let r' and r'' be the number of red edges in H_a and H'_b, respectively. Similarly, let b' and b'' be the number of blue edges in H_a and H'_b, respectively. Then $j = j_1 + j_2, r = r' + r'', b = b' + b''$. Let h_a and h_b denote the color-preserving homomorphism obtained from h by restricting the domain to $V(H_a)$ and $V(H_b)$, respectively. Now, considering the connected relaxed-subgraphs $S_1 = (H_a, h_a)$ and $S_2 = (H_b, h_b)$, by induction there is a monomial M_1 in $P_{j_1}(e', r', b')$ that corresponds to S_1 and there is a monomial M_2 in $P_{j_2}(e, r'', b'')$ that corresponds to S_2. Then, the monomial $M_1 M_2$ which is in $P_j(e, r, b)$ corresponds to S. Finally, let $P = \sum_{e \in E(G)} P_k(e, \frac{k}{2}, \frac{k}{2})$. Every monomial in P has degree k. Then from the arguments above, P is the desired polynomial. As these polynomials can be represented as a polynomial-sized arithmetic circuit, the reduction runs in polynomial time. $\square$

Next, we move on to EXACT EDGE BALANCED TREE. We define a polynomial over the variable set corresponding to vertices (as opposed to variable set corresponding to edges as in EXACT EDGE BALANCED CONNECTED SUBGRAPH) leading to the following result.

Theorem 9. [⋆] EXACT EDGE BALANCED TREE *admits a randomized* $\mathcal{O}^*(2^k)$-*time algorithm.*

Proof. **(sketch)** We define a polynomial $P = \sum_{e \in E(G)} P_k(e, \frac{k}{2}, \frac{k}{2})$ over the variable set $\{y_v : v \in V(G)\}$ with subtle differences from the one defined for balanced connected subgraphs where $P_j(e, r, b)$ is defined as follows with $e = \{u, v\}$.

$$P_1(e, 1, 0) = \begin{cases} y_u y_v, & \text{if } e \in E_R \\ 0 & \text{otherwise.} \end{cases} \quad \text{and} \quad P_1(e, 0, 1) = \begin{cases} y_u y_v, & \text{if } e \in E_B \\ 0 & \text{otherwise.} \end{cases}$$

For $j > 1$, if $e \in E_R$ (the polynomial for $e \in E_B$ is similar),

$$P_j(e, r, b) = \sum_{\substack{e', e'' \in N_G(e) \\ u \in V(e'), v \in V(e'') \\ r' < r, b' \le b, \ell < j}} P_\ell(e', r', b') P_{j-1-\ell}(e'', r - 1 - r', b - b')$$

$$+ \sum_{\substack{e' \in N_G(e) \\ v \in V(e')}} y_u P_{j-1}(e', r - 1, b) + \sum_{\substack{e' \in N_G(e) \\ u \in V(e')}} y_v P_{j-1}(e', r - 1, b).$$

The properties of $P = \sum_{e \in E(G)} P_k(e, \frac{k}{2}, \frac{k}{2})$ then lead to the claimed result. $\square$

Finally, we define a (simpler) polynomial for EXACT EDGE BALANCED TREE satisfying certain properties leading to the following result.

Theorem 10. [⋆] EXACT EDGE BALANCED PATH *admits a randomized* $\mathcal{O}^*(2^k)$-*time algorithm.*

4.2 Deterministic Algorithms

We first describe deterministic algorithms for EXACT EDGE BALANCED CONNECTED SUBGRAPH and EXACT EDGE BALANCED TREE using the color-coding technique [1,8,19]. Consider an instance (G, k) of EXACT EDGE BALANCED CONNECTED SUBGRAPH/TREE. Let E_R denote the set of red edges and E_B denote the set of blue edges in G. Let $\sigma : E(G) \to [k]$ be a coloring of edges of G and $\tau : V(G) \to [k+1]$ be a coloring of vertices of G. For $L \subseteq [k+1]$, a subgraph $H \subseteq G$ is said to be L-edge-colorful if $|E(H)| = |L|$ and $\bigcup_{e \in E(H)} \sigma(e) = L$. Similarly, H is said to be L-vertex-colorful if $|V(H)| = |L|$ and $\bigcup_{v \in V(H)} \tau(v) = L$. We describe dynamic programming algorithms to find a $[k]$-edge-colorful balanced connected subgraph and a $[k+1]$-vertex-colorful balanced tree in G (if they exist) in $\mathcal{O}^*(4^k)$ time. Then, a standard derandomization technique using *perfect hash families* [1,8,19] leads to the following results.

Theorem 11. [⋆] EXACT EDGE BALANCED CONNECTED SUBGRAPH *can be solved in* $\mathcal{O}^*((4e)^k)$ *time.*

Theorem 12. [⋆] EXACT EDGE BALANCED TREE *can be solved in* $\mathcal{O}^*((4e)^k)$ *time.*

Analogous to Theorems 11 and 12, we can show that EXACT EDGE BALANCED PATH can be solved in $\mathcal{O}^*((2e)^k)$ time. Subsequently, we describe a faster algorithm using representative sets. We begin with some definitions and results related to representative sets. For a finite set U, let $\binom{U}{p}$ denote the set of all subsets of size p of U. Given two families $\mathcal{S}_1, \mathcal{S}_2 \subseteq 2^U$, the *convolution* of $\mathcal{S}_1$ and $\mathcal{S}_2$ is the new family defined as $\mathcal{S}_1 * \mathcal{S}_2 = \{X \cup Y \mid X \in \mathcal{S}_1, Y \in \mathcal{S}_2, X \cap Y = \emptyset\}$.

Definition 3. *Let U be a set and $\mathcal{S} \subseteq \binom{U}{p}$. A subfamily $\widehat{\mathcal{S}} \subseteq \mathcal{S}$ is said to be q-represent $\mathcal{S}$ (denoted as $\widehat{\mathcal{S}} \subseteq_{rep}^q \mathcal{S}$) if for every set $Y \subseteq U$ of size at most q such that there is a set $X \in \mathcal{S}$ with $X \cap Y = \emptyset$, there is a set $\widehat{X} \in \widehat{\mathcal{S}}$ with $\widehat{X} \cap Y = \emptyset$. If $\widehat{\mathcal{S}} \subseteq_{rep}^q \mathcal{S}$, then $\widehat{\mathcal{S}}$ is called a q-representative family for $\mathcal{S}$.*

Representative families (also called representative sets) are transitive and have nice union and convolution properties [8, Lemmas 12.26, 12.27 and 12.28]. A classical result due to Bollobás states that small representative families exist [5] and a result due to [12] and [21] (see also [8]) shows that such families can be efficiently computed.

Theorem 13. [⋆] EXACT EDGE BALANCED PATH *can be solved in* $\mathcal{O}^*(2.619^k)$ *time.*

Proof. (sketch) Consider an instance (G, k). Let E_R and E_B denote the sets of red and blue edges of G. For a pair of vertices $u, v \in V(G)$ and non-negative integers r and b with $r + b \geq 1$, define the family $\mathcal{P}_{uv}^{(r,b)}$ as follows.

$$\mathcal{P}_{uv}^{(r,b)} = \{X : X \subseteq V(G), |X| = r + b + 1 \text{ and there is a } uv\text{-path } P \text{ with}$$
$$V(P) = X, |E_R \cap E(P)| = r \text{ and } |E_B \cap E(P)| = b\}.$$

Now, it suffices to determine if $\mathcal{P}_{uv}^{(\frac{k}{2},\frac{k}{2})}$ is non-empty for some $u, v \in V(G)$. The families $\mathcal{P}_{uv}^{(r,b)}$ can be computed using the following formula. For $r + b = 1$,

$$\mathcal{P}_{uv}^{(1,0)} = \begin{cases} \{\{u,v\}\}, & \text{if } \{u,v\} \in E_R \\ \emptyset, & \text{otherwise} \end{cases} \quad \text{and} \quad \mathcal{P}_{uv}^{(0,1)} = \begin{cases} \{\{u,v\}\}, & \text{if } \{u,v\} \in E_B \\ \emptyset, & \text{otherwise} \end{cases}$$

For $r + b > 1$, $\mathcal{P}_{uv}^{(r,b)} = (\bigcup_{\{w,v\} \in E_R} (\mathcal{P}_{uw}^{(r-1,b)} * \{\{v\}\})) \bigcup (\bigcup_{\{w,v\} \in E_B} (\mathcal{P}_{uw}^{(r,b-1)} * \{\{v\}\}))$.

Clearly, a naive computation of $\mathcal{P}_{uv}^{(r,b)}$ is not guaranteed to result in an FPT (in k) algorithm. Therefore, instead of computing $\mathcal{P}_{uv}^{(r,b)}$, we compute $\widehat{\mathcal{P}}_{uv}^{(r,b)} \subseteq_{rep}^{k-(r+b)}$ $\mathcal{P}_{uv}^{(r,b)}$ and use the fact that $\widehat{\mathcal{P}}_{uv}^{(\frac{k}{2},\frac{k}{2})} \subseteq_{rep}^{0} \mathcal{P}_{uv}^{(\frac{k}{2},\frac{k}{2})}$. Now, we describe a dynamic programming algorithm to compute $\widehat{\mathcal{P}}_{uv}^{(\frac{k}{2},\frac{k}{2})}$ for every $u, v \in V(G)$. For $r + b = 1$, set $\widehat{\mathcal{P}}_{uv}^{(1,0)} = \mathcal{P}_{uv}^{(1,0)}$ and $\widehat{\mathcal{P}}_{uv}^{(0,1)} = \mathcal{P}_{uv}^{(0,1)}$. Clearly, $\widehat{\mathcal{P}}_{uv}^{(1,0)} \subseteq_{rep}^{k-1} \mathcal{P}_{uv}^{(1,0)}$ and $\widehat{\mathcal{P}}_{uv}^{(0,1)} \subseteq_{rep}^{k-1} \mathcal{P}_{uv}^{(0,1)}$. Further, $|\widehat{\mathcal{P}}_{uv}^{(1,0)}|, |\widehat{\mathcal{P}}_{uv}^{(0,1)}| \leq 1$ and this computation is polynomial time. Now, we proceed to computing $\widehat{\mathcal{P}}_{uv}^{(r,b)} \subseteq_{rep}^{k-(r+b)} \mathcal{P}_{uv}^{(r,b)}$ for $k \geq r+b > 1$ in the increasing order of $r + b$. Towards this, we compute a new family $\widetilde{\mathcal{P}}_{uv}^{(r,b)}$ as follows.

$$\widetilde{\mathcal{P}}_{uv}^{(r,b)} = (\bigcup_{\{w,v\} \in E_R} (\widehat{\mathcal{P}}_{uw}^{(r-1,b)} * \{\{v\}\})) \bigcup (\bigcup_{\{w,v\} \in E_B} \widehat{\mathcal{P}}_{uw}^{(r,b-1)} * \{\{v\}\})$$

Using the union and convolution properties of representative sets, $\widetilde{\mathcal{P}}_{uv}^{(r,b)} \subseteq_{rep}^{k-(r+b)}$ $\mathcal{P}_{uv}^{(r,b)}$. Further, $|\widetilde{\mathcal{P}}_{uv}^{(r,b)}| = \mathcal{O}^*(|\widehat{\mathcal{P}}_{uv}^{(r-1,b)}| + |\widehat{\mathcal{P}}_{uv}^{(r,b-1)}|)$. Then, we use the result of [12] and [21] to compute a family $\widehat{\mathcal{P}}_{uv}^{(r,b)} \subseteq_{rep}^{k-(r+b)} \widetilde{\mathcal{P}}_{uv}^{(r,b)}$. By the transitivity property of representative sets, it follows that $\widehat{\mathcal{P}}_{uv}^{(r,b)} \subseteq_{rep}^{k-(r+b)} \mathcal{P}_{uv}^{(r,b)}$. Further, from the running time analysis given in [12] and [21], the overall running time of the algorithm is $\mathcal{O}^*(2.619^k)$.   □

5   Concluding Remarks

To summarize our work, we studied the complexity of finding balanced connected subgraphs, trees and paths in red-blue graphs. We gave fixed-parameter tractability results using color-coding, representative sets and reductions to MULTILINEAR MONOMIAL DETECTION. En route, we showed combinatorial results on the existence of small balanced connected subgraphs, trees and paths. We observe that these results also extend to vertex-balanced connected subgraphs, trees and paths. As a result the algorithms described in this work also generalize to solve the vertex-analogue of the problems. Note that using line graphs, one can reduce EDGE BALANCED CONNECTED SUBGRAPH to VERTEX BALANCED CONNECTED SUBGRAPH, however, when the solution is required to be a path or a tree, this reduction is not useful. An interesting next direction of research is determining the complexity of finding other balanced substructures. Also, studying the problems on graphs that are colored using more than two colors and on colored weighted graphs are natural questions in this context.

References

1. Alon, N., Yuster, R., Zwick, U.: Color-coding. J. ACM **42**(4), 844–856 (1995). https://doi.org/10.1145/210332.210337. https://doi.org/10.1145/210332.210337
2. Bhore, S., Chakraborty, S., Jana, S., Mitchell, J.S.B., Pandit, S., Roy, S.: The balanced connected subgraph problem. In: Pal, S.P., Vijayakumar, A. (eds.) CALDAM 2019. LNCS, vol. 11394, pp. 201–215. Springer, Cham (2019). https://doi.org/10.1007/978-3-030-11509-8_17
3. Bhore, S., Jana, S., Pandit, S., Roy, S.: Balanced connected subgraph problem in geometric intersection graphs. In: Li, Y., Cardei, M., Huang, Y. (eds.) COCOA 2019. LNCS, vol. 11949, pp. 56–68. Springer, Cham (2019). https://doi.org/10.1007/978-3-030-36412-0_5
4. Bodlaender, H.L., Downey, R.G., Fellows, M.R., Hermelin, D.: On problems without polynomial kernels. J. Comput. Syst. Sci. **75**(8), 423–434 (2009). https://doi.org/10.1016/j.jcss.2009.04.001. https://doi.org/10.1016/j.jcss.2009.04.001
5. Bollobás, B.: On generalized graphs. Acta Math. Hungar. **16**(3–4), 447–452 (1965)
6. Brewster, R.C., Dedic, R., Huard, F., Queen, J.: The recognition of bound quivers using edge-coloured homomorphisms. Discret. Math. **297**(1-3), 13–25 (2005). https://doi.org/10.1016/j.disc.2004.10.026. https://doi.org/10.1016/j.disc.2004.10.026
7. Caro, Y., Hansberg, A., Lauri, J., Zarb, C.: On zero-sum spanning trees and zero-sum connectivity. Electron. J. Comb. **29**(1), P1.9 (2022). https://doi.org/10.37236/10289. https://doi.org/10.37236/10289
8. Cygan, M., et al.: Parameterized Algorithms. Springer, Cham (2015). https://doi.org/10.1007/978-3-319-21275-3
9. Darties, B., Giroudeau, R., Jean-Claude, K., Pollet, V.: The balanced connected subgraph problem: complexity results in bounded-degree and bounded-diameter graphs. In: Li, Y., Cardei, M., Huang, Y. (eds.) COCOA 2019. LNCS, vol. 11949, pp. 449–460. Springer, Cham (2019). https://doi.org/10.1007/978-3-030-36412-0_36
10. Diestel, R.: Graph Theory. GTM, vol. 173. Springer, Heidelberg (2017). https://doi.org/10.1007/978-3-662-53622-3
11. Dom, M., Lokshtanov, D., Saurabh, S.: Kernelization lower bounds through colors and ids. ACM Trans. Algorithms **11**(2), 1–20 (2014). https://doi.org/10.1145/2650261. https://doi.org/10.1145/2650261
12. Fomin, F.V., Lokshtanov, D., Panolan, F., Saurabh, S.: Efficient computation of representative families with applications in parameterized and exact algorithms. J. ACM **63**(4), 1–60 (2016). https://doi.org/10.1145/2886094. https://doi.org/10.1145/2886094
13. Garey, M.R., Johnson, D.S.: The rectilinear steiner tree problem in NP complete. J. SIAM Appl. Math. **32**, 826–834 (1977)
14. Garey, M.R., Johnson, D.S.: Computers and Intractability: a Guide to the Theory of NP-Completeness. W. H. Freeman (1979)
15. Kobayashi, Y., Kojima, K., Matsubara, N., Sone, T., Yamamoto, A.: Algorithms and hardness results for the maximum balanced connected subgraph problem. In: Li, Y., Cardei, M., Huang, Y. (eds.) COCOA 2019. LNCS, vol. 11949, pp. 303–315. Springer, Cham (2019). https://doi.org/10.1007/978-3-030-36412-0_24
16. Koutis, I.: Faster algebraic algorithms for path and packing problems. In: Aceto, L., Damgård, I., Goldberg, L.A., Halldórsson, M.M., Ingólfsdóttir, A., Walukiewicz, I. (eds.) ICALP 2008. LNCS, vol. 5125, pp. 575–586. Springer, Heidelberg (2008). https://doi.org/10.1007/978-3-540-70575-8_47

17. Koutis, I., Williams, R.: LIMITS and applications of group algebras for parameterized problems. ACM Trans. Algorithms **12**(3), 1–18 (2016). https://doi.org/10.1145/2885499. https://doi.org/10.1145/2885499

18. Martinod, T., Pollet, V., Darties, B., Giroudeau, R., König, J.: Complexity and inapproximability results for balanced connected subgraph problem. Theor. Comput. Sci. **886**, 69–83 (2021). https://doi.org/10.1016/j.tcs.2021.07.010. https://doi.org/10.1016/j.tcs.2021.07.010

19. Naor, M., Schulman, L.J., Srinivasan, A.: Splitters and near-optimal derandomization. In: 36th Annual Symposium on Foundations of Computer Science, Milwaukee, Wisconsin, USA, 23–25 October 1995, pp. 182–191 IEEE Computer Society (1995). https://doi.org/10.1109/SFCS.1995.492475. https://doi.org/10.1109/SFCS.1995.492475

20. Nederlof, J.: Fast polynomial-space algorithms using inclusion-exclusion. Algorithmica **65**(4), 868–884 (2013). https://doi.org/10.1007/s00453-012-9630-x. https://doi.org/10.1007/s00453-012-9630-x

21. Shachnai, H., Zehavi, M.: Representative families: a unified tradeoff-based approach. J. Comput. Syst. Sci. **82**(3), 488–502 (2016). https://doi.org/10.1016/j.jcss.2015.11.008. https://doi.org/10.1016/j.jcss.2015.11.008

22. White, K., Farber, M., Pulleyblank, W.R.: Steiner trees, connected domination and strongly chordal graphs. Networks **15**(1), 109–124 (1985). https://doi.org/10.1002/net.3230150109. https://doi.org/10.1002/net.3230150109

23. Williams, R.: Finding paths of length k in $O^*(2^k)$ time. Inf. Process. Lett. **109**(6), 315–318 (2009). https://doi.org/10.1016/j.ipl.2008.11.004. https://doi.org/10.1016/j.ipl.2008.11.004

On the Complexity of Scheduling Problems with a Fixed Number of Parallel Identical Machines

Klaus Jansen$^{(\boxtimes)}$ (ID) and Kai Kahler$^{(\boxtimes)}$ (ID)

Department of Computer Science, Kiel University, Kiel, Germany
{kj,kka}@informatik.uni-kiel.de

Abstract. In parallel machine scheduling, we are given a set of jobs, together with a number of machines and our goal is to decide for each job, when and on which machine(s) it should be scheduled in order to minimize some objective function. Different machine models, job characteristics and objective functions result in a multitude of scheduling problems and many of them are NP-hard, even for a fixed number of identical machines. In this work, we give conditional running time lower bounds for a large number of scheduling problems, indicating the optimality of some classical algorithms. Most notably, we show that the algorithm by Lawler and Moore for $1||\sum w_j U_j$ and $Pm||C_{\max}$, as well as the algorithm by Lee and Uzsoy for $P2||\sum w_j C_j$ are probably optimal. There is still small room for improvement for the $1|Rej \leq Q|\sum w_j U_j$ algorithm by Zhang et al., the algorithm for $1||\sum T_j$ by Lawler and the FPTAS for $1||\sum w_j U_j$ by Gens and Levner. We also give a lower bound for $P2|any|C_{\max}$ and improve the dynamic program by Du and Leung from $\mathcal{O}(nP^2)$ to $\mathcal{O}(nP)$, matching this new lower bound. Here, P is the sum of all processing times. The same idea also improves the algorithm for $P3|any|C_{\max}$ by Du and Leung from $\mathcal{O}(nP^5)$ to $\mathcal{O}(nP^2)$. While our results suggest the optimality of some classical algorithms, they also motivate future research in cases where the best known algorithms do not quite match the lower bounds.

Keywords: SETH · Subset sum · Scheduling · Fine-grained complexity · Pseudo-polynomial algorithms

1 Introduction

Consider the problem of working on multiple research papers. Each paper j has to go to some specific journal or conference and thus has a given due date d_j. Some papers might be more important than others, so each one has a weight w_j. In order to not get distracted, we may only work on one paper at a time and this work may not be interrupted. If a paper does not meet its due date, it is not important by how much it misses it; it is either late or on time. If it is late, we

Supported by the German Research Foundation (DFG) project JA 612/25-1.

must pay its weight w_j. In the literature, this problem is known as $1||\sum w_j U_j$ and it is one of Karp's original 21 NP-hard problems [16]. The naming of $1||\sum w_j U_j$ and the problems referred to in the abstract will become clear when we review the three-field notation by Graham et al. [10] in Sect. 2. Even when restricted to a fixed number of identical machines, many combinations of job characteristics and objective functions lead to NP-hard problems. For this reason, a lot of effort has been put towards finding either pseudo-polynomial exact or polynomial approximation algorithms. Sticking to our problem $1||\sum w_j U_j$, where we aim to minimize the weighted number of late jobs on a single machine, there are e.g. an $\mathcal{O}(nW)$ algorithm by Lawler and Moore [21] and an FPTAS by Gens and Levner [9]. Here, W is the sum of all weights w_j and n is the number of jobs.

In recent years, research regarding scheduling has made its way towards parameterized and fine-grained complexity (see e.g. [2,12,18,26,27]), where one goal is to identify parameters that make a problem difficult to solve. If those parameters are assumed to be small, parameterized algorithms can be very efficient. Similarly, one may consider parameters like the total processing time P and examine how fast algorithms can be in terms of these parameters, while maintaining a sub-exponential dependency on n. That is our main goal in this work. Most of our lower bounds follow from a lower bound for SUBSET SUM:

Problem 1. SUBSET SUM

Instance: Items $a_1,\ldots,a_n \in \mathbb{N}$, integer target $T \in \mathbb{N}$.
Task: Decide whether there is a subset $S \subseteq [n]$ such that $\sum_{i \in S} a_i = T$.

Fine-grained running time lower bounds are often based on the Exponential Time Hypothesis (ETH) or the Strong Exponential Time Hypothesis (SETH). Intuitively, the ETH conjectures that 3-SAT cannot be solved in sub-exponential time and the SETH conjectures that the trivial running time of $\mathcal{O}(2^n)$ is optimal for k-SAT, if k tends to infinity. For details, see the original publication by Impagliazzo and Paturi [13]. A few years ago, Abboud et al. gave a beautiful reduction from k-SAT to SUBSET SUM [1]. Previous results based on the ETH excluded $2^{o(n)}T^{o(1)}$-time algorithms [15], while this new result based on the SETH suggests that we cannot even achieve $\mathcal{O}(2^{\delta n}T^{1-\varepsilon})$:

Theorem 1 (Abboud et al. [1]). *For every $\varepsilon > 0$, there is a $\delta > 0$ such that* SUBSET SUM *cannot be solved in time $\mathcal{O}(2^{\delta n}T^{1-\varepsilon})$, unless the SETH fails.*[1]

By revisiting many classical reductions in the context of fine-grained complexity, we transfer this lower bound to scheduling problems like $1||\sum w_j U_j$. Although lower bounds do not have the immediate practical value of an algorithm, it is clear from the results of this paper how finding new lower bounds can push research into the right direction: Our lower bound for the scheduling problem $P2|any|C_{\max}$ indicated the possibility of an $\mathcal{O}(nP)$-time algorithm, but the best known algorithm (by Du and Leung [8]) had running time $\mathcal{O}(nP^2)$. A modification of this algorithm closes this gap.

It should be noted that all lower bounds in this paper are conditional, that is, they rely on some complexity assumption. However, all of these assumptions

---

[1] Though it might seem unintuitive at first, it is not required that $\varepsilon < 1$.

are reasonable in the sense that a lot of effort has been put towards refuting them. And in the unlikely case that they are indeed falsified, this would have big complexity theoretical implications.

This paper is organized as follows: We first give an overview on terminology, the related lower bounds by Abboud et al. [2] and our results in Sect. 2. Then we examine scheduling problems with a single machine in Sect. 3 and problems with two or more machines in Sect. 4. Finally, we give a summary as well as open problems and promising research directions in Sect. 5.

2   Preliminaries

In this section, we first introduce the PARTITION problem, a special case of SUBSET SUM from which many of our reductions start. Then we recall common terminology from scheduling theory and finally, we give a short overview of the recent and closely related work [2] by Abboud et al. and briefly state our main results.

Throughout this paper, log denotes the base 2 logarithm. Moreover, we write $[n]$ for the set of integers from 1 to n, i.e. $[n] := \{1, \ldots, n\}$. If we consider a set of items or jobs $[n]$ and a subset $S \subseteq [n]$, we use $\bar{S} = [n] \setminus S$ to denote the complement of S. The $\tilde{\mathcal{O}}$-notation hides poly-logarithmic factors.

2.1   Subset Sum and Partition

In this work, we provide lower bounds for several scheduling problems; our main technique are *fine-grained reductions*, which are like polynomial-time reductions, but with more care for the exact sizes and running times. With these reductions, we can transfer the (supposed) hardness of one problem to another. Most of the time, our reductions start with an instance of SUBSET SUM or PARTITION and construct an instance of some scheduling problem. PARTITION is the special case of SUBSET SUM, where the sum of all items is exactly twice the target value:

Problem 2 PARTITION

Instance: Items $a_1, \ldots, a_n \in \mathbb{N}$.
Task: Decide whether there is a subset $S \subseteq [n]$ such that $\sum_{i \in S} a_i = \sum_{i \in \bar{S}} a_i$.

In the following, we always denote the total size of all items by $A := \sum_{i=1}^{n} a_i$ for SUBSET SUM and PARTITION. Note that we can always assume that $T \leq A$, since otherwise the target cannot be reached, even by taking all items. Moreover, in the reduction by Abboud et al. [1], A and T are quite close, in particular, we can assume that $A = \text{poly}(n)T$. Hence, if we could solve SUBSET SUM in time $\mathcal{O}\left(2^{\delta n} A^{1-\varepsilon}\right)$ for some $\varepsilon > 0$ and every $\delta > 0$, this would contradict Theorem 1 for large enough n. For details on this, we refer to the full version [14].

Corollary 1. *For every $\varepsilon > 0$, there is a $\delta > 0$ such that SUBSET SUM cannot be solved in time $\mathcal{O}\left(2^{\delta n} A^{1-\varepsilon}\right)$, unless the SETH fails.*

Using a classical reduction from SUBSET SUM to PARTITION that only adds two large items, we also get the following lower bound for PARTITION (for a detailed proof, see [14]):

Theorem 2. *For every $\varepsilon > 0$, there is a $\delta > 0$ such that* PARTITION *cannot be solved in time $\mathcal{O}\left(2^{\delta n} A^{1-\varepsilon}\right)$, unless the SETH fails.*

2.2 Scheduling

In all scheduling problems we consider, we are given a number of machines and a set of n jobs with processing times p_j, $j \in [n]$; our goal is to assign each job to (usually) one machine such that the resulting *schedule* minimizes some objective.[2] So these problems all have a similar structure: A machine model, some (optional) job characteristics and an objective function. This structure motivates the use of the three-field notation introduced by Graham et al. [10]. Hence, we denote a scheduling problem as a triple $\alpha|\beta|\gamma$, where α is the machine model, β is a list of (optional) job characteristics and γ is the objective function. As is usual in the literature, we leave out job characteristics like due dates that are implied by the objective function, e.g. for $1||\sum w_j U_j$. In this work, we mainly consider the decision variants of scheduling problems (as opposed to the optimization variants). In the decision problems, we are always given a threshold denoted by y and the task is to decide whether there is a solution with value at most y. Note that the optimization and the decision problems are – at least in our context – equivalent: An algorithm for the decision problem can be used to find a solution of the optimization problem with a binary search over the possible objective values (which are always integral and bounded, here). Vice versa, an algorithm for the optimization problem can also solve the decision problem.

In order to have a unified notation, given some job-dependent parameters $g_1, \ldots, g_n$ (e.g. processing times), we let $g_{max} := \max_{i \in [n]} g_i$, $g_{min} := \min_{i \in [n]} g_i$ and $G := \sum_{i \in [n]} g_i$. We now briefly go over the considered machine models, job characteristics and objective functions.

As the title of this work suggests, we consider problems with a fixed number of m parallel identical machines, denoted by 'Pm' if $m > 1$ or simply '1' if $m = 1$. In this setting, a job has the same processing time on every machine.

In the case of *rigid* and *moldable jobs*, each job has a given '*size*' and must be scheduled on that many machines or it may be scheduled on '*any*' number of machines, respectively, needing a possibly different (usually lower) processing time when scheduled on multiple machines. Sometimes, not all jobs are available at time 0, but instead each job j arrives at its release date 'r_j'.[3] Similarly, jobs might have deadlines d_j (i.e. due dates that may not be missed) and we must assure that '$C_j \leq d_j$' holds for every job j, where C_j is the *completion time* of j. Additionally, every job j might have a weight w_j and we are allowed to reject

---

[2] Depending on the scheduling problem, it may also be important in which order the jobs of a machine are scheduled or whether there are gaps between the execution of consecutive jobs.

[3] This is not to be confused with *online* scheduling; we know the r_j's in advance.

(i.e., choose not to schedule) jobs of total weight at most Q; this constraint is denoted by '$Rej \leq Q$'.[4]

The arguably most popular objective in scheduling is to minimize the so-called *makespan* '$C_{\max}$', which is the largest completion time C_j among all jobs j, i.e. the time at which all jobs are finished. In order to give the jobs different priorities, we can minimize the *total (weighted) completion time* '$\sum C_j$' ('$\sum w_j C_j$'). If there is a due date d_j for each job, we might be concerned with minimizing the *(weighted) number of late jobs* '$\sum U_j$' ('$\sum w_j U_j$'), where $U_j = 1$ if j is late, i.e. $C_j > d_j$ and $U_j = 0$ otherwise. Similar objectives are the *maximum lateness* '$L_{\max}$' and the *maximum tardiness* '$T_{\max}$' of all jobs, where the lateness L_j of job j is the (uncapped) difference $C_j - d_j$ and the tardiness T_j is the (capped) difference $\max\{C_j - d_j, 0\}$. Another objective, the *total tardiness* '$\sum T_j$', measures the tardiness of all jobs together and the *total late work* '$\sum V_j$' is the late work $V_j := \min\{p_j, C_j - d_j\}$ summed over all jobs. Both objectives may also appear in combination with weights. Lastly, if release dates r_j are present, we might be interested in minimizing the *maximum flow time* '$F_{\max}$', the total flow time '$\sum F_j$' or the weighted total flow time '$\sum w_j F_j$'. These objectives are similar to the previous ones; F_j, the flow time of job j, is defined as $F_j := C_j - r_j$, i.e. the time that passes between j's release and completion.

2.3   The Scheduling Lower Bounds by Abboud et al.

In their more recent work [2], Abboud et al. show lower bounds for the problems $1||\sum w_j U_j$, $1|Rej \leq Q|\sum U_j$, $1|Rej \leq Q|T_{\max}$, $1|r_j, Rej \leq Q|C_{\max}$, $P2||T_{\max}$, $P2||\sum U_j$, $P2|r_j|C_{\max}$ and $P2|level\text{-}order|C_{\max}$.[5] From those problems, only $1||\sum w_j U_j$ appears in this version; the full version [14] also contains results for $1|Rej \leq Q|\sum U_j$, $1|Rej \leq Q|T_{\max}$, $P2||T_{\max}$ and $P2||\sum U_j$. As we will see however, the results by Abboud et al. are not directly comparable to our results.

Standard dynamic programming approaches often give running times like $\mathcal{O}(nP)$; on the other hand, it is usually possible to try out all subsets of jobs, yielding an exponential running time like $\mathcal{O}(2^n \text{polylog}(P))$ (see e.g. the work by Jansen et al. [15]). The intuitive way of thinking about our lower bounds is that we cannot have the best of both worlds, i.e.: '*An algorithm cannot be sub-exponential in n and sub-linear in P at the same time.*' To be more specific, most of our lower bounds have this form: For every $\varepsilon > 0$, there is a $\delta > 0$ such that the problem cannot be solved in time $\mathcal{O}(2^{\delta n} P^{1-\varepsilon})$.

However, note that algorithms with running time $\tilde{\mathcal{O}}(n + P)$ or $\tilde{\mathcal{O}}(n + p_{\max})$ are not excluded by our bounds, as they are not sub-linear in P. But in a setting where n and P (resp. $p_{\max}$) are roughly of the same order, such algorithms would be much more efficient than the dynamic programming approaches. In particular, they would be near-linear in n instead of quadratic. This is where

---

[4] This is usually denoted by $Rej \leq R$, but since we will use R for the sum of all release dates, we denote the total rejection weight by Q.

[5] In 'level-order' problems, the jobs are ordered hierarchically and all jobs of one level have to be finished before jobs of higher levels can be scheduled.

the lower bounds from the more recent paper [2] by Abboud et al. come into play, as they have the following form: There is no $\varepsilon > 0$ such that the problem can be solved in time $\tilde{\mathcal{O}}\left(n + p_{\max}n^{1-\varepsilon}\right)$, unless the $\forall\exists$-SETH fails. The $\forall\exists$-SETH is similar to the SETH, but assuming yet another assumption (the NSETH), $\forall\exists$-SETH is a strictly stronger assumption than SETH. However, these lower bounds by Abboud et al. [2] can exclude algorithms with an additive-type running time $\tilde{\mathcal{O}}\left(n + p_{\max}\right)$. Algorithms with running time $\tilde{\mathcal{O}}\left(n + p_{\max}n\right)$ may still be possible, but they would only be near-quadratic instead of near-linear in the $n \approx p_{\max}$ setting. It should be noted that our lower bounds also include parameters other than $p_{\max}$, e.g. the largest due date $d_{\max}$ or the threshold for the objective value y.

2.4 Our Results

The main contribution of this work is two-fold: On the one hand, we give plenty of lower bounds for classical scheduling problems with a fixed number of machines. These lower bounds all either rely on the ETH, SETH or the (min, +)-conjecture[6] and are shown by revisiting classical reductions in the context of fine-grained complexity, i.e., we pay much attention to the parameters of the constructed instances. On the other hand, we show how the dynamic programming algorithms for $P2|any|C_{\max}$ and $P3|any|C_{\max}$ by Du and Leung [8] can be improved. Most notably, we show the following (for the precise statements, we refer to the upcoming sections):

- The algorithm by Lawler and Moore [21] is probably optimal for $1||\sum w_j U_j$ and $Pm||C_{\max}$.
- The algorithm by Lee and Uzsoy [23] is probably optimal for $P2||\sum w_j C_j$.
- The algorithm by Zhang et al. [30] for $1|Rej \leq Q|\sum w_j U_j$, the algorithm by Lawler [19] for $1||\sum T_j$ and the FPTAS by Gens and Levner [9] for $1||\sum w_j U_j$ are nearly optimal, but there is still some room for improvement.
- $P2|any|C_{\max}$ can be solved in time $\mathcal{O}(nP)$ *and* this is probably optimal.
- $P3|any|C_{\max}$ can be solved in time $\mathcal{O}(nP^2)$, which greatly improves upon the $\mathcal{O}(nP^5)$-time algorithm by Du and Leung [8].

Due to space restrictions, this version does not include the following content, which can be found in the full version [14]:

- Lower bounds for strongly NP-hard problems,
- implications of our lower bounds for other scheduling problems using classical reductions between objective functions (see Fig. 1),
- detailed correctness proofs of the classical reductions from the literature and
- proofs of some of our (less prominent or more technical) results.

---

[6] Under the (min, +)-conjecture, the (min, +)-convolution problem cannot be solved in sub-quadratic time, see [6] for details.

Note that our SETH-based lower bounds mainly show that improvements for some pseudo-polynomial algorithms are unlikely. For problems that are strongly NP-hard, pseudo-polynomial algorithms cannot exist, unless P = NP [4]. However, the lower bounds for strongly NP-hard problems may be of independent interest, e.g. in the context of parameterized algorithms.

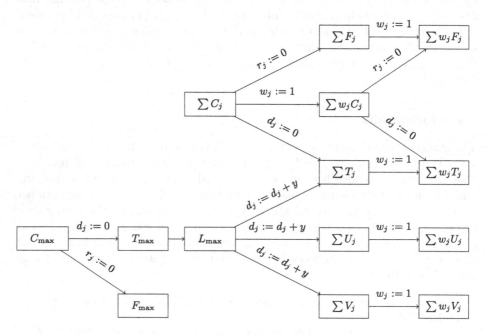

Fig. 1. Classical reductions between objective functions (see e.g. [20] and the very useful website http://schedulingzoo.lip6.fr/about.php).

3   Problems with One Machine

In this section, we consider problems on a single machine. For these problems, the main task is to order the jobs. First, consider again the problem $1||\sum w_j U_j$ of minimizing the weighted number of late jobs on a single machine. With a reduction very similar to the one by Karp [16], we get the following lower bound:[7]

Theorem 3. *For every $\varepsilon > 0$, there is a $\delta > 0$ such that $1||\sum w_j U_j$ cannot be solved in time $\mathcal{O}\left(2^{\delta n}(d_{\max} + y + P + W)^{1-\varepsilon}\right)$, unless the SETH fails.*

---

[7] It should be noted that some of the parameters in our lower bounds could be omitted, as they are overshadowed by others. For example, we can assume w.l.o.g. that $d_{\max} \leq P$ for $1||\sum w_j U_j$, since we can assume a schedule to be gap-less and hence due dates larger than P could be set to P. But having all the parameters in the lower bound makes the comparison with known upper bounds easier.

Proof. Let $a_1, \ldots, a_n$ be a PARTITION instance and let $T = \frac{1}{2} \sum_{i=1}^{n} a_i$. Construct an instance of $1 || \sum w_j U_j$ by setting $p_j = w_j = a_j, d_j = T$ for each $j \in [n]$ and $y = T$. The idea is that the jobs corresponding to items in one of the partitions can be scheduled early (i.e. before the uniform due date T).

With this reduction, we get $N := n$ jobs. We have $P = \sum_{i=1}^{n} a_i = A$ and hence $K := d_{max} + y + P + W = T + T + A + A = \text{poly}(n)A = n^c A$. The reduction itself takes time $\mathcal{O}(N)$. Assuming that we can solve $1 || \sum w_j U_j$ in time $\mathcal{O}\left(2^{\delta N} K^{1-\varepsilon}\right)$ for some $\varepsilon > 0$ and every $\delta > 0$, we could also solve PARTITION in time:

$$\mathcal{O}(N) + \mathcal{O}\left(2^{\delta N} K^{1-\varepsilon}\right) = \mathcal{O}(n) + \mathcal{O}\left(2^{\delta n}(n^c A)^{1-\varepsilon}\right) \leq \mathcal{O}\left(2^{\delta n} n^c A^{1-\varepsilon}\right)$$

$$= \mathcal{O}\left(2^{\delta n + c \log(n)} A^{1-\varepsilon}\right)$$

$$\leq \mathcal{O}\left(2^{2\delta n} A^{1-\varepsilon}\right)$$

The last step holds for large enough n; for smaller n, we can solve the problem efficiently, anyway, as n is then bounded by a constant. Now, to contradict Theorem 2, we can set $\varepsilon' := \varepsilon$ and for every $\delta' > 0$, we have $\delta = \frac{\delta'}{2} > 0$. So by assumption, we can solve PARTITION in time $\mathcal{O}\left(2^{2\delta n} A^{1-\varepsilon}\right) = \mathcal{O}\left(2^{\delta' n} A^{1-\varepsilon'}\right)$. $\square$

Using the algorithm by Lawler and Moore [21], $1 || \sum w_j U_j$ is solvable in time $\mathcal{O}(nW)$ or $\mathcal{O}(n \min\{d_{max}, P\})$. Our $\mathcal{O}\left(2^{\delta n}(d_{max} + y + P + W)^{1-\varepsilon}\right)$-time lower bound suggests the optimality of both variants, as we cannot hope to reduce the linear dependency on W, d_{max} or P without getting a super-polynomial dependency on n. As noted above, Abboud et al. [2] exclude $\tilde{\mathcal{O}}\left(n + p_{max} n^{1-\varepsilon}\right)$-time algorithms; Hermelin et al. [12] exclude algorithms with running time $\tilde{\mathcal{O}}\left(n + w_{max} n^{1-\varepsilon}\right)$, $\tilde{\mathcal{O}}\left(n + w_{max}^{1-\varepsilon} n\right)$ and $\tilde{\mathcal{O}}\left(n^{\mathcal{O}(1)} + d_{max}^{1-\varepsilon}\right)$ (all under the stronger $\forall\exists$-SETH).

One interesting property of $1 || \sum w_j U_j$ is that its straightforward formulation as an Integer Linear Program has a triangular structure that collapses to a single constraint when all due dates are equal (see e.g. Lenstra and Shmoys [25]). This shows that the problem is closely related to KNAPSACK:

Problem 3. KNAPSACK

Instance: Item values $v_1, \ldots, v_n \in \mathbb{N}$, item sizes $a_1, \ldots, a_n \in \mathbb{N}$, knapsack capacity $T \in \mathbb{N}$ and threshold y.

Task: Decide whether there is a subset S of items with $\sum_{j \in S} a_j \leq T$ and $\sum_{j \in S} v_j \geq y$.

Cygan et al. [6] conjectured that the $(\min, +)$-CONVOLUTION problem cannot be solved in sub-quadratic time (this is known as the $(\min, +)$-conjecture) and showed that this conditional lower bound transfers to KNAPSACK, excluding $\mathcal{O}\left((n + T)^{2-\delta}\right)$ algorithms. As noted by Mucha et al. [28], these results also hold when we swap the role of sizes and values. As we can discard items with too large value v_i, a lower bound depending on the largest item value v_{max} directly follows from Corollary 9.6 in [28]:

Corollary 2. *For any constant $\delta > 0$, there is no $\mathcal{O}\left((n + v_{\max})^{2-\delta}\right)$-time exact algorithm for* KNAPSACK, *unless the* $(\min, +)$-*conjecture fails.*

We show that the conditional hardness of KNAPSACK transfers to $1||\sum w_j U_j$:

Theorem 4. *For any constant $\delta > 0$, the existence of an exact algorithm for $1||\sum w_j U_j$ with running time $\mathcal{O}\left((n + w_{\max})^{2-\delta}\right)$ refutes the $(\min, +)$-conjecture.*

Proof. We give a reduction from KNAPSACK to $1||\sum w_j U_j$. Consider an instance $v_1, \ldots, v_n, a_1, \ldots, a_n, T, y$ of KNAPSACK. We construct jobs with $p_j = a_j$, $w_j = v_j$ and $d_j = T$ for every $j \in [n]$. The threshold is set to $y' = \sum_{j=1}^{n} v_j - y$.

Suppose that there is an $\mathcal{O}\left((n + w_{\max})^{2-\delta}\right)$-time algorithm for $1||\sum w_j U_j$. Since $w_{\max} = v_{\max}$ in the reduction and the reduction takes time $\mathcal{O}(n)$, we could then solve KNAPSACK in time $\mathcal{O}(n) + \mathcal{O}\left((n + w_{\max})^{2-\delta}\right) = \mathcal{O}\left((n + v_{\max})^{2-\delta}\right)$, which is a contradiction to Corollary 2, unless the $(\min, +)$-conjecture fails. $\square$

Lower bounds such as this one also imply lower bounds for approximation schemes, as setting the accuracy parameter ε small enough yields an exact solution. The above result implies the following (see the full version [14] for the proof):

Corollary 3. *For any constant $\delta > 0$, the existence of an $\mathcal{O}\left((n + \frac{1}{2n\varepsilon})^{2-\delta}\right)$-time approximation scheme for the optimization version of $1||\sum w_j U_j$ refutes the $(\min, +)$-conjecture.*

As the currently fastest FPTAS by Gens and Levner [9] has running time $\mathcal{O}\left(n^2(\log(n) + \frac{1}{\varepsilon})\right)$, there is still a small gap. This relation between exact and approximation algorithms might also be an interesting subject of further investigation, as many other scheduling problems admit approximation schemes and exact lower bounds.

We wish to mention two other results that follow from examining classical reductions, the proofs of which can also be found in the full version [14]. The first result concerns $1||\sum T_j$:

Theorem 5. *For every $\varepsilon > 0$, there is a $\delta > 0$ such that $1||\sum T_j$ cannot be solved in time $\mathcal{O}\left(2^{\delta n} P^{1-\varepsilon}\right)$, unless the SETH fails.*

There is an $\mathcal{O}\left(n^4 P\right)$-time algorithm by Lawler [19] and while we can derive no statement about the exponent of n, our lower bound suggests that an improvement of the linear factor P is unlikely without getting a super-polynomial dependency on n. We have a similar situation for the problem $1|Rej \leq Q|C_{\max}$:

Theorem 6. *For every $\varepsilon > 0$, there is a $\delta > 0$ such that $1|Rej \leq Q|C_{\max}$ cannot be solved in time $\mathcal{O}\left(2^{\delta n}(y + P + Q + W)^{1-\varepsilon}\right)$, unless the SETH fails.*

The lower bound can also be shown to hold for $1|Rej \leq Q|\sum w_j U_j$ using reductions between objective functions (see Fig. 1) and this problem can be solved in time $\mathcal{O}(nQP)$ with the algorithm by Zhang et al. [30]. This almost matches our lower bound: An algorithm with running time $\mathcal{O}(n(Q + P))$ might still be possible, for example.

4   Problems with Multiple Machines

We now turn our attention to problems on two or more machines. For **standard jobs**, a straightforward reduction from PARTITION yields the following result (for a formal proof, see [14]):

Theorem 7. *For every $\varepsilon > 0$, there is a $\delta > 0$ such that $P2||C_{\max}$ cannot be solved in time $\mathcal{O}\left(2^{\delta n}(y + P)^{1-\varepsilon}\right)$, unless the SETH fails.*

This lower bound also applies to the harder objectives (e.g. $T_{\max}$) and in particular to $P2||\sum w_j U_j$ (using the reductions in Fig. 1); the dynamic program by Lawler and Moore [21] (which is also sometimes attributed to Rothkopf [29]) solves most common objectives like $C_{\max}$ and $T_{\max}$ in time $\mathcal{O}(ny)$ but needs $\mathcal{O}(ny^2)$ for $P2||\sum w_j U_j$ (see [25], in particular exercise 8.10). So the gap is likely closed in the $C_{\max}$, $T_{\max}$, ... cases, but there is still a factor-y-gap for the $\sum w_j U_j$-objective.

In general, the dynamic program by Lawler and Moore [21] solves $Pm||C_{\max}$ in a running time of $\mathcal{O}\left(nmy^{m-1}\right) \leq \mathcal{O}\left(nmP^{m-1}\right)$. Our matching lower bound for $m = 2$ gives rise to the question whether the running time is optimal for general $m > 1$. Chen et al. [5] showed a $2^{\mathcal{O}\left(m^{\frac{1}{2}-\delta}\sqrt{|I|}\right)}$-time lower bound for $Pm||C_{\max}$ and with a careful analysis, one can also show the following lower bound (see the full version [14] for a proof):

Theorem 8. *There is no $\mathcal{O}\left(nmP^{o\left(\frac{m}{\log^2(m)}\right)}\right)$-time algorithm for $Pm||C_{\max}$, unless the ETH fails.*

So the algorithm by Lawler and Moore [21] is indeed almost optimal, as we can at best hope to shave off logarithmic factors in the exponent (assuming the weaker assumption ETH). Since the algorithm not only works for $C_{\max}$, one might ask whether we can find similar lower bounds for other objectives as well. For most common objective functions, we answer this question positively using the reductions in Fig. 1 (see the full version [14]), but it remains open for $\sum w_j C_j$. Note that the unweighted $Pm||\sum C_j$ is polynomial-time solvable [3].

An alternative dynamic program by Lee and Uzsoy [23] solves $Pm||\sum w_j C_j$ in time $\mathcal{O}\left(mnW^{m-1}\right)$. In order to get a matching lower bound (i.e. one that depends on the weights) for $m = 2$, we examine another classical reduction:

Theorem 9. *For every $\varepsilon > 0$, there is a $\delta > 0$ such that $P2||\sum w_j C_j$ cannot be solved in time $\mathcal{O}\left(2^{\delta n}(\sqrt{y} + P + W)^{1-\varepsilon}\right)$, unless the SETH fails.*

Proof. We show that the lower bound for PARTITION can be transferred to $P2||\sum w_j C_j$ using the reduction by Lenstra et al. [24] and Bruno et al. [3].

Given a PARTITION instance $a_1, \ldots, a_n$, we construct a $P2||\sum w_j C_j$ instance in the following way: Define $p_j = w_j = a_j$ for all $j \in [n]$ and set the limit $y = \sum_{1 \leq i \leq j \leq n} a_j a_i - \frac{1}{4}A^2$. Of course, the idea of the reduction is that the limit y forces the jobs to be equally distributed among the two machines (regarding the processing time).

Assume that there is an algorithm that solves an instance of $P2||\sum w_j C_j$ in time $\mathcal{O}\left(2^{\delta N} K^{1-\varepsilon}\right)$ for some $\varepsilon > 0$ and every $\delta > 0$, where $N := n$ and $K := \sqrt{y} + P + W$. By the choice of y, we can see that

$$y = \sum_{1 \le i \le j \le n} a_j a_i - \frac{1}{4} A^2 \le \left(\sum_{j \in [n]} a_j\right)^2 - \frac{1}{4} A^2 = \frac{3}{4} A^2 = \mathcal{O}\left(A^2\right).$$

Since $w_j = p_j = a_j$, we also have $P = W = A$. Hence, we have $K = \sqrt{y} + P + W = \mathcal{O}(A + A + A) = \mathcal{O}(A)$ and an algorithm with running time

$$\mathcal{O}\left(2^{\delta N} K^{1-\varepsilon}\right) = \mathcal{O}\left(2^{\delta n} \mathcal{O}(A)^{1-\varepsilon}\right) = \mathcal{O}\left(2^{\delta n} c^{1-\varepsilon} A^{1-\varepsilon}\right) = \mathcal{O}\left(2^{\delta n} A^{1-\varepsilon}\right)$$

would contradict the lower bound for PARTITION from Theorem 2. Here, c covers the constants in the $\mathcal{O}$-term and the running time $\mathcal{O}(N)$ of the reduction vanishes.   □

So the $\mathcal{O}(nW)$-time algorithm by Lee and Uzsoy [23] is probably optimal for $P2||\sum w_j C_j$, as we cannot hope to reduce the linear dependency on W without getting a super-polynomial dependency on n.

We briefly turn our attention towards **rigid jobs**. Clearly, $P2|size|C_{\max}$ is a generalization of $P2||C_{\max}$ (the latter problem simply does not have two-machine jobs), so we get the following lower bound (for a formal proof, see the full version [14]):

Theorem 10. *For every $\varepsilon > 0$, there is a $\delta > 0$ such that $P2|size|C_{\max}$ cannot be solved in time $\mathcal{O}\left(2^{\delta n}(y + P)^{1-\varepsilon}\right)$, unless the SETH fails.*

Similarly, the algorithm by Lawler and Moore [21] can be used to find a feasible schedule for the one-machine jobs and the two-machine jobs can be scheduled at the beginning. This gives an $\mathcal{O}(ny)$-time algorithm for $P2|size|C_{\max}$, and the linear dependency on y cannot be improved without getting a super-polynomial dependency on n, unless the SETH fails. For other objectives, the problem quickly becomes more difficult: Already $P2|size|L_{\max}$ is strongly NP-hard, as well as $P2|size|\sum w_j C_j$ (for both results, see Lee and Cai [22]). It is still open whether the unweighted version $P2|size|\sum C_j$ is also strongly NP-hard or whether there is a pseudo-polynomial algorithm; this question has already been asked by Lee and Cai [22], more than 20 years ago.

It is not hard to see that the hardness of $P2||C_{\max}$ also transfers to **moldable jobs** (i.e. $P2|any|C_{\max}$); we simply create an instance where it does not make sense to schedule any of the jobs on two machines (again, for a formal proof, see the full version [14]):

Theorem 11. *For every $\varepsilon > 0$, there is a $\delta > 0$ such that $P2|any|C_{\max}$ cannot be solved in time $\mathcal{O}\left(2^{\delta n}(y + P)^{1-\varepsilon}\right)$, unless the SETH fails.*

The problems $P2|any|C_{\max}$ and $P3|any|C_{\max}$ can be solved via dynamic programming, as shown by Du and Leung [8] (a nice summary is given in the book by Drozdowski [7]). We show that these programs can be improved to match our new lower bound for the two-machine case:

Theorem 12. *The problem $P2|any|C_{\max}$ can be solved in time $\mathcal{O}(nP)$ via dynamic programming.*

Proof. Assume that we are given processing times $p_j(k)$, indicating how long it takes to run job j on k machines. The main difficulty is to decide whether a job is to be processed on one or on two machines. Our dynamic program fills out a table $F(j, t)$ for every $j \in [n]$ and $t \in [y]$, where the entry $F(j, t)$ is the minimum load we can achieve on machine 2, while we schedule all the jobs in $[j]$ and machine 1 has load t. To fill the table, we use the following recurrence formula:

$$F(j, t) = \min \begin{cases} F(j - 1, t - p_j(1)) \\ F(j - 1, t) + p_j(1) \\ F(j - 1, t - p_j(2)) + p_j(2) \end{cases}$$

Intuitively speaking, job j is executed on machine 1 in the first case, on machine 2 in the second case and on both machines in the third case. The initial entries of the table are $F(0, 0) = 0$ and $F(0, t) = \infty$ for every $t \in [y]$.

There are $ny \leq n \sum_{j=1}^{n} \max\{p_j(1), p_j(2)\} = \mathcal{O}(nP)$ entries we have to compute.[8] Then, we can check for every $t \in [y]$ whether $F(n, t) \leq y$. If we find such an entry, this directly corresponds to a schedule with makespan at most y, so we can accept. Otherwise, there is no such schedule and we can reject. The actual schedule can be obtained by traversing backwards through the table; alternatively, we can store the important bits of information while filling the table (this works exactly like, e.g., in the standard knapsack algorithm). Note that we might have to reorder the jobs such that the jobs executed on two machines are run in parallel. But it can be easily seen that all two-machine jobs can be executed at the beginning of the schedule. Computing the solution and reordering does not change the running time in $\mathcal{O}$-notation, so we get an $\mathcal{O}(nP)$ algorithm. $\square$

As Theorem 11 shows, improving the dependency on P to sub-linear is only possible if we get a super-polynomial dependency on n, unless the SETH fails. Using a similar recurrence formula and the fact that information about an optimal placement of jobs directly leads to an optimal schedule (i.e. there is a canonical schedule), one can show a similar result for $P3|any|C_{\max}$ (see the full version [14] for a proof):

Theorem 13. *The problem $P3|any|C_{\max}$ can be solved in time $\mathcal{O}(n^2 P)$ via dynamic programming.*

This improves upon the $\mathcal{O}(nP^5)$-algorithm by Du and Leung [8]. Even though the same approach could be applied to an arbitrary number of machines m in time $\mathcal{O}(nmP^{m-1})$, the strong NP-hardness of $Pm|any|C_{\max}$ for $m \geq 4$ shows that the information on which machine each job is scheduled is not enough to directly construct an optimal schedule in those cases, unless P=NP (see Henning et al. [11] as well as Du and Leung [8]).

---

[8] The precise definition of P in this context does not matter for the running time in $\mathcal{O}$-notation; we can either add $p_j(1)$ *and* $p_j(2)$ to the sum or just the larger of the two.

5   Conclusion

In this work, we examined the complexity of scheduling problems with a fixed number of machines. Our conditional lower bounds indicate the optimality of multiple well-known classical algorithms. For the problems $P2|any|C_{\max}$ and $P3|any|C_{\max}$, we managed to improve the currently best known algorithm, closing the gap in the case of two machines.

As we have seen at the example of $1||\sum w_j U_j$, lower bounds for exact algorithms can be quite easily used to obtain lower bounds for approximation schemes. We strongly believe that the same technique can be used for other problems, either to show tightness results or to indicate room for improvement.

For exact algorithms, there is a number of open problems motivated by our results: First of all, there is still a gap between our lower bound for $Pm||C_{\max}$ (and other objectives) and the algorithm by Lawler and Moore [21]. So an interesting question is where the 'true' complexity lies between $m-1$ and $o\left(\frac{m}{\log^2(m)}\right)$ in the exponent. Zhang et al. give an $\mathcal{O}\left(n(r_{\max}+P)\right)$-time algorithm for $1|r_j, Rej \leq Q|C_{\max}$ in their work [30]. Since $r_{\max}+P \geq y$ w.l.o.g., it would be interesting to find an $\mathcal{O}\left(2^{\delta n}(r_{\max}+P)^{1-\varepsilon}\right)$ or $\mathcal{O}\left(2^{\delta n}y^{1-\varepsilon}\right)$ lower bound for this problem. As noted by Lenstra and Shmoys [25], the algorithm by Lawler and Moore [21] cannot be improved to $\mathcal{O}\left(mny^{m-1}\right)$ for the objective $\sum w_j U_j$. So this algorithm would be quadratic in y for two machines, while our lower bound excludes anything better than linear (and still polynomial in n). Hence, it would be interesting to see whether there is a different algorithm with running time $\mathcal{O}(ny)$. Similarly, there is an algorithm for $1|Rej \leq Q|\sum w_j U_j$ with running time $\mathcal{O}(nQP)$ [30], while our lower bound suggests that an $\mathcal{O}(n(Q+P))$-time algorithm could be possible.

On another note, it would be interesting to extend the sub-quadratic equivalences by Cygan et al. [6] and Klein [17] to scheduling problems. Finally, the question by Lee and Cai [22] whether $P2|size|\sum C_j$ is strongly NP-hard or not is still open since 1999.

Acknowledgements. The authors wish to thank Sebastian Berndt, Max Deppert, Sören Domrös, Lena Grimm, Leonie Krull, Marten Maack, Niklas Rentz and anonymous reviewers for very helpful comments and ideas.

References

1. Abboud, A., Bringmann, K., Hermelin, D., Shabtay, D.: Seth-based lower bounds for subset sum and bicriteria path. In: Proceedings of the Thirtieth Annual ACM-SIAM Symposium on Discrete Algorithms, SODA 2019, San Diego, California, USA, 6–9 January, 2019, pp. 41–57 (2019). https://doi.org/10.1137/1.9781611975482.3
2. Abboud, A., Bringmann, K., Hermelin, D., Shabtay, D.: Scheduling lower bounds via and subset sum. J. Comput. Syst. Sci. **127**, 29–40 (2022). https://doi.org/10.1016/j.jcss.2022.01.005

3. Bruno, J.L., Coffman, Jr., E.G., Sethi, R.: Scheduling independent tasks to reduce mean finishing time. Commun. ACM **17**(7), 382–387 (1974). https://doi.org/10.1145/361011.361064

4. Chen, B., Potts, C.N., Woeginger, G.J.: A review of machine scheduling: Complexity, algorithms and approximability. In: Du, D.Z., Pardalos, P.M. (eds.) Handbook of Combinatorial Optimization: Volume 1–3, pp. 1493–1641. Springer, US, Boston, MA (1998). https://doi.org/10.1007/978-1-4613-0303-9_25

5. Chen, L., Jansen, K., Zhang, G.: On the optimality of approximation schemes for the classical scheduling problem. In: Chekuri, C. (ed.) Proceedings of the Twenty-Fifth Annual ACM-SIAM Symposium on Discrete Algorithms, SODA 2014, Portland, Oregon, USA, 5–7 January, 2014. pp. 657–668. SIAM (2014). https://doi.org/10.1137/1.9781611973402.50

6. Cygan, M., Mucha, M., Wundefinedgrzycki, K., Włodarczyk, M.: On problems equivalent to (min,+)-convolution. ACM Trans. Algorithms **15**(1) (2019). https://doi.org/10.1145/3293465

7. Drozdowski, M.: Scheduling for Parallel Processing. Springer Publishing Company, Incorporated, 1st edn. (2009). https://doi.org/10.1007/978-1-84882-310-5

8. Du, J., Leung, J.Y.T.: Complexity of scheduling parallel task systems. SIAM J. Discret. Math. **2**(4), 473–487 (1989). https://doi.org/10.1137/0402042

9. Gens, G., Levner, E.: Fast approximation algorithm for job sequencing with deadlines. Discret. Appl. Math. **3**(4), 313–318 (1981). https://doi.org/10.1016/0166-218X(81)90008-1

10. Graham, R.L., Lawler, E.L., Lenstra, J.K., Kan, A.H.R.: Optimization and approximation in deterministic sequencing and scheduling: a survey. In: Hammer, P., Johnson, E., Korte, B. (eds.) Discrete Optimization II, Annals of Discrete Mathematics, vol. 5, pp. 287–326. Elsevier (1979). https://doi.org/10.1016/S0167-5060(08)70356-X

11. Henning, S., Jansen, K., Rau, M., Schmarje, L.: Complexity and Inapproximability Results for Parallel Task Scheduling and Strip Packing. Theory of Computing Systems **64**(1), 120–140 (2019). https://doi.org/10.1007/s00224-019-09910-6

12. Hermelin, D., Molter, H., Shabtay, D.: Minimizing the weighted number of tardy jobs via (max,+)-convolutions (2022). https://doi.org/10.48550/ARXIV.2202.06841

13. Impagliazzo, R., Paturi, R.: On the complexity of k-sat. J. Comput. Syst. Sci. **62**(2), 367–375 (2001). https://doi.org/10.1006/jcss.2000.1727

14. Jansen, K., Kahler, K.: On the complexity of scheduling problems with a fixed number of parallel identical machines (2022). https://doi.org/10.48550/ARXIV.2202.07932, https://arxiv.org/abs/2202.07932

15. Jansen, K., Land, F., Land, K.: Bounding the running time of algorithms for scheduling and packing problems. SIAM J. Discret. Math. **30**(1), 343–366 (2016). https://doi.org/10.1137/140952636

16. Karp, R.M.: Reducibility among combinatorial problems. Complexity of Computer Computations (1972). https://doi.org/10.1007/978-1-4684-2001-2_9

17. Klein, K.M.: On the Fine-Grained Complexity of the Unbounded SubsetSum and the Frobenius Problem, pp. 3567–3582. SIAM (2022). https://doi.org/10.1137/1.9781611977073.141

18. Knop, D., Koutecký, M.: Scheduling meets n-fold integer programming. J. Sched. **21**(5), 493–503 (2017). https://doi.org/10.1007/s10951-017-0550-0

19. Lawler, E.L.: A "pseudopolynomial" algorithm for sequencing jobs to minimize total tardiness. In: Hammer, P., Johnson, E., Korte, B., Nemhauser, G. (eds.) Studies in Integer Programming, Annals of Discrete Mathematics, vol. 1, pp. 331–342. Elsevier (1977). https://doi.org/10.1016/S0167-5060(08)70742-8

20. Lawler, E.L., Lenstra, J.K., Rinnooy Kan, A.H., Shmoys, D.B.: Chapter 9 sequencing and scheduling: algorithms and complexity. In: Logistics of Production and Inventory, Handbooks in Operations Research and Management Science, vol. 4, pp. 445–522. Elsevier (1993). https://doi.org/10.1016/S0927-0507(05)80189-6

21. Lawler, E.L., Moore, J.M.: A functional equation and its application to resource allocation and sequencing problems. Manage. Sci. **16**(1), 77–84 (1969). https://doi.org/10.1287/mnsc.16.1.77

22. Lee, C.Y., Cai, X.: Scheduling one and two-processor tasks on two parallel processors. IIE Trans. **31**(5), 445–455 (1999). https://doi.org/10.1080/07408179908969847

23. Lee, C.Y., Uzsoy, R.: A new dynamic programming algorithm for the parallel machines total weighted completion time problem. Operations Research Letters **11**(2), 73–75 (mar 1992). https://doi.org/10.1016/0167-6377(92)90035-2

24. Lenstra, J.K., Rinnooy Kan, A.H., Brucker, P.: Complexity of machine scheduling problems. Ann. Discrete Math. **1**, 343–362 (1977). https://doi.org/10.1016/S0167-5060(08)70743-X

25. Lenstra, J.K., Shmoys, D.B.: Elements of scheduling (2020). https://doi.org/10.48550/ARXIV.2001.06005

26. Mnich, M., van Bevern, R.: Parameterized complexity of machine scheduling: 15 open problems. Comput. Oper. Res. **100**, 254–261 (2018). https://doi.org/10.1016/j.cor.2018.07.020

27. Mnich, M., Wiese, A.: Scheduling and fixed-parameter tractability. Mathematical Programming **154**(1–2), 533–562 (dec 2015). https://doi.org/10.1007/s10107-014-0830-9

28. Mucha, M., Wundefinedgrzycki, K., Włodarczyk, M.: A subquadratic approximation scheme for partition. In: Chan, T.M. (ed.) Proceedings of the Thirtieth Annual ACM-SIAM Symposium on Discrete Algorithms, SODA 2019, pp. 70–88. Society for Industrial and Applied Mathematics, USA (2019). https://doi.org/10.1137/1.9781611975482.5

29. Rothkopf, M.H.: Scheduling independent tasks on parallel processors. Manage. Sci. **12**(5), 437–447 (1966). https://doi.org/10.1287/mnsc.12.5.437

30. Zhang, L., Lu, L., Yuan, J.: Single-machine scheduling under the job rejection constraint. Theoret. Comput. Sci. **411**(16–18), 1877–1882 (2010). https://doi.org/10.1016/j.tcs.2010.02.006

SOFSEM 2023 Best Student Papers

On the 2-Layer Window Width Minimization Problem

Michael A. Bekos[1] , Henry Förster[2] , Michael Kaufmann[2] ,
Stephen Kobourov[3] , Myroslav Kryven[3] , Axel Kuckuk[2(✉)] ,
and Lena Schlipf[2]

[1] Department of Mathematics, University of Ioannina, Ioannina, Greece
bekos@uoi.gr
[2] Department of Computer Science, University of Tübingen, Tübingen, Germany
{henry.foerster,michael.kaufmann,axel.kuckuk,
lena.schlipf}@uni-tuebingen.de
[3] Department of Computer Science, University of Arizona, Tucson, AZ, USA
{kobourov,kryven}@cs.arizona.edu

Abstract. When interacting with a visualization of a bipartite graph,
one of the most common tasks requires identifying the neighbors of a
given vertex. In interactive visualizations, selecting a vertex of interest
usually highlights the edges to its neighbors while hiding/shading the rest
of the graph. If the graph is large, the highlighted subgraph may not fit in
the display window. This motivates a natural optimization task: find an
arrangement of the vertices along two layers that reduces the size of the
window needed to see a selected vertex and all its neighbors. We consider
two variants of the problem; for one we present an efficient algorithm,
while for the other we show NP-hardness and give a 2-approximation.

Keywords: Graph drawing · Bipartite graphs · 2-layer drawings ·
Window width

1 Introduction

Two-layer networks model relationships between two disjoint sets of entities in various applications. Such networks are naturally modeled by bipartite graphs and are usually visualized with 2-layer drawings, where vertices are drawn as points on two distinct parallel lines ℓ_t and ℓ_b, and edges are straight-line segments [5]. Such drawings occur as components in layered drawings of directed graphs [15] and also as final drawings, e.g., in tanglegrams for phylogenetic trees [1,2,6,14] or in network layouts highlighting relationships between two communities [4,10,13].

A common task in the exploration of such networks is to identify the neighbors of a vertex of interest. A typical approach is to click on this vertex and

H. Förster, M. Kaufmann and A. Kuckuk are supported by DFG grant Ka512/18-2.
L. Schlipf is supported by the Ministry of Science, Research and the Arts Baden-Württemberg (Germany).

© The Author(s), under exclusive license to Springer Nature Switzerland AG 2023
L. Gąsieniec (Ed.): SOFSEM 2023, LNCS 13878, pp. 209–221, 2023.
https://doi.org/10.1007/978-3-031-23101-8_14

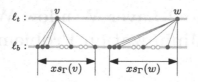

Fig. 1. The x-spans of vertices v and w.

highlight the edges to its neighbors, while hiding/shading the rest of the graph. Of course, the highlighted edges should fit in the display window. This motivates a natural optimization task: *find permutations of the vertices that minimize the size of the window needed to see any vertex and all its neighbors*. Related is the problem of minimizing the number of crossings instead, which is an NP-complete problem [5,7,11] and does not always result in easy-to-read drawings.

In applications, the vertex orders cannot always be treated as permutations; the vertices may have specific coordinates in one of the two layers ℓ_t or ℓ_b. For instance, the *ASCT+B Reporter* [8], a tool for displaying anatomical structures, cell types, and biomarkers, exemplifies this issue; by selecting a cell type its related biomarkers are highlighted. Minimizing the actual window width makes the tool easier to use. Note that in this use-case, the window widths of cell types are very important while the corresponding widths for biomarkers are negligible.

Our Contribution. Motivated by the discussion above, we study the following problem. The input consists of a bipartite graph $G = (A \cup B, E)$. The output is a 2-layer drawing Γ of G, that is, one in which the vertices in A and B are placed at distinct integer coordinates on two parallel lines ℓ_t and ℓ_b, respectively (w.l.o.g., $\ell_t : y = 1$ and $\ell_b : y = 0$; *top* and *bottom layers*). The objective is to minimize the *window width* of Γ, i.e., the maximum taken over all vertices v in A of the maximum x-distance between all neighbors of v along ℓ_b including the projection of v to ℓ_b. Motivated by common assumptions in layered graph drawing [3,9] we consider two variants, where the x-coordinates of the vertices of either A or B on ℓ_t or ℓ_b, respectively, are fixed. The former is NP-complete (Theorem 3); the latter is efficiently solvable (Theorem 1).

Preliminaries. For a vertex v in a drawing Γ denote by $x_\Gamma(v)$ and $y_\Gamma(v)$ the x- and y-coordinate of v in drawing Γ; when the reference drawing is clear, we simplify the notation to $x(v)$ and $y(v)$. Given a bipartite graph $G = (A \cup B, E)$ with $n_A = |A|$ and $n_B = |B|$, the *x-span* $xs_\Gamma(v)$ *of a vertex* $v \in A$ in a 2-layer drawing Γ of G is the maximum x-distance of all neighbors of v in B including v itself. To be more formal, $xs_\Gamma(v) = \max_{u,w \in N[v]}\{|x_\Gamma(u) - x_\Gamma(w)|\}$ where $N[v] = \{v\} \cup \{w|(v,w) \in E\}$ is the *closed neighborhood* of v. We define the *window width* $ww(\Gamma)$ *of the drawing* Γ as the maximum of the x-spans over all vertices in A, that is, $ww(\Gamma) = \max_{v \in A}\{xs_\Gamma(v)\}$, see Fig. 1. In the *2-layer window width minimization problem*, we seek to determine the *window width* $ww(G)$ *of a graph* G, which is the minimum window width of all of its 2-layer drawings.

2   Window Width Minimization with Bottom Layer Fixed

We present an efficient algorithm to find a 2-layer drawing of minimum window width when the x-coordinates of the vertices of B along ℓ_b are fixed.

Theorem 1. *Given a bipartite graph $G = (A \cup B, E)$ and a function $\xi_B \colon B \to \mathbb{Z}$, there is an $O(n_A \log n_A + |E|)$-time algorithm to determine a 2-layer drawing Γ of G with minimum window width $k^\star$ and $x_\Gamma(b) = \xi_B(b)$ for each $b \in B$.*

Proof. For each vertex $v \in A$ it suffices to focus on its leftmost neighbor $\ell(v)$ in ξ_B and rightmost neighbor $r(v)$ in ξ_B (ignoring intermediate ones). Note that $\ell(v) = r(v)$ is possible. This preprocessing, which can be done in $O(|E|)$ time, allows us to continue with a graph of $O(n_A)$ vertices and edges, called the *critical part* of G. We now determine the x-coordinate of each vertex v in A.

Let k_0 be the maximum x-distance between $\ell(v)$ and $r(v)$ over all vertices v in A and note that k_0 is a lower bound for $k^\star$. We describe an $O(n_A \log n_A)$-time algorithm to compute $k^\star$ and a corresponding solution. In this process, we start by attempting to find a drawing with window width $k = k_0$. If at some point, we conclude that the current value of k is too small, we increase k by 1 and proceed. When the algorithm terminates it will hold that $k = k^\star$.

Let $I(v) = [x(r(v)) - k, x(\ell(v)) + k]$ be the *interval* of $v \in A$; the x-distance of v to $\ell(v)$ and $r(v)$ is at most k if and only if its x-coordinate is in $I(v)$.

We sweep the intervals of the vertices from left-to-right by a vertical sweep line L, which is a data-structure maintaining a set of *active intervals* (i.e., those intersected by L whose vertices in A have not been placed yet) assumed to be *sorted by their right endpoints*. In this process, we distinguish three different types of events: *start*, *placement* and *end*. If during the sweep multiple events occur at the same x-coordinate i we first perform all start events at i, followed by a possible placement event at i before finally performing the end events at i.

Start Event. It occurs at the left endpoint i of each interval $I(v)$. Here, the interval $I(v)$ is inserted into L. We add a placement event at i, if there is none.

Placement Event at i. We remove the first active interval $I(v)$ from L, set $x(v) := i$ and mark $I(v)$ as *inactive*. If L is not empty, we add a placement event at $i + 1$. Note that placement events always place a vertex, hence there is only a linear number of placement events in total.

End Event. It occurs at the right endpoint i of each interval $I(v)$. We check if $I(v)$ is marked as inactive. If this is the case, we proceed. If not, we failed to place v at a position within $I(v)$. We increase k by 1 (i.e., all start events and already placed vertices are moved by -1 and all end events by $+1$ on the x-axis) and replace the already existing placement event with a new placement event at i.

Correctness. We begin with two useful observations. First, once our algorithm failed to place a vertex v within $I(v)$, the partial solution obtained by increasing k by one and shifting all placed vertices one unit to the left is identical to the one that would be obtained by restarting the algorithm with window width $k + 1$.

Second, by increasing k the ordering of the start events of the intervals remains the same and the same holds true for the end events. Consequently, the following property holds. Assume that we increased k by 1 at x-coordinate i after failing to place a vertex v with $I(v) = [\ell, r]$. Note that $r = i$ before increasing k and $r = i + 1$ after increasing k. Now let P_i denote the set of vertices that has been placed by our algorithm so far and let S_i denote the set of vertices whose start event occurs at i after increasing k to $k + 1$. Then, after handling the end event, for each $p \in P_i$ with interval $I(p) = [\ell_p, r_p]$ it holds for each $s \in S_i$ with $I(s) = [\ell_s, r_s]$ that $r_p \le r_s$ since $r_p \le r = i + 1$ and $i = \ell_s < r_s$.[1]

To complete the correctness proof, we show that we increase k only if it is necessary. Recall that we increase k if a vertex v cannot be placed within $I(v) = [\ell, r]$. Hence, all x-coordinates of $I(v)$ have been assigned to previously placed vertices. Let $\ell' < \ell$ be the largest x-coordinate our algorithm assigned no vertex from A and let $A_v \subset A$ be the vertices placed in $I'(v) = [\ell' + 1, r]$. We prove that in each solution with window width k, all vertices in A_v have to be placed in $I'(v)$. Assume for a contradiction that there is a vertex $a \in A_v$ that can be placed outside of $I'(v)$ such that its x-span is at most k. To this end, recall that a has x-span at most k if and only if it is placed within $I(a)$. First, a cannot be placed at an x-coordinate greater than x_r, since a has been placed before v by the algorithm, i.e., the right end of $I(a)$ is at an x-coordinate of at most x_r. Second, a cannot be placed at an x-coordinate smaller than $x'_\ell + 1$ as our algorithm would have placed a at coordinate x'_ℓ (or even beforehand) if its interval would have started at an x-coordinate smaller or equal to x'_ℓ; contradiction.

Time Complexity. We store the start and end events in two left-to-right sorted lists, while we maintain at most one placement event (with associated x-coordinate). The active intervals are stored in a binary min heap (the keys are the right endpoints). By keeping offset values for start and end events, as well as for the last placed vertex, the performed shifts can be done in $O(n_A)$ time with one additional right-to-left pass. Since L maintains at most $O(n_A)$ intervals the running time is $O(n_A \log n_A)$, after computing the critical part of G in $O(|V| + |E|)$ time. □

Remark 1. The core of the algorithm, given sorted start and end events, can be completed in $O(n_A \log k^\star)$ time since the number of intervals in L is actually bounded by $2k^\star$.

Proof. Consider some x-coordinate i at which there are $2k^\star + 2$ intervals maintained in L. Since there can only be one vertex placed on each integer coordinate, there must be one placed on x-coordinate $i + 2k^\star + 1$, let this be vertex v with interval $I(v) = [\ell, r]$. Note that since this interval is active at i it must hold that $\ell \le i$. With the definition of $I(v)$ it follows $x(r(v)) \le \ell + k^\star \le i + k$. The interval is maximal if $r(v) = \ell(v)$, thus $x(\ell(v)) + k^\star \le r(v)) + k^\star \le i + 2k^\star$ which contradicts the placement of v at $i + 2k^\star + 1$. □

---

[1] We point out that the latter relation $\ell_s < r_s$ does not hold if $k = 0$, but since we increased k by 1, it holds $k \ge 1$.

Next, we show that a variant of our algorithm can be used to optimize the *maximum edge-length*.

Theorem 2. *Given a bipartite graph $G = (A \cup B, E)$ and a function $\xi_B \colon B \to \mathbb{Z}$, there is an $O(n_A \log n_A + |E|)$-time algorithm to determine a 2-layer drawing Γ of G that minimizes the maximum x-distance $k^\star$ between any vertex in A and any adjacent vertex in B and $x_\Gamma(b) = \xi_B(b)$ for each $b \in B$.*

Proof. As in the proof of Theorem 1, we first identify the critical part of G which has $O(n_A)$ vertices and edges. In the following, we determine the x-coordinate of each vertex v in A such that the maximum x-distance between adjacent vertices, denoted by k, is minimized in the critical part, which implies that it is minimized in G as well. As in the proof of Theorem 1, for a sufficiently large value of k, we define for each vertex $v \in A$ an *interval* $I(v)$ such that v is placed on any x-coordinate in $I(v)$ if and only if its x-distance to any neighbor of v is at most k. More precisely, $I(v) = [x(r(v)) - k, x(\ell(v)) + k]$. We start the algorithm of Theorem 1 with $k = k_0$, where $k_0 := \lceil \frac{k_{\max}}{2} \rceil$ and $k_{\max}$ denotes the maximum x-distance between $\ell(v)$ and $r(v)$ over all vertices v in A (that is, k_0 is the trivial lower bound for $k^\star$). During the algorithm, we might conclude that the current value of k is not sufficient, thus k is increased by 1 before proceeding.

Since the rest of the algorithm of Theorem 1 consists of finding placements of all vertices within their intervals and increasing the intervals if necessary, this part of the algorithm can be completely adopted. Both the correctness and the time complexity of the algorithm follow analogously to Theorem 1.   □

3   Window Width Minimization with Top Layer Fixed

In contrast to the positive result from Theorem 1, we prove here that the problem is NP-complete when the order of the vertices A on the top layer ℓ_t is fixed.

Theorem 3. *Given a bipartite graph $G = (A \cup B, E)$, a function $\xi_A \colon A \to \mathbb{Z}$ and an integer k, it is NP-complete to test whether a 2-layer drawing Γ of G exists, such that $ww(\Gamma) = k$ and $x_\Gamma(a) = \xi_A(a)$ for each $a \in A$.*

Proof. Membership in NP is obvious. To prove NP-hardness, we adapt a reduction by Papadimitriou from the EXACT-3-SAT problem to the BANDWIDTH problem [12]. Let φ be an instance of EXACT-3-SAT, that is, a Boolean formula with n variables and m clauses (each with 3 different literals). We assume w.l.o.g. that $n \geq 5$ and reduce the problem of determining whether φ is satisfiable to an instance of our problem consisting of a bipartite graph $G = (A \cup B, E)$, a function $\xi_A \colon A \to \mathbb{Z}$ and the integer $k = 6n + 3$. We first sketch the general idea of the reduction by Papadimitriou and discuss the relation to our construction; for an example illustration see Fig. 2.

Introduction to the Reduction. A central concept in the reduction for the BAND-WIDTH problem[2] is a subgraph $\mathcal{H}$ that contains a *literal-vertex* for each possible

---

[2] Given $k \in \mathbb{N}$ and a graph $G = (V, E)$ the BANDWIDTH problem asks for an ordering $\prec$ of V so that for each $(u, v) \in E$ there are at most k vertices between u and v in $\prec$.

literal (i.e., for each variable x_i, it contains vertices ℓ_{x_i} and $\ell_{\neg x_i}$) and two additional vertices denoted by M and M'. By fixing the value of the bandwidth, it can be ensured that in any layout of $\mathcal{H}$ exactly n of the literal-vertices appear in a sequence P to the left of M and M' whereas the remaining n literal-vertices appear to the right of M and M' in a sequence Q. The vertices placed in P correspond to the satisfied literals, while the vertices placed in Q correspond to unsatisfied literals. In our reduction, we achieve the same behavior using *block-gadgets* and *$\mathcal{H}$-gadgets* where our B_2-*blocks* correspond to vertices M and M' in Papadimitriou's reduction.

In the reduction for the bandwidth problem, there are $n + m$ consecutive copies of $\mathcal{H}$ that are "synchronized" via the bandwidth restriction. Namely, additional edges ensure that each literal consistently occurs either in every sequence P or in every sequence Q. We achieve the same behavior using the *propagation gadgets*. Papadimitriou associates each of the first n copies of $\mathcal{H}$ with a *variable-gadget* that checks that only one of the literal-vertices corresponding to x and $\neg x$ occurs within Q, namely, as the leftmost vertex in Q. Finally, each of the last m copies of $\mathcal{H}$ is associated with a *clause-gadget* that ensures that at most two literals of a given clause can occur within sequence Q, namely, as the leftmost two vertices. In our construction, we use similar gadgets exploiting this idea.

Finally, it is worth remarking that in contrast to the bandwidth problem, in the window width minimization problem vertices in B are restricted to certain positions along ℓ_b by inputs ξ_A and k. With these additional restrictions fixing vertices to certain intervals (e.g., one copy of each literal in each $\mathcal{H}$-gadget) is simplified, however, it also becomes less apparent that the model still allows for enough flexibility to show NP-hardness (as for instance required in the propagation between consecutive $\mathcal{H}$-gadgets).

Next, we provide a description of the gadgets of our construction. The functionality of each gadget is ensured by introducing one or two vertices at appropriate coordinates along ℓ_t. We start by introducing the basic structure of our construction consisting of block- and $\mathcal{H}$-gadgets.

Block-Gadget. The purpose of the block-gadget is to fix a certain number β of vertices of B to be consecutive at fixed x-coordinates $i, \ldots, i + \beta - 1$ so that no other vertex can be placed there; see Fig. 3a. Hence, these β *block vertices* occupy a block of x-coordinates where no other vertex of B may be placed. To achieve this property, we introduce two vertices $a_\ell, a_r \in A$ with $\xi_A(a_\ell) = i - (k - \beta + 1)$ and $\xi_A(a_r) = i + k$ which both are connected to all β block vertices. It is easy to verify that each block vertex has x-distance at most k to both a_ℓ and a_r if and only if it is located inside the interval $[i, i + \beta - 1]$ in B (the order of the β vertices inside the interval is free).

We use two types of blocks, namely, one with $\beta_1 = 2n+3$ vertices of B (empty dark gray circles in Fig. 2a) and one with $\beta_2 = n + 1$ vertices of B (filled dark gray circles in Fig. 2a). We call the B-vertices of such blocks B_1- and B_2-*blocks*, respectively. Further, we assume that the vertices of a B_1-block are partitioned into three parts B_1^ℓ, B_1^m and B_1^r. Part B_1^m has exactly n vertices, while B_1^ℓ and B_1^r have $\lfloor (n + 3)/2 \rfloor$ and $\lceil (n + 3)/2) \rceil$ vertices, respectively; see Fig. 3b.

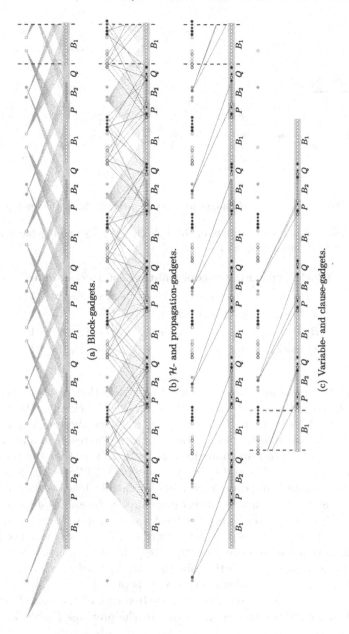

Fig. 2. Example of our NP-hardness reduction for $\varphi = (\neg x_1 \vee x_2 \vee x_3) \wedge (\neg x_2 \vee x_4 \vee x_5) \wedge (x_1 \vee \neg x_4 \vee x_5)$ with satisfying assignment $x_1 = x_3 = x_5 = \top$ and $x_2 = x_4 = \bot$. Literal-vertices of x_1, x_2, x_3, x_4 and x_5 are colored blue, red, green, yellow and pink, respectively, and filled white or black, if associated with the positive or negative literal, respectively. Subfigures (a) and (b) show only part of the $\mathcal{H}$-gadgets, in subfigure (c) all $\mathcal{H}$-gadgets are shown (split into two lines at the B_1-block separated with dashed lines). Subfigures (b) and (c) also show the vertices of the gadgets introduced in the previous subfigures. (Color figure online)

216   M. A. Bekos et al.

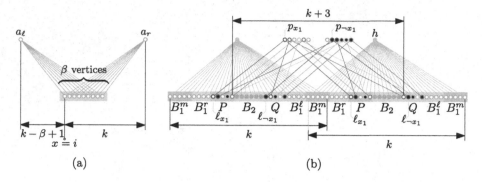

Fig. 3. (a) Block-gadget. (b) $\mathcal{H}$-gadget and propagation-gadget. (Color figure online)

Table 1. First and last x-coordinate of the i-th B_1-, P-, B_2-, and Q-block as enumerated from left-to-right starting at 1.

Block Type	First x-coordinate	Last x-coordinate
B_1-block	$p \cdot (i-1) + 1$	$p \cdot (i-1) + 2n + 3$
P-block	$p \cdot (i-1) + 2n + 4$	$p \cdot (i-1) + 3n + 3$
B_2-block	$p \cdot (i-1) + 3n + 4$	$p \cdot (i-1) + 4n + 4$
Q-block	$p \cdot (i-1) + 4n + 5$	$p \cdot (i-1) + 5n + 4$

B_1- and B_2-blocks alternate from left-to-right so that in total there are $n + m + 1$ B_1-blocks and $n + m$ B_2-blocks. Between a B_1-block and a B_2-block there is a *P-block* while between a B_2-block and a B_1-block there is a *Q-block*. Both P- and Q-blocks are intervals supporting n x-coordinates each and correspond to sequences P and Q in Papadimitriou's reduction. Note that the total number of vertices in a B_1-, P-, B_2- and Q-block is $p = 5n + 4$. We let the first B_1-block start at x-coordinate 1 and obtain intervals for the block types shown in Table 1.

More precisely, we ensure the correct positions of B_1- and B_2-blocks as follows. For the i-th B_1-block, vertex a_ℓ is placed at $\xi_A(a_\ell) = p \cdot (i-2) + n + 4$ while vertex a_r is placed at $\xi_A(a_r) = p \cdot i + n$. In other words, a_ℓ is placed above the $(n+4)$-th vertex (left-to-right) of the previous B_1-block and a_r is placed above the n-th vertex of the next B_1-block. Further, for the i-th B_2-block, vertex a_ℓ is placed at $\xi_A(a_\ell) = p \cdot (i-2) + 3n + 5$ whereas vertex a_r is placed at $\xi_A(a_r) = p \cdot i + 4n + 3$. Intuitively, vertex a_r is placed above the n-th vertex of the next B_2-block and vertex a_ℓ is placed above the second vertex of the previous B_2-block.

$\mathcal{H}$-Gadget. The purpose of the $\mathcal{H}$-gadget is to introduce *literal-vertices* for all literals of φ, that is, literals ℓ_{x_i} and $\ell_{\neg x_i}$ for each variable x_i ($2n$ in total; see red, blue, green, yellow and pink vertices in Figs. 2b and 3b). Each $\mathcal{H}$-gadget is associated with a B_2-block b and ensures that each of its $2n$ literal-vertices is placed either in the P-block preceding b (containing all satisfied literals) or in the Q-block succeeding b (containing all unsatisfied literals); Fig. 3b depicts two consecutive copies of the $\mathcal{H}$-gadget; note that there is a shared part of n vertices, denoted by B_1^m.

More precisely, there exists one $\mathcal{H}$-gadget H for each B_2-block b. H contains a vertex h in A that is incident to all vertices of b, to the B_1^m- and B_1^r-vertices of the B_1-block preceding b and to the B_1^ℓ- and B_1^m-vertices of the B_1-block succeeding b, i.e., $\mathcal{H}$-gadgets corresponding to consecutive B_2-blocks share $n = |B_1^m|$ vertices. If H corresponds to the i-th B_2-block, vertex h is placed at $\xi_A(h) = p \cdot (i-1) + 3n + 4$, that is, above the first B-vertex of its associated B_2-block. Further, h is connected to a literal-vertex for each literal of φ. Since the leftmost vertex of the B_1^m-block preceding b and the rightmost vertex of the B_1^m-block succeeding b are at distance k, all literal-vertices connected to h must be placed between these two blocks. The only available positions in this range are covered by the P-block preceding b and the Q-block succeeding b. Note that in the following, no further edges incident to vertices in a B_1-block are introduced, i.e., the vertex-order inside a B_1-block is only restricted by h-vertices. Finally, observe that the h-vertices have x-span k if the vertices in B_1^ℓ precede (left-to-right) those in B_1^m, which precede those in B_1^r.

In the following, we assume literal-vertices in P-blocks and Q-blocks to correspond to satisfied and unsatisfied literals, respectively. Next, we ensure consistency.

Propagation-Gadget. The propagation-gadgets (see red, blue, green, yellow and pink vertices and edges in Figs. 2b and 3b) ensure consistency, that is, literals in P-blocks are satisfied, while literals in Q-blocks are unsatisfied in φ. Namely, the propagation gadget for x_i ensures that the literal-vertex $\ell_\lambda \in \{\ell_{x_i}, \ell_{\neg x_i}\}$ occurring in the P-block of an $\mathcal{H}$-gadget H_1 will also occur in the P-block of the next $\mathcal{H}$-gadget H_2 in their left-to-right order. Since all vertices from P-blocks are propagated, literal-vertices in the Q-blocks are also propagated from H_1 to H_2. Note that literal-vertices do not necessarily have the same order in H_1 and H_2.

More formally, for each B_1-block b and for each variable x_i there is a copy of the propagation-gadget containing two *propagation-vertices* p_{x_i} and $p_{\neg x_i}$. Let H_1 and H_2 be the two (consecutive) $\mathcal{H}$-gadgets incident to the B_1^m-vertices of b. Then, vertex p_{x_i} is connected to the literal-vertices ℓ_{x_i} of H_1 and H_2 while $p_{\neg x_i}$ is connected to the literal-vertices $\ell_{\neg x_i}$ of H_1 and H_2. If b is the j-th B_1-block, we set $\xi_A(p_{x_i}) = p \cdot (j-1) + (i-1)$ and $\xi_A(p_{\neg x_i}) = p \cdot (j-1) + (n+4) + i$, i.e., all propagation-vertices with positive literals are to the left of the a_r-vertex above b while all propagation-vertices with negative literals are to the right of the a_ℓ-vertex above b. Note that p_{x_1} is above the last vertex in the Q-block preceding b while $p_{\neg x_n}$ is above the first vertex in the P-block succeeding b; the remaining literal-vertices are placed on unique x-coordinates above b. Since the distance between the leftmost literal-vertex in the P-block of H_1 and the rightmost literal-vertex in the P-block of H_2 is $k - n + 1$, we can reorder all literal-vertices freely in the P-blocks of H_1 and H_2; the same holds for the corresponding Q-blocks. On the other hand, the rightmost literal-vertex ℓ_λ of the P-block of H_1 cannot occur in the Q-block of H_2 as otherwise their connecting vertex p_λ has x-span at least $k + 3$; see Fig. 3b. As already mentioned above, since all literals from

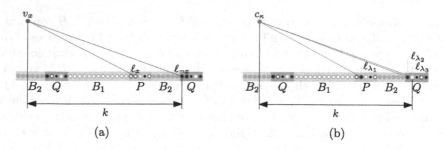

Fig. 4. (a) Variable-gadget. (b) Clause-gadget.

the P-block are propagated from H_1 to H_2, all literals from the Q-block are propagated as well.

Now each literal is either consistently satisfied (in P-blocks) or unsatisfied (in Q-blocks). It remains to encode the logic of φ with variable- and clause-gadgets.

Variable-Gadget. The variable-gadget for variable x_i ensures that only one of the literal-vertices ℓ_{x_i} and $\ell_{\neg x_i}$ can be placed within Q-blocks. Since these gadgets guarantee that at most one literal for each variable is false, it is only possible to place all $2n$ literals if exactly one literal per variable is true while the other is false. Hence, each variable is either true or false consistently in all $\mathcal{H}$-gadgets.

More precisely, the first n (in left-to-right-order) $\mathcal{H}$-gadgets are augmented with a variable-gadget. Namely, each variable gadget is associated with a unique variable x and ensures that one of the literals x and $\neg x$ must be true. The variable gadget associated with $\mathcal{H}$-gadget H consists of both literal-vertices ℓ_x and $\ell_{\neg x}$ of H and an additional *variable-vertex* v_x connected to ℓ_x and $\ell_{\neg x}$; see Fig. 4a and purple vertices and edges in Fig. 2c. We set the x-coordinate of v_x so that it is at distance k to the left of the leftmost vertex in the Q-block of H, i.e., if H is the i-th $\mathcal{H}$-gadget we have $\xi_A(v_x) = p \cdot (i - 2) + 3n + 6$. As a result, v_x is placed above the third vertex of the B_2-block preceding the B_2-block of H. Clearly, the x-span of v_x is at most k if at most one of ℓ_x and $\ell_{\neg x}$ is in the Q-block of H. As mentioned above, since for each variable the variable gadget guarantees this property, it is only possible to place all $2n$ literals if exactly one literal per variable is true while the other is false. Thus, each variable is either consistently true or consistently false.

Clause-Gadget. There are m clause-gadgets associated with the last m copies of $\mathcal{H}$-gadgets. The clause-gadget for a clause $\kappa = (\lambda_1 \vee \lambda_2 \vee \lambda_3)$ ensures that at most two of the literal-vertices ℓ_{λ_1}, ℓ_{λ_2} and ℓ_{λ_3} can be placed within Q-blocks. At least one literal must be placed inside the P-blocks and thus, κ contains at least one satisfied literal.

To this end, the clause gadget for clause $\kappa = (\lambda_1 \vee \lambda_2 \vee \lambda_3)$ consists of a vertex $c_\kappa \in A$ connected to the three literal-vertices $\ell_{\lambda_1}, \ell_{\lambda_2}, \ell_{\lambda_3}$; see Fig. 4b and brown vertices and edges in Fig. 2c. We assign the x-coordinate of c_κ so that it is at distance $k - 1$ to the left of the leftmost vertex in the Q-block of H, i.e., if H

Table 2. Placements of vertices in A above the j-th vertex within B_1- and B_2-blocks. If a vertex x of A is above the j-th vertex of the i-th B_1-block, $\xi_A(x) = p \cdot (i-1) + j$, while $\xi_A(x) = p \cdot (i-1) + 3n + 3 + j$ if x is above the j-th vertex of the i-th B_2-block.

B_1-block	j	B_2-block
	1	h of associated $\mathcal{H}$-gadget
	2	a_ℓ of next B_2-block
$p_{x_{j+1}}$ of prop.-gadget	3	v_x associated with the next $\mathcal{H}$-gadget
	4	c_κ associated with the next $\mathcal{H}$-gadget
	$[5, n-1]$	——
a_r of previous B_1-block	n	a_r of previous B_2-block
——	$[n+1, n+3]$	
a_ℓ of next B_1-block	$n+4$	——
$p_{\neg x_{j-(n-4)}}$ of prop.-gadget	$[n+5, 2n+3]$	

is the i-th H-gadget, then $\xi_A(c_\kappa) = p(i-2) + 3n + 7$. Hence, c_κ is placed above the fourth vertex of the B_2-block preceding H. Since the distance between the second vertex in the Q-block of H and c_κ is k, the x-span of c_κ is at most k if at most two of $\ell_{\lambda_1}, \ell_{\lambda_2}, \ell_{\lambda_3}$ are in the Q-block, i.e., one of $\lambda_1, \lambda_2, \lambda_3$ is true.

Polynomial Time of the Reduction and Equivalence. The construction can clearly be done in $O(n \cdot (n+m))$ time. Next, we prove that no two vertices of A share the same x-coordinate, i.e., ξ_A is injective.

Recall that we have placed vertices in A only on coordinates that are covered by B_1- and B_2-blocks (or are located to the left of the first B_1-block or to the right of the last B_1-block); see Fig. 2. Table 2 summarizes the positioning described in the construction. The two exceptions to this are vertices p_{x_1} and $p_{\neg x_n}$ of propagation gadgets which are located above the last vertex of P-blocks and above the first vertex of Q-blocks, respectively. For $n \geq 5$ (as assumed at the beginning) indeed no x-coordinate is assigned twice by ξ_A.

It remains to prove that φ is satisfiable if and only if there is a drawing Γ with $ww(\Gamma) \leq k$ and $x_\Gamma(a) = \xi_A(a)$ for $a \in A$. First, assume that φ is satisfiable. We can construct a drawing with window width at most k by placing all satisfied literals in the P-block and each unsatisfied literal in the Q-block of each $\mathcal{H}$-gadget. The literal-vertices are sorted so that the unsatisfied literals in a variable- or clause-gadget are the leftmost ones in the corresponding Q-block. Second, assume that there is a drawing Γ with $ww(\Gamma) \leq k$ and $x_\Gamma(a) = \xi_A(a)$ for $a \in A$. As discussed above, each variable is either true or false while each clause contains a satisfied literal. Thus, a satisfying truth assignment for φ can be read from any P-block of Γ. □

Next, we prove that our algorithm from Theorem 2 can be used for a 2-approximation algorithm for the window width minimization problem with fixed top layer.

Theorem 4. *Given a bipartite graph $G = (A \cup B, E)$ and a function $\xi_A \colon A \to \mathbb{Z}$, there is an $O(n_B \log n_B + |E|)$-time 2-approximation algorithm for computing the minimum value $k^\star$ such that there is a 2-layer drawing Γ of G with $ww(\Gamma) = k^\star$ and $x_\Gamma(a) = \xi_A(a)$ for each $a \in A$ that also produces a corresponding solution.*

Proof. The idea is to use the optimization algorithm from the proof of Theorem 2, where the vertex sets A and B are interchanged, to compute a placement of the vertices of B in time $O(n_B \log n_B + |E|)$, so that the length k' of the longest edge is minimized. Let k denote the window width of the obtained 2-layer drawing Γ.

Let $k^\star$ be the minimum window width of a 2-layer drawing $\Gamma^\star$ of G with $x_{\Gamma^\star}(a) = \xi_A(a)$. We show that $k \leq 2k^\star$. First, recall that the longest edge in Γ has length k'. Thus $k \leq 2k'$ as in the worst case, a vertex $v \in A$ has distance k' to both its leftmost and its rightmost neighbor. Second, consider the 2-layer drawing $\Gamma^\star$. Since the longest edge in $\Gamma^\star$ has length at most $k^\star$ and k' is chosen optimally, we obtain $k' \leq k^\star$. Combining both arguments, we obtain $k \leq 2k' \leq 2k^\star$.

Finally, drawing Γ is a corresponding solution as stated in the theorem. □

4   Open Problems

We conclude with some open problems. First, the case where all vertices can be freely positioned along ℓ_t and ℓ_b may be investigated in future work. Second, the setting of Theorem 3 with the additional constraint that the vertices in A are degree-restricted, is of interest. Third, other optimization criteria could be useful in practice. For instance, one may try to minimize the average x-span while potentially also weighting spans of important vertices differently.

References

1. Buchin, K., Buchin, M., Byrka, J., Nöllenburg, M., Okamoto, Y., Silveira, R., Wolff, A.: Drawing (complete) binary tanglegrams. Algorithmica **62**(1–2), 309–332 (2012). https://doi.org/10.1007/s00453-010-9456-3
2. Czabarka, É., Székely, L.A., Wagner, S.G.: A tanglegram Kuratowski theorem. J. Graph Theory **90**(2), 111–122 (2019). https://doi.org/10.1002/jgt.22370
3. Di Battista, G., Eades, P., Tamassia, R., Tollis, I.G.: Graph Drawing: Algorithms for the Visualization of Graphs. Prentice-Hall (1999)
4. Dumas, M., McGuffin, M.J., Robert, J.-M., Willig, M.-C.: Optimizing a radial layout of bipartite graphs for a tool visualizing security alerts. In: van Kreveld, M., Speckmann, B. (eds.) GD 2011. LNCS, vol. 7034, pp. 203–214. Springer, Heidelberg (2012). https://doi.org/10.1007/978-3-642-25878-7_20
5. Eades, P., Wormald, N.C.: Edge crossings in drawings of bipartite graphs. Algorithmica **11**(4), 379–403 (1994). https://doi.org/10.1007/BF01187020
6. Fernau, H., Kaufmann, M., Poths, M.: Comparing trees via crossing minimization. J. Comput. Syst. Sci. **76**(7), 593–608 (2010). https://doi.org/10.1016/j.jcss.2009.10.014
7. Garey, M.R., Johnson, D.S.: Crossing number is NP-complete. SIAM J. Algebraic Discrete Methods **4**(3), 312–316 (1983)

8. HuBMAP Consortium: CCF ASCT+B Reporter. https://hubmapconsortium. github.io/ccf-asct-reporter/
9. Kaufmann, M., Wagner, D. (eds.): Drawing Graphs. LNCS, vol. 2025. Springer, Heidelberg (2001). https://doi.org/10.1007/3-540-44969-8
10. Meulemans, W., Schulz, A.: A tale of two communities: assessing homophily in node-link diagrams. In: Di Giacomo, E., Lubiw, A. (eds.) GD 2015. LNCS, vol. 9411, pp. 489–501. Springer, Cham (2015). https://doi.org/10.1007/978-3-319-27261-0_40
11. Muñoz, X., Unger, W., Vrt'o, I.: One sided crossing minimization is NP-hard for sparse graphs. In: Mutzel, P., Jünger, M., Leipert, S. (eds.) GD 2001. LNCS, vol. 2265, pp. 115–123. Springer, Heidelberg (2002). https://doi.org/10.1007/3-540-45848-4_10
12. Papadimitriou, C.H.: The NP-completeness of the bandwidth minimization problem. Computing **16**(3), 263–270 (1976). https://doi.org/10.1007/BF02280884
13. Pezzotti, N., Fekete, J.D., Höllt, T., Lelieveldt, B.P.F., Eisemann, E., Vilanova, A.: Multiscale visualization and exploration of large bipartite graphs. Comput. Graph. Forum **37**(3), 549–560 (2018). https://doi.org/10.1111/cgf.13441
14. Scornavacca, C., Zickmann, F., Huson, D.H.: Tanglegrams for rooted phylogenetic trees and networks. Bioinformatics **27**(13), i248–i256 (2011). https://doi.org/10.1093/bioinformatics/btr210
15. Sugiyama, K., Tagawa, S., Toda, M.: Methods for visual understanding of hierarchical system structures. IEEE Trans. Syst. Man Cybern. **11**(2), 109–125 (1981). https://doi.org/10.1109/TSMC.1981.4308636

Sequentially Swapping Tokens: Further on Graph Classes

Hironori Kiya[1]🆔, Yuto Okada[2]🆔, Hirotaka Ono[2]🆔, and Yota Otachi[2](✉)🆔

[1] Kyushu University, Fukuoka, Japan
h-kiya@econ.kyushu-u.ac.jp
[2] Nagoya University, Nagoya, Japan
okada.yuto.b3@s.mail.nagoya-u.ac.jp, {ono,otachi}@nagoya-u.jp

Abstract. We study the following variant of the 15 puzzle. Given a graph and two token placements on the vertices, we want to find a walk of the minimum length (if any exists) such that the sequence of token swappings along the walk obtains one of the given token placements from the other one. This problem was introduced as SEQUENTIAL TOKEN SWAPPING by Yamanaka et al. [JGAA 2019], who showed that the problem is intractable in general but polynomial-time solvable for trees, complete graphs, and cycles. In this paper, we present a polynomial-time algorithm for block-cactus graphs, which include all previously known cases. We also present general tools for showing the hardness of problem on restricted graph classes such as chordal graphs and chordal bipartite graphs. We also show that the problem is hard on grids and king's graphs, which are the graphs corresponding to the 15 puzzle and its variant with relaxed moves.

Keywords: Sequential token swapping · The (generalized) 15 puzzle · Block-cactus graph · Grid graph · King's graph

1 Introduction

Let $G = (V, E)$ be an undirected graph and $f, f' \colon V \to \{1, \ldots, c\}$ be colorings of G.[1] We call a sequence $\langle f_1, \ldots, f_p \rangle$ of colorings of G a *swapping sequence* of length $p - 1$ from f to f' if $f_1 = f$, $f_p = f'$, and there is a walk $\langle w_1, w_2, \ldots, w_p \rangle$ such that for $2 \le i \le p$, f_i is obtained from f_{i-1} by *swapping* the colors of w_{i-1} and w_i; that is, $f_i(w_i) = f_{i-1}(w_{i-1})$, $f_i(w_{i-1}) = f_{i-1}(w_i)$, and $f_i(v) = f_{i-1}(v)$ for $v \notin \{w_{i-1}, w_i\}$. See Fig. 1. Now the problem can be formulated as follows.

Problem: SEQUENTIAL TOKEN SWAPPING
Input: A graph $G = (V, E)$, colorings f, f' of G, and an integer k.

---

[1] By a coloring, we mean a mapping from the vertex set to a color set, which is not necessarily a proper coloring.

Partially supported by JSPS KAKENHI Grant Numbers JP17H01698, JP17K19960, JP18H04091, JP20H05793, JP20H05967, JP21K11752, JP21K19765, JP21K21283, JP22H00513. The full version is available at https://arxiv.org/abs/2210.02835.

L. Gąsieniec (Ed.): SOFSEM 2023, LNCS 13878, pp. 222–235, 2023.
https://doi.org/10.1007/978-3-031-23101-8_15

Question: Is there a swapping sequence of length at most k from f to f'?

We assume that f and f' color the same number of vertices for each color since otherwise it becomes a trivial no-instance. We also assume that the input graph G is connected as a swapping sequence affects only one connected component.

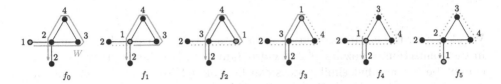

Fig. 1. An example of a swapping sequence.

The intuition behind its name, SEQUENTIAL TOKEN SWAPPING, is as follows: we consider a coloring as an assignment of colored tokens (or pebbles) to the vertices; we proceed along a walk; and when we visit an edge in the walk, we swap the tokens on the endpoints. For the ease of presentation, we often use the concept of tokens in this paper. For example, we call the token on the first vertex of the walk the *moving token* as it will always be the one exchanged during the swapping sequence. In other words, $f_i(w_i) = f_1(w_1)$ holds for all i.

Yamanaka et al. [17] introduced SEQUENTIAL TOKEN SWAPPING as a variant of the (generalized) 15 puzzle (Fig. 2), in which the first and last vertices in a swapping sequence are given as part of input. They showed that SEQUENTIAL TOKEN SWAPPING is polynomial-time solvable in some restricted cases such as trees, complete graphs, and cycles. They also showed that there is a constant $\varepsilon > 0$ such that the shortest length of a swapping sequence is NP-hard to approximate within a factor $1 + \varepsilon$.

Our Results. We unify and extend the positive results in [17] by showing that SEQUENTIAL TOKEN SWAPPING is polynomial-time solvable on block-cactus graphs, which include the classes of trees, complete graphs, and cycles. To this end, we first show that SEQUENTIAL TOKEN SWAPPING on a graph is reducible to a generalized problem (called SUB-STS) on its biconnected components, which may be of independent interest. We then show that the generalized problem SUB-STS can be solved in polynomial time on complete graphs and cycles. As a byproduct, we also show that the generalized 15 puzzle is polynomial-time solvable on the same graph class.

To complement the positive results, we show negative results on several classes of graphs. We first present two general tools for showing the NP-hardness of SEQUENTIAL TOKEN SWAPPING on restricted graph classes. One is for the *few-color case*, where we use only a fixed number of colors, and the other is for the *colorful case*, where we use a unique color for each vertex. The graph classes covered by the general tools include chordal graphs and chordal bipartite graphs.

Fig. 2. The 15 puzzle. Each step can be seen as a move of the vacant cell.

We also show the hardness on grids and king's graphs that play important roles in the connection to puzzles [9] and video games [5]. For them, our general tools cannot be applied, but similar ideas can be tailored. Also for split graphs, our general tools cannot be applied, but the NP-completeness of the few-color case follows as a corollary to some discussions for grid-like graphs. The complexity of the colorful case on split graphs remains unsettled.

Related Results. SEQUENTIAL TOKEN SWAPPING can be seen as a variant of the famous *15 puzzle*. The 15 puzzle is played on a 4 × 4 board with 16 cells. On the board, there are 15 pieces numbered from 1 to 15 and one vacant cell. In each turn, we can slide an adjacent piece to the vacant cell. The goal is to place the pieces at the right positions (see Fig. 2). By regarding the vacant cell (instead of an adjacent piece) as the piece moving in each step, we can see the sliding process in the 15 puzzle as a swapping sequence on the 4 × 4 grid that starts at the vacant cell. If we define the *generalized 15 puzzle* as the same problem considered on general graphs with arbitrary colorings, then it is almost the same as SEQUENTIAL TOKEN SWAPPING, and the difference is whether the first and last vertices in the walk corresponding to a swapping sequence are specified in the input (for the generalized 15 puzzle) or not (for SEQUENTIAL TOKEN SWAPPING).

The generalized 15 puzzle has been extensively studied with respect to the "reachability", i.e., under the setting where the question is the existence of a swapping sequence (not the minimum length). It was shown by Johnson and Story [9] that the reachability in the original 15 puzzle can be decided from the parity of the total distance from the initial to final token placements. This was later generalized further and a characterization of the reachability was given. For example, it is easy to see that the characterization given by Trakultraipruk [15] is polynomial-time testable. On the other hand, the problem of finding a swapping sequence of the minimum length has been studied only for a couple of cases. It was shown by Ratner and Warmuth [14] that the generalized 15 puzzle is NP-complete on $n \times n$ grids, in which case the problem is called the $(n^2 - 1)$ *puzzle*. A short proof for the same result was presented later by Demaine and Rudoy [2].

Although SEQUENTIAL TOKEN SWAPPING is quite close to the generalized 15 puzzle, its concept comes also from its non-sequential variant TOKEN SWAP-PING, which does not ask for the existence of a walk consistent with a swapping sequence but allows to swap the tokens on the endpoints of any edge in each step. The complexity of TOKEN SWAPPING has shown to be quite different from its sequential variant. For example, it is recently shown that TOKEN SWAPPING is NP-complete even on trees [1].

The generalized 15 puzzle and (SEQUENTIAL) TOKEN SWAPPING are sometimes considered in *combinatorial reconfiguration* as well. See the surveys [6, 13] for the background and related results in this context.

2 Preliminaries

We use standard terminologies for graphs (see e.g., [3] for the terms not defined here). Let $G = (V, E)$ be an undirected graph. For $S \subseteq V$, the subgraph of G induced by S is denoted by $G[S]$. A sequence $W = \langle w_1, \ldots, w_{|W|} \rangle$ of vertices is a *walk* of length $|W| - 1$ in G if $\{w_i, w_{i+1}\} \in E$ for $1 \leq i < |W|$. A vertex of a connected graph is a *cut vertex* if the removal of the vertex makes the graph disconnected. A connected graph is *biconnected* if it contains no cut vertex. A maximal induced biconnected subgraph of a graph is called a *biconnected component* of the graph. Let $\mathcal{B}_G$ denote the set of biconnected components of G. It is known that $\mathcal{B}_G$ can be computed in linear time [7].

A graph is a *cactus* if each biconnected component is a cycle or a 2-vertex complete graph. A graph is a *block graph* if each biconnected component is a complete graph. A graph is a *block-cactus graph* if each biconnected component is a cycle or a complete graph. A *chordal graph* is a graph with no induced cycle of length 4 or more. A *chordal bipartite graph* is a bipartite graph with no induced cycle of length 6 or more. A graph is a *split graph* if its vertex set can be partitioned into a clique and an independent set.

The $h \times w$ *grid* has the vertex set $V = \{1, \ldots, h\} \times \{1, \ldots, w\}$ and the edge set $\{\{(x, y), (x', y')\} \mid (x, y), (x', y') \in V, |x - x'| + |y - y'| = 1\}$. A graph is a *grid* if it is the $h \times w$ grid for some integers h and w. A graph is a *grid graph* if it is an induced subgraph of some grid. We say that a grid graph $G = (V, E)$ is given with a *grid representation* if $V \subseteq \mathbb{Z}^2$ and $E = \{\{(x, y), (x', y')\} \mid (x, y), (x', y') \in V, |x - x'| + |y - y'| = 1\}$. The $h \times w$ *king's graph* is obtained from the $h \times w$ grid by adding all diagonal edges of the unit squares (4-cycles) in the grid; that is, the vertex set is $V = \{1, \ldots, h\} \times \{1, \ldots, h\}$ and the edges set is $\{\{(x, y), (x', y')\} \mid (x, y), (x', y') \in V, \max\{|x - x'|, |y - y'|\} = 1\}$. A graph is a *king's graph* if it is the $h \times w$ king's graph for some integers h and w. We call a vertex (x, y) of a king's graph *even* if $x + y$ is even and *odd* if $x + y$ is odd. In passing, the name of a king's graph comes from the legal moves of the king chess piece on a chessboard.

As mentioned in Sect. 1, the generalized 15 puzzle can be seen as a variant of SEQUENTIAL TOKEN SWAPPING with the first and last vertices specified. In the following, we call it (s, t)-STS.

Problem: (s, t)-STS
Input: A graph $G = (V, E)$, colorings f, f' of G, $s, t \in V$, and an integer k.
Question: Is there a swapping sequence of length at most k from f to f' such that the corresponding walk starts at s and ends at t?

Note that s and t in an instance of (s, t)-STS are not necessarily distinct.

3   Polynomial-Time Algorithm for Block-Cactus Graphs

In this section, we present a polynomial-time algorithm for SEQUENTIAL TOKEN SWAPPING on block-cactus graphs. We prove the following theorem.

Theorem 3.1. SEQUENTIAL TOKEN SWAPPING *on block-cactus graphs can be solved in* $O(n^3)$ *time, where n is the number of vertices.*

Note that although Theorem 3.1 is stated for SEQUENTIAL TOKEN SWAPPING, which is a decision problem, the algorithm presented below actually solves the optimization version of the problem in the same running time. That is, it computes the minimum length of a swapping sequence from f to f' in $O(n^3)$ time.

The main part of the algorithm is the subroutine for solving (s, t)-STS. Given that subroutine, the algorithm just tries all pairs of vertices as the first and last vertices. In the following, we focus on this subroutine.

We show that the problem on a graph can be reduced to a generalized problem on its biconnected components. Then it suffices to show that the generalized problem can be solved in polynomial time on complete graphs and cycles. We prove this in a way similar to Yamanaka et al. [17] but the proofs here are much more involved because of the generality of the problem.

3.1   Reduction to a Generalized Problem on Biconnected Components

We generalize (s, t)-STS by adding a subset P of vertices to be visited as follows.

Problem: SUB-STS
Input: A graph $G = (V, E)$, colorings f, f' of G, $s, t \in V$, and $P \subseteq V$.
Task: Find the minimum length of a swapping sequence from f to f' (if any exists) such that the corresponding walk $W = \langle w_1, w_2, \ldots, w_{|W|} \rangle$ satisfies that $w_1 = s$, $w_{|W|} = t$, and $P \subseteq \{w_1, w_2, \ldots, w_{|W|}\}$.

Let $\lambda(G, f, f', s, t, P)$ denote the answer for the instance $\langle G, f, f', s, t, P \rangle$ of SUB-STS. We set it to ∞ if no swapping sequence from f to f' exists. Note that $\lambda(G, f, f', s, t, \emptyset)$ is the minimum k such that $\langle G, f, f', s, t, k \rangle$ is a yes-instance of (s, t)-STS.

Let $\langle G, f, f', s, t, k \rangle$ be an instance of (s, t)-STS and let H be a biconnected component of G. Let us see how a solution to this instance passes through H. If $s \notin V(H)$, then the first vertex visited in H is the cut vertex closest to s. Similarly, if $t \notin V(H)$, then the last vertex visited in H is the cut vertex closest to t. Also, a cut vertex u of G belonging to H has to be visited if at least one vertex in H is visited and there is a vertex $v \notin V(H)$ such that $f(v) \neq f'(v)$ and u is the closest vertex in H to v. With these observations, we construct an instance $\langle H, f_H, f'_H, s_H, t_H, P_H \rangle$ of SUB-STS as follows, where c_v is the cut vertex in H that separates v and $V(H)$.

Fig. 3. A swapping sequence on a cycle.

- Set $f_H = f|_{V(H)}$. If $s \notin V(H)$, then update f_H as $f_H(c_s) := f(s)$.
- Set $f'_H = f'|_{V(H)}$. If $t \notin V(H)$, then update f'_H as $f'_H(c_t) := f(s)$.
- Set $s_H = s$ if $s \in V(H)$. Otherwise, set $s_H = c_s$.
- Set $t_H = t$ if $t \in V(H)$. Otherwise, set $t_H = c_t$.
- Set P_H to the set of cut vertices c_v of G belonging to $V(H)$ such that c_v separates v and H for some $v \notin V(H)$ with $f(v) \neq f'(v)$.

The following lemma says that this instance correctly captures how a solution to (s, t)-STS on G affects H.[2]

Lemma 3.2 (★). *For a graph G, colorings f, f' of G, and $s, t \in V$,*

$$\lambda(G, f, f', s, t, \emptyset) = \sum_{H \in \mathcal{B}_G} \lambda(H, f_H, f'_H, s_H, t_H, P_H).$$

3.2 SUB-STS on cycles

Lemma 3.3 SUB-STS *on cycles can be solved in linear time.*

Proof Let $\langle C, f, f', s, t, P \rangle$ be an instance of SUB-STS, where C is a cycle of n vertices. We assume that $f(s) = f'(t)$ since otherwise it is a trivial no-instance. We arbitrarily fix a cyclic orientation on C and call it the clockwise direction (and the other one the counterclockwise direction).

Observe that if the moving token goes in one direction on the cycle, then the other tokens passed are shifted to the other direction (see Fig. 3). Observe also that if the moving token goes one step in one direction and goes back in the other direction immediately, then these moves cancel out and the coloring stays the same. Thus, if $P = \emptyset$, then an optimal solution never goes back and forth. Based on these observations, Yamanaka et al. [17] presented a polynomial-time algorithm for SEQUENTIAL TOKEN SWAPPING on cycles. We also use these facts, but since $P \neq \emptyset$ in general, we need some new ideas.

Let $W = \langle u_1, \ldots, u_p (= v_1), \ldots, v_q (= w_1), \ldots, w_r \rangle$ be a walk corresponding to a desired swapping sequence of the minimum length, where

- v_1 is the last vertex in W such that $v_1 = s$ and the coloring after executing the swapping sequence up to v_1 is f, and
- v_q is the first vertex in W such that $v_q = t$ and the coloring after executing the swapping sequence up to v_q is f'.

---

[2] The proofs of the statements marked with ★ are omitted in this short version and can be found in the full version.

We show that there is a direction ←, which is clockwise or counterclockwise, such that the following properties hold:

- the moves along $\langle u_1, \ldots, u_p \rangle$ first go in the direction ← some number of steps and then go back in the opposite direction → the same number of steps;
- the moves along $\langle v_1, \ldots, v_q \rangle$ go in the direction → only;
- the moves along $\langle w_1, \ldots, w_r \rangle$ first go in the direction → some number of steps and then go back in the direction ← the same number of steps.

To show the property of $\langle v_1, \ldots, v_q \rangle$, assume that $q \geq 2$ and that v_2 is the clockwise neighbor of v_1. If $v_i = s$ for some $i \neq 1$, then $V(C) = \{v_1, \ldots, v_q\}$ holds as the coloring after executing the swapping sequence up to v_i is not f. Similarly, if $v_i = t$ for some $i \neq q$, then $V(C) = \{v_1, \ldots, v_q\}$ holds as the coloring after executing the swapping sequence up to v_i is not f'. Otherwise, $\{v_1, \ldots, v_q\}$ is the set of consecutive vertices on C from s to t in the clockwise direction. In all cases, if there is a counterclockwise move, then the first such move $v_j \to v_{j+1}$ can be removed with the previous one $v_{j-1} \to v_j$ without changing the set of visited vertices. Since W is of the minimum length, we can conclude that there is no such move. Now the properties of $\langle u_1, \ldots, u_p \rangle$ and $\langle w_1, \ldots, w_r \rangle$ follows easily as they are necessary only for visiting more vertices (in P).

We compute the minimum length for each of the cases $V(C) \neq \{v_1, \ldots, v_q\}$ and $V(C) = \{v_1, \ldots, v_q\}$.

The case of $V(C) \neq \{v_1, \ldots, v_q\}$. In this case, $\langle v_1, \ldots, v_q \rangle$ is a simple path from s to t. We assume that this is a clockwise path. (The other case is symmetric.) Since only $\langle v_1, \ldots, v_q \rangle$ changes the coloring and the other parts of W cancel out, we first check that we get f' by applying $\langle v_1, \ldots, v_q \rangle$ to f and then compute the other parts that visit $P \setminus \{v_1, \ldots, v_q\}$. Let $\langle x_1, \ldots, x_k \rangle$ be the sequence of the vertices of $P \setminus \{v_1, \ldots, v_q\}$ ordered in the counterclockwise order from s to t. Observe that if $p \geq 2$, then the first part $\langle u_1, \ldots, u_p \rangle$ of W starts at s in the counterclockwise direction, visits some vertices $x_1, \ldots, x_{k'}$, and comes back to s. Thus, its length is the twice of the distance from s to $x_{k'}$ in the counterclockwise direction. Similarly, if $r \geq 2$, then the last part $\langle w_1, \ldots, w_r \rangle$ of W starts at t in the clockwise direction, visits the remaining vertices $x_{k'+1}, \ldots, x_k$, and comes back to t. Its length is the twice of the distance from t to $x_{k'+1}$ in the clockwise direction. The index $k' \in \{0, \ldots, k\}$ that minimizes the sum $p + r$ can be found in linear time by precomputing the counterclockwise distances from s to $x_1, \ldots, x_k$ and the clockwise distances from t to $x_1, \ldots, x_k$ in linear time.

The case of $V(C) = \{v_1, \ldots, v_q\}$. In this case, $W = \langle v_1, \ldots, v_q \rangle$ as there is no other vertex to visit. Now it is easy to compute the minimum length in polynomial time: guess the direction of the walk; go in the guessed direction $n-1$ steps from s; and then further proceed in the same direction until we get the desired coloring. Since the minimum length is $O(n^3)$ (if not ∞) in general [17], this algorithm runs in polynomial time.

To do it in linear time, we reduce the problem to a substring matching problem that can be solved in linear time by the KMP algorithm [10].

Assume that W goes in the clockwise direction. (The other case is symmetric.) Let g be the coloring obtained from f by executing the swapping sequence along W up to the first point where all vertices are visited and the moving token is placed at t. The remaining of the walk we are looking for repeats the (clockwise) walk from t to t some number of times. Observe that if we repeat it i times, then the coloring we get is the one obtained from g by shifting the non-moving tokens i steps in the counterclockwise direction (see Fig. 3). Thus it suffices to compute the minimum number of shifts to obtain f'.

Let t_{next} and t_{prev} be the clockwise and counterclockwise neighbors of t, respectively. Let $S_g = \langle c_1, \ldots, c_{n-1} \rangle$ be the sequence of the colors under g of vertices from t_{next} to t_{prev} in the clockwise ordering. Similarly, let $S_{f'}$ be the same sequence but under f'. Observe that if $S_{f'}$ can be obtained from S_g by i cyclic shifts (in the counterclockwise direction, or to the left in this context), then $S_{f'} = \langle c_{i+1}, \ldots, c_{n-1}, c_1, \ldots, c_i \rangle$ holds. The minimum i satisfying this can be found by finding the first index such that $S_{f'}$ starts in $S_g \cdot S_g = \langle c_1, \ldots, c_{n-1}, c_1, \ldots, c_{n-1} \rangle$ as a substring, which can be done in linear time [10].   □

3.3   Sub-STS on complete graphs

Let $I = \langle K, f, f', s, t, P \rangle$ be an instance of Sub-STS, where $K = (V, E)$ is a complete graph. As before, we assume that $f(s) = f'(t)$. Furthermore, we assume that s has a unique color under f (and so does t under f'). We set the unique color to 0. That is, we assume that $f(s) = f'(t) = 0$, $f(v) \neq 0$ if $v \neq s$, and $f'(v) \neq 0$ if $v \neq t$. Observe that this does not change the instance since the moving token anyway moves from s to t.

Let $R = \{v \in V \mid f(v) \neq f'(v)\} \cup \{s, t\} \cup P$. We define a directed multigraph $D = (V_D, E_D)$, possibly with self-loops and parallel edges, as $V_D = \{f(v) \mid v \in R\}$ and $E_D = \{(f(v), f'(v)) \mid v \in R\}$. This graph D is almost the same as the conflict graph defined in [17]. The difference here is the self-loops corresponding to the vertices in $P \setminus \{v \in V \mid f(v) \neq f'(v)\}$ (and s when $s = t$). Thus the assumption that f and f' use the same number of vertices for each color implies that the indegree and the outdegree are the same for each vertex (or, color) in D. This implies that each connected component of D is strongly connected and has an Eulerian circuit. Let $\text{cc}(D)$ denote the number of (strongly) connected components of D.

We show that the following equation holds (the proof is omitted in this short version):

$$\lambda(I) = |R| + \text{cc}(D) - 2. \tag{1}$$

Lemma 3.4 (★). $\lambda(I) \leq |R| + \text{cc}(D) - 2$.

Lemma 3.5 (★). $\lambda(I) \geq |R| + \text{cc}(D) - 2$.

3.4   The Whole Algorithm

Let $\langle G = (V, E), f, f', k \rangle$ be an instance of Sequential Token Swapping, such that $G = (V, E)$ is a block-cactus graph with $|V| = n$ and $|E| = m$. We first compute the set $\mathcal{B}_G$ of the biconnected components. For each $H \in \mathcal{B}_G$, we mark all cut vertices, check whether H contains a vertex v with $f(v) \neq f(v')$, and check whether H is a cycle or a complete graph. If H is a complete graph, then we construct an implicit representation so that we do not have to store the redundant information $E(H)$. These preprocessing can be done in $O(m+n)$ time in total.

Let $s, t \in V$. We compute the instance $I_H = \langle H, f_H, f'_H, s_H, t_H, P_H \rangle$ of SUB-STS for all $H \in \mathcal{B}_G$. We can do it in $O(m+n)$ time in a bottom-up manner over the tree structure of the biconnected components. Let $H \in \mathcal{B}_G$. If H is a cycle, then we compute $\lambda(I_H)$ in $O(|V(H)|)$ time using the algorithm in Lemma 3.3. If H is a complete graph, then we compute $\lambda(I_H)$ using Eq. (1), which can be done in $O(|V(H)|)$ time from the implicit representation of H. Thus, by Lemma 3.2, we can solve (s, t)-STS in $O(n)$ time, given that the aforementioned preprocessing is done. Since we have n^2 candidates for the pair s, t, the total running time is $O(n^3)$. This completes the proof of Theorem 3.1.

Note that we only need $O(m+n)$ time to solve (s, t)-STS.

Corollary 3.6. *(s, t)-STS on block-cactus graphs can be solved in linear time.*

4   Hardness of the Few-Color and Colorful Cases

Since SEQUENTIAL TOKEN SWAPPING clearly belongs to NP, in the following we only show the NP-hardness for each case.

4.1   General Tools for Showing Hardness

The first tool uses the hardness of HAMILTONIAN PATH to show the hardness of SEQUENTIAL TOKEN SWAPPING with few colors.

A path (a cycle) in a graph is a *Hamiltonian path* (a *Hamiltonian cycle*, resp.) if it visits every vertex in the graph exactly once. The problems HAMILTONIAN PATH and HAMILTONIAN CYCLE ask whether a given graph has a Hamiltonian path or a Hamiltonian cycle, respectively. In the problem (s, t)-HAMILTONIAN PATH, the first and last vertices s and t are fixed.

By *attaching a cycle* of length q at a vertex v, we mean the operation of adding $q - 1$ new vertices and q edges that form a cycle with v.

Theorem 4.1 (★). *Let $\mathcal{C}$ be a graph class. For every fixed $c \geq 2$, SEQUENTIAL TOKEN SWAPPING with c colors is NP-complete on $\mathcal{C}$ if the following conditions are satisfied:*

1. *(s, t)-HAMILTONIAN PATH is NP-complete on $\mathcal{C}$;*
2. *there is an integer $q \geq 3$ such that $\mathcal{C}$ is closed under the operation that attaches a cycle of length q at a vertex.*

The second tool uses the hardness of STEINER TREE to show the hardness of the colorful case of SEQUENTIAL TOKEN SWAPPING. In this case, we ask f to be injective and call this condition the *colorful condition*.

For a walk W, let $V(W)$ be the set of vertices in W.

Lemma 4.2 (★). *Let f be an injective coloring of a graph G. If a walk W in G corresponds to a swapping sequence from f to f itself, then $|V(W)| \leq (|W|+1)/2$.*

For a graph $G = (V, E)$ and a set $K \subseteq V$, the subgraph T of G is a *Steiner tree* if it is a tree and contains all vertices in K.

Problem: STEINER TREE
Input: A graph $G = (V, E)$, a set $K \subseteq V$ with $|K| \geq 2$, and an integer ℓ.
Question: Is there a connected subgraph T of G such that $K \subseteq V(T)$ and $|E(T)| \leq \ell$?

Theorem 4.3 (★). SEQUENTIAL TOKEN SWAPPING *with the colorful condition is NP-complete on a graph class C if the following conditions are satisfied:*

1. STEINER TREE *is NP-complete on C;*
2. *there is an integer $q \geq 3$ such that C is closed under the operation that attaches a cycle of length q at a vertex.*

It is known that STEINER TREE is NP-complete on chordal graphs [16] and chordal bipartite graphs [12]. It is also known that (s, t)-HAMILTONIAN PATH is NP-complete on chordal graphs and chordal bipartite graphs [11]. Observe that chordal graphs and chordal bipartite graphs are closed under the operations that attach a cycle of length 3 and 4, respectively. Thus, Theorems 4.1 and 4.3 implies the hardness on them.

Corollary 4.4. SEQUENTIAL TOKEN SWAPPING *is NP-complete on chordal graphs and on chordal bipartite graphs in both the colorful and few-color cases.*

4.2 The Few-Color Case on Grid-Like Graphs

Recall that a graph is a *grid graph* if it is an induced subgraph of a grid. A bipartite graph is *balanced* if it admits a proper 2-coloring such that the color classes have the same size. Note that a grid graph is bipartite.

It is known that HAMILTONIAN CYCLE is NP-complete on grid graphs [8]. The next lemma follows easily from this fact.

Lemma 4.5 (★). HAMILTONIAN PATH *is NP-complete on balanced grid graphs given with grid representations.*

Theorem 4.6. *For every fixed constant $c \geq 2$,* SEQUENTIAL TOKEN SWAPPING *with c colors is NP-complete on king's graphs.*

Proof. We prove the theorem only for the case where $c = 2$. For $c > 2$, we add $c - 2$ new vertices to G defined below and for each new vertex, set a new color as its initial and target colors. Then the proof works as it is.

We prove the NP-hardness by a reduction from HAMILTONIAN PATH on balanced grid graphs (see Lemma 4.5).

Let $G = (V, E)$ be a balanced grid graph given with a grid representation. We assume that G is connected. From G, we construct an instance $\langle H, f, f', k \rangle$ of SEQUENTIAL TOKEN SWAPPING. We set $k = |V| - 1$.

Let $\min_x = \min\{x \in \mathbb{Z} \mid (x, y) \in V\}$ and $\min_y = \min\{y \in \mathbb{Z} \mid (x, y) \in V\}$. We also define $\max_x$ and $\max_y$ in the analogous ways. Let $U = \{(x, y) \in \mathbb{Z}^2 \setminus V \mid \min_x \leq x \leq \max_x, \min_y \leq x \leq \max_y\}$. The grid graph represented by $U \cup V$ is a grid and has size polynomial in $|V|$. From this grid, we obtain H by adding all diagonal edges of the unit squares. Note that H is a king's graph.

Let f be the coloring of H that maps the odd vertices to 1 and the even vertices to 2. Let f' be the coloring obtained from f by reversing the colors of the vertices in the original grid graph G. That is, $f'(v) = f(v)$ for $v \in U$, $f'(v) = 1$ for $v \in V$ with $f(v) = 2$, and $f'(v) = 2$ for $v \in V$ with $f(v) = 1$.

We show that G has a Hamiltonian path if and only if $\langle H, f, f', k \rangle$ is a yes-instance of SEQUENTIAL TOKEN SWAPPING.

The Only-if Direction. Let $P = \langle v_1, \ldots, v_{|V|} \rangle$ be a Hamiltonian path of G. Let $S = \langle f_1, \ldots, f_{|V|} \rangle$ be a swapping sequence corresponding to P, where $f_1 = f$. Since each vertex in the walk is visited only once, we have $f_{|V|}(v_i) = f(v_{i+1})$ for $1 \leq i \leq |V| - 1$, and $f_{|V|}(v_{|V|}) = f(v_1)$. Since P is a path of G, v_i and v_{i+1} have different parities. Also, since G is balanced, $v_{|V|}$ and v_1 have different parities. Therefore, $f_{|V|}$ is obtained from f_1 by changing the color of each vertex in V to the other one. That is, $f_{|V|} = f'$.

The if Direction. Let $\langle f_1, \ldots, f_{k'+1} \rangle$ be a swapping sequence between f and f' with $k' \leq k$. Let $W = \langle w_1, \ldots, w_{k'+1} \rangle$ be the corresponding walk in H. Since $f(v) \neq f'(v)$ for every $v \in V$, the moving token has to visit all vertices in V. Furthermore, since $|W| = k' + 1 \leq k + 1 = |V|$, indeed the moving token visits each vertex in V exactly once and does not visit other vertices (and thus, $k' = k$). Hence, it suffices to show that W is a walk also in G; that is, each edge $\{w_i, w_{i+1}\}$ is not diagonal. Suppose to the contrary that an edge $\{w_i, w_{i+1}\}$ in W is diagonal. This implies that w_i and w_{i+1} have the same parity, and thus $f(w_i) = f(w_{i+1})$. Since W visits a vertex at most once, $f(w_{i+1}) = f'(w_i)$ has to hold. Thus we have that $f(w_i) = f'(w_i)$. This contradicts the assumption that $f'(v) \neq f(v)$ for each $v \in V$.   □

In the proof above, we showed that no diagonal edge is used in shortest swapping sequences. Therefore, the proofs work without the diagonal edges.

Corollary 4.7. *For every fixed constant $c \geq 2$, SEQUENTIAL TOKEN SWAPPING with c colors is NP-complete on grids.*

Observe further that the proof of Theorem 4.6 works even if we add or remove an arbitrary set of edges connecting vertices of the same parity since we can just ignore them. Now consider the graph obtained from a king's graph by removing all edges connecting even vertices and adding all possible edges connecting odd vertices. Such a graph is a split graph since the even vertices form an independent set and the odd vertices form a clique. Thus it is hard on split graphs as well.

Corollary 4.8. *For every fixed constant* $c \geq 2$, SEQUENTIAL TOKEN SWAPPING *with* c *colors is NP-complete on split graphs.*

4.3   The Colorful Case on Grid-Like Graphs

We now consider SEQUENTIAL TOKEN SWAPPING with the colorful condition on grid-like graphs.

We first show the hardness on the ordinary grids. Recall that (s, t)-STS with the colorful condition on grids are known as the generalized 15 puzzle (or the $(n^2 - 1)$ puzzle) and shown to be NP-complete [2,14]. For SEQUENTIAL TOKEN SWAPPING, we can use the reduction by Demaine and Rudoy [2] almost directly with a small change. Their reduction is from the following problem.

Problem: RECTILINEAR STEINER TREE
Input: A set $P \subseteq \mathbb{Z}_+^2$ of integer points in the plane and an integer ℓ.
Question: Is there a tree T on the plane that satisfies the following conditions?
T contains all points in P; every edge of T is horizontal or vertical; the total length of the edges in T is at most ℓ.

RECTILINEAR STEINER TREE is known to be strongly NP-hard [4], and thus we assume that the maximum coordinate of the points in P is bounded from above by a polynomial in $|P|$.

The high-level idea of the reduction in [2] is to represent the integer points in the plane by a grid and then each point in P by some local changes of the colors around the vertex corresponding to the point. Then, a swapping sequence for this instance forms a rectilinear Steiner tree on the plane. The difference between their setting and ours is that they can fix the starting and ending vertices, but we cannot. This difference actually does not affect the correctness of their proof applied to our case. To not repeat their argument here, we only give a proof sketch.

Theorem 4.9 (★). SEQUENTIAL TOKEN SWAPPING *with the colorful condition is NP-complete on grids.*

Before showing the hardness of SEQUENTIAL TOKEN SWAPPING with the colorful condition on king's graphs, we first show that STEINER TREE is NP-complete on king's graphs since the proofs are similar and this one is easier than the one for SEQUENTIAL TOKEN SWAPPING. Also, the result itself might be useful for connecting some graph problems and geometric problems.

Theorem 4.10 (★). STEINER TREE *is NP-complete on king's graphs.*

We now show the hardness of SEQUENTIAL TOKEN SWAPPING with the colorful condition on king's graphs. Although the proof is similar to the one for grids, the presence of diagonal edges makes it a little more complicated.

Theorem 4.11 (★). SEQUENTIAL TOKEN SWAPPING *with the colorful condition is NP-complete on king's graphs.*

5 Concluding Remarks

We have studied SEQUENTIAL TOKEN SWAPPING from the view point of restricted graph classes and shown several positive and negative results. We note that the complexity of the problem with the colorful condition remained unsettled for split graphs.

As another direction, it would be interesting to study the parameterized complexity of the problem. Lemma 3.2 and the $O(n^3)$ upper bound of the minimum length of a swapping sequence [17] together imply that SEQUENTIAL TOKEN SWAPPING is fixed-parameter tractable parameterized by the maximum size of a biconnected component. To the best of our knowledge, nothing is known for other structural graph parameters.

References

1. Aichholzer, O., et al.: Hardness of token swapping on trees. In: ESA 2022. LIPIcs, vol. 244, pp. 3:1–3:15 (2022). https://doi.org/10.4230/LIPIcs.ESA.2022.3
2. Demaine, E.D., Rudoy, M.: A simple proof that the $(n^2 - 1)$-puzzle is hard. Theor. Comput. Sci. **732**, 80–84 (2018). https://doi.org/10.1016/j.tcs.2018.04.031
3. Diestel, R.: Graph Theory. GTM, vol. 173. Springer, Heidelberg (2017). https://doi.org/10.1007/978-3-662-53622-3
4. Garey, M.R., Johnson, D.S.: The rectilinear Steiner tree problem is NP-complete. SIAM J. Appl. Math. **32**(4), 826–834 (1977). https://doi.org/10.1137/0132071
5. GungHo Online Entertainment America, Inc.: Puzzle & Dragons, official website. https://www.puzzleanddragons.us/. Accessed 22 July 2022
6. van den Heuvel, J.: The complexity of change. In: Blackburn, S.R., Gerke, S., Wildon, M. (eds.) Surveys in Combinatorics 2013, London Mathematical Society Lecture Note Series, vol. 409, pp. 127–160. Cambridge University Press (2013). https://doi.org/10.1017/CBO9781139506748.005
7. Hopcroft, J.E., Tarjan, R.E.: Algorithm 447: efficient algorithms for graph manipulation. Commun. ACM **16**(6), 372–378 (1973). https://doi.org/10.1145/362248.362272
8. Itai, A., Papadimitriou, C.H., Szwarcfiter, J.L.: Hamilton paths in grid graphs. SIAM J. Comput. **11**(4), 676–686 (1982). https://doi.org/10.1137/0211056
9. Johnson, W.W., Story, W.E.: Notes on the "15" puzzle. Am. J. Math. **2**(4), 397–404 (1879). https://doi.org/10.2307/2369492
10. Knuth, D.E., Jr., J.H.M., Pratt, V.R.: Fast pattern matching in strings. SIAM J. Comput. **6**(2), 323–350 (1977). https://doi.org/10.1137/0206024
11. Müller, H.: Hamiltonian circuits in chordal bipartite graphs. Discret. Math. **156**(1–3), 291–298 (1996). https://doi.org/10.1016/0012-365X(95)00057-4
12. Müller, H., Brandstädt, A.: The NP-completeness of steiner tree and dominating set for chordal bipartite graphs. Theor. Comput. Sci. **53**, 257–265 (1987). https://doi.org/10.1016/0304-3975(87)90067-3
13. Nishimura, N.: Introduction to reconfiguration. Algorithms **11**(4), 52 (2018). https://doi.org/10.3390/a11040052
14. Ratner, D., Warmuth, M.K.: The $(n^2 - 1)$-puzzle and related relocation problems. J. Symb. Comput. **10**(2), 111–138 (1990). https://doi.org/10.1016/S0747-7171(08)80001-6

15. Trakultraipruk, S.: Connectivity properties of some transformation graphs, Ph. D. thesis, London School of Economics and Political Science, London, UK (2013)
16. White, K., Farber, M., Pulleyblank, W.R.: Steiner trees, connected domination and strongly chordal graphs. Networks **15**(1), 109–124 (1985). https://doi.org/10.1002/net.3230150109
17. Yamanaka, K., et al.: Sequentially swapping colored tokens on graphs. J. Graph Algorithms Appl. **23**(1), 3–27 (2019). https://doi.org/10.7155/jgaa.00482

Communication and Temporal Graphs

On the Preservation of Properties When Changing Communication Models

Olav Bunte[1]([✉]), Louis C. M. van Gool[2], and Tim A. C. Willemse[1]

[1] Eindhoven University of Technology, Eindhoven, The Netherlands
{o.bunte,t.a.c.willemse}@tue.nl
[2] Canon Production Printing, Venlo, The Netherlands
louis.vangool@cpp.canon

Abstract. In a system of processes that communicate asynchronously by means of FIFO channels, there are many options in which these channels can be laid out. In this paper, we compare channel layouts in how they affect the behaviour of the system using an ordering based on splitting and merging channels. This order induces a simulation relation, from which the preservation of safety properties follows. Also, we identify conditions under which the properties reachability, deadlock freedom and confluence are preserved when changing the channel layout.

Keywords: Asynchronous communication · Communication models · Property preservation · Confluence

1 Introduction

In asynchronous communication, sending and receiving a message are two separate actions, which makes it possible for messages to be received in a different order than they were sent. What orderings are possible, depends on the asynchronous communication model(s) used within the system, for which many flavours are possible. We consider communication models that are implemented by means of a layout of (unbounded) FIFO (First In First Out) channels, which defines how messages in transit are stored. For instance, using a channel per message implements a fully asynchronous model, while having a single input channel per process enforces that messages that are sent to the same process are received in the same order in which they are sent.

While (re)designing or refactoring a software system of asynchronously communicating processes, it may be desirable to change (part of) the channel layout. This can, for instance, be the case when design choices are still being explored, when the performance of the system needs to be improved, when the behaviour of the system has grown too complex due to the additions of new processes, or when the channel implementation is part of legacy software. However, changing

---

This work was carried out as part of the VOICE-B project, which is funded by Canon Production Printing.

L. Gąsieniec (Ed.): SOFSEM 2023, LNCS 13878, pp. 239–253, 2023.
https://doi.org/10.1007/978-3-031-23101-8_16

the channel layout may impact the behaviour of the system in unexpected ways, possibly violating desired properties. In this paper, we investigate the extent of this impact.

We use the notion of a FIFO system [1,7] to represent a software system of asynchronously communicating processes. Firstly, we define an ordering on FIFO systems based on whether one can be created from the other by merging channels. We then analyse the difference in the behaviour between related FIFO systems and show that it induces a simulation order, from which the preservation of safety properties follows. Secondly, we analyse whether reachability, deadlock freedom and confluence are preserved when changing the channel layout. Reachability is particularly relevant in practice, since changing the method of communication should typically not cause previously possible process behaviour to become impossible. If deadlock freedom is preserved, it is ensured that changing the method of communication does not introduce undesired situations where all processes are stuck waiting for each other. Confluence is related to the independence of actions, which is often expected between actions of different processes. A violation of confluence between actions of different processes indicates a possible race condition, where the faster process determines how the system progresses. We identify conditions under which these properties are guaranteed to be preserved when merging or splitting channels.

Related Work. In [6], seven distinct channel-based asynchronous communication models are related to each other in a hierarchy based on trace and MSC implementability. The authors of [4] also consider the causal communication model [10], and show a similar hierarchy. They prove this hierarchy correct in [3] using automated proof techniques. Compared to these works, we consider mixed (channel-based) communication models, which is more realistic for complex software systems. For communication models that can be defined by FIFO systems, the hierarchies in these works relate these models the same way as our relation does.

Property preservation is closely related to the field of incremental model checking [8,12,14], which is an efficient method for rechecking a property on a system that has undergone some changes. Under some conditions, one can actually prove that a property will be preserved, as shown in multiple contexts [5,9,11,13,15]. To our knowledge, no such work exists in the context of asynchronously communicating processes however.

Outline. We first introduce the necessary definitions to reason about FIFO systems in Sect. 2. We define an ordering between FIFO systems in Sect. 3 and show that it induces a simulation relation. Then in Sect. 4 we identify conditions under which the aforementioned properties are preserved when changing the channel layout. Lastly, we conclude in Sect. 5. The proofs for all lemmas and theorems in Sect. 3 and 4 can be found in [2].

2   The FIFO System

Let P be a set of processes that make up a software system and let M be the set of messages that can be communicated between these processes. We represent the behaviour of each process $p \in P$ with a Labelled Transition System (LTS) $B_p = \langle Q_p, q_p^0, L_p, \rightarrow_p \rangle$ where Q_p is its set of states, q_p^0 its initial state, $L_p \subseteq (\{?, !\} \times M) \cup \{\tau\}$ its set of actions and $\rightarrow_p \subseteq Q_p \times L_p \times Q_p$ its transition relation. An action $?m$ indicates the receiving of m, $!m$ the sending of m and τ is an internal action. We assume that processes do not share non-internal actions, that is $L_p \cap L_{p'} \subseteq \{\tau\}$ for all distinct $p, p' \in P$. We write $q \xrightarrow{a}_p q'$ iff $(q, a, q') \in \rightarrow_p$. A FIFO system then describes how these processes communicate with each other via FIFO channels.

Definition 1. *A FIFO system is a tuple $\langle P, C, M \rangle$ where $C \subseteq \mathcal{P}(M)$ is a set of (FIFO) channels defined as a partition of M.*

Each channel in C is defined as a set of messages, which represents the messages that this channel can hold. Note that because C partitions M, we assume that each message can only be sent to and received from exactly one channel. For a message $m \in M$, we define $[m]_C$ as the channel in C that m belongs to, that is $m \in [m]_C$ and $[m]_C \in C$. We write $m \simeq_C o$ iff $[m]_C = [o]_C$ for messages $m, o \in M$.

We define M^* as the set of all finite sequences of messages, also known as *words*. We use ϵ as the empty word and concatenate two words with $+\!\!+$. Given a word $m +\!\!+ w$ for message $m \in M$ and word $w \in M^*$, we define its head as $hd(m +\!\!+ w) = m$ and its tail as $tl(m +\!\!+ w) = w$.

Example 1. Imagine two vending machines, one for healthy snacks and one for unhealthy snacks, and some user who can interact with these vending machines. After receiving a "healthy voucher" ⬛, the healthy vending machine can supply apples ● and bananas ⟋. After receiving an "unhealthy voucher ⬛, the unhealthy vending machine can supply chocolate ▰▰ and donuts ◉. The user decides to use ⬛ before ⬛ and can receive the snacks whenever they are ready.

Let $P_V = \{hvm, uvm, user\}$ be processes that represent the two vending machines and the user. Their LTSs are visualised in Fig. 1. The set of messages is $M_V = \{⬛, ⬛, ●, ⟋, ▰▰, ◉\}$. The realistic case where both vending

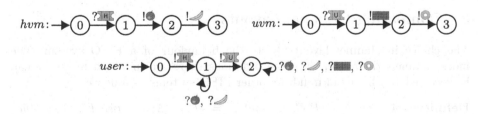

Fig. 1. Processes *hvm*, *uvm* and *user* for Example 1 and 2.

machines have their own voucher slot and output slot is represented by the channel set $C_V = \{\{\blacksquare\}, \{\blacksquare\}, \{\bullet, \mathcal{A}\}, \{\blacksquare, \mathbf{O}\}\}$, resulting in the FIFO system $\langle P_V, C_V, M_V \rangle$. Note that $\bullet \simeq_{C_V} \mathcal{A}$ and $\blacksquare \simeq_{C_V} \mathbf{O}$.

Semantics. A FIFO system induces an LTS that represents the communication behaviour between all processes. A state in this LTS consists of two parts: the states of the individual processes and the contents of the channels. For a set of processes P, $\mathbf{P} = \{\kappa \in P \to \bigcup_{p \in P} Q_p \mid \forall_{p \in P} : \kappa(p) \in Q_p\}$ denotes the set of functions that map processes to their current states. For a set of channels C, $\mathbf{C} = \{\zeta \in C \to M^* \mid \forall_{c \in C} : \zeta(c) \in c^*\}$ denotes the set of functions that map channels to their contents. In case of a set of channels C', we write $\mathbf{C'}$. Note that we assume unbounded channels.

Definition 2. *Let* $F = \langle (Q_p, q_p^0, \to_p)_{p \in P}, C, M \rangle$ *be a FIFO system. The semantics of* F *is an LTS* $B_F = \langle S, s_0, L, \to \rangle$ *where*

- $S = \mathbf{P} \times \mathbf{C}$,
- $s_0 = (\kappa_0, \zeta_\epsilon)$, *where* $\kappa_0(p) = q_p^0$ *for all* $p \in P$ *and* $\zeta_\epsilon(c) = \epsilon$ *for all* $c \in C$,
- $L = (\{?, !\} \times M) \cup \{\tau\}$,
- $\to \subseteq S \times L \times S$ *such that for all* $(\kappa, \zeta) \in S$, $p \in P$, $q \in Q_p$ *and* $m \in M$, *with* $c = [m]_C$:

$$(\kappa, \zeta) \xrightarrow{\tau} (\kappa[p \mapsto q], \zeta) \text{ iff } \kappa(p) \xrightarrow{\tau}_p q$$

$$(\kappa, \zeta) \xrightarrow{?m} (\kappa[p \mapsto q], \zeta[c \mapsto tl(\zeta(c))]) \text{ iff } \kappa(p) \xrightarrow{?m}_p q \wedge hd(\zeta(c)) = m$$

$$(\kappa, \zeta) \xrightarrow{!m} (\kappa[p \mapsto q], \zeta[c \mapsto \zeta(c) + \!\!+ m]) \text{ iff } \kappa(p) \xrightarrow{!m}_p q$$

We write $s \xrightarrow{a} s'$ iff $(s, a, s') \in \to$. We lift the transition relation to one over sequences of actions $\to^* \subseteq S \times L^* \times S$ in the usual way. In the context of a FIFO system F, we refer to the semantics of the FIFO system as defined above as "the LTS of F".

Two LTSs can be compared by means of a simulation relation [11].

Definition 3. *Let* $B = \langle S, s_0, L, \to \rangle$ *and* $B' = \langle S', s_0', L, \to' \rangle$ *be two LTSs. We say that* B *simulates* B' *iff there exists a simulation relation* $R \subseteq S' \times S$ *such that* $s_0' R s_0$ *and for all* $s \in S$ *and* $s' \in S'$, *if* $s'Rs$ *and* $s' \xrightarrow{a}' t'$ *for some* $t' \in S'$ *and* $a \in L$, *then there must exist a* $t \in S$ *such that* $s \xrightarrow{a} t$ *and* $t'Rt$.

3   Comparing Channel Layouts

The choice in channel layout affects the behaviour of a FIFO system. The more channels there are, the more orderings there are in which messages can be received. With this in mind, we order FIFO systems as follows:

Definition 4. *Let* $F = \langle P, C, M \rangle$ *and* $F' = \langle P, C', M \rangle$ *be two FIFO systems. We define the relation* $\succ$ *on FIFO systems such that* $F \succ F'$ *iff* $C \neq C'$ *and* $\forall_{c \in C} : \exists_{c' \in C'} : c \subseteq c'$ *(that is, C is a more refined partition of M than C' is).*

One can create F' from F by merging a number of channels (splitting channels in the opposite direction). We first illustrate how this affects the behaviour of the system with an example.

Example 2. Continuing from Example 1, consider the FIFO systems $F_m = \langle P_V, \{\{\blacksquare\}, \{\blacksquare\}, \{\bullet\}, \{\diagup\}, \{\blacksquare\}, \{\bigcirc\}\}, M_V\rangle$ (one channel per message), $F_o = \langle P_V, \{\{\blacksquare, \blacksquare\}, \{\bullet, \diagup\}, \{\blacksquare, \bigcirc\}\}, M_V\rangle$ (one output channel per process) and $F_g = \langle P_V, \{M_V\}, M_V\rangle$ (one global channel). Observe that $F_m \succ F_o \succ F_g$.

In F_m, the trace !\[■\]!\[■\]?\[■\] is possible, but in F_o it is not. This is because in F_o, both vouchers sent by *user* are put in the same channel, so *hvm* has to receive its voucher before *uvm* can. In F_o, the trace !\[■\]!\[■\]?\[■\]?\[■\]!\[●\]!\[■\]?\[■\] is possible, but in F_g it is not. This is because in F_g, both vending machines send their snacks to the same channel, which fixes the order in which *user* receives the snacks.

In the remainder of this section, to avoid duplication in definitions, lemmas and theorems, we universally quantify over FIFO systems $F = \langle P, C, M\rangle$ and $F' = \langle P, C', M\rangle$ such that $F \succ F'$, with $B_F = \langle S, s_0, L, \rightarrow\rangle$ and $B_{F'} = \langle S', s'_0, L, \rightarrow'\rangle$.

The effect on the LTS when changing the channel layout is visualised in Fig. 2. When the channels $\{m\}$ and $\{o\}$ are merged into one channel $\{m, o\}$, state s results in states s_1 and s_2, one for every interleaving of the contents of the two channels in s. Conversely, when channel $\{m, o\}$ is split into channels $\{m\}$ and $\{o\}$, the channel contents of states s_1 and s_2 are split as well, making them coincide, resulting in s. We say that state s *generalises* states s_1 and s_2 and that states s_1 and s_2 *specialise* state s.

To define this formally, we first define the interleavings of words. Given a message $m \in M$ and a set of words W, let $m + W = \{m + w \mid w \in W\}$. Then for a set of words W, we define the set of possible interleavings of these words $\|\, W$ as $\|\, W = \{\epsilon\}$ if $W = \{\epsilon\}$, else $\|\, W = \bigcup_{m + w \in W} m + \|((W \setminus \{m + w\}) \cup \{w\})$.

Example 3. Continuing from Example 2, let $W = \{\epsilon, \bullet\diagup, \blacksquare\bigcirc\}$. Then $\|\, W = \{\bullet\diagup\blacksquare\bigcirc, \bullet\blacksquare\diagup\bigcirc, \bullet\blacksquare\bigcirc\diagup, \blacksquare\bullet\diagup\bigcirc, \blacksquare\bullet\bigcirc\diagup, \blacksquare\bigcirc\bullet\diagup\}$.

With this, we can define generalisation/specialisation of states as follows:

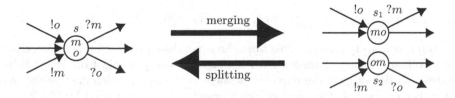

Fig. 2. A visualisation of the effect on the LTS of a FIFO system when merging or splitting channels. Transitions without a label cover any transition that is not already represented by other incoming or outgoing transitions.

Definition 5. *Let $\zeta \in \mathbf{C}$. For channels C' we define the set of functions $\mathbf{C}'_\zeta$, each representing possible interleavings of channel contents in ζ, as:*

$$\mathbf{C}'_\zeta = \left\{ \zeta' \in \mathbf{C}' \mid \forall_{c' \in C'} : \zeta'(c') \in \big|\big|\{\zeta(c) \mid c \in C \wedge c \subseteq c'\} \right\}$$

Definition 6. *Let $s = (\kappa, \zeta) \in S$ and $s' = (\kappa', \zeta') \in S'$. We say that s generalises s' and s' specialises s, written as $s \rhd s'$, iff $\kappa = \kappa' \wedge \zeta' \in \mathbf{C}'_\zeta$.*

Example 4. Continuing from Example 2, take the LTSs of FIFO systems F_o and F_g. Let $\kappa \in \mathbf{P}_V$ such that $\kappa(hvm) = 3$, $\kappa(uvm) = 3$ and $\kappa(user) = 2$ (the vending machines have supplied their snacks). Assume that $user$ has not retrieved any snack from the channels yet. In F_o there only exists one state $s = (\kappa, \zeta)$ that represents this situation, namely where $\zeta(\{\text{⬛}, \text{⬛}\}) = \epsilon$, $\zeta(\{\text{⬤}, \text{⬗}\}) = \text{⬤⬗}$ and $\zeta(\{\text{⬛}, \text{⬤}\}) = \text{⬛⬤}$. In F_g there are 6 such states, because the two vending machines use the same channel for output (the only channel M), so their outputs get interleaved. Let S'_κ be the set of these 6 states. Let $(\kappa, \zeta') \in S'_\kappa$, then the possible values for $\zeta'(M_V)$ are the interleavings mentioned in Example 3. The states in S'_κ specialise state s since they are stricter in how the snacks are ordered in the channel(s). Vice versa, state s generalises the states in S'_κ.

For every state in the LTS of a FIFO system, specialising or generalising states exist in the LTS of $\succ$-related FIFO systems.

Lemma 1. $\forall_{s' \in S'} : \exists_{s \in S} : s \rhd s'$ and $\forall_{s \in S} : \exists_{s' \in S'} : s \rhd s'$

As shown in Fig. 2, after merging channels $\{m\}$ and $\{o\}$, the action $?m$ is only possible from s_1, since it has the interleaving where m is at the head of the channel. Action $!m$ can only result in s_2, since it has the interleaving where m is at the back of the channel. Similar arguments can be made for $?o$ and $!o$. Any other transitions to and from s are possible for both s_1 and s_2. When splitting channel $\{m, o\}$ the opposite happens: the incoming and outgoing transitions for s are all transitions to and from s_1 and s_2 combined.

We show which transitions are preserved in the LTS when changing the channel layout formally in the below four lemmas, for any $\kappa_1, \kappa_2 \in \mathbf{P}$, $\zeta \in \mathbf{C}$, $\zeta' \in \mathbf{C}'_\zeta$ and $m \in M$, with $c = [m]_C$ and $c' = [m]_{C'}$. Firstly, internal actions are always possible from specialising or generalising states after merging or splitting channels.

Lemma 2. $(\kappa_1, \zeta) \xrightarrow{\tau} (\kappa_2, \zeta)$ iff $(\kappa_1, \zeta') \xrightarrow{\tau}' (\kappa_2, \zeta')$.

Input actions remain possible from the generalising state after splitting channels. When merging channels, such actions are not possible from specialising states that do not have the required message at the head, which may be the case when the channel of this message was merged, as was illustrated by Fig. 2.

Lemma 3. *If $c = c'$, then $(\kappa_1, \zeta) \xrightarrow{?m} (\kappa_2, \zeta[c \mapsto tl(\zeta(c))])$ iff $(\kappa_1, \zeta') \xrightarrow{?m}' (\kappa_2, \zeta'[c' \mapsto tl(\zeta'(c'))])$.*

Lemma 4. *If $c \neq c'$, then $(\kappa_1, \zeta) \xrightarrow{?m} (\kappa_2, \zeta[c \mapsto tl(\zeta(c))]) \wedge hd(\zeta'(c')) = m$ iff $(\kappa_1, \zeta') \xrightarrow{?m}' (\kappa_2, \zeta'[c' \mapsto tl(\zeta'(c'))])$.*

Output actions are always possible from specialising or generalising states after merging or splitting channels. Note however that sending a message to a merged channel increases the number of possible interleavings, so not all specialising target states are reached, as was illustrated by Fig. 2.

Lemma 5. $(\kappa_1, \zeta) \xrightarrow{!m} (\kappa_2, \zeta[c \mapsto \zeta(c) + m])$ *iff* $(\kappa_1, \zeta') \xrightarrow{!m}' (\kappa_2, \zeta'[c' \mapsto \zeta'(c') + m])$.

Note that for each of the above four lemmas, the target state of the $\rightarrow$-transition generalises the target state of the $\rightarrow'$-transition. Since the structure of the transitions in these lemmas is the same as in Definition 2, it follows that the above lemmas cover all transitions in $\rightarrow$ and $\rightarrow'$. In general, merging channels reduces the behaviour that a FIFO system allows. This can be formalised with the simulation preorder.

Lemma 6. $\triangleright^{-1}$ *is a simulation relation.*

Theorem 1. B_F *simulates* $B_{F'}$.

4   Property Preservation

In the previous section we have formally shown how the LTS of a FIFO system is affected when changing the channel layout. In this section we investigate how properties of a system are affected by such changes. Here the question is: if a property ϕ holds on the LTS of a FIFO system F, denoted by $B_F \models \phi$, under which conditions does it still hold after changing the channel layout? For this, we define the following notions.

Definition 7. *Let F and F' be two FIFO systems such that $F \succ F'$ and let ϕ be some property on FIFO systems. We say that:*

- *ϕ is merge-preserved iff $B_F \models \phi \Rightarrow B_{F'} \models \phi$.*
- *ϕ is split-preserved iff $B_F \models \phi \Leftarrow B_{F'} \models \phi$.*

In [11] it has already been shown that simulation preserves safety properties, that is properties of the form "some bad thing is not reachable", so from Theorem 1 we can derive the following:

Theorem 2. *Safety properties are merge-preserved.*

In the remainder of this section, we analyse the preservation of reachability, deadlock freedom and confluence. To avoid duplication in definitions, lemmas and theorems, we again universally quantify over FIFO systems $F = \langle P, C, M \rangle$ and $F' = \langle P, C', M \rangle$ such that $F \succ F'$, with $B_F = \langle S, s_0, L, \rightarrow \rangle$ and $B_{F'} = \langle S', s_0', L, \rightarrow' \rangle$.

4.1 Reachability

Reachability asks whether a state can be reached in the LTS of a FIFO system by a sequence of transitions, starting from the initial state.

Definition 8. *Let $B = \langle S, s_0, L, \rightarrow \rangle$ be an LTS and let $L' \subseteq L$. A state $s \in S$ is L'-reachable in B iff there exists a sequence of actions $\alpha \in L'^*$ such that $s_0 \xrightarrow{\alpha}^* s$. We define $Reach_{L'}(S)$ as the set of all L'-reachable states. We omit L' if $L' = L$. We define $Reach(B)$ as the LTS B restricted to only reachable states and the transitions between them.*

Preservation of reachability depends on whether a state's specialising or generalising states are still reachable after changing the channel layout. When splitting channels this is the case, as it follows from Theorem 1.

Lemma 7. *Let $s \in S$ and $s' \in S'$. Assume that $s \rhd s'$. Then $s \in Reach(S) \Leftarrow s' \in Reach(S')$.*

When merging channels however, reachability is only guaranteed to be preserved when only transitions have been taken with actions that do not use merged channels. Formally, we define the set of such actions as $IL(F, F') = \{\tau\} \cup \{?m, !m \mid m \in M \wedge [m]_C = [m]_{C'}\}$.

Lemma 8. *Let $s \in S$ and $s' \in S'$. Assume that $s \rhd s'$. Then $s \in Reach_{IL(F,F')}(S) \Rightarrow s' \in Reach_{IL(F,F')}(S')$.*

We argue using Fig. 2 why other actions violate merge-preservation of reachability. The transition with action $!m$ can be done to s and to s_2, but not to s_1, because it does not have m at the end of its channel. If s would not have any other incoming transitions, s_2 is possibly unreachable. The transition with action $?m$ can be done from s to some state t (not depicted in the figure) and from s_1, but not from s_2 since it does not have m at the head of its channel. Due to this, some states that specialise t are possibly unreachable.

Lifting these results to the full system, we will only focus on the reachability of process states. For a $\kappa \in \mathbf{P}$ and $L' \subseteq L$, we say that κ is $(L'-)$reachable in F iff there exists a $\zeta \in \mathbf{C}$ such that (κ, ζ) is $(L'-)$reachable in B_F.

Theorem 3. *For all $\kappa \in \mathbf{P}$, reachability of κ is split-preserved.*

Theorem 4. *For all $\kappa \in \mathbf{P}$, $IL(F, F')$-reachability of κ is merge-preserved.*

Example 5. See Fig. 3 for an example that shows that reachability of process states is not merge-preserved in general. In $Reach(B_F)$, state 2 of process p_2 is reachable, but in $Reach(B_{F'})$ it is not. This is because in $Reach(B_{F'})$, the messages m and o can only be received by p_2 in the order in which they are sent by p_1. Note that the actions $!m$, $!o$ and $?o$ that are necessary to reach state 2 of p_2 in $Reach(B_F)$ are not in $IL(F, F')$.

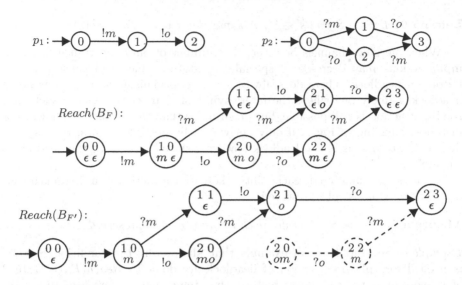

Fig. 3. Processes p_1 and p_2 and LTSs $Reach(B_F)$ and $Reach(B_{F'})$ for Example 5, with $F = \langle\{p_1, p_2\}, \{\{m\}, \{o\}\}, \{m, o\}\rangle$ and $F' = \langle\{p_1, p_2\}, \{\{m, o\}\}, \{m, o\}\rangle$. The dashed states are unreachable states that specialise states in $Reach(B_F)$.

The preservation of reachability is not only interesting on its own, but also for property preservation in general, because one is typically only interested in the preservation of a property within reachable behaviour. Thanks to Theorem 3 and Lemma 1, we know that for merge-preservation of a property that needs to hold for all reachable states, it suffices to check whether if it holds for a state, then it also holds for its specialising states. We cannot give such local arguments for split preservation, because after splitting channels, states may be reachable that do not generalise any reachable state in the original LTS. To be able to claim split-preservation of a property, we need to assume that all reachable states in the new system generalise some state in the original. We call this assumption *unaltered reachability* and represent it formally with $S \blacktriangleright S'$, which is true iff for all $s \in Reach(S)$ there exists an $s' \in Reach(S')$ such that $s \triangleright s'$.

4.2 Deadlock Freedom

A deadlock is a state in an LTS from which no action is possible.

Definition 9. *Let $B = \langle S, s_0, L, \rightarrow \rangle$ be an LTS. A state $s \in S$ is a deadlock, denoted as $\delta(s)$, iff there does not exist an $a \in L$ and $t \in S$ such that $s \xrightarrow{a} t$. We say that B is deadlock free iff for all $s \in Reach(S)$ it holds that $\neg\delta(s)$.*

The preservation of deadlock freedom comes down to whether for every non-deadlock state, its generalising or specialising states are not deadlocks as well. Whether this is the case can be easily derived from Fig. 2. When splitting channels, the number of outgoing transitions cannot decrease, so s cannot become a deadlock if s_1 or s_2 were not deadlocks already.

Lemma 9. *Let $s \in S$ and $s' \in S'$. Assume that $s \triangleright s'$. Then $\neg\delta(s) \Leftarrow \neg\delta(s')$.*

When merging channels however, the number of outgoing transitions with input actions may decrease in specialising states, which can cause some to become a deadlock. There does always exist a specialising state that is not a deadlock, namely one where the interleaving of channel contents is such that the input action is still possible, but much more than this cannot be shown. For instance, referring to Fig. 2, if only $?m$ would be possible from s, then s_2 is a deadlock. State s_1 is not a deadlock however, since it has m at the head of its channel.

If we assume unaltered reachability, then we can derive from Lemma 9 that deadlock freedom is split-preserved.

Theorem 5. *If $S \blacktriangleright S'$, then deadlock freedom is split-preserved.*

Example 6. See Fig. 4 for an example that shows why the condition $S \blacktriangleright S'$ is needed for split-preservation of deadlock freedom. In $Reach(B_{F'})$, state 2 of process p_2 is not reachable, because the single channel forces p_2 to receive m and o in the order that they are sent. In $Reach(B_F)$, m and o are put in different channels, so p_2 is free to choose which it receives first. This makes 2 of p_2 reachable, which violates unaltered reachability. The corresponding state in $Reach(B_F)$ is a deadlock, because p_2 expects another o which is never supplied.

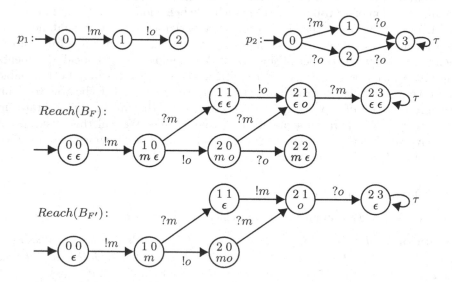

Fig. 4. An example that shows that deadlock freedom is not split-preserved without assuming unaltered reachability ($S \blacktriangleright S'$), with $F = \langle \{p_1, p_2\}, \{\{m\}, \{o\}\}, \{m, o\} \rangle$ and $F' = \langle \{p_1, p_2\}, \{\{m, o\}\}, \{m, o\} \rangle$.

4.3   Confluence

Confluence of two actions indicates a form of independence between them. Since a FIFO system consists of multiple processes acting mostly independently of each other, confluence in a FIFO system is common.

Definition 10. *Let $B = \langle S, s_0, L, \rightarrow \rangle$ be an LTS. For $a, b \in L$ and $s \in S$, a and b are confluent from s, denoted as $Conf_b^a(s)$, iff for all $t, u \in S$ we have that $(s \xrightarrow{a} t \wedge s \xrightarrow{b} u) \Rightarrow (\exists_{v \in S} : t \xrightarrow{b} v \wedge u \xrightarrow{a} v)$. Note that $Conf_b^a(s) = Conf_a^b(s)$. We say that a and b are confluent in B iff $Conf_b^a(s)$ for all $s \in Reach(S)$.*

Again, we will first look at the preservation on state level. The relation between confluence and independence is reflected in its preservation: confluence of two actions from a state is preserved when merging channels, if the two actions do not use the same channel. This is the case when at least one of two actions is τ and when both actions use different channels in both FIFO systems. When splitting channels, there is an exception when an input action a is involved that uses a split channel. If a choice between a and another action exists from a state s after splitting channels, there may be some specialising states in the original LTS from which a is not possible due to the interleaving of channel contents. This makes confluence trivially true from these states, while confluence may be false from s. We represent this case with the condition $?SC_{C'}^C(a)$, which is true iff $a = ?m \Rightarrow [m]_C = [m]_{C'}$ for some $m \in M$.

Lemma 10. *Let $s \in S$, $s' \in S'$ and $a \in L$. Assume that $s \rhd s'$. Then $Conf_a^\tau(s) \Rightarrow Conf_a^\tau(s')$ and if $?SC_{C'}^C(a)$, then $Conf_a^\tau(s) \Leftarrow Conf_a^\tau(s')$.*

Lemma 11. *Let $s \in S$, $s' \in S'$ and $m, o \in M$. Assume that $s \rhd s'$ and $m \not\sim_{C'} o$. Let $a \in \{?m, !m\}$ and $b \in \{?o, !o\}$. Then $Conf_b^a(s) \Rightarrow Conf_b^a(s')$ and if $?SC_{C'}^C(a)$ and $?SC_{C'}^C(b)$, then $Conf_b^a(s) \Leftarrow Conf_b^a(s')$.*

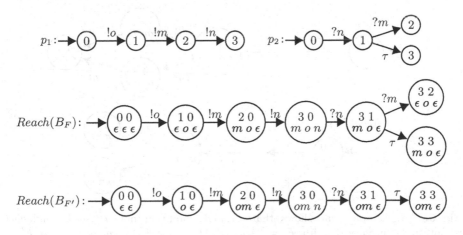

Fig. 5. An example that shows that without condition $?SC_{C'}^C(a)$ confluence is not split-preserved, with $F = \langle \{p_1, p_2\}, \{\{m\}, \{o\}, \{n\}\}, \{m, o, n\} \rangle$ and $F' = \langle \{p_1, p_2\}, \{\{m, o\}, \{n\}\}, \{m, o, n\} \rangle$.

Example 7. See Fig. 5 for an example why condition $?SC_{C'}^{C}(a)$ is necessary for the split-preservation of confluence. In $Reach(B_F)$ confluence of $?m$ and τ is not met, because there is a choice between the two that does not result in a confluence diamond. In $Reach(B_{F'})$ confluence of $?m$ and τ is trivially met because the choice between the two is never possible. Compared to $Reach(B_F)$, the choice was made impossible in $Reach(B_{F'})$ because the channels for m and o have now merged. Because o is sent before m, p_2 is forced in $Reach(B_{F'})$ to receive o first, but it never does. Note that $?SC_{C'}^{C}(?m)$ is not met.

If both actions a and b use the same channel in both FIFO systems, there is an edge case where confluence is not preserved when merging channels, namely when both are the exact same input action. In this case, two messages m are required at the head of the channel of m to create the confluence diamond. However, if the channel of m is merged with another channel, there are specialised states with an interleaving of channel contents without both messages m in front. We represent this case with $a \equiv_? b$ for actions $a, b \in L$, which is true iff $a = ?m = b$ for some $m \in M$.

Lemma 12. *Let $s \in S$, $s' \in S'$ and $m, o \in M$. Assume that $s \triangleright s'$ and $m \simeq_C o$. Let $a \in \{?m, !m\}$ and $b \in \{?o, !o\}$. If not $a \equiv_? b$, then $Conf_b^a(s) \Rightarrow Conf_b^a(s')$ and if $?SC_{C'}^{C}(a)$ and $?SC_{C'}^{C}(b)$, then $Conf_b^a(s) \Leftarrow Conf_b^a(s')$.*

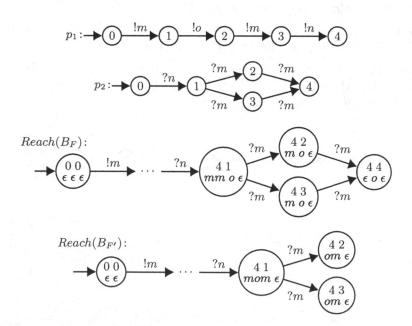

Fig. 6. An example that shows that without condition not $a \equiv_? b$ confluence is not merge-preserved, with $F = \langle\{p_1, p_2\}, \{\{m\}, \{o\}, \{n\}\}, \{m, o, n\}\rangle$ and $F' = \langle\{p_1, p_2\}, \{\{m, o\}, \{n\}\}, \{m, o, n\}\rangle$.

Example 8. See Fig. 6 for an example why condition not $a \equiv_? b$ is necessary for merge-preservation of confluence. In $Reach(B_F)$ confluence of $?m$ and $?m$ is met, because the only choice between $?m$ and $?m$ results in a confluence diamond. In $Reach(B_{F'})$ confluence of $?m$ and $?m$ is not met, because there is a choice between $?m$ and $?m$ that does not result in a confluence diamond. This is because p_2 first needs to receive the o before it can receive the second m. In $Reach(B_F)$ this was not an issue, because m and o both had their own channels. Note that $?m \equiv_? ?m$.

If both actions use channels that are distinct in F but equal in F', confluence of two actions is merge-preserved if at least one of the actions is an input action. In case one action is an input action and the other an output action, merge-preservation follows from the fact that the actions touch different ends of the channel and are therefore in some sense independent. In case both actions are input actions, the choice between the two actions is not possible from any state in $B_{F'}$, since they use the same channel. This makes their confluence hold trivially, from which merge-preservation trivially follows. Confluence is not split-preserved in these cases for the same reason as for the example in Fig. 5 (for instance, replace τ with $?o$).

In case both actions are output actions, confluence is only split-preserved. This is because from any state s' in $B_{F'}$, the two different orders of these actions produce different orders of channel contents, since both actions use the same channel. This implies that confluence cannot hold from s', and therefore it is trivially preserved when splitting channels. Confluence is still possible in B_F, so confluence is not necessarily preserved when merging channels. For actions $a, b \in L$, we formally represent this last case with $!AC_{C'}^C(a, b)$, which is true iff $a = !m$, $b = !o$, $m \not\simeq_C o$ and $m \simeq_{C'} o$ for some $m, o \in M$.

Lemma 13. *Let $s \in S$, $s' \in S'$ and $m, o \in M$. Assume that $s \rhd s'$, $m \not\simeq_C o$ and $m \simeq_{C'} o$. Let $a \in \{?m, !m\}$ and $b \in \{?o, !o\}$. If not $!AC_{C'}^C(a, b)$, then $Conf_b^a(s) \Rightarrow Conf_b^a(s')$.*

Lemma 14. *Let $s \in S$, $s' \in S'$ and $a, b \in L$ Assume that $s \rhd s'$. If $!AC_{C'}^C(a, b)$, then $Conf_b^a(s) \Leftarrow Conf_b^a(s')$.*

Lifting these state-based results to confluence in the LTSs of FIFO systems, we can derive the following theorems:

Theorem 6. *Let $a, b \in L$. If not $!AC_{C'}^C(a, b)$ and not $a \equiv_? b$, then confluence of a and b is merge-preserved.*

Theorem 7. *Let $a, b \in L$. If $?SC_{C'}^C(a)$, $?SC_{C'}^C(b)$ and $S \blacktriangleright S'$, then confluence of a and b is split-preserved.*

4.4   Summary of Results

The results of this section are summarised in the table below. Remember that $IL(F, F') = \{\tau\} \cup \{?m, !m \mid m \in M \wedge [m]_C = [m]_{C'}\}$, that $S \blacktriangleright S'$ iff $\forall_{s \in Reach(S)} : \exists_{s' \in Reach(S')} : s \triangleright s'$, that $?SC_{C'}^C(a)$ iff $a = ?m \Rightarrow [m]_C = [m]_{C'}$ for some $m \in M$, that $a \equiv_? b$ iff $a = ?m = b$ for some $m \in M$, and that $!AC_{C'}^C(a, b)$ iff $a = !m$, $b = !o$, $m \not\simeq_C o$ and $m \simeq_{C'} o$ for some $m, o \in M$.

	Merge-preserved	Split-preserved
L'-reachability of κ	If $L' = IL(F, F')$ (Theorem 4)	If $L' = L$ (Theorem 3)
Deadlock freedom	No	If $S \blacktriangleright S'$ (Theorem 5)
Confluence of a and b	If not $!AC_{C'}^C(a, b)$ and not $a \equiv_? b$ (Theorem 6)	If $?SC_{C'}^C(a)$, $?SC_{C'}^C(b)$ and $S \blacktriangleright S'$ (Theorem 7)

5   Conclusion

We have studied asynchronously communicating systems and their channel layouts by modelling them as FIFO systems and ordered them based on whether one can be created from the other by merging channels. We have shown that the LTS that describes the behaviour of a split FIFO system simulates the LTS of the original system. As a consequence of this, safety properties are merge-preserved. We have also identified conditions under which reachability, deadlock freedom and confluence are preserved when changing the channel layout.

For most conditions that are required for a property to be preserved, their truth can be derived easily. An exception of this is the unaltered reachability assumption, for which it should be investigated how feasible it is to check them. It is also the question how likely the conditions are met in practice, given that some are rather strict. Using more detailed information from the processes of the FIFO systems could lead to less strict conditions, but they can be more difficult to check. Another option would be to find whether sufficient property-specific conditions exist.

The properties mentioned in this paper are of course not the only interesting properties one could want to be preserved. Other options would be preservation of maximum queue length, of eventual termination, of (lack of) starvation and of behavioural equivalence between two systems.

Acknowledgements. We thank the reviewers for their helpful feedback.

References

1. Bollig, B., Finkel, A., Suresh, A.: Bounded reachability problems are decidable in FIFO machines. In: CONCUR. LIPIcs, vol. 171, pp. 49:1–49:17. Schloss Dagstuhl - Leibniz-Zentrum für Informatik (2020)
2. Bunte, O., van Gool, L.C., Willemse, T.A.: On the preservation of properties when changing communication models (2022). https://doi.org/10.48550/ARXIV.2210.06196
3. Chevrou, F., Hurault, A., Nakajima, S., Quéinnec, P.: A map of asynchronous communication models. In: Sekerinski, E., et al. (eds.) FM 2019. LNCS, vol. 12233, pp. 307–322. Springer, Cham (2020). https://doi.org/10.1007/978-3-030-54997-8_20
4. Chevrou, F., Hurault, A., Quéinnec, P.: On the diversity of asynchronous communication. Form. Asp. Comput. **28**(5), 847–879 (2016)
5. Derrick, J., Smith, G.: Temporal-logic property preservation under Z refinement. Form. Asp. Comput. **24**(3), 393–416 (2012)
6. Engels, A., Mauw, S., Reniers, M.A.: A hierarchy of communication models for message sequence charts. Sci. Comput. Program. **44**(3), 253–292 (2002)
7. Finkel, A., Praveen, M.: Verification of flat FIFO systems. Log. Methods Comput. Sci. **16**(4) (2020)
8. Henzinger, T.A., Jhala, R., Majumdar, R., Sanvido, M.A.A.: Extreme model checking. In: Dershowitz, N. (ed.) Verification: Theory and Practice. LNCS, vol. 2772, pp. 332–358. Springer, Heidelberg (2003). https://doi.org/10.1007/978-3-540-39910-0_16
9. Huang, J., Voeten, J., Geilen, M.: Real-time property preservation in approximations of timed systems. In: MEMOCODE, pp. 163–171. IEEE Computer Society (2003)
10. Lamport, L.: Time, clocks, and the ordering of events in a distributed system. Commun. ACM **21**(7), 558–565 (1978)
11. Loiseaux, C., Graf, S., Sifakis, J., Bouajjani, A., Bensalem, S.: Property preserving abstractions for the verification of concurrent systems. Form. Methods Syst. Des. **6**(1), 11–44 (1995)
12. Sokolsky, O.V., Smolka, S.A.: Incremental model checking in the modal mu-calculus. In: Dill, D.L. (ed.) CAV 1994. LNCS, vol. 818, pp. 351–363. Springer, Heidelberg (1994). https://doi.org/10.1007/3-540-58179-0_67
13. Wehrheim, H.: Behavioural subtyping and property preservation. In: FMOODS. IFIP Conference Proceedings, vol. 177, pp. 213–231. Kluwer (2000)
14. Wijs, A., Engelen, L.: Efficient property preservation checking of model refinements. In: Piterman, N., Smolka, S.A. (eds.) TACAS 2013. LNCS, vol. 7795, pp. 565–579. Springer, Heidelberg (2013). https://doi.org/10.1007/978-3-642-36742-7_41
15. Xia, C., Li, C.: Property preservation of petri synthesis net based representation for embedded systems. IEEE CAA J. Autom. Sinica **8**(4), 905–915 (2021)

Introduction to Routing Problems with Mandatory Transitions

Christian Laforest and Timothée Martinod$^{(\boxtimes)}$

Université Clermont Auvergne, CNRS, Mines de Saint-Étienne,
Clermont-Auvergne-INP, LIMOS, 63000 Clermont-Ferrand, France
{christian.laforest,timothee.martinod}@uca.fr

Abstract. A sequence P of vertices $v_1, \ldots, v_k$ from a directed graph (or digraph) $D = (V, A)$ is a directed *route* (or diroute) if $(v_i, v_{i+1}) \in A$ for $1 \leq i \leq k - 1$. In this paper, we study a generalisation of this classical notion. Namely, an instance of our problem contains a digraph $D = (V, A)$ and a set of *mandatory transitions* Π from A^2. Each mandatory transition is composed of two contiguous arcs. A diroute P in the instance (D, Π) is a diroute if $\forall i = 1, \ldots, k-2, \ (v_i = a, v_{i+1} = b) \implies (v_{i+2} = c)$ for each mandatory transition $\langle a, b, c \rangle \in \Pi$ *and* $(a, b)\ (b, c) \in A$ (we say that P *respects* the mandatory transitions). We show that finding a shortest diroute between two vertices of V, respecting the mandatory transitions Π, is solvable in polynomial time. On the other side, we show that finding an elementary shortest diroute between two vertices of V, and respecting the mandatory transitions Π is not approximable by any ratio unless $\mathcal{P} = \mathcal{NP}$, even in the very particular case of a dense digraph (each vertex has at least $|V| - 2$ outgoing arcs). We also show that finding a shortest diroute containing several target vertices (or all the vertices at least once) is not approximable by any ratio unless $\mathcal{P} = \mathcal{NP}$, even in very restricted classes of digraphs.

Keywords: Complexity · Obligation · Routing problem · Shortest path · Directed graph

1 Introduction

Many variants of routing problems according to the additional properties are considered, namely periodic vehicle routing problem, dynamic vehicle routing problem, stochastic vehicle routing problem, among others (see [11]). These problems are central in graph theory and operational research. They are also useful and daily used for delivery problems, among others. It is well-known that routing problems are mostly $\mathcal{NP}$-complete (see [4]). However, some polynomial approximation algorithms exist (see [10]).

This work was sponsored by a public grant overseen by the French National Research Agency as part of the "Investissements d'Avenir" through the IMobS3 Laboratory of Excellence (ANR-10-LABX-0016) and the IDEX-ISITE initiative CAP 20–25 (ANR-16-IDEX-0001).

L. Gąsieniec (Ed.): SOFSEM 2023, LNCS 13878, pp. 254–266, 2023.
https://doi.org/10.1007/978-3-031-23101-8_17

Let $D = (V, A)$ be a directed graph (or *digraph*), with V its set of *vertices* and A its set of *arcs*. A sequence P of vertices $v_1, \ldots, v_k$ from V (or equivalently a sequence of arcs $(v_1, v_2), \ldots, (v_{k-1}, v_k)$ from A) is a directed route (or *diroute*) if $(v_i, v_{i+1}) \in A$ for $1 \le i \le k - 1$. A diroute P is *elementary* if $v_i \neq v_j$ for $1 \le i \neq j \le k$. A diroute P is *simple* if $(v_i, v_{i+1}) \neq (v_j, v_{j+1})$ for $1 \le i \neq j \le k - 1$. An elementary diroute P is *Hamiltonian* if $\forall v_i \in V, v_i \in P$. A simple diroute P is *eulerian* if $\forall (v_i, v_j) \in V, (v_i, v_j) \in P$. Note that we could use 'directed path' instead of 'directed diroute'. We choose this terminology to avoid confusing between a solution for routing problems and all the defined paths in it.

In this paper, we study a generalisation of this classical notion. Namely, an instance of our problem contains a digraph $D = (V, A)$ and a set of *mandatory transitions* Π from A^2 (each mandatory transition is composed of two contiguous arcs). A diroute P in the instance (D, Π) is a diroute if $\forall i = 1, \ldots, k - 2$, $(v_i = a, v_{i+1} = b) \implies (v_{i+2} = c)$ (except if v_{i+1} is the end of the diroute) for each mandatory transition $\langle a, b, c \rangle \in \Pi$ and (a, b) $(b, c) \in A$ (we say that P *respects* the mandatory transitions). Given an instance of a routing problem, a solution is a diroute passing through all the target vertices and respecting each mandatory transition. The length of a solution $(v_1, v_2), \ldots, (v_{k-1}, v_k)$ is its number of arcs ($k - 1$ in this case). A mandatory transition $\langle a, b, c \rangle$ is composed by an *head* arc (a, b) and a *tail* arc (b, c). So, a diroute respects a mandatory transition if its tail is the next arc each time the head appears (except if there is no next arc in the diroute). An arc (a, b) can be the head of at most one mandatory transition. An arc (b, c) can be the tail of at most the incoming degree of b. A tail arc in a mandatory transition can be a head arc in another mandatory transition. Therefore, a digraph has at most as many mandatory transitions as arcs. Two mandatory transition are *consecutive* if the tail of the first is the head of the second (consecutiveness is ordered: $\langle a, b, c \rangle$ and $\langle b, c, d \rangle$ are consecutive but $\langle b, c, d \rangle$ and $\langle a, b, c \rangle$ are not consecutive). Consecutive mandatory transitions form a directed *consecutive path* (for example $\langle a, b, c \rangle$, $\langle b, c, d \rangle$ and $\langle c, d, e \rangle$). A directed *consecutive circuit* is a directed consecutive path such that the last and the first mandatory transitions are consecutive. Note that if a diroute passes through an arc in a directed consecutive circuit, the diroute *must* follow this circuit: the diroute cannot pass through an arc outside the circuit anymore. In the instance of Fig. 1, there is no diroute from s to t respecting the mandatory transitions, even though $P = s, u, t$ is a diroute.

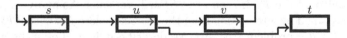

Fig. 1. Illustration of a consecutive circuit, with four vertices $\{s, u, v, t\}$ and three mandatory transitions $\Pi = \{ \langle s, u, v \rangle, \langle u, v, s \rangle, \langle v, s, u \rangle \}$.

Mandatory transitions are placed in a broader research context: the obligations. An obligation can represent a subset of devices that must be used jointly, not individually or a team of people that must all be present to make an action. Routing problems use graphs modeling real roads. An obligation can represent a 'vehicles must turn right/left ahead' sign. Obligations were recently introduced in [3] for many graph problems such as vertex cover, connected vertex cover, dominating set, total dominating set, independent dominating set, spanning tree, matching, and Hamiltonian path. Some properties of obligations have been studied in the independent dominating set problem in [8]. The authors mainly obtained hardness results. Studying classical graph problems with additional constraints is not new. The introduction of obligations as an object of study was motivated by considering them as the converse to *conflict* constraints. They have been studied for the last years. The general context of conflict constraint is the following. Given a graph $G = (V, E)$, we add a set of pairs of elements of G (vertices or edges) that *cannot* appear together in the same solution (which can be a path, a tree, a dominating set, etc. depending on the goal). Unlike obligations, a conflict models that two elements of a system cannot be used simultaneously because they are incompatible for example. Most problems with conflicts are hard, even if the underlying classical version is solvable in polynomial time. See the following recent publications on the subject for examples: [1,2,5–7,9].

In this paper, we study the Multi-target Routing problem with Mandatory Transitions ($MRMT$), which we define as follows:

Instance: A directed graph $D = (V, A)$, a set of mandatory transitions Π, a source vertex $s \in V$, target vertices $t_1, \ldots, t_k \in V$ and an integer l.

Objective: Find a shortest diroute P starting from s, containing $t_1, \ldots, t_k$, and respecting Π.

Question: Does P contain at most l arcs?

We show that the $MRMT$ problem cannot admit an approximation by any ratio if all the vertices are targets (denoted $FRMT$ version), even in a symmetric directed complete graph. On the other side, we show it is solvable in polynomial time if only one vertex is a target (denoted $SPMT$ version) but cannot admit an approximation by any ratio if the solution diroute has to be elementary (denoted elementary $SPMT$ version), even if each vertex of the digraph $D = (V, A)$ has $|V| - 2$ outgoing arcs. Note that we study the complexity of the $MRMT$ problem only though two special cases: its complexity can be extrapolated from the non approximability of the $FRMT$ and the $SPMT$ versions. As many results in this paper are obtained by a reduction from the $\mathcal{NP}$-complete *Restricted Exact Cover by 3-Sets (RX3C)* problem (see [4]), we describe it here. Let $\mathcal{X}$ be a finite set of $3q$ *elements* and $\mathcal{C}$ a collection of $3q$ *triplets* (sets of three elements) of $\mathcal{X}$. Given such an instance $(\mathcal{X}, \mathcal{C})$ it is $\mathcal{NP}$-complete to decide whether there is a sub-collection $C' \subseteq \mathcal{C}$ such that each element of $\mathcal{X}$ occurs in exactly one triplet of C'. Such a sub-collection C' has exactly q disjoint triplets (each triplet covers exactly three elements, and each element is covered by exactly one triplet).

2   Shortest Path Problems with Mandatory Transitions

In this section, we show that the $MRMT$ problem is solvable in polynomial time in general if we want to find the diroute to a single vertex (only one vertex is a target), but it is not approximable by any ratio unless $\mathcal{P} = \mathcal{NP}$ if the solution diroute has to be elementary, even if each vertex of the digraph $D = (V, A)$ has $|V| - 2$ outgoing arcs. For more clarity we denote these special cases as the Shortest Path with Mandatory Transitions ($SPMT$) and the elementary Shortest Path with Mandatory Transitions (elementary $SPMT$). Note that these $SPMT$ problems may admit no solutions, even in a symmetric dense digraph. For example, take a symmetric directed completed graph with vertices $u_1, \ldots, u_k$ for any $k > 1$, add the vertex v, the arcs (v, u_1) and (u_1, v) and the mandatory transition $\langle v, u_1, v \rangle$. There are no solutions if v is the source vertex and u_1 is not the target vertex.

We will show that the $MRMT$ problem with only one vertex as a target (equivalent to find the path to a single vertex) is solvable in polynomial time. To do so, we will make a polynomial construction to convert an instance of the $SPMT$ problem into an equivalent instance of the Weighted Shortest Path problem. This construction will replace directed consecutive paths by weighted arcs between their extremities. If there is a solution at a cost c in an instance, there is a solution with the same cost in the other instance.

2.1   Polynomiality of $SPMT$

The Shortest Path with Mandatory transitions problem is a special case of the $MRMT$ problem where only one vertex is a target. As seen in Fig. 1, some existing diroutes do not respect mandatory transitions. These diroutes cannot be solutions. Since optimal solutions without mandatory transitions can be forbidden in $SPMT$, we cannot use classical algorithms (such as Dijkstra's algorithm) to find a diroute between two vertices. To solve this problem, we will make a reduction to the Weighted Shortest Path problem to remove the mandatory transitions without adding or removing any solution. Since there cannot be any solution respecting mandatory transitions if there are no diroutes in the digraph, we only need to look for instances where a diroute from the source to the destination exists. Note that directed consecutive circuits can be found and removed in polynomial time (simply follow consecutive mandatory transitions until passing through the same arc more than once, these arcs cannot be in a solution).

Construction 1. *Let* $(D = (V, A), \Pi, s, t)$ *be an instance of SPMT (or MRMT) without directed consecutive circuits (arcs from directed consecutive circuits containing t are cut after t, all the arcs for the others circuits are deleted), where $s, t \in V$. We construct $(D' = (V, A'), s, t)$ of the Weighted Shortest Path problem and L the list of each arc of A used by each arc of A' as follows (we number the arcs of A' as we add them):*

We add the arc (u, v) of weight 1 $(\forall u, v \in V)$ if $(u, v) \in A$ (the arc exists) and $\forall w \in V, \langle u, v, w \rangle \notin \Pi$ (the arc is no constraint by a mandatory transition).

Let i be the arc number, $L_i = [u, v]$.
We add the arc (u, v) of weight $k+1$ ($\forall u, v \in V$) if there is a directed consecutive path $u, u_1, \ldots, u_k, v \in (D, \Pi)$ and there are no mandatory transitions $\langle u_k, v, w \rangle$ for any vertex $w \in V$. Let i be the arc number, $L_i = [u, u_1, \ldots, u_k, v]$.
We add the arc (u, t) of weight $j + 1$ ($\forall u, v \in V$) if there is a directed consecutive path $u, u_1, \ldots, u_j, t, \ldots, u_k, v \in (D, \Pi)$. Let i be the arc number, $L_i = [u, u_1, \ldots, u_j, t]$.

The weighted multi-digraph D' contains at most $|V|$ vertices and $2(|V|^2)$ arcs. This digraph is constructed in polynomial time $(O(|V|^2 * |A|^2))$.

Lemma 1. *Let $(D = (V, A), \Pi, s, t)$ be an instance of SPMT (or MRMT) without directed consecutive circuits, where $s, t \in V$. Let $(D' = (V, A'), s, t)$ be the instance of the Weight Shortest Path problem following Construction 1. There is a solution for $(D = (V, A), \Pi, s, t)$ if and only if there is a solution for $(D' = (V, A'), s, t)$.*

Proof. Suppose that $P = u_0, u_1, \ldots, u_k$ with $u_0 = s$ and $u_k = t$ is a diroute respecting mandatory transitions in (D, Π). Let P' be the diroute in D' defined as follows: a) P' starts at vertex s, let $i = 1$; b) if there is a mandatory transition $\langle u_{i-1}, u_i, u_{i+1} \rangle$, let u_j be the extremity of the longest directed consecutive path starting at the arc (u_{i-1}, u_i), $i = j$; c) add the vertex u_i at the end of P'; d) if $i \leq k$ go to step b with $i = i + 1$. P' is the shortest diroute between s and t in D'.

Suppose that $P' = u_0, u_1, \ldots, u_k$ with $u_0 = s$ and $u_k = t$ is a diroute in D'. Let P be the diroute in (D, Π), respecting the mandatory transitions, defined as follows: a) $P = P'$; b) for each arc (u_i, u_{i+1}) ($0 \leq i < k$), numbered as j, substitute in P the vertices u_i and u_{i+1} by the vertices defined in L_j. P is the shortest diroute between s and t in (D, Π), respecting the mandatory transitions. □

Since we can use a straightforward generalisation of Dijkstra's algorithm for extraction of Shortest Paths in weighted directed multi-graphs, we can state the following theorem.

Theorem 1. *The SPMT problem is solvable in polynomial time.*

In a solution of Shortest Path problems, the diroute is elementary and simple. In the *SPMT* problem, a solution is still simple but not necessarily elementary. Next, we will show that the elementary *SPMT* problem is not approximable by any ratio unless $\mathcal{P} = \mathcal{NP}$.

2.2 Non Approximation of Elementary *SPMT* and *MRMT*

As said in the Introduction, we will use a reduction from the *RX3C* problem. Each element (resp. triplet) in *RX3C* is related to exactly three triplets (resp. elements). To represent the membership of the elements and the intersection between triplets, we will create one gadget per triplet (Construction 2): a triplet covers either all its elements or none of them. We show how gadgets behave in a complexity proof of the $MRMT$ problem (Theorem 2). We have to ensure that a) each element is covered by only one triplet and b) each triplet covers either all its elements or none of them. Next, we will duplicate the instance such that each copy allows choosing exactly one triplet (Construction 3). Copies form a directed path between the source and the target vertex. If an instance of $RX3C$ (size q) has a solution, the associated digraph has a diroute going through q triplets and all elements before ending at the target vertex. We show the elementary $SPMT$ problem is not approximable by any ratio unless $\mathcal{P} = \mathcal{NP}$ (Theorem 4). In the last step, we will propose a gadget to get a dense directed graph (Construction 4). This gadget is a way to add missing arcs while forbidding them. We show that the elementary $SPMT$ problem is not approximable by any ratio unless $\mathcal{P} = \mathcal{NP}$, even if each vertex of the digraph $D = (V, A)$ has $|V| - 2$ outgoing arcs (Theorem 5). In all constructions, we associate each element and triplet of $RX3C$ to one or more vertices in the associated digraph. In the rest of our paper, we will call them element vertices and triplet vertices. Let us start with the first step, the $MRMT$ problem.

Construction 2. *Let $(\mathcal{X}, \mathcal{C})$ be an instance of RX3C where $\mathcal{X} = \{x_1, \ldots, x_{3q}\}$ and $\mathcal{C} = \{c_1, \ldots, c_{3q}\}$. We construct $(D = (V, A), \Pi)$ of the MRMT problem as follows:*
For each $c_i \in \mathcal{C}$ (resp. $x_i \in \mathcal{X}$), we add the vertices $c_{i,a}, c_{i,b}, c_{i,c}, c_{i,d}$ (resp. x_i). We add the vertex s and the arcs $(s, c_{1,a})$, $\ldots$, $(s, c_{3q,a})$ and $(c_{1,d}, s)$, $\ldots$, $(c_{3q,d}, s)$.
For each $c_i = \{x_j, x_k, x_h\} \in \mathcal{C}$, we add the arcs $(c_{i,a}, x_j)$, $(x_j, c_{i,b})$, $(c_{i,b}, x_k)$, $(x_k, c_{i,c})$, $(c_{i,c}, x_h)$, and $(x_h, c_{i,d})$ as well as the seven mandatory transitions $\langle s, c_{i,a}, x_j \rangle$, $\langle c_{i,a}, x_j, c_{i,b} \rangle$, $\langle x_j, c_{i,b}, x_k \rangle$, $\langle c_{i,b}, x_k, c_{i,c} \rangle$, $\langle x_k, c_{i,c}, x_h \rangle$, $\langle c_{i,c}, x_h, c_{i,d} \rangle$, and $\langle x_h, c_{i,d}, s \rangle$.

The instance (D, Π) contains $15q + 1$ vertices, $24q$ arcs and $21q$ mandatory transitions. Mandatory transitions are represented in Fig. 2 (top) for one triplet. A big picture of the construction is represented in Fig. 2 (bottom). Links between triplet vertices and their elements vertices are condensed and shown as edges, mandatory transitions are not shown.

In Construction 2, mandatory transitions ensure that each triplet covers either all its elements or none of them. Note that the vertex s is the only one where an arc can be chosen freely. For each other vertex, a mandatory transition constrains the diroute to only one possible next arc. To make sure each element is covered exactly one time, we prevent the choice of more than q triplets. To do that, we choose the size of the solution diroute wisely.

Theorem 2. *The MRMT problem is $\mathcal{NP}$-complete.*

Proof. Clearly $MRMT$ is in $\mathcal{NP}$.

Let $(\mathcal{X}, \mathcal{C})$ be an instance of $RX3C$ with $3q$ elements and $3q$ triplets. Let (D, Π) be the associated instance of $MRMT$ following Construction 2. Let s be

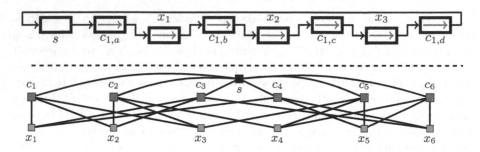

Fig. 2. Illustration of Construction 2 for the instance $(\mathcal{X}, \mathcal{C})$ of *RX3C* where $\mathcal{X} = \{1, \ldots, 6\}$ and $\mathcal{C} = \{\{1, 2, 3\}, \{2, 3, 4\}, \{1, 2, 6\}, \{1, 5, 6\}, \{3, 4, 5\}, \{4, 5, 6\}\}$.

the source vertex, and $x_1, \ldots, x_{3q}$ the target vertices. Let P be a shortest diroute starting from s, going through $x_1, \ldots, x_{3q}$ and respecting Π.

Due to the mandatory transitions structure, we can see P as a succession of element collectors. Each *element collector* p^f is a directed consecutive path composed of eight arcs. Path p^f starts at the vertex s, heads towards a triplet vertex $c_{i,a}$, followed by its first element vertex x_j, alternates between the triplet vertices and its element vertices, and ends at vertex s: $p^f = (s, c_{i,a}), (c_{i,a}, x_j), (x_j, c_{i,b}), (c_{i,b}, x_k), (x_k, c_{i,c}), (c_{i,c}, x_h), (x_h, c_{i,d}), (c_{i,d}, s)$, for each $c_i = \{x_j, x_k, x_h\} \in \mathcal{C}$. Thus $P = p^1, \ldots, p^\lambda$ deprived of the two last arcs. Indeed, the diroute ends as soon as the last target is reached. Each element collector p^f contains exactly three element vertices among the $3q$. The shortest diroute cannot have less than $8q - 2$ arcs. Let l be $8q - 2$.

Suppose that P is a diroute with exactly $8q - 2$ arcs. Each element vertex is reached exactly one time. Exactly q triplet vertices have been reached. The triplet vertices crossed in P give a solution for the instance $(\mathcal{X}, \mathcal{C})$ of *RX3C*.

Suppose there is a solution S for the instance of *RX3C*. Let P' be the diroute defined as follows: a) P' starts at vertex s; b) for each triplet $c_i = \{x_j, x_k, x_h\} \in S$, we add the corresponding element collector at the end of P' except for the last two vertices of the last element collector. The diroute P' contains $8q - 2$ arcs, starts at s, goes through all the targets and respects the mandatory transitions.

Thus, there is a solution for the instance $(\mathcal{X}, \mathcal{C})$ of *RX3C* if and only if there is, in the associated instance of *MRMT*, a diroute of size $8q - 2$ starting at s, going through all the targets and respecting the mandatory transitions.   □

The *MRMT* problem contains the core of the proofs: the link between a solution in *RX3C* and a solution in our routing problems. The choice of the triplet to cover the elements is already defined. We now place ourselves in the version of *MRMT* in which only one vertex is a target: the elementary Shortest Path problem with Mandatory Transitions. In the next constructions, we will focus on the associated digraph. We copy and connect our instance from Construction 2. The diroute between the source and the destination will pass through q copies of the previous instance. In each copy three new elements are chosen, such that there is a diroute only if there is a solution for *RX3C*. After, in Construction 4, we add arcs (that cannot be in a solution diroute) to obtain a dense digraph. The choice of triplet for $RX3C$ will remain the same (they depend on choice vertices $s^1, \ldots s^q$).

Construction 3. *Let $(\mathcal{X}, \mathcal{C})$ be an instance of RX3C where $\mathcal{X} = \{x_1, \ldots, x_{3q}\}$ and $\mathcal{C} = \{c_1, \ldots, c_{3q}\}$. We construct $(D = (V, A), \Pi)$ of the SPMT problem as follows.*

- *We add the vertices $s^1, \ldots s^{q+1}$ and for each $x_i \in \mathcal{X}$, we add the vertices x_i.*
- *For $1 \leq l \leq q$:*
  - *For each $c_i \in \mathcal{C}$, we add the vertices $c_{i,a}^l, c_{i,b}^l, c_{i,c}^l, c_{i,d}^l$.*
  - *We add the arcs $(s^l, c_{1,a}^l), \ldots, (s^l, c_{3q,a}^l)$ and $(c_{1,d}^l, s^{l+1}), \ldots, (c_{3q,d}^l, s^{l+1})$.*
  - *For each $c_i = \{x_j, x_k, x_h\} \in \mathcal{C}$, we add the arcs $(c_{i,a}^l, x_j)$, $(x_j, c_{i,b}^l)$, $(c_{i,b}^l, x_k)$, $(x_k, c_{i,c}^l)$, $(c_{i,c}^l, x_h)$, and $(x_h, c_{i,d}^l)$, as well as the seven mandatory transitions $\langle s^l, c_{i,a}^l, x_j \rangle$, $\langle c_{i,a}^l, x_j, c_{i,b}^l \rangle$, $\langle x_j, c_{i,b}^l, x_k \rangle$, $\langle c_{i,b}^l, x_k, c_{i,c}^l \rangle$, $\langle x_k, c_{i,c}^l, x_h \rangle$, $\langle c_{i,c}^l, x_h, c_{i,d}^l \rangle$, and $\langle x_h, c_{i,d}^l, s^{l+1} \rangle$.*

The instance (D, Π) contains $12q^2 + 4q + 1$ vertices, $24q^2$ arcs and $21q^2$ mandatory transitions. A big picture of the construction is represented in Fig. 3, where the red arcs represent directed consecutive paths. For each copy, only one gadget is shown.

Theorem 3. *The elementary SPMT problem is $\mathcal{NP}$-complete.*

Proof. Clearly elementary $SPMT$ is in $\mathcal{NP}$.

Let $(\mathcal{X}, \mathcal{C})$ be an instance of $RX3C$ with $3q$ elements and $3q$ triplets. Let (D, Π) be the associated instance of $SPMT$ following Construction 3. Let s^1 be the source vertex, and s^{q+1} the target vertex. Let P be an elementary shortest diroute starting from s^1, ending at s^{q+1} and respecting Π.

Due to the mandatory transitions structure, we can see P as a succession of element collectors. Each *element collector* p^l is a directed consecutive path composed of 8 arcs. Path p^l starts at a vertex s^l, heads towards a triplet vertex $c_{i,a}^l$, followed by its first element vertex x_j, alternates between the triplet vertices and its element vertices, and ends at vertex s^{l+1}: $p^l = (s^l, c_{i,a}^l), (c_{i,a}^l, x_j), (x_j, c_{i,b}^l), (c_{i,b}^l, x_k), (x_k, c_{i,c}^l), (c_{i,c}^l, x_h), (x_h, c_{i,d}^l), (c_{i,d}^l, s^{l+1})$, for each $c_i = \{x_j, x_k, x_h\} \in \mathcal{C}$. Thus $P = p^1, \ldots, p^q$. Each element collector p^l contains exactly three element vertices among the $3q$. The shortest diroute cannot have less than $8q$ arcs. Let l be $8q$.

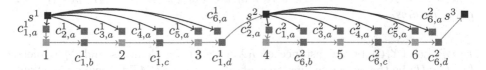

Fig. 3. Partial illustration of Construction 3 for the instance $(\mathcal{X}, \mathcal{C})$ of $RX3C$ where $\mathcal{X} = \{1, \ldots, 6\}$ and $\mathcal{C} = \{\{1, 2, 3\}, \{4, 5, 6\}, \{1, 2, 6\}, \{1, 5, 6\}, \{3, 4, 5\}, \{2, 3, 4\}\}$.

Suppose that P is a diroute with exactly $8q$ arcs. Since the diroute is elementary, each element vertex is reached exactly once. Exactly q triplet vertices have been reached. The triplet vertices crossed in P give a solution for the RX3C instance.

Suppose there is a solution S for the instance of RX3C. We arbitrarily order each triplet c_i of S from 1 to q. We denoted these triplets as c_i^l. Let P' be the diroute defined as follows: a) P' starts at vertex s^1; b) for each triplet $c_i^l = \{x_j, x_k, x_h\} \in S$ with $1 \leq l \leq q$, we add the corresponding element collector (starting at s^l and ending at s^{l+1}) at the end of P'. The elementary diroute P' contains $8q$ arcs, starts at s^1, ends at s^{q+1} and respects the mandatory transitions.

Thus, there is a solution for the instance $(\mathcal{X}, \mathcal{C})$ of RX3C if and only if there is, in the associated instance of elementary SPMT, a diroute of size $8q$ starting at s^1, ending at s^{q+1} and respecting the mandatory transitions.   □

Note that the vertices $s^1, \ldots s^q$ are the only ones where an arc can be chosen freely. For each other vertex, a mandatory transition constrains the diroute to only one possible next arc. In Construction 3, all the possible solutions have the same size. There are no diroutes from s^1 to s^{q+1} respecting the mandatory transitions with more (or less) than $8q$ arcs. Since the optimal solution is the only solution, no approximation algorithm can exist.

Theorem 4. *The elementary SPMT problem is not approximable by any ratio unless $\mathcal{P} = \mathcal{NP}$.*

We have stated that the elementary SPMT problem cannot be approximate by any ratio in general cases. The last step is to extend the result to dense digraphs. We add the missing arcs but prevent any solution from taking them. The Construction 4 shows the general way to add these arcs. The exact number of new arcs depends on the problem studied.

Construction 4. *Let $(D = (V, A), \Pi)$ be a directed graph and its associated mandatory transitions. We add an arc and a mandatory transition as follows. For an absent arc of D $(u, v \in V, (u, v) \notin A)$, we add the arc (u, v) in A and the mandatory transition $\langle u, v, u \rangle$ in Π.*

The new arcs are represented in Fig. 4. We can see two configurations with missing arcs on the left, and all the arcs on the right.

Theorem 5. *The elementary SPMT problem is not approximable by any ratio unless $\mathcal{P} = \mathcal{NP}$, even if each vertex of the digraph $D = (V, A)$ has $|V| - 2$ outgoing arcs.*

Proof. Let $(\mathcal{X}, \mathcal{C})$ be an instance of RX3C with $3q$ elements and $3q$ triplets. Let (D, Π) be the associated instance of elementary SPMT following Construction 3. We add all the missing arcs following Construction 4 except the arcs (s^i, s^{q+1}) for $1 \leq i \leq q$ and the arc (s^{q+1}, s^1). Note that each vertex has exactly

Fig. 4. Illustration of new arcs of Construction 4.

$|V| - 2$ outgoing arcs. Let s^1 be the source vertex and s^{q+1} the target vertex. Let P be an elementary shortest diroute starting from s^1, ending at s^{q+1} and respecting Π.

As we mentioned before, there are only q vertices where an arc can be freely chosen: $s^1, \ldots, s^q$. Since no arcs can be chosen for other vertices, new outcoming arcs cannot be passed through either. Due to the mandatory transitions structure, new outcoming arcs from $s^1, \ldots, s^q$ cannot be chosen in a diroute. If one of these arcs is chosen, the vertex s^{q+1} cannot be reached. Since no new arcs can be used in P, the proof of Theorem 4 also proves Theorem 5.   □

3   Routing Problems with Mandatory Transitions

In this section, we extend our results on the *MRMT* problem and the *FRMT* problem thanks to our result on the elementary *SPMT* problem. Some results can be obtained with a construction very similar to Construction 3 and the same arguments of proof. Thus, we do not recall all these proofs here. Note that these problems may admit no solutions, even in a symmetric complete digraph. For example, take a symmetric directed completed graph with vertices $u_1, \ldots, u_k$ for any $k > 2$ and add the mandatory transitions $\langle u_1, u_i, u_1 \rangle$ and $\langle u_i, u_1, u_i \rangle$ for $2 \leq i \leq k$. There are no solutions if u_1 is the source vertex and there is more than one target vertices.

We can see element vertices are indirect targets in the elementary *SPMT* problem. If we add the vertex s^{q+2}, the arcs (s^{q+1}, s^{q+2}) and (s^{q+2}, s^{q+1}), the mandatory transitions $\langle s^{q+1}, s^{q+2}, s^{q+1} \rangle$ and $\langle c^q_{i,d}, s^{q+1}, s^{q+2} \rangle$ for $1 \leq i \leq 3q$, and the missing arcs following Construction 4, we can state the following theorem (the proof is obtained with $x_1, \ldots, x_{3q}, s^{q+1}, s^{q+2}$ as targets in *MRMT* problem).

Theorem 6. *The MRMT problem (and its elementary version) are not approximable by any ratio unless $\mathcal{P} = \mathcal{NP}$, even if the digraph is a symmetric directed complete graph.*

Since the *SPMT* problem is solvable in polynomial time (Theorem 1) but the elementary *SPMT* problem is not approximable by any ratio (Theorem 5), we can investigate the role of the uniqueness of each vertex in a diroute. However, if we allow passing k times through the same vertex, the complexity stays the same. Let k be the number of times a vertex can appear in a solution diroute. We can make a first directed path starting at s^0, passing through the element vertices $k - 1$ times following Construction 5 and connect it at the vertex s^1.

Construction 5. *Let $(\mathcal{X}, \mathcal{C})$ be an instance of RX3C where $\mathcal{X} = \{x_1, \ldots, x_{3q}\}$ and $\mathcal{C} = \{c_1, \ldots, c_{3q}\}$. Let $(D = (V, A), \Pi)$ be the digraph and its associated*

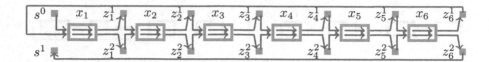

Fig. 5. Illustration of Construction 5 with $k = 3$ and $q = 2$.

mandatory transitions following Construction 3. Let $k > 1$ be the constant number of time a vertex can appear in a solution diroute. We add some vertices, arcs and mandatory transitions as follows.

- *We add the vertex s^0. For $1 \leq l \leq k - 1$:*
  - *We add the vertices z_i^l and the arcs (x_i, z_i^l) for $1 \leq i \leq 3q$.*
  - *We add the arcs (z_i^l, x_{i+1}) and the mandatory transitions $\langle x_i, z_i^l, x_{i+1} \rangle$ and $\langle z_i^l, x_{i+1}, z_{i+1}^l \rangle$ for $1 \leq i \leq 3q - 1$.*
- *We add the arcs (s^0, x_1), (z_{3q}^{k-1}, s^1), and (z_{3q}^l, x_1) and the mandatory transitions $\langle s^0, x_1, z_1^1 \rangle$, $\langle x_{3q}, z_{3q}^{k-1}, s^1 \rangle$, and $\langle x_{3q}, z_{3q}^l, x_1 \rangle$ for $1 \leq l \leq k - 2$.*

The Construction 5 is represented in Fig. 5. Since each arc is constraint by a directed consecutive path, the red arcs represents arcs and mandatory transitions. For more clarity, the mandatory transitions in element vertices are still represented. Note that the source vertex becomes s^0. Since each element vertex appears $k - 1$ times in the diroute between s^0 and s^1, the diroutes from s^1 to s^{q+1} for the k-elementary *SPMT* problem are the same as for the elementary *SPMT* problem. We can add all the missing arcs following Construction 4 except the arcs (s^i, s^{q+1}) for $0 \leq i \leq q$ and the arc (s^{q+1}, s^0). Thus we can state the following theorem.

Theorem 7. *The k-elementary SPMT problem is not approximable by any ratio unless $\mathcal{P} = \mathcal{NP}$, even if each vertex of the digraph $D = (V, A)$ has $|V| - 2$ outgoing arcs.*

The *MRMT* problem is not approximable by any ratio, even if there is only one target or if the diroute is elementary. The solution diroutes uses only few vertices. We now investigate the complexity of the Full-target Routing problem with Mandatory Transitions. The *FRMT* problem is a special case of the *MRMT* problem where all vertices are targets. We show that the *FRMT* problem is not approximable by any ratio unless $\mathcal{P} = \mathcal{NP}$, even if the digraph is a symmetric directed complete graph. For that we will complete Construction 3 such that each vertex appears in a diroute if and only if there is a solution for the *RX3C* problem. We will create an additional gadget (Construction 6). This gadget will construct a directed path between triplets without going through elements. If an instance of *RX3C* (size q) has a solution, the associated digraph has a diroute going through q triplets and all elements before going through the remaining triplets. We will then use Construction 4 to add all missing arcs while forbidding them. We show that the *FRMT* problem is not approximable by any ratio unless $\mathcal{P} = \mathcal{NP}$, even if the digraph is a symmetric directed complete graph (Theorem 8). Let us start construction.

Construction 6. *Let $(\mathcal{X}, \mathcal{C})$ be an instance of RX3C where $\mathcal{X} = \{x_1, \ldots, x_{3q}\}$ and $\mathcal{C} = \{c_1, \ldots, c_{3q}\}$. We construct $(D = (V, A), \Pi)$ of the FRMT problem as follows. Let (D', Π') be the obtained instance following Construction 3.*

We add D' to D and Π' to Π.
We add the vertex F and the directed consecutive circuit passing through s^{q+1}, all the triplets vertices $(c_{1,a}^1, \ldots c_{3q,d}^1 \cdots c_{1,a}^q, \ldots c_{3q,d}^q)$, and F in that order.

The instance (D, Π) contains $12q^2 + 4q + 2$ vertices. A solution for (D, Π) will always pass through the vertex s^{q+1} only after all the element vertices. We call the diroute passing through all the element vertices (following the element collectors) the *solution collector*. This diroute starts at the vertex s^1 and ends at the vertex s^{q+1}. We call the diroute which passes through all the remaining vertices the *remaining collector*. This diroute starts at the vertex s^{q+1} and ends at the vertex F. The triplet vertices in the solution collector define a solution for *RX3C*. The remaining collector does not affect the link between a solution in *RX3C* and a solution in *FRMT*. Indeed, the remaining collector is trapped in the directed consecutive circuit, without any element vertex. If q triplets are not enough, there is still no way to pass through all the element vertices. Since the proof is still the same, we do not recall it here. In Construction 6, all the possible solutions have the same size. There are no diroutes from s^1, going through all the vertices and respecting the mandatory transitions with more (or less) arcs than $12q^2 + 8q + 1$. Since the optimal solution is the only solution, no approximation algorithm can exist. The last step is to extend the digraph following Construction 4 to obtain a complete directed graph.

Theorem 8. *The FRMT problem is not approximable by any ratio unless $\mathcal{P} = \mathcal{NP}$, even if the digraph is a symmetric directed complete graph.*

4 Conclusion

In this paper, we have introduced a problem: the Multi-target Routing problem with Mandatory Transitions (*MRMT*). We have shown it cannot admit an approximation by any ratio if we want to pass through all the vertices (*FRMT* version), even if the digraph is a symmetric directed complete graph. On the other side, we have shown it is solvable in polynomial time if there is only one target (*SPMT* version) but cannot admit an approximation by any ratio if the solution diroute has to be elementary, even if each vertex of the digraph $D = (V, A)$ has $|V| - 2$ outgoing arcs. As a perspective we plan to study the hardness of these problems by considering a limited number of mandatory transitions and the influence of the place of targets in mandatory transitions. We can already see we can rewrite arcs and mandatory transitions like in Construction 1 to prove that the *MRMT* problem is equivalent to the Multi-target Routing problem unless a target appears in the tail of a mandatory transition's head arc.

Acknowledgements. The authors thank the anonymous referees for their advice on their work.

References

1. Cornet, A., Laforest, C.: Total domination, connected vertex cover and steiner tree with conflicts. Discret. Math. Theor. Comput. Sci. 19(3) (2017)
2. Cornet, A., Laforest, C.: Domination problems with no conflicts. Discret. Appl. Math. **244**, 78–88 (2018)
3. Cornet, A., Laforest, C.: Graph problems with obligations. In: Kim, D., Uma, R.N., Zelikovsky, A. (eds.) COCOA 2018. LNCS, vol. 11346, pp. 183–197. Springer, Cham (2018). https://doi.org/10.1007/978-3-030-04651-4_13
4. Garey, M.R., Johnson, D.S.: Computers and Intractability: A Guide to the Theory of NP-Completeness. W. H. Freeman & Co., New York (1979)
5. Kanté, M.M., Laforest, C., Momège, B.: An exact algorithm to check the existence of (elementary) paths and a generalisation of the cut problem in graphs with forbidden transitions. In: van Emde Boas, P., Groen, F.C.A., Italiano, G.F., Nawrocki, J., Sack, H. (eds.) SOFSEM 2013. LNCS, vol. 7741, pp. 257–267. Springer, Heidelberg (2013). https://doi.org/10.1007/978-3-642-35843-2_23
6. Kanté, M.M., Laforest, C., Momège, B.: Trees in graphs with conflict edges or forbidden transitions. In: Chan, T.-H.H., Lau, L.C., Trevisan, L. (eds.) TAMC 2013. LNCS, vol. 7876, pp. 343–354. Springer, Heidelberg (2013). https://doi.org/10.1007/978-3-642-38236-9_31
7. Kanté, M.M., Moataz, F.Z., Momège, B., Nisse, N.: Finding paths in grids with forbidden transitions. In: Mayr, E.W. (ed.) WG 2015. LNCS, vol. 9224, pp. 154–168. Springer, Heidelberg (2016). https://doi.org/10.1007/978-3-662-53174-7_12
8. Laforest, C., Martinod, T.: On the complexity of independent dominating set with obligations in graphs. Theoret. Comput. Sci. **904**, 1–14 (2022)
9. Laforest, C., Momège, B.: Some hamiltonian properties of one-conflict graphs. In: Kratochvíl, J., Miller, M., Froncek, D. (eds.) IWOCA 2014. LNCS, vol. 8986, pp. 262–273. Springer, Cham (2015). https://doi.org/10.1007/978-3-319-19315-1_23
10. Laporte, G.: The traveling salesman problem: an overview of exact and approximate algorithms. Eur. J. Oper. Res. **59**(2), 231–247 (1992)
11. Rachid, M.H., Cherif, W.R., Bloch, C., Chatonnay, P.: Proposition de notation pour les problèmes de tournées. In: 9ème congrès de la ROADEF: ROADEF 2008, pp. 63–77. Presses universitaires de l'Université Blaise Pascal (2008)

Payment Scheduling in the Interval Debt Model

Tom Friedetzky[ID], David C. Kutner[✉][ID], George B. Mertzios[ID],
Iain A. Stewart[ID], and Amitabh Trehan[ID]

Department of Computer Science, Durham University, Durham, UK
{tom.friedetzky,david.c.kutner,george.mertzios,i.a.stewart,
amitabh.trehan}@durham.ac.uk

Abstract. The networks-based study of financial systems has received considerable attention in recent years, but seldom explicitly incorporated the dynamic aspects of such systems. We consider this problem setting from the temporal point of view, and we introduce the *Interval Debt Model (IDM)* and some scheduling problems based on it, namely: BANKRUPTCY MINIMIZATION/MAXIMIZATION, in which the aim is to produce a schedule with at most/at least k bankruptcies; PERFECT SCHEDULING, the special case of the minimization variant where $k = 0$; and BAILOUT MINIMIZATION, in which a financial authority must allocate a smallest possible bailout package to enable a perfect schedule.

In this paper we investigate the complexity landscape of the various variants of these problems. We show that each of them is NP-complete, in many cases even on very restrictive input instances. On the positive side, we provide for PERFECT SCHEDULING a polynomial-time algorithm on (rooted) out-trees. In wide contrast, we prove that this problem is NP-complete on directed acyclic graphs (DAGs), as well as on instances with a constant number of nodes (and hence also constant treewidth). When the problem definition is relaxed to allow *fractional* payments, we show by a linear programming argument that BAILOUT MINIMIZATION can be solved in polynomial time.

Keywords: Temporal graph · Financial network · Payment scheduling · NP-complete · Polynomial-time algorithm

1 Introduction

In the study of financial systems, network-based paradigms were introduced to model behaviors associated with the connectedness and complexity exhibited in real-world financial systems. We introduce the *Interval Debt Model*, focusing on the *choices* that real-world financial entities have in times at which they pay their debts by applying temporal graphs to this setting. Previous work in the study of financial networks had seldom explicitly incorporated the temporal aspects inherent to real-world debt.

G.B. Mertzios—Partially supported by the EPSRC grant EP/P020372/1.

L. Gąsieniec (Ed.): SOFSEM 2023, LNCS 13878, pp. 267–282, 2023.
https://doi.org/10.1007/978-3-031-23101-8_18

Financial Networks have been studied by applying concepts from ecology [9], statistical physics [3], and Boolean networks [6]. In 2001, Eisenberg and Noe (EN) [7] introduced a paradigm which has been the basis for much work in the network-based analysis of financial systems, see also e.g. the survey [11]. In this model, financial entities all operate within a single clearing system. The paradigm has been extended to include default costs [18], Credit Default Swaps (CDSs) (derivatives through which banks can bet on the default of another bank in the system) [19], and the sequential behavior of bank defaulting in real-world financial networks [16].

A core motivation of financial network analysis is to inform central banks' and regulators' policies. The concepts of *solvency* and *liquidity* are core to this task: a bank is said to be *solvent* if it has enough assets (including non-liquid assets) to meet all its obligations, and is said to be *liquid* if it has enough liquid assets (e.g. cash) to meet its obligations on time. An illiquid but solvent bank may exist even in modern interbank markets [17]. In such cases, a central bank may act as a *lender of last resort* and extend loans to such banks to prevent their defaulting on debts [2,17]. The optimal allocation of bailouts to a system in order to minimize damage has also been studied as an extension of Eisenberg and Noe's model [15]. Here, bailouts refer to funds provided by a third party (such as the government) to entities to help them avoid bankruptcy.

Temporal Graphs are graphs whose underlying network structure changes over time. These allow us to model real-world networks which have inherent dynamic properties, such as transportation networks [20], contact networks in an epidemic [8], and communication networks; for an overview see [4,5,10]. Most commonly, following the formulation introduced by Kempe, Kleinberg and Kumar [13], a temporal graph has a fixed set of vertices, and edges which appear and disappear at integer times up to an (integer) lifetime T. In such cases the temporal graph can be thought of as a static graph $G = (V, E)$ in which the edges are labeled with the times at which they appear. Often, a natural extension of a problem on static graphs to the temporal setting yields a computationally harder problem; for example, finding node-disjoint paths in a temporal graph remains NP-complete even when the underlying graph is itself a path [14], and finding a temporal vertex cover remains NP-complete even on star temporal graphs [1].

Our Contribution. In this paper we present a new framework for considering problems of bailout allocation and payment scheduling in financial networks by taking into account the temporal aspect of debts between financial entities, the *Interval Debt Model* (IDM). We introduce several natural problems and problem variants in this model, and show that the tractability of such problems depends greatly on the network topology and on the restrictions on payments (i.e. the admission or exclusion of partial and fractional payments on debts). While previous work has mainly focused on static financial networks, we go further and introduce the time dimension in financial networks to account for the temporal nature of real-world debts. In particular, the IDM offers the capability to represent the flexibility that entities have in paying debts earlier or later, within some *interval*. In Sect. 2 we present our new model IDM in detail. The formal definitions of our problems, as well as a summary of our results (see Table 1) are given in Sect. 2.3. All our results are formally presented in Sect. 3. Due to space constraints, some proofs are deferred to more complete versions of this paper.

2   The Interval Debt Model

In this section, we introduce (first by example, then formally) the Interval Debt Model, a framework in which temporal graphs are used to represent the system of debts in a financial network.

As an example, consider a tiny financial network consisting of 3 banks u, v, w, with €30, €20 and €10 respectively in initial assets, and the following inter-bank financial obligations. Bank u owes bank v €20, which it must pay by time 3, and €15, which it must pay between times 4 and 5. Bank v has agreed to lend bank w €25 at time 2 exactly, which bank w must repay to bank v between times 4 and 6. A graphical representation of this system is shown in Fig. 1.

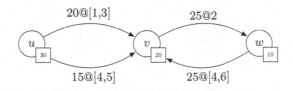

Fig. 1. A simple instance of the IDM

Several points can be made about this system: node u is insolvent as its €30 in assets are insufficient to pay all its debts; node v may be illiquid (it may default on part of its debt to w, e.g. if u pays all of its first debt at time 3) or may remain liquid (e.g. if it receives at least €5 from u by time 2); and node w is solvent and certain to remain liquid in any case.

One may ask several questions about this system: Are partial payments admitted (i.e. u paying €18 of the €20 debt at time 1, and the rest later)? If so, are non-integer payments admitted? Can money received be immediately forwarded (e.g. u paying v €20 at time 2, and v paying w €25 at time 2)?

We now specify in detail the setting we consider in the remainder of the paper.

2.1   Formal Setting

Formally, an Interval Debt Model (IDM) instance is a 3-tuple (G, D, A^0), where:

- $G = (V, E)$ is a finite digraph with n nodes (or, alternatively, banks) from $V = \{v_i : i = 1, 2, \ldots, n\}$ and directed labelled multi-edges (but no loops) from $E \subseteq V \times V \times \mathbb{N}$, with the edge $(u, v, id) \in E$ denoting that there is a *debt*, whose label is id, from the *debtor* u to the *creditor* v; moreover, the labels of some pair (u, v) (appearing in at least one triple $(u, v, id) \in E$) form a non-empty contiguous integer sequence $0, 1, 2, \ldots$ We refer to the subset of edges directed out of or in to some specific node v by $E_{\text{out}}(v)$ and $E_{\text{in}}(v)$, respectively.

- $D : E \to \{(a, t_1, t_2) : a, t_1, t_2 \in \mathbb{N} \setminus \{0\}, t_1 \leq t_2\}$ is the *debt function* which associates *terms* to every debt (ordinarily, we abbreviate $D((u, v, n))$ as $D(u, v, n)$). Here, if e is a debt with terms $D(e) = (a, t_1, t_2)$ then a is the monetary amount to be paid and t_1 (resp. t_2) is the first (resp. last) time at which any portion of this amount can be paid; also, for any debt $e \in E$, we write $D(e) = (D_a(e), D_{t_1}(e), D_{t_2}(e))$. For simplicity of notation, we sometimes denote the terms $D(e) = (a, t_1, t_2)$ by $a@[t_1, t_2]$, and $a@t_1$ when $t_1 = t_2$.
- $A^0 = (e^0_{v_1}, e^0_{v_2}, \ldots e^0_{v_n}) \in \mathbb{N}^n$ is a tuple with $e^0_{v_i}$ denoting the *initial external assets* of bank v_i.

We refer to the greatest timestamp that appears in any debt for a given instance as *lifetime*. The instance shown in Fig. 1, which has lifetime $T = 6$, is given by: $V = \{u, v, w\}$, $E = \{(u, v, 0), (u, v, 1), (v, w, 0), (w, v, 0)\}$, $D(u, v, 0) = (20, 1, 3)$, $D(u, v, 1) = (15, 4, 5)$, $D(v, w, 0) = (25, 2, 2)$, $D(w, v, 0) = (25, 4, 6)$, and $A^0 = (e^0_u, e^0_v, e^0_w)$, where $e^0_u = 30$, $e^0_v = 20$, and $e^0_w = 10$.

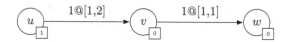

Fig. 2. An instance of the IDM with exactly two schedules.

Similarly, the instance shown in Fig. 2 has lifetime $T = 2$ and is given by $V = \{u, v, w\}$, $E = \{(u, v, 0), (v, w, 0)\}$, $D(u, v, 0) = (1, 1, 2)$, $D(v, w, 0) = (1, 1, 1)$, and $A^0 = (e^0_u, e^0_v, e^0_w)$, where $e^0_u = 1$, $e^0_v = 0$, $e^0_w = 0$.

2.2 Schedules

Given an IDM instance (G, D, A^0), a *schedule* describes at what times the banks transfer *assets* to one another via *payments*. Formally, a schedule is a set of $|E| * T$ *payment* values $p^t_e \geq 0$, one for each time-edge pair. Equivalently, a schedule can be expressed as an $|E| \times T$ matrix S, and the variables p^t_e are the entries of that matrix. The value of p^t_e is the monetary amount of the debt e paid at time t. Our intention is that at any time $t \in [1, T]$, every payment $p^t_e > 0$ of a schedule S is paid by the debtor of e to the creditor of e, not necessarily for the full amount $D_a(e)$ but for the amount p^t_e. A schedule for the instance of Fig. 2 consists of the four payments $p^1_{(u,v,0)}$, $p^1_{(v,w,0)}$, $p^2_{(u,v,0)}$ and $p^2_{(v,w,0)}$. Note that, using the above representations of a schedule S, we might have a large number of zero payments. Therefore, for simplicity of presentation, in the remainder of the paper we specify schedules by only specifying the non-zero payments. An example schedule for the instance in Fig. 2 is then $p^1_{(u,v,0)} = 1$, $p^1_{(v,w,0)} = 1$.

We now introduce some auxiliary variables which are not strictly necessary but help us to concisely express constraints on and properties of schedules (for nodes $u, v \in V(G)$ and time $t \in [T]$):

- Denote by I_v^t the total monetary amount of incoming payments of node v at time t.
- Denote by O_v^t the total monetary amount of outgoing payments (expenses) of node v at time t.
- We write $p_{u,v}^t$ to denote the total amount of all payments made from u to v at time t in reference to *all* debts from u to v. That is, $p_{u,v}^t = \sum_i p_{(u,v,i)}^t$.
- We denote by e_v^t node v's external assets at time t. Then $e_v^t = e_v^{t-1} + I_v^t - O_v^t$ for every v and t.

Note that, whenever there is only one edge from a node u to a node v, we have $p_{u,v}^t = p_{(u,v,0)}^t$; we use this in proofs for conciseness where possible. Recall the example schedule for Fig. 2, which we can then represent as $p_{u,v}^1 = 1$, $p_{v,w}^1 = 1$. As we shall see, the payments in this schedule can be legitimately discharged in order to satisfy the terms of all debts but in general this need not be the case. In fact, there might be schedules that are invalid, as well as schedules in which banks default on debts (go bankrupt). We deal with the notions of validity and bankruptcy now.

Definition 1 (Valid schedule; payable, due, overdue debts). *A schedule is valid if it satisfies the following properties (for any edge e and debt $D(e) = (a, t_1, t_2)$):*

- *All payment variables are nonnegative. That is, $p_e^t \geq 0$ for every e and t.*
- *All asset variables (as derived from payment variables and initial assets) are nonnegative. That is, $e_v^t \geq 0$ for every v and t.*
- *No debts are overpaid. That is, $\sum_{t \in [t_1, T]} p_e^t \leq a$.*
- *No debts are paid early. $\sum_{t \in [0, t_1 - 1]} p_e^t = 0$.*

Given some IDM instance and schedule, a debt $D(e) = a@[t_1, t_2]$ is said to be payable for the interval $[t_1, t_2 - 1]$. At time t_2, $D(e)$ is said to be due. At every time $t \geq t_2$, if the full amount a has not yet been paid with reference to e, then $D(e)$ is said to be overdue at time t. A debt is active whenever it is payable, due, or overdue.

A bank is said to be withholding if, at some time t, it has an overdue debt and sufficient assets to pay (part of, where fractional or partial payments (see below) are permitted) the debt. If any bank is withholding in the schedule, then the schedule is not valid.

Definition 2 (Bankrupt). *A bank is said to be bankrupt (at time t) if it is the debtor of an overdue debt (at time t). We say a schedule has k bankruptcies if k distinct banks go bankrupt at any point in the schedule. A bank may recover from bankruptcy if it receives sufficient income to pay off all its overdue debts.*

Definition 3 (Insolvent). *A bank v is said to be* insolvent *if all its assets (the sum of all debts due to v and of v's initial assets) are insufficient to cover all its obligations (the sum of all debts v owes). Formally, v is insolvent if*

$$e_v^0 + \sum_{e \in E_{in}(v)} D_a(e) < \sum_{e \in E_{out}(v)} D_a(e)$$

A bank which is insolvent will necessarily be bankrupt in any schedule.

We emphasize that the timing of bankruptcy and recovery or not of the banks is not considered.

We consider three natural variants of the model, in which different natural constraints are imposed on the payment variables:

- In the **Fractional Payments (FP)** variant, the payment variables may take rational values. That is, $p_e^t \in \mathbb{Q}$ for every e and t.
- In the **Partial Payments (PP)** variant, the payment variables may take only integer values. That is, $p_e^t \in \mathbb{N}$ for every e and t, and we allow payments for a smaller amount than the total debt.
- In the **All-or-Nothing (AoN)** variant, every payment must fully cover the relevant debt; every payment variable must be for the full amount, or zero. That is, $p_e^t \in \{D_a(e), 0\}$ for every edge e.

For example, the instance of Fig. 2 has the following valid schedules:

- (In all variants) the one above in which $p_{u,v}^1 = p_{v,w}^1 = 1$ (all debts are paid in full at time 1).
- (In all variants) one in which $p_{u,v}^2 = p_{v,w}^2 = 1$ (all debts are paid in full at time 2). Under this schedule, node v is **bankrupt** at time 1, as €1 of the debt $D(v, w, 0)$ is unpaid and that debt is **due**.
- (In the FP variant only) for every $a \in \mathbb{Q}$, where $0 < a < 1$, the schedule in which $p_{u,v}^1 = p_{v,w}^1 = a$ and $p_{u,v}^2 = p_{v,w}^2 = 1 - a$. Under each of these, node v is **bankrupt** at time 1, as €$1 - a$ of the debt $D(v, w, 0)$ is unpaid and that debt is **due**.

Instant Forwarding and Cycles. We emphasize that we allow a bank to instantly spend income received. Note that in all valid schedules for the instance in Fig. 2 above, v *instantly forwards* money received from u to w; the assets of v never exceed 0 in any valid schedule. This behavior is consistent with the EN model [7] in which financial entities operate under a single clearing authority. Indeed, in such cases a payment chain of any length is admitted and the payment takes place in unit time regardless of chain length. Furthermore, still consistent with the EN model is the possibility of a payment cycle.

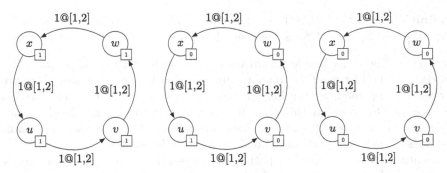

(a) Every node starts with €1. (b) Only u starts with €1. (c) Every node starts with €0.

Fig. 3. Examples illustrating the behavior of cycles in the IDM.

Figure 3 shows three cyclic IDM instances, all with lifetime $T = 2$. By our definition of a valid schedule, the schedule $p^1_{u,v} = p^1_{v,w} = p^1_{w,x} = p^1_{x,u} = 1$ is valid in all three instances. In Fig. 3b we may imagine the €1 moves from node u along the cycle, satisfying every debt at time 1. This is a useful abstraction, but not strictly accurate - rather, we may imagine that all 4 banks simultaneously order payments forward under a single clearing system. The clearing system calculates the balances that each bank would have with those payments executed, ensures they are all nonnegative (one of our criteria for schedule validity) and then executes the transfer by updating all accounts simultaneously. This distinction is significant when we consider Fig. 3c, in which no node has assets. A clearing system ordered to simultaneously pay all debts would have no problem doing so in the EN model, and in our model this constitutes a valid schedule. We highlight that there also exist valid schedules for the instance in Fig. 3c in which all 4 banks go bankrupt, namely the schedule in which every payment variable is set to zero; then no bank is withholding (they all have zero assets), so the schedule is valid, and every bank has an overdue debt.

Lemma 1. *For any given IDM instance and a (FP/PP/AoN) schedule, it is possible to check in polynomial time whether the schedule is valid for that instance, and to compute the number of bankruptcies under the schedule.*

Proof Sketch. It is possible to iterate over the schedule once and calculate: the assets of every node at every time; the number of debts which are overdue; the number of nodes which have overdue debts. The validity of the schedule (no withholding banks, nonnegative assets at every time, no overpaid debts, and no debts paid early, FP/PP/AoN constraint) can similarly be verified in a single iteration over the schedule. □

Definition 4. *Let* (G, D, A^0) *be an instance. Then the set of timestamps* $\{t :$ $D_{t_1}(e) = t$ *or* $D_{t_2}(e) = t$ *for some edge* $e\}$ *is the set of* extremal *timestamps.*

Remark 1. There is a simple preprocessing step such that we can assume afterwards that the lifetime T is polynomially bounded in the input size. This preprocessing step modifies the instance such that every $t \in [T]$ is an extremal timestamp. Observe that this procedure does not make any previously impossible schedule outcome (number of bankruptcies and finishing assets) possible, nor does it make any previously possible outcome impossible.

Hence we need not consider pathological cases in which the lifetime (and so the size of schedules) is exponential in the size of the input.

2.3 Problem Definitions

Here, we define some problems with natural real-world applications in the IDM.

IDM BANKRUPTCY MINIMIZATION

> **Input:** an IDM instance (G, D, A^0) and integer k
> **Question:** does there exist a valid schedule S for the input such that at most k banks go bankrupt (have overdue debts) at any point in the schedule?

IDM PERFECT SCHEDULE

> **Input:** an IDM instance (G, D, A^0)
> **Question:** does there exist a valid schedule S for the input such that no debt is ever overdue?

This problem is equivalent to IDM BAILOUT MINIMIZATION where $b = 0$ and to IDM BANKRUPTCY MINIMIZATION where $k = 0$.

IDM BAILOUT MINIMIZATION

> **Input:** an IDM instance (G, D, A^0) and integer b
> **Question:** does there exist a positive bailout vector $B = (b_1, b_2, \ldots b_{|V|})$ with $\sum_{i \in |V|} b_i \leq b$ and schedule S such that S is a perfect schedule for the instance $(G, D, A^0 + B)$?

IDM BANKRUPTCY MAXIMIZATION

> **Input:** an IDM instance (G, D, A^0) and integer k
> **Question:** does there exist a valid schedule S for the input such that **at least** k banks go bankrupt (have overdue debts) at any point in the schedule?

This problem is interesting to consider for quantifying a "worst-case" schedule, where banks' behavior is unconstrained beyond the terms of their debts.

All of the problems above exist in the **All-or-Nothing (AoN)** variant, where an AoN schedule is required; in the **Partial Payments (PP)** variant, in which a PP schedule is required; and in the **Fractional Payments (FP)** variant, in which an FP schedule is required.

All of the above problems, in all three variants, are in NP. For every yes-instance, there exists a witness schedule polynomial in the size of the input the validity of which can be verified in polynomial time by Lemma 1.

Every valid **PP** schedule is a valid **FP** schedule. Not every valid **AoN** schedule is a valid **PP** schedule. In an **AoN** schedule, a bank may go bankrupt while still having assets (insufficient to pay off any of its debts) - this is prohibited in any **PP** schedule as that bank would be **withholding**. If we restrict the input to only those in which for every edge e $D_a(e) = 1$ then every valid **AoN** schedule for that instance is a valid **PP** schedule and a valid **FP** schedule.

We call a graph *multiditree* whenever the underlying undirected graph (i.e. the undirected graph that is obtained by replacing each directed multiedge with an undirected edge) is a tree. We call *rooted out-tree* (or *out-tree*) a multiditree in which every edge is directed away from the root. By an *out-path* we mean an out-tree where the underlying undirected graph is a path, and the root is either of the endpoints.

3   Our Results

In this section we investigate the complexity of the problems presented. We first present our hardness results in Subsect. 3.1, and then show in Subsect. 3.2 that under certain constraints the problem of Bailout Minimization becomes tractable.

3.1   Hardness Results

Here we show that every problem introduced is NP-complete in the PP and AoN variants, even in various special cases, and that Bankruptcy Minimization is NP-complete and para-NP-Hard in all three variants for a variety of possible parameters.

Theorem 1. *For each of the variants* AoN, PP, *and* FP, BANKRUPTCY MIN-IMIZATION *is (i) NP-complete, even when the underlying graph G has $O(1)$ vertices, and (ii) NP-complete, even when $T = 1$, the underlying graph G is a DAG with a longest path of length 4, out-degree at most 2, in-degree at most 3, debt at most €3 per edge, and starting assets at most €3 per bank.*

Table 1. Summary of results. Note that PERFECT SCHEDULING is a subproblem of both BAILOUT MINIMIZATION and BANKRUPTCY MINIMIZATION.

Problem \ Constraint on graph G	Out-tree	Multiditree	DAG	General case
FP BAILOUT MINIMIZATION and FP PERFECT SCHEDULING	P (Theorems 8,9)	P (Theorem 9)	P (Theorem 9)	P (Theorem 9)
FP BANKRUPTCY MINIMIZATION	?	?	NP-C (Theorem 1)	NP-C (Theorem 1)
PP BAILOUT MINIMIZATION and PP PERFECT SCHEDULING	P (Theorem 8)	NP-C (Theorem 3)	NP-C (Theorems 2,4)	NP-C (Theorems 2,3,4)
PP BANKRUPTCY MINIMIZATION	?	NP-C (Theorem 3)	NP-C (Theorem 1)	NP-C (Theorems 1-3,4)
PP BANKRUPTCY MAXIMIZATION	?	?	NP-C (Theorem 5)	NP-C (Theorem 5)
AoN, all problems	NP-C (Theorems 6+7)	NP-C (Theorems 6+7)	NP-C (Theorems 6+7)	NP-C (Theorems 6+7)

By Theorem 1(i) it follows that each of the AoN, PP, and FP variants of BANKRUPTCY MINIMIZATION are *para-NP-hard*, when parameterized by any parameter that is upper-bounded by the number of vertices, such as e.g. the number of bankruptcies k, or the treewidth of the underlying graph.

Theorem 2. AoN PERFECT SCHEDULING *and* PP PERFECT SCHEDULING *are both NP-complete even when* $T \leq 3$, *the underlying graph G is a DAG with out-degree at most 3, in-degree at most 3, debt at most €2 per edge, and starting assets at most €3 per bank.*

Theorem 3. PP PERFECT SCHEDULING *is NP-complete even when the input is restricted to multiditrees with diameter 6, to €1 debts, and to a maximum of 6 multiedges between any two nodes.*

In all the above results, the input is allowed to have unlimited (i.e. unbounded) total assets in the system, which might be unrealistic in practically relevant financial systems. We now show that, even in the highly restricted case where just €1 in liquid assets exists in the system, PP PERFECT SCHEDULING still remains NP-complete.

Theorem 4. AoN PERFECT SCHEDULING *and* PP PERFECT SCHEDULING *are both NP-complete even when* $sum(A^0) = 1$, *i.e. the total of all external assets is €1.*

Proof. We proceed by reduction from DIRECTED HAMILTONIAN CYCLE (DHC), an NP-complete problem [12]:

Input: a digraph $G = (V, E)$.

Question: does there exist a DHC on G (a directed cycle which visits every vertex exactly once)?

For this reduction, we introduce the **at-least-once** gadget shown in Fig. 4. The intuition of the proof is that there is only €1 in the system, and that in any perfect schedule that €1 must pass through each gadget at least once, and therefore exactly once since there are n such gadgets and n timesteps the €1 can "rest" at a **center node** v_C.

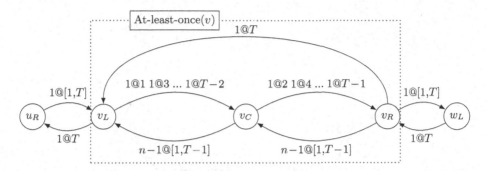

Fig. 4. The **at-least-once** gadget for node $v \in V(G_{DHC})$ where $\{(u,v),(v,w)\} \subseteq E(G_{DHC})$. Note the lifetime $T = 2|V(G_{DHC})| + 1$.

Given a DHC instance G_{DHC}, construct a PP PERFECT SCHEDULING instance (G_{PS}, D, A^0) by introducing a copy of the *at-least-once* gadget for each node $v \in G_{DHC}$ and then connecting gadgets within G_{PS} iff there exists a directed edge from one of the corresponding nodes in G_{DHC} to the other, as shown in Fig. 4. We then give €1 in assets to exactly onearbitrarily-chosen *right node* v_R in G_{PS}, and €0 in assets to every other node.

Claim 1. *If the IDM instance G_{PS}, D, A^0 admits a perfect schedule, then the DHC instance G_{DHC} admits a directed Hamiltonian cycle.*

Claim 2. *If the DHC instance G_{DHC} admits a directed Hamiltonian cycle, then G_{PS} admits a perfect schedule.*

Proof Sketch. Given a DHC v_1, v_2, ..., v_n, v_1, we describe the order in which the €1 in "real" assets moves through the network in our constructed perfect schedule. All other payments in the schedule can be efficiently found by debt cancellation, i.e. some node u pays some node v some amount €1 at time t and v pays u the same amount €1 also at time t, resulting in no "real" asset movement. This is possible by construction of the instance (G_{PS}, D, A^0) - in general, there is no guarantee that v would also have an active debt to u at time t. The €1 starts in node v_{1R}, then is paid forward in the order indicated by the edge labels in Fig. 5. We emphasize once more that the "payment" around the entire cycle at time $2n + 1$ does not result in any "real" asset movement – all balances remain unchanged.   □

Hence we have that there exists a perfect schedule in (G_{PS}, D, A^0) iff there exists a Hamiltonian cycle in G_{DHC}. This completes the proof of Theorem 4. □

Theorem 5. AON BANKRUPTCY MAXIMIZATION *and* PP BANKRUPTCY MAXIMIZATION *are both NP-complete even when $T = 2$, the underlying graph G is a DAG with out-degree at most 2, in-degree at most 3, debt at most €2 per edge, and assets at most €3 per bank.*

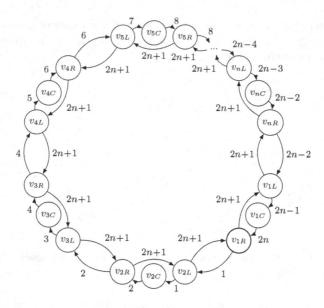

Fig. 5. The path that the "real" €1 takes in our constructed PP PERFECT SCHEDULING instance, if the input graph G contained a Hamiltonian cycle.

The constraints imposed by AoN schedules rapidly increase the problem complexity. Indeed, every problem considered is NP-complete, even whenever the input graph is an out-path on at most 4 vertices and with lifetime $T \leq 2$.

Theorem 6. *When the underlying graph G is an out-path on 4 vertices* AoN PERFECT SCHEDULING *is weakly NP-Hard when the lifetime $T = 2$, and strongly NP-Hard when T is unbounded.*

Theorem 7. AoN BANKRUPTCY MAXIMIZATION *is NP-Hard even when the underlying graph G is an out-path on 3 vertices and the lifetime T of the graph is at most 2.*

3.2 Polynomial-Time Algorithms

In this section we show that the PP variant of BAILOUT MINIMIZATION is solvable in polynomial time on out-trees, while its FP variant is always polynomial-time solvable. Our algorithm for the PP variant contrasts with the NP-completeness of its subproblem PP PERFECT SCHEDULING on both DAGs (Theorem 2) and multiditrees (Theorem 3); note that out-trees are a subclass of both DAGs and multiditrees.

Theorem 8. PP BAILOUT MINIMIZATION *is in P when the input is restricted to out-trees.*

Proof Sketch. We show that, given an instance of PP BAILOUT MINIMIZATION G, D, A^0, b in which G is an out-tree on at least 3 nodes, it is always possible to produce an updated instance which is equivalent (a yes-instance iff the original instance was a yes-instance) and in which G is strictly smaller (has fewer nodes). The process is illustrated in Fig. 6, and is as follows:

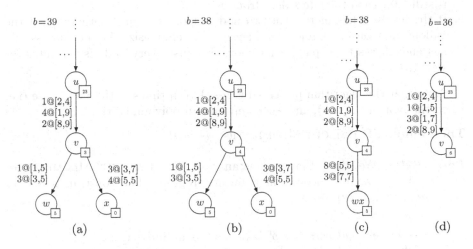

Fig. 6. Example of shrinking an out-tree while preserving (non)existence a bailout schedule.

1. While any node v is insolvent, increment v's assets and decrement the bailout amount b. In Fig. 6a, for instance, node v has €7 in income and €3 in external assets, but €11 in debt, so is insolvent. In Fig. 6b v's assets have increased by €1 and the bailout amount has decreased by €1.
2. Update every debt from a parent to its leaf children to be a 1-interval at the end timestep as, so a debt due in the interval [3,7] is updated to be due at time 6 exactly (i.e. at [7,7]).
3. While any parent has two leaf children (i.e. in Fig. 6b v has two leaf children w and x), **merge** the two sibling leaves into a single leaf (i.e. in Fig. 6c w and x are combined into wx). Sum node assets, and combine debts due at the same time into a single debt (i.e. v's debts 1@[6,6] to w and 4@[6,6] to x are combined into a single debt 5@[6,6] to wx).
4. While there exist 3 nodes j, k, l with l a leaf and the only child of k, and k a child of j **prune** l. In Fig. 6c $j = u$, $k = v$, and $l = wx$, and we prune wx. That is, update the debts from j to k to reflect exactly the constraints imposed by k's debts to l, then remove l from the graph. If payments from j cannot cover all payments to l in time (i.e. k is necessarily illiquid), increment k's assets and decrement the bailout amount b until they do, then update the debts. (Binary search may be applied where debt amounts are very large.)

In Fig. 6d, the 8@[5,5] debt to wx will be paid using €6 of v's external assets (including €1 of "solvency" bailout and €2 of "liquidity" bailout), €1 from u's debt 1@[2,4], and €1 of u's debt at [1,9], now constrained to the interval [1,5]. The 3@[7,7] debt to wx will be paid using €3 of u's debt at [1,9], now constrained to the interval [1,7]. The money received at v in the interval [8,9] cannot be usefully directed to wx, as the earliest it could be received ($t = 8$) is still later than the latest debt from v ($t = 7$).

5. If the instance has more than two nodes, loop to step 2. Otherwise, if the bailout amount $b < 0$, reject, else accept. We emphasize the root necessarily has enough assets to pay all its debts, because every node is solvent after step 1.

Note that the algorithm presented loops $|V(G)|$ times in the worst case (i.e. when the input is a path), and each step runs in polynomial time.   □

Theorem 9. FP BAILOUT MINIMIZATION *is in P.*

Proof Sketch. We show that an instance of FP BAILOUT MINIMIZATION G, D, A^0, b can be encoded in a linear program. The constraints are all expressed as linear expressions:

- $\mathrm{sum}(B) \leq b$.
- The set of *payment variables* p_e^t is defined as in Subsect. 2.2.
- The definitions of I_v^t, O_v^t, and e_v^t are all linear combinations of *payment variables*. Set $e_v^0 := A^0[v] + B[v]$; that is, the starting assets of v are its *external assets* combined with the *bailouts received* at v under the vector B.
- The constraints on valid schedules in Definition 1 can all also be expressed as linear combinations.
- We additionally impose that no banks are bankrupt under the schedule, or equivalently that no debt is ever overdue. That is, for every edge e and debt $D(e) = (a, t_1, t_2)$, $\sum_{t \in [t_1, t_2]} p_e^t = a$.

Any assignment to B and to the payment variables satisfying the above is necessarily a valid perfect schedule on an instance in which starting assets of nodes were supplemented by at most €b in total.

Linear programs can be efficiently solved when fractional solutions are admitted, hence all instances of FP BAILOUT MINIMIZATION are tractable. We emphasize that this is in contrast with PP BAILOUT MINIMIZATION and FP BANKRUPTCY MINIMIZATION, both of which are NP-complete. The method above solves neither of these: the former would correspond to an integer linear program (which are NP-complete in general) and it is not possible to express a constraint on the *number of bankruptcies* through a linear combination on the payment variables.   □

4   Conclusion and Open Problems

This paper introduces the *Interval Debt Model (IDM)*, a new model seeking to capture the temporal aspects of debts in financial networks. We investigate the computational complexity of various problems involving debt scheduling, bankruptcy and bailout with different payment options (All-or-nothing (AoN), Partial (PP), Fractional (FP)) in this setting. We prove that many variants are hard even on very restricted inputs but certain special cases are tractable. For example, we present a polynomial time algorithm for PP BAILOUT MINIMIZA-TION where the IDM graph is an out-tree. However, for a number of other classes (DAGs, multitrees, total assets are €1), we show that the problem remains NP-hard. This leaves open the intriguing question of the complexity status of problems which are combinations of two or more of these constraints, most naturally on multitrees which are also DAGs, an immediate superclass of our known tractable case.

We prove that FP BAILOUT MINIMIZATION is polynomial-time solvable by expressing it as a Linear Program. Can a similar argument be applied to some restricted version of FP Bankruptcy Minimization (which is NP-Complete, in general)? A natural generalization is simultaneous Bailout and Bankruptcy minimization i.e. can we allocate €b in bailouts such that a schedule with at most k bankruptcies becomes possible. Variations of this would be of practical interest. For example, if regulatory authorities can allocate bailouts as they see fit, but not impose specific payment times, it would be useful to consider the problem of allocation of €b in bailouts such that the maximum number of bankruptcies in any valid schedule is at most k. Conversely, where financial authorities can impose specific payment times, the combination of the problems Bankruptcy Minimization and Bailout Minimization would be more applicable.

Finally, can we make our models even more realistic and practical? How well do our approaches perform on real-world financial networks? Can we identify topological and other properties of financial networks that may be leveraged in designing improved algorithms?

Acknowledgements. The authors are thankful to Roger Wattenhofer for insightful discussions and suggestions and to Nina Klobas and Tamio-Vesa Nakajima for technical discussions early in the project.

References

1. Akrida, E.C., Mertzios, G.B., Spirakis, P.G., Zamaraev, V.: Temporal vertex cover with a sliding time window. J. Comput. Syst. Sci. **107**, 108–123 (2020)
2. Bagehot, W.: Lombard Street: A Description of the Money Market. King, London (1873)
3. Bardoscia, M., et al.: The physics of financial networks. Nat. Rev. Phys. **3**, 1–18 (2021)
4. Casteigts, A., Flocchini, P.: Deterministic Algorithms in Dynamic Networks: Formal Models and Metrics. Tech. rep., Defence R&D Canada CR 2013-020, April 2013. https://hal.archives-ouvertes.fr/hal-00865762
5. Casteigts, A., Flocchini, P.: Deterministic Algorithms in Dynamic Networks: Problems, Analysis, and Algorithmic Tools. Tech. rep., Defence R&D Canada CR 2013-021, April 2013. https://hal.archives-ouvertes.fr/hal-00865764

6. Eisenberg, L.: A summary: Boolean networks applied to systemic risk. Neural Networks Financ. Eng. 436–449 (1996)
7. Eisenberg, L., Noe, T.H.: Systemic risk in financial systems. Manage. Sci. **47**(2), 236–249 (2001)
8. Enright, J., Meeks, K., Mertzios, G.B., Zamaraev, V.: Deleting edges to restrict the size of an epidemic in temporal networks. J. Comput. Syst. Sci. **119**, 60–77 (2021)
9. Haldane, A.G., May, R.M.: Systemic risk in banking ecosystems. Nature **469**(7330), 351–355 (2011)
10. Holme, P., Saramäki, J. (eds.): Temporal Networks. Springer Berlin, Heidelberg (2013). https://doi.org/10.1007/978-3-642-36461-7
11. Jackson, M.O., Pernoud, A.: Systemic risk in financial networks: a survey. Annu. Rev. Econ. **13**(1), 171–202 (2021)
12. Karp, R.M.: Reducibility among combinatorial problems. In: Miller, R.E., Thatcher, J.W., Bohlinger, J.D. (eds.) Complexity of Computer Computations. The IBM Research Symposia Series. Springer, Boston (1972). https://doi.org/10.1007/978-1-4684-2001-2_9
13. Kempe, D., Kleinberg, J.M., Kumar, A.: Connectivity and inference problems for temporal networks. In: Proceedings of the 32nd Annual ACM Symposium on Theory of computing (STOC), pp. 504–513 (2000)
14. Klobas, N., Mertzios, G.B., Molter, H., Niedermeier, R., Zschoche, P.: Interference-free walks in time: temporally disjoint paths. In: Proceedings of the 30th International Joint Conference on Artificial Intelligence (IJCAI), pp. 4090–4096 (2021)
15. Papachristou, M., Kleinberg, J.: Allocating stimulus checks in times of crisis. In: Proceedings of the ACM Web Conference 2022, pp. 16–26 (2022)
16. Papp, P.A., Wattenhofer, R.: Sequential defaulting in financial networks. In: 12th Innovations in Theoretical Computer Science Conference ITCS, pp. 52:1–52:20 (2021)
17. Rochet, J.C., Vives, X.: Coordination failures and the lender of last resort: was Bagehot right after all? J. Eur. Econ. Assoc. **2**(6), 1116–1147 (2004)
18. Rogers, L.C., Veraart, L.A.: Failure and rescue in an interbank network. Manage. Sci. **59**(4), 882–898 (2013)
19. Schuldenzucker, S., Seuken, S., Battiston, S.: Finding clearing payments in financial networks with credit default swaps is PPAD-complete. In: Proceedings of the 8th Innovations in Theoretical Computer Science (ITCS) Conference, vol. 67, pp. 32:1–32:20 (2017)
20. Tesfaye, B., Augsten, N., Pawlik, M., Böhlen, M., Jensen, C.: Speeding up reachability queries in public transport networks using graph partitioning. Inf. Syst. Front. **24**, 11–29 (2022)

Multi-Parameter Analysis of Finding Minors and Subgraphs in Edge-Periodic Temporal Graphs

Emmanuel Arrighi[1][ID], Niels Grüttemeier[2][ID], Nils Morawietz[2],
Frank Sommer[2][✉][ID], and Petra Wolf[1][ID]

[1] University of Bergen, Bergen, Norway
emmanuel.arrighi@uib.no, mail@wolfp.net
[2] Philipps-Universität Marburg, Marburg, Germany
{niegru,morawietz,fsommer}@informatik.uni-marburg.de

Abstract. We study the computational complexity of determining structural properties of edge-periodic temporal graphs (EPGs). EPGs are time-varying graphs that compactly represent periodic behavior of components of a dynamic network, for example, train schedules on a rail network. In EPGs, for each edge e of the graph, a binary string s_e determines in which time steps the edge is present, namely e is present in time step t if and only if s_e contains a 1 at position $t \mod |s_e|$. Due to this periodicity, EPGs serve as very compact representations of complex periodic systems and can even be exponentially smaller than classic temporal graphs representing one period of the same system, as the latter contain the whole sequence of graphs explicitly. In this paper, we study the computational complexity of fundamental questions of the new concept of EPGs such as is there a time step or a sliding window of size Δ in which the graph (1) is minor-free; (2) contains a minor; (3) is subgraph-free; (4) contains a subgraph; with respect to a given minor or subgraph. We give a detailed parameterized analysis for multiple combinations of parameters for the problems stated above including several algorithms.

Keywords: Temporal graphs · Minor-free · Minor containment · Subgraph-free · Subgraph containment · Parameterized complexity · FPT-algorithm

1 Introduction

In general, a *time-varying graph* describes a graph that changes over time. For most applications, this change can be reduced to the availability or weight of

E. Arrighi—Supported by Research Council of Norway (no. 274526 and 329745) and IS-DAAD (no. 309319).
N. Morawietz—Supported by DFG project OPERAH (KO 3669/5–1).
F. Sommer—Supported by DFG project EAGR (KO 3669/6–1).
P. Wolf—The research was mainly performed when the author was associated with University of Trier, Germany, and supported by DFG project FE 560/9–1 and DAAD PPP (no. 57525246).

L. Gąsieniec (Ed.): SOFSEM 2023, LNCS 13878, pp. 283–297, 2023.
https://doi.org/10.1007/978-3-031-23101-8_19

edges, meaning that edges are only present at certain time steps or the time needed to cross an edge changes over time. Time varying graphs are of great interest in the area of *dynamic networks* [8,14–16] such as *mobile ad hoc networks* [35] and *vehicular networks* [4,11] as in those networks, the topology naturally changes over time. There are plenty of representations for time-varying graphs in the literature which are not equivalent in general, see [6–8] for some overview. In general, a time-varying graph $\mathcal{G}$ consists of an underlying graph G and functions describing how the availability or weights of edges change over time. Thereby, settings with *discrete* and *continuous* time steps are considered [8,21,24,27]. In this work, we only deal with the discrete time setting. Usually, in the field of time-varying graphs, for each time step t of the *lifetime* of the graph, the *snapshot* graph $\mathcal{G}(t)$, i.e., the graph present in time step t, is explicitly given in the input [5,26,33]. This implies that the lifetime of $\mathcal{G}$ is linear in the input size and that the input is mostly dominated by the sequence of snapshot graphs $(\mathcal{G}(t))_t$. In addition, the lifetime might be superpolynomially in the size of the underlying graph G. We will call graphs where the whole sequence of snapshot graphs is explicitly given *temporal graphs*.

Knowing the whole sequence of snapshot graphs of the temporal graph requires a detailed knowledge of the usually complex system that is modelled by the graph. On the other hand, describing a system by its components is a natural concept in computer science [10,17,30] and requires only individual knowledge of the components. In the context of time-varying graphs, this approach is realized by so called *edge-periodic (temporal) graphs*, EPGs for short, categorized as Class 8 in [8] and considered for instance in [12,28,29]. An edge-periodic (temporal) graph $\mathcal{G} = (V, E, \tau)$ consists of an underlying graph $G = (V, E)$ and a function τ that assigns each edge with a binary string, the *edge label*, that indicates in which time step the edge is present. Thereby, the time step is considered modulo the length of the edge label. As the length of the edge labels can differ, the sequence of snapshot graphs only repeats after the least common multiple of the individual edge label lengths. Hence, an EPG can compactly represent an exponentially longer sequence of snapshot graphs without explicitly describing each snapshot graph individually. This implies that the lifetime can be exponential in the input size. Figure 1 shows an example of an EPG together with some snapshot graphs.

As humans tend to follow a daily routine and the systems that are to be described by time-varying graphs are mostly influenced by human behavior, they naturally exhibit a periodic behavior. For instance, in social networks describing the dynamics of people meeting, see [22,32], the whole network will be quite complex, but every person individually follows mostly a daily routine. Another example is to model a train network. There, the underlying graph represents the railway system, while an edge is present in a time step if and only if a train is scheduled to run on the respective rail segment at that time. A major advantage of modelling a time-varying system with EPGs is that, if we are only interested in a part of the temporal graph (for instance, we are only interested in the train schedule of a commune and not of the whole state), then we can first extract

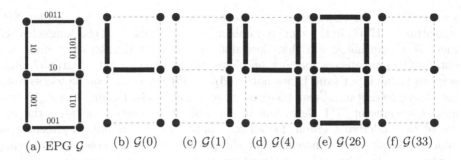

(a) EPG $\mathcal{G}$ (b) $\mathcal{G}(0)$ (c) $\mathcal{G}(1)$ (d) $\mathcal{G}(4)$ (e) $\mathcal{G}(26)$ (f) $\mathcal{G}(33)$

Fig. 1. Example EPG $\mathcal{G}$ and snapshot graphs for $t \in \{0, 1, 4, 26, 33\}$. $\mathcal{G}$ has a period of length 60. It illustrates the blow-up in complexity due to the compact representation. For example, the first K_2-free snapshot graph is at time step 33.

the corresponding subgraph of $\mathcal{G}$ and then compute the sequence of snapshot graphs. Thereby, the size of the individual snapshots will be smaller, and the sequence of considered snapshot graphs might be shorter as the period of the sequence might be smaller. Hence, we avoid considering the complete huge and complicated system if we are only interested in a part of the system.

So far, to the best of our knowledge, the class of edge-periodic (temporal) graphs is not studied in detail yet. We counter this by giving a fundamental analysis of the parameterized complexity of essential graph-theoretical problems on EPGs such as being minor- or subgraph-free, containing a minor or subgraph, and the fundamental short traversal problem [1,34] from the theory of time-varying graphs. The theory on graph minors, established by Robertson and Seymour in a series of over 20 publications [23], is one of the most fundamental results in graph theory. They showed that minor closed properties of a given graph can be checked in polynomial time. Hence, it is natural to ask, whether the toolbox of minors carries over to EPGs. For those, one could be interested in two questions: (1) Do all snapshot graphs obey a minor closed property? (2) Is there some snapshot graph that obeys a minor closed property? As those properties are proved by excluding certain minors, question (1) relates to a no-answer to the question whether there exists a snapshot graph containing a certain minor and question (2) relates to a yes-answer to the question whether there exists a snapshot graph being minor-free. Note that for EPGs, the underlying graph may not be contained in a minor closed graph class while each snapshot is contained.

While classically, both problems of being minor-free and finding a minor are FPT in the size of the sought minor, we observe that for EPGs, both problems are NP-hard even if the minor is fixed and very simple, such as a triangle, or a a a star with four leaves. This implies that the graph minor toolbox does not translate to EPGs. In fact, our NP-hardness results hold even in the case of topological minors. On the other hand, the problem of finding a subgraph is not getting harder when we shift from classic graphs to EPGs. This problem is classically W[1]-hard for the size of the subgraph (consider cliques as subgraphs) [9] and in XP for the same parameter. Surprisingly, we can obtain a similar XP-algorithm for EPGs in the same parameter. For the problem of checking whether there is a snapshot graph that does not contain a fixed subgraph/minor, we obtain NP-completeness for both problems, while if the sought subgraph/minor is given in the input (and hence not fixed), we lift the coNP-completeness from the classic setting to Σ_2^P-completeness concerning EPGs. Despite the high complexity, we present FPT-algorithms in a combined parameter, including the size $|G|$ of the underlying graph, for all four problems of containment/freeness of minors/subgraphs. We indicate that the parameter $|G|$ is necessary by giving hardness results when $|G|$ is replaced by smaller structural parameters such as vertex cover number, treewidth, and pathwidth of the underlying graph.

Proofs of statements annotated with (*) are deferred to the full version.

2 Preliminaries

For a string $w = w_0 w_1 \ldots w_{n-1}$ with $w_i \in \{0,1\}$, for $0 \leq i \leq n-1$, we denote with $w[i]$ the symbol w_i at position i in w. Let $|w| = n$ be the *length* of w. We write the concatenation of strings u and v as $u \cdot v$. For non-negative integers $i \leq j$ we denote with $[i,j]$ the interval of natural numbers n with $i \leq n \leq j$. Given two graphs $H = (V,E)$ and $G = (V',E')$, a *(graph) monomorphism* $\varphi \colon V \to V'$ is an injective mapping such that $\{\varphi(v), \varphi(w)\} \in E'$ if and only if $\{v,w\} \in E$ for all $v \in V$ and $w \in V$. For a set $S = \{s_1, s_2, \ldots, s_n\}$, we might denote the set $\{\varphi(s_1), \varphi(s_2), \ldots, \varphi(s_n)\}$ by $\varphi(S)$. Given a graph $G = (V,E)$, we let $|G| = |V| + |E|$ denote the size of the graph.

An *edge-periodic (temporal) graph*, EPG for short, $\mathcal{G} = (V, E, \tau)$ (see also [12]) consists of a graph $G = (V, E)$ (called the *underlying graph*) and a function $\tau \colon E \to \{0,1\}^*$ where τ maps each edge e to a string $\tau(e) \in \{0,1\}^*$ such that e exists in a time step $t \geq 0$ if and only if $\tau(e)[t]^\circ = 1$, where $\tau(e)[t]^\circ := \tau(e)[t \bmod |\tau(e)|]$. Let $|\mathcal{G}| = |V| + |E| + \sum_{e \in E} |\tau(e)|$ be the size of $\mathcal{G}$. For an edge e and non-negative integers $i \leq j$, we inductively define $\tau(e)[[i,j]]^\circ = \tau(e)[i]^\circ \cdot \tau(e)[[i+1,j]]^\circ$ and $\tau(e)[[j,j]]^\circ = \tau(e)[j]^\circ$. Every edge e exists in at least one time step. We might abbreviate i repetitions of the same symbol σ in $\tau(e)$ as σ^i. We call $\#1_{\max}$ the maximal number of ones appearing in an edge label $\tau(e)$ over all edges $e \in E$. Similarly, we call $\#0_{\max}$ the maximal number of zeros appearing in some $\tau(e)$. Let $L_{\mathcal{G}} = \{|\tau(e)| \mid e \in E\}$ be the set of all edge-periods of some edge-periodic graph $\mathcal{G} = (V, E, \tau)$ and let $\mathrm{lcm}(L_{\mathcal{G}})$ be the *least common multiple* of all periods in $L_{\mathcal{G}}$. We denote with $\mathcal{G}(t)$ the subgraph of G present in time step t. We do not assume that $\mathcal{G}$ is connected in any time step. If not stated otherwise, we assume an edge-periodic graph to be undirected.

For an EPG $\mathcal{G} = (V, E, \tau)$ and an integer $\Delta \in \mathbb{N}$, we let a Δ-expansion of $\mathcal{G}$ be an EPG $\mathcal{G}^\Delta = (V, E, \tau^\Delta)$ such that for every time step t, we have $\mathcal{G}^\Delta(t) = \mathcal{G}(t) \cup \mathcal{G}(t+1) \cup \ldots \cup \mathcal{G}(t+\Delta-1)$. In other words, for every time step, $\mathcal{G}^\Delta(t)$ represents the union of the snapshot of $\mathcal{G}$ in a sliding window of size Δ starting at position t.

3 Temporal Extension of Graph Problems

There exists several ways to generalize problems on graphs to temporal graphs. In this work, we are interested in problems that ask for the existence or non existence of some sub-structures in a graph (subgraphs, minors). Such problems can be generalized in two natural ways. One can ask if there exists a snapshot that contains the sub-structure. Or, for a more temporal setting, given an integer $\Delta \in \mathbb{N}$, one can ask if there exists a sliding window of size Δ that contains the sub-structure. Such generalizations have been considered for other graph problems such as vertex cover [2] and graph coloring [25]. Using the Δ-expansion defined earlier, we can rephrase the sliding window question as follow: given an EPG $\mathcal{G}$, does a snapshot of $\mathcal{G}^\Delta$ contain the sub-structure? If we take as example the existence of a minor, we have those two generalizations to temporal graphs:

EPG MINOR
Input: EPG $\mathcal{G} = (V, E, \tau)$ and graph $H = (V_H, E_H)$.
Question: Is there a time step t, s.t. H is a minor of $\mathcal{G}(t)$?

SLIDING WINDOW EPG MINOR
Input: EPG $\mathcal{G} = (V, E, \tau)$, graph $H = (V_H, E_H)$, and integer Δ.
Question: Is there a time step t, s.t. H is a minor of $\mathcal{G}^\Delta(t)$?

The next lemma shows that both approaches are equivalent.

Lemma 1. *Given an EPG $\mathcal{G}$ and an integer $\Delta \in \mathbb{N}$, one can construct in polynomial time a Δ-expansion $\mathcal{G}^\Delta$ of $\mathcal{G}$ such that $|\mathcal{G}^\Delta| \leq |\mathcal{G}|$.*

Proof. Let $\mathcal{G} = (V, E, \tau)$ be an EPG, and $\Delta \in \mathbb{N}$ an integer. We let $\mathcal{G}^\Delta = (V, E, \tau^\Delta)$ such that for all edges $e \in E$, $|\tau^\Delta(e)| = |\tau(e)|$ and for all time step $t \in [0; |\tau(e)|)$, $\tau^\Delta[t] = 1$ if and only if there exist a time step $t' \subset [t; t+\Delta)$ such that $\tau(e)[t']^\circ = 1$. By construction of τ^Δ, $\mathcal{G}^\Delta$ is a Δ-expansion of $\mathcal{G}$. □

This allows us to reduce instances of problems defined with a sliding window to instances of problems looking for a single time step. In the other direction, looking for a single time step correspond to looking for a sliding window of size 1. Hence, in the rest of the paper, we focus only on finding one time step containing the sub-structure. By using the Δ-expansion construction (Lemma 1), all our results can be extended to work with a sliding window.

4 Periodic Character Alignment

Most of our hardness results presented in this work will be based on the PERIODIC CHARACTER ALIGNMENT problem which was shown to be NP-complete in [28]. This problem builds a bridge between the modern setting of edge-periodic temporal graphs and the classical field of automata theory as it is closely related to the INTERSECTION NON-EMPTINESS problem of deterministic finite automata over a unary alphabet.

PERIODIC CHARACTER ALIGNMENT (PCA)
Input: A finite set $X \subseteq \{0, 1\}^*$ of binary strings.
Question: Is there a position i, such that $x[i]^\circ = 1$ for all $x \in X$?

The parameterized complexity of PCA was already considered in [28] where W[1]-hardness was shown for the parameter $|X|$ and FPT-algorithms were given for the total number of runs of 1's, in all strings, the combined parameter $|X|$ plus the greatest common divisor of any pair of lengths of strings of X, and the length of the longest string in X. Here, a *run* is a nonextendable repetition of one symbol in a string. As the reductions from PCA, presented in this work, are parameter preserving, we inherit several W[1]-hardness results from PCA for the different problems on EPGs. Due to this tight connection, we begin with a more detailed analysis of the parameterized complexity of the PCA problem.

Theorem 1 (*). PCA *is NP-hard even if* $\#0_{max} = 1$ *or* $\#1_{max} \leq 9$.

Corollary 1 (*). PCA *is W[1]-hard with respect to the number of different prime numbers in the prime factorizations of the integers in* $L_{\mathcal{G}}$.

The core task of the problems introduced in the next sections is to determine whether a certain graph structure exists in one time step or over a sequence of consecutive time steps. As the existence of an edge e in an EPG is determined by a binary string $\tau(e)$, we associate each EPG with a corresponding PCA instance. Hence, if the location of the sought graph structure in the underlying graph of the EPG is known, the problem of finding a time step in which the structure exists is equivalent to finding a time step in which the 1's of the corresponding PCA-instance align. We refer to Theorem 11 for details.

Definition 1. *Let* X *be an instance of PCA. A triple* $(\mathcal{G}, H, \varphi)$, *where* $\mathcal{G} = (V, E, \tau)$ *is an EPG,* $H = (V_H, E_H)$ *is a subgraph of the underlying graph* $G = (V, E)$ *of* $\mathcal{G}$, *and* $\varphi : V_H \to V$ *is a monomorphism that identifies* H *in* G, *is called an* X-embedding *if* $\tau(E_H) = X$.

Lemma 2 (*). *Let* X *be an instance of PCA and let* $(\mathcal{G}, H, \varphi)$ *be an* X-embedding. *Then, there exists a time step* t *in which* $\varphi(H)$ *exists in* $\mathcal{G}(t)$ *if and only if* X *is a yes-instance of PCA.*

5   Minors and Subgraphs

We now come to the main part of this paper considering the existence and non existence of sub-structures in an EPG such as induced subgraphs and minors. Recall that $G' = (V', E')$ is a *subgraph* of a graph $G = (V, E)$ if $V' \subseteq V$ and $E' \subseteq E$. If further for all $u, v \in V'$ it holds that $\{u, v\} \in E'$ if and only if $\{u, v\} \in E$, we call G' an *induced subgraph* of G. In the following, we see subgraphs as induced subgraphs. We call G' a *minor* of G if G' can be obtained from G, by deletion of vertices, deletion of edges, and contraction of edges. Here, we consider the following questions: Does there exists a time step t, such that $\mathcal{G}(t)$ has a subgraph/minor or is subgraph-/minor-free.

5.1 Subgraphs

EPG SUBGRAPH
Input: EPG $\mathcal{G} = (V, E, \tau)$ and graph $H = (V_H, E_H)$.
Question: Is there a time step t, s.t. H is a subgraph of $\mathcal{G}(t)$?

EPG SUBGRAPH-FREE
Input: EPG $\mathcal{G} = (V, E, \tau)$ and graph $H = (V_H, E_H)$.
Question: Is there a time step t, s.t. H is *not* a subgraph of $\mathcal{G}(t)$?

Theorem 2. *The* EPG SUBGRAPH *problem is* NP-*complete and* W[1]-*hard parameterized by* $|\mathcal{G}|$. *This holds even if H is a path and $G = H$.*

Proof. EPG SUBGRAPH belongs to NP, since we may non-deterministically choose a time step t of size at most $\operatorname{lcm}(L_\mathcal{G})$ and an embedding $\varphi : V_H \to V$ and check, whether φ identifies H in $\mathcal{G}(t)$. Since $t \leq \max(L_\mathcal{G})^{(n^2)}$, this certificate can be encoded polynomially in the input size.

It remains to show that EPG SUBGRAPH is NP-hard. Let $X := \{x_1, \ldots, x_{|X|}\}$ be an instance of PCA. We define an equivalent instance $(\mathcal{G}, H)$ of EPG SUBGRAPH. First, we define $H := (V_H, E_H)$ to be a path on $|X|$ edges $e_1, \ldots, e_{|X|}$. Second, we define $\mathcal{G} := (V_H, E_H, \tau)$ with $\tau(e_i) := x_i$ for every $i \in [1, |X|]$.

We next use Lemma 2 to show that X is a yes-instance of PCA if and only if $(\mathcal{G}, H)$ is a yes-instance of EPG SUBGRAPH. Observe that $\varphi : V_H \to V_H$ with $\varphi(v) := v$ is a trivial monomorphism that identifies H in the underlying graph of $\mathcal{G}$. Note that ϕ, and the isomorphic monomorphism reversing the path, are the only monomorphisms that identify H, as H and $\mathcal{G}$ share their vertices and edges. Furthermore, by the definition of τ we have $\tau(E_H) = X$. Thus, $(\mathcal{G}, H, \varphi)$ is an X-embedding according to Definition 1. Then, by Lemma 2 we have that X is a yes-instance of PCA if and only if there is a time step t in which $\varphi(H)$ exists in $\mathcal{G}(t)$. Consequently, X is a yes-instance of PCA if and only if $(\mathcal{G}, H)$ is a yes-instance of EPG SUBGRAPH. □

Note that the length of the paths in the construction behind Theorem 2 corresponds to the size of the PCA instance. Thus, these paths might be arbitrarily long. If we—in contrast—assume that the size of sought subgraph H is bounded by some constant h, we obtain a polynomial time algorithm for EPG SUBGRAPH. In other words, EPG SUBGRAPH is XP when parameterized by h as we show in the following theorem.

Theorem 3. EPG SUBGRAPH *can be solved in time* $\mathcal{O}(n^h \cdot \max(L_\mathcal{G})^{(h^2)}) \cdot 2^{\mathcal{O}(\sqrt{h \log h})}$, *where h is the number of vertices in H.*

Proof. We prove the theorem by describing the algorithm. Let $(\mathcal{G} = (V, E, \tau), H)$ be an instance of EPG SUBGRAPH. The algorithm is straightforward: We iterate over all possible subsets $W \subseteq V$ of size h. For each of these sets we check whether there is a time step $t \in [1, \max(L_\mathcal{G})^{(h^2)}]$ such that $\mathcal{G}(t)[W]$ is isomorphic to H. If such a time step exists, return *yes*. Otherwise, return *no*.

The algorithm runs within the claimed running time since there are $\binom{n}{h} \in \mathcal{O}(n^h)$ possible choices of W. For each choice, we consider $\max(L_\mathcal{G})^{(h^2)}$ distinct graphs $\mathcal{G}(t)$ and check whether one of these graphs is isomorphic to H in $2^{\mathcal{O}(\sqrt{h \log h})}$ time [3].

We next show that the algorithm is correct. Suppose that the algorithm returns *yes*. Then, for one choice of W and one time step t, the graph $\mathcal{G}(t)[W]$ is isomorphic to H and therefore, $(\mathcal{G}, H)$ is a yes-instance.

Conversely, suppose that $(\mathcal{G}, H)$ is a yes-instance. Let $W \subseteq V$ be the subset of size h such that $\mathcal{G}(t)[W]$ is isomorphic to H at some time step t. Let $e_1, \ldots, e_k \in E$ be all edges between vertices of W in (V, E). Since $|W| = h$ we have $k \leq h^2$. Thus, the least common multiple of all string lengths $|\tau(e_1)|, \ldots, |\tau(e_k)|$ is at most $\max(L_\mathcal{G})^{(h^2)}$. Therefore, we may assume that $t \in [1, \max(L_\mathcal{G})^{(h^2)}]$. Consequently, the algorithm returns *yes*.   □

Recall that Theorem 3 reveals that the NP-hardness of EPG SUBGRAPH crucially relies on the fact that the size of H is unbounded. In contrast, we show next that EPG SUBGRAPH-FREE is NP-hard for every fixed size of H.

Theorem 4 (*). EPG SUBGRAPH-FREE *is* NP-*complete and* W[1]-*hard parameterized by* $|G|$ *for every fixed subgraph* H *containing at least two vertices.*

The containment in Theorem 4 follows by guessing the time step t in which $\mathcal{G}$ is H-free and checking for each subset V' of V whether $\mathcal{G}(t)[V']$ is not isomorphic to H. As H is fixed, this can be done in polynomial time. For the hardness, we distinguish whether H contains an edge ore not. If H is edgeless, having H as a subgraph correspond to having an independent set of size $|H|$. The key idea is that, by putting the label of a PCA instance on the edges of a clique, the obtained EPG has a time step with a maximum independent set of size 1 if and only if the PCA instance is a yes-instance. On the other hand, if H contains at least one edge, then we can take the disjoint union of several copies of H. Given a PCA instance, by putting the complement of each string (that is, replacing each 1 by a 0 and vice versa) of the PCA instance to an edge in each copy, we obtain that the PCA instance is a yes-instance if and only if there exists a time step where H is not a subgraph.

In contrast, if the subgraph H is not fixed, then the problem becomes even harder. Intuitively, the following theorem is based on a construction from the problem $\exists\forall$3UNSAT, where in the resulting EPG, we first have to guess a time step t and then need to check that each selection of k vertices is not a clique in $\mathcal{G}(t)$. More precisely, the underlying graph is the k-partite graph from the classic reduction by Karp [20] where each partite set corresponds to a clause of a given 3SAT-formula. We use edge labels to model which variables are existentially and which variables are universal.

Theorem 5 (*). EPG SUBGRAPH-FREE *is* Σ_P^2-*complete even if* H *is a clique.*

5.2   Minors

EPG MINOR
Input: EPG $\mathcal{G} = (V, E, \tau)$ and graph $H = (V_H, E_H)$.
Question: Is there a time step t, s.t. H is a minor of $\mathcal{G}(t)$?

EPG MINOR-FREE
Input: EPG $\mathcal{G} = (V, E, \tau)$ and graph $H = (V_H, E_H)$.
Question: Is there a time step t, s.t. H is *not* a minor of $\mathcal{G}(t)$?

As in the subgraph variant, we obtain Σ_2^P-completeness for EPG MINOR-FREE.

Theorem 6 (*). *The* EPG MINOR-FREE *problem is* Σ_2^P*-complete.*

If we fix the minor, the complexity falls to NP-completeness, which is still significantly harder than the poly-time solvability in the case of static graphs [31].

Theorem 7 (*). EPG MINOR-FREE *is* NP-*complete and* W[1]-*hard parameterized by* $|\mathcal{G}|$ *for every fixed H containing at least one edge.*

In the case that H is edgeless, in contrast to the subgraph variant, we only need to compare the number of vertices in H and $\mathcal{G}$.

Proposition 1 (*). EPG MINOR-FREE *can be solved in linear time using logarithmic space for every fixed edgeless graph H.*

We now turn to the related problem EPG MINOR, in which we ask whether a graph H exists as a minor in some time step t in an EPG. For finding an H-minor, the problem is already NP-complete for very simple minors. More precisely, we provide a dichotomy for minors of constant sizes into cases which are NP-complete and those which are solvable in polynomial time. First, we provide NP-completeness for the case that H contains at least one cycle.

Theorem 8 (*). EPG MINOR *is* NP-*complete and* W[1]-*hard parameterized by* $|\mathcal{G}|$ *for every fixed H containing a cycle.*

Since Theorem 8 shows hardness for each fixed minor containing a cycle, it remains to consider fixed minors H which are forests. Second, we provide NP-hardness for forests containing a tree with some minimum-degree vertices.

Theorem 9 (*). EPG MINOR *is* NP-*complete and* W[1]-*hard parameterized by* $|\mathcal{G}|$ *for every fixed forest H with a connected component that contains a) two vertices of degree at least 3 or b) one vertex of degree at least 4.*

For all remaining cases, that is, each connected component of H is either a path, or a tree with exactly one vertex of degree 3 and no vertex of degree at least 4, we provide an XP-algorithm for the parameter h, the number of vertices of H. The algorithm works completely analogously to the algorithm of Theorem 3 for EPG SUBGRAPH. This algorithm also works for minors, since for this structure of H the minor must already be contained as a subgraph.

Corollary 2. *The* EPG MINOR *problem can be solved in* $\mathcal{O}(n^h \cdot \max(L_\mathcal{G})^{(h^2)}) \cdot 2^{\mathcal{O}(\sqrt{h \log h})}$ *time if H is a forest such that each connected component of H contains no vertex of degree at least 4 and at most one vertex of degree 3.*

5.3   Further Parameterized Analysis

Finally, we take a closer look on the parameterized complexity of the 4 problems concerning minors and subgraphs.

Corollary 3 (*). *The problems* EPG SUBGRAPH, EPG SUBGRAPH-FREE, EPG MINOR, *and* EPG MINOR-FREE *are*

- NP-*hard even if G is a disjoint union of paths and* $\#1_{max} \in \mathcal{O}(1)$ *and*
- NP-*hard even if G is a disjoint union of paths and* $\#0_{max} \in \mathcal{O}(1)$.

Theorem 10 (*). *The problems* EPG SUBGRAPH, EPG SUBGRAPH-FREE, EPG MINOR, *and* EPG MINOR-FREE *are* W[1]-*hard when parameterized by the vertex cover number of the underlying graph even if* $\#1_{max} = 1$.

Theorem 11. *The problems* EPG SUBGRAPH, EPG SUBGRAPH-FREE, EPG MINOR, *and* EPG MINOR-FREE *are FPT with respect to the combined parameter* $\min(\#1_{max}, \#0_{max})$ *plus the number of vertices* $|V|$ *of G.*

Proof. We prove the theorem by providing a class of FPT-algorithms solving the four considered problems. Intuitively, our algorithms iterate over all possible graphs that can be present in some time step and check, whether these graphs (not) contain H as an induced subgraph or as a minor, respectively. To this end, we introduce an auxiliary problem that asks whether there exists a time step where $\mathcal{G}(t)$ consists of a specific edge set.

> EPG PRESENT EDGES
> **Input:** An EPG $\mathcal{G} = (V, E, \tau)$ and an edge set $E' \subseteq E$
> **Question:** Is there a time step t such that $\mathcal{G}(t) = (V, E')$?

Claim 1 (*). EPG PRESENT EDGES *is FPT for* $|V| + \min(\#1_{max}, \#0_{max})$.

We prove the claim by providing a parameterized reduction to PCA parameterized by the total number of runs of 1's, that is, the number of groups of consecutive 1's, in all strings, which is known to be FPT [28].

We next describe the FPT-algorithms for EPG SUBGRAPH, EPG SUBGRAPH-FREE, EPG MINOR, and EPG MINOR-FREE. Let $(\mathcal{G} = (V, E, \tau), H)$ be an instance of one of these problems. We iterate over every possible $E' \subseteq E$ and check if (V, E') (not) contains an induced H or (not) contains H as a minor, respectively. If this is the case, we check whether $(\mathcal{G}, E')$ is a yes-instance of EPG PRESENT EDGES and return *yes* or *no* accordingly.

The correctness follows by the fact that we consider every possible graph (V, E') that might be present in some time step. It remains to consider the running time. Due to the previous claim, checking whether $(\mathcal{G}, E')$ is a yes-instance of EPG PRESENT EDGES can be performed in FPT time parameterized by $|V| + \min(\#1_{\max}, \#0_{\max})$. Checking whether H is a minor of (V, E') or H is an induced subgraph of (V, E') can clearly be done in a running time only depending on the graph size $|V|$. Consequently, the four considered problems are FPT when parameterized by $|V| + \min(\#1_{\max}, \#0_{\max})$.   □

6   Short Traversal

One class of subgraphs plays a special role in temporal settings: paths. Our general positive results for EPG SUBGRAPH generalize the temporal setting of finding paths of fixed length that are present in a sliding window in an EPG. If we consider paths as subgraphs in temporal graphs, one may also ask for other temporal aspects: The most natural question in this context is to ask if there is a path connecting two vertices a and b such that the edges of this path can be traversed in a consecutive manner in our temporal setting. More precisely, we want to know the most favorable time step t to start the traversal from a in order to have the shortest traversal time. Note that, in contrast to EPG SUBGRAPH, we do not ask for a path of fixed length. Furthermore, the ordering in which the edges appear in our graph becomes important.

EPG SHORT TRAVERSAL (EPG-ST)
Input: Edge-periodic graph $\mathcal{G} = (V, E, \tau)$, vertices $a, b \in V$, and $k \in \mathbb{N}$.
Question: Is there a time step t such that starting from vertex a at time step t, we can reach vertex b at the beginning of time step $t + k$ while traversing at most one edge per time step?

Theorem 12 (*). EPG SHORT TRAVERSAL *is* NP-*hard and* W[1]-*hard with respect to the combined parameter* $|G| + k$ *even if* G *is a path.*

Theorem 13 (*). EPG SHORT TRAVERSAL *is* W[1]-*hard when parameterized by the vertex cover number of the underlying graph and* k, *even if* $\#1_{max} = 1$.

In contrast, if we combine the size of the underlying graph and the maximal number of ones per edge label, we can obtain an FPT-algorithm. Note that the length of each edge label $\tau(e)$, and therefore $\mathrm{lcm}(L_{\mathcal{G}})$, is not restricted by the combination of parameters.

Theorem 14. EPG SHORT TRAVERSAL *is* FPT *with respect to the combined parameter* $|G| + \#1_{max}$ *and can be solved in* $\mathcal{O}(|G| \cdot \#1_{max})^{\mathcal{O}(|G| \cdot \#1_{max})} \cdot |\mathcal{G}|^{\mathcal{O}(1)}$ *time.*

Proof. Let $I = (\mathcal{G} = (V, E, \tau), a, b, k)$ be an instance of EPG SHORT TRAVERSAL. To obtain an FPT-algorithm, we perform two steps: First, we iterate over all possible (a, b)-paths $P = (v_0, \ldots, v_r)$ in the underlying graph G, where $v_0 = a$ and $v_r = b$. Since we can assume that the temporal walk with the shortest traversal time is vertex simple, that is, each vertex is visited at most once, it remains to show that there is a time step t and an (a, b)-path in the underlying graph, such that at time step t one can start at vertex a and reach vertex b in at most k time steps by only traversing edges of the path P. To check if such a time step exists for a given path P, we present the following ILP-formulation.

For each edge $e_i := \{v_{i-1}, v_i\}$, we use a variable t_i which is equal to the time step in which the considered temporal walk with shortest traversal time traverses edge e_i. Since at most one edge can be traversed at a time step, we need to ensure that $t_i + 1 \leq t_{i+1}$. Moreover, an edge e_i can only be traversed at time step t_i, if $\tau(e_i)[t_i \mod |\tau(e_i)|] = 1$. Hence, we first introduce two additional variables c_i and m_i for each edge e_i, where $m_i \in [0, |\tau(e_i)| - 1]$ and $|\tau(e_i)| \cdot c_i + m_i = t_i$. That is, m_i stores the value of $t_i \mod |\tau(e_i)|$. Finally, we have to ensure that $\tau(e_i)[m_i] = 1$. Let $J_i := \{j \in [0, |\tau(e_i)|] \mid \tau(e_i)[j] = 1\}$ denote the set of positions where $\tau(e_i)$ is equal to one. We introduce for each $i \in [1, r]$ and each $j \in J_i$ a new binary variable $\ell_{i,j} \in \{0, 1\}$ which is equal to zero if and only if $m_i = j$. To make sure that $\tau(e_i)[m_i] = 1$, the value of exactly one $\ell_{i,j}$ has to be zero, which can be achieve by adding the constraint $\sum_{j \in J_i} \ell_{i,j} = |J_i| - 1$. The complete ILP formulation now reads as follows:

$$t_i, c_i \in \mathbb{N} \qquad \text{for each } i \in [1, r]$$
$$m_i \in \{0, |\tau(e_i)| - 1\} \qquad \text{for each } i \in [1, r]$$
$$\ell_{i,j} \in \{0, 1\} \qquad \text{for each } i \in [1, r], j \in J_i$$

Minimize $t_r - t_1$ subject to

$$t_i + 1 \leq t_{i+1} \qquad \text{for each } i \in [1, r - 1]$$
$$c_i \cdot |\tau(e_i)| + m_i = t_i \qquad \text{for each } i \in [1, r]$$
$$-\ell_{i,j} \cdot 2|\tau(e_i)| + j \leq m_i \qquad \text{for each } i \in [1, r], j \in J_i$$
$$\ell_{i,j} \cdot 2|\tau(e_i)| + j \geq m_i \qquad \text{for each } i \in [1, r], j \in J_i$$
$$\sum_{j \in J_i} \ell_{i,j} = |J_i| - 1 \qquad \text{for each } i \in [1, r]$$

Note that the number of variables in this ILP-formulation is bounded by $\mathcal{O}(r \cdot \max_{i \in [1,r]} |J_i|)$. Since $r \leq |G|$ and $\max_{i \in [1,r]} |J_i| \leq \#1_{\max}$, this ILP can be solved in $\mathcal{O}(|G| \cdot \#1_{\max})^{\mathcal{O}(|G| \cdot \#1_{\max})} \cdot |I|^{\mathcal{O}(1)}$ time [13, 18, 19].

Since there are at most $2^{|G|}$ many possible (a, b)-paths in G and we can solve for each such path the corresponding ILP in $\mathcal{O}(|G| \cdot \#1_{\max})^{\mathcal{O}(|G| \cdot \#1_{\max})} |I|^{\mathcal{O}(1)}$ time, EPG SHORT TRAVERSAL can be solved in the stated running time. $\square$

7   Conclusion

We studied the computational and parameterized complexity of multiple essential graph-theoretical problems on edge-periodic temporal graphs (EPGs). Interestingly, it turned out to be equivalent whether we ask for the (non)existence of a minor or subgraph in a single time step or in a sliding window of size Δ.

We emphasize that EPGs can trivially be converted into temporal graphs by unrolling the whole sequence of snapshot graphs in exponential time and space. Hence, the apparent complexity blow-up comes from the compact representation via periodic edge labels. Intuitively, as for encoding a problem in binary instead of unary, we do not need more time than for temporal graphs, we are just measuring in a smaller input size. But we can exploit the additional structure of EPGs to obtain better algorithms than with the naive approach of unrolling the EPG. For example, we presented an FPT-algorithm for the parameter $|G| + \#1_{max}$ for EPG-ST that relies on an ILP-formulation which checks if at time step t one can start at vertex a and reach vertex b in at most k time steps. The idea of this ILP-formulation might be useful for other problems on EPGs.

Another direction of research is to evaluate to which extent the model of EPGs is suitable to model real-world time-varying graphs by only using little space. Furthermore, one could also consider a model where each edge label corresponds to a compression that is even smaller than the periodic representation (for example, a run length encoding). Is it possible to adapt our positive results for EPGs to this model? Note for instance that in the current encoding, the label of an edge can be interpreted as the encoding of a deterministic unary permutation finite automaton that accepts the unary encodings of all time steps in which the edge is present (for details, see [28]). A more powerful model could be obtained by considering deterministic finite automata over a binary alphabet. Here, the time step is encoded in binary and the automaton corresponding to an edge accepts all time steps in which the edge is present. This would allow, for example, for edges being present in time steps that are powers of two. Another approach is to simplify edge labels by going to multi-graphs and splitting a label into multiple labels. This would correspond to decomposing the respective finite automaton as it was considered in [17].

References

1. Akrida, E.C., Mertzios, G.B., Nikoletseas, S.E., Raptopoulos, C.L., Spirakis, P.G., Zamaraev, V.: How fast can we reach a target vertex in stochastic temporal graphs? J. Comput. Syst. Sci. **114**, 65–83 (2020)
2. Akrida, E.C., Mertzios, G.B., Spirakis, P.G., Zamaraev, V.: Temporal vertex cover with a sliding time window. J. Comput. Syst. Sci. **107**, 108–123 (2020)
3. Babai, L., Luks, E.M.: Canonical labeling of graphs. In: Johnson, D.S., et al. eds, Proceedings of the 15th Annual ACM Symposium on Theory of Computing, Boston, Massachusetts, USA, pp. 171–183. ACM (1983)
4. Berman, K.A.: Vulnerability of scheduled networks and a generalization of Menger's theorem. Networks **28**(3), 125–134 (1996)
5. Bhadra, S., Ferreira, A.: Complexity of connected components in evolving graphs and the computation of multicast trees in dynamic networks. In: Pierre, S., Barbeau, M., Kranakis, E. (eds.) ADHOC-NOW 2003. LNCS, vol. 2865, pp. 259–270. Springer, Heidelberg (2003). https://doi.org/10.1007/978-3-540-39611-6_23
6. Casteigts, A., Flocchini, P.: Deterministic algorithms in dynamic networks: formal models and metrics. Technical Report (2013)
7. Casteigts, A., Flocchini, P.: Deterministic algorithms in dynamic networks: problems, analysis, and algorithmic tools. Technical Report (2013)
8. Casteigts, A., Flocchini, P., Quattrociocchi, W., Santoro, N.: Time-varying graphs and dynamic networks. Int. J. Parallel Emergent Distrib. Syst. **27**(5), 387–408 (2012)
9. Cygan, M., Fomin, F.V., Kowalik, L., Lokshtanov, D., Marx, D., Pilipczuk, M., Pilipczuk, M., Saurabh, S.: Parameterized Algorithms. Springer, Cham (2015). https://doi.org/10.1007/978-3-319-21275-3

10. de Roever, W.-P., Langmaack, H., Pnueli, A. (eds.): COMPOS 1997. LNCS, vol. 1536. Springer, Heidelberg (1998). https://doi.org/10.1007/3-540-49213-5
11. Ding, B., Yu, J.X., Qin, L.: Finding time-dependent shortest paths over large graphs. In: Kemper, A., et al. eds, Proceedings of the 11th International Conference on Extending Database Technology, vol. 261 of ACM International Conference Proceeding Series, pp. 205–216. ACM (2008)
12. Erlebach, T., Spooner, J.T.: A game of cops and robbers on graphs with periodic edge-connectivity. In: Chatzigeorgiou, A., Dondi, R., Herodotou, H., Kapoutsis, C., Manolopoulos, Y., Papadopoulos, G.A., Sikora, F. (eds.) SOFSEM 2020. LNCS, vol. 12011, pp. 64–75. Springer, Cham (2020). https://doi.org/10.1007/978-3-030-38919-2_6
13. Frank, A., Tardos, É.: An application of simultaneous diophantine approximation in combinatorial optimization. Combinatorica 7(1), 49–65 (1987)
14. Ganguly, N., Deutsch, A., Mukherjee, A.: Dyn. Complex Netw. Computer Science, and the Social Sciences, Applications to Biology (2009)
15. Holme, P.: Modern temporal network theory: a colloquium. Eur. Phys. J. B 88(9), 1–30 (2015)
16. Holme, P., Saramäki, J.: Temporal networks. Phys. Rep. 519(3), 97–125 (2012)
17. Jecker, I., Mazzocchi, N., Wolf,P.: Decomposing permutation automata. In: Haddad, S., Varacca, D., eds, 32nd International Conference on Concurrency Theory, CONCUR 2021, Virtual Conference, vol. 203 of LIPIcs, pp. 18:1–18:19. Schloss Dagstuhl - Leibniz-Zentrum für Informatik (2021)
18. Hendrik, W., Lenstra: Integer programming with a fixed number of variables. Math. Oper. Res. 8(4), 538–548 (1983)
19. Kannan, R.: Minkowski's convex body theorem and integer programming. Math. Oper. Res. 12(3), 415–440 (1987)
20. Karp, R.M.: Reducibility among combinatorial problems. In: Complexity of Computer Computations, pp. 85–103 (1972)
21. Kempe, D., Kleinberg, J.M., Kumar, A.: Connectivity and inference problems for temporal networks. J. Comput. Syst. Sci. 64(4), 820–842 (2002)
22. Leskovec, J., Krevl, A.: SNAP datasets: stanford large network dataset collection. http://snap.stanford.edu/data/ (2014)
23. Lovász, L.: Graph minor theory. Bull. Am. Math. Soc. 43(1), 75–86 (2006)
24. Mertzios, G.B., Michail, O., Spirakis, P.G.: Temporal network optimization subject to connectivity constraints. Algorithmica 81(4), 1416–1449 (2019)
25. Mertzios, G.B., Molter, H., Zamaraev, V.: Sliding window temporal graph coloring. J. Comput. Syst. Sci. 120, 97–115 (2021)
26. Michail, O., Spirakis, P.G.: Traveling salesman problems in temporal graphs. Theoret. Comput. Sci. 634, 1–23 (2016)
27. Michail, O., Spirakis, P.G.: Elements of the theory of dynamic networks. Commun. ACM 61(2), 72 (2018)
28. Morawietz, N., Rehs, C., Weller, M.: A timecop's work is harder than you think. In: Esparza, J., Král', D., eds, 45th International Symposium on Mathematical Foundations of Computer Science, MFCS 2020, Prague, Czech Republic, vol. 170 of LIPIcs, pp. 71:1–71:14. Schloss Dagstuhl - Leibniz-Zentrum für Informatik (2020)
29. Morawietz, N., Wolf, P.: A timecop's chase around the table. In: Bonchi, F., Puglisi, S.J., eds, 46th International Symposium on Mathematical Foundations of Computer Science, MFCS 2021, Tallinn, Estonia, vol. 202 of LIPIcs, pp. 77:1–77:18. Schloss Dagstuhl - Leibniz-Zentrum für Informatik (2021)
30. Pagin, P.: Compositionality, computability, and complexity. Rev. Symbolic Logic 14(3), 551–591 (2021)

31. Robertson, N., Seymour, P.D.: Graph Minors. XIII. The Disjoint Paths Problem. J. Comb. Theory Ser. B **63**(1), 65–110 (1995)
32. Sapiezynski, P., Stopczynski, A., Gatej, R., Lehmann, S.: Tracking human mobility using WiFi signals. PLoS ONE **10**(7), 1–11 (2015)
33. Wehmuth, K., Ziviani, A., Fleury, E.: A unifying model for representing time-varying graphs. In: 2015 IEEE International Conference on Data Science and Advanced Analytics, DSAA 2015, Campus des Cordeliers, Paris, France, pp. 1–10. IEEE (2015)
34. Huanhuan, W., Cheng, J., Huang, S., Ke, Y., Yi, L., Yanyan, X.: Path problems in temporal graphs. Proc. VLDB Endowment **7**(9), 721–732 (2014)
35. Zhang, Z.: Routing in intermittently connected mobile ad hoc networks and delay tolerant networks: overview and challenges. IEEE Commun. Surv. Tutorials **8**(1–4), 24–37 (2006)

Complexity and Learning

Lower Bounds for Monotone
q-Multilinear Boolean Circuits

Andrzej Lingas[(✉)]

Department of Computer Science, Lund University, 22100 Lund, Sweden
Andrzej.Lingas@cs.lth.se

Abstract. A monotone Boolean circuit is composed of OR gates, AND gates and input gates corresponding to the input variables and the Boolean constants. It is multilinear if for any AND gate the two input functions have no variable in common. We consider a generalization of monotone multilinear Boolean circuits to include monotone q-multilinear Boolean circuits. Roughly, a sufficient condition for the q-multilinearity is that in the formal Boolean polynomials at the output gates of the circuit no variable has degree larger than q. First, we study a relationship between q-multilinearity and the conjunction depth of a monotone Boolean circuit, i.e., the maximum number of AND gates on a path from an input gate to an output gate. As a corollary, we obtain a trade-off between the lower bounds on the size of monotone q-multilinear Boolean circuits for semi-disjoint bilinear forms and the parameter q. Next, we study the complexity of the monotone Boolean function $Isol_{k,n}$ verifying if a k-dimensional matrix has at least one 1 in each line (e.g., each row and column when $k = 2$) in terms of monotone k-multilinear Boolean circuits. We show that the function admits Π_2 monotone k-multilinear circuits of $O(n^k)$ size. On the other hand, we demonstrate that any Π_2 monotone Boolean circuit for $Isol_{k,n}$ is at least k-multilinear. Also, we show under an additional assumption that any Σ_3 monotone Boolean circuit for $Isol_{k,n}$ is not $(k - 1)$-multilinear or it has an exponential in n size.

Keywords: Monotone Boolean circuit · Monotone multilinear Boolean circuit · Monotone arithmetic circuit · Circuit size

1 Introduction

The derivation of superlinear lower bounds on the size of Boolean circuits for natural problems appeared extremely hard. Therefore, already at the end of the 70s and the beginning of the 80s, several researches started to study the complexity of monotone arithmetic circuits or monotone Boolean circuits for natural multivariate arithmetic polynomials and natural Boolean functions, respectively. The monotone arithmetic circuits are composed of addition gates, multiplication gates and input gates for variables and non-negative real constants. Similarly, monotone Boolean circuits are composed of OR gates, AND gates, and the input

L. Gąsieniec (Ed.): SOFSEM 2023, LNCS 13878, pp. 301–312, 2023.
https://doi.org/10.1007/978-3-031-23101-8_20

gates for variables and Boolean constants. In the case of monotone arithmetic circuits, one succeeded to derive even exponential lower bounds relatively easily [3,18] while in the case of monotone Boolean circuits, the derivation of exponential lower bounds for natural problems required much more effort [1,16].

The problem of computing the permanent of an $n \times n$ matrix equivalent to counting the number of perfect matchings in a bipartite graph is an example of a problem for which the gap between lower bounds in the models of monotone arithmetic circuits and monotone Boolean circuits remains very large up to today. Namely, Jerrum and Snir established an exponential lower bound on the size of monotone arithmetic circuits for this problem [3] while the best known lower bound on the size of a monotone Boolean circuit computing the Boolean variant of the permanent shown by Razborov [17] is only superpolynomial. In order to tackle the gap, Ponnuswami and Venkateswaran considered the concept of monotone multilinear Boolean circuits and showed an exponential lower bound on the size of the restricted monotone Boolean circuits for the Boolean permanent [12]. A Boolean circuit is multilinear if for any AND gate the two input functions have no variable in common. To be more precise, the authors of [12] used a semantic version of multilinearity by forbidding minimal representations of the two input functions to share a variable [12]. This means for example that if the first function is represented by $x \vee x \wedge y$ and the second one by y then the AND gate represented by $(x \wedge y) \vee (x \wedge y \wedge y)$ is allowed in their multilinear Boolean circuit. In fact, soon later, Krieger claimed exponential lower bounds on the number of OR gates in multilinear Boolean circuits for among other things a clique function [6]. He used a much more restricted syntactic version of multilinearity, where the function computed at a gate is declared to be dependent on a variable if there is a path from the input gate with the variable or its negation to the gate in the circuit. This syntactic version directly makes impossible for a multilinear Boolean circuit to produce terms with two or more occurrences of a variable or its negation. Hence, in particular term cancellation, i.e., terms with a variable x and its negation $\bar{x}$, is not possible in syntactically multilinear Boolean circuits.

Note however that the concept of multilinearity in the case of unrestricted Boolean circuits does not make too much sens as any Boolean circuit can be easily turned into a multilinear one. It is just enough to apply the DeMorgan rule $f_1 \wedge f_2 = \neg(\neg f_1 \vee \neg f_2)$ in order to eliminate all AND gates increasing the total number of gates at most by the factor 4. Therefore, it seems that Krieger used implicitly a restriction of Boolean circuits to DeMorgan Boolean circuits, a generalization of monotone Boolean circuits, where negation can be applied solely to input gates. This seems to be evident in the proof Lemma 3 in [6] stating that any optimal multilinear Boolean circuit for a monotone Boolean function is monotone. The negation seems to occur solely at the input gates in the proof. So, the proof of the lower bound of $\binom{n}{k} - 1$ on the number of OR gates in any multilinear Boolean circuit for the k-clique function in [12] seems to work solely for DeMorgan multilinear Boolean circuits[1].

---

[1] A similar interpretation of Krieger's results can be found in [5].

In our recent report [7], we have used a simple argument to obtain a more general result than the lower bounds of Ponnuswami and Venkateswaran or Krieger in [12] and [6], respectively. We have shown that the known lower bounds on the size of monotone arithmetic circuits for multivariate polynomials that are sums of monomials consisting of the same number of distinct variables [3,18] yield almost the analogous lower bounds on the size of monotone multilinear Boolean circuits computing the functions represented by the corresponding multivariate Boolean polynomials. Our result can be slightly improved to yield exactly analogous lower bounds by using the lower envelope argument from [3] as observed in [5].

On the other hand, Raz and Widgerson showed that monotone Boolean circuits for the Boolean permanent require linear depth [15] and Raz proved that multilinear Boolean formulas for this problem have superpolynomial size [14].

The concept of circuit multilinearity is also natural for circuits over other semi-rings beside the Boolean one ($\{0, 1\} \vee, \wedge$) such as the arithmetic one ($R_+, +, \times$) or the tropical one ($R_+, \min, +$), where R_+ stands for the set of nonnegative real numbers [5]. In particular, Jukna observed that the classical dynamic programming algorithms for shortest paths and traveling salesman problems can be expressed as multilinear circuits over the tropical semi-ring [5].

In this paper, we consider a generalization of monotone multilinear Boolean circuits to include monotone q-multilinear Boolean circuits. Roughly, a sufficient condition for the q-multilinearity is that in the formal Boolean polynomials at the output gates of the circuit, no variable has degree larger than q. The requirement can be relaxed to hold only for monom representatives of prime implicants of the computed functions (see Sect. 3). Monotone q-multilinear circuits correspond to the so called monotone read-q Boolean circuits introduced in [5].

The central question is how restrictive is the requirement of q-multilinearity in monotone Boolean circuits and whether or not there are substantial gaps between the sizes of monotone q-multilinear Boolean circuits and those of the ($q + 1$)-multilinear ones computing the same Boolean functions. Of course, Razborov's approximation method [16] and several other proofs of lower bounds on the size of monotone Boolean circuits (e.g., for the clique function [1] or matrix multiplication [9,11,13]) work for any q. It seems that the lowest envelope argument from [3] can be solely used to show the gap between monotone 1-multilinear Boolean circuits and the q-multilinear ones, where $q \geq 2$ [5].

First, we study a relationship between the q-multilinearity and the conjunction depth of a monotone Boolean circuit, i.e., the maximum number of AND gates on a path from an input gate to an output gate. As a corollary, we obtain a general lower bound trade-off between the size of monotone q-multilinear Boolean circuits computing a semi-disjoint bilinear form and the parameter q. For instance, in case of Boolean convolution of two n-dimensional Boolean vectors, our lower bound is higher than the best known lower bound of $\Omega(n/\log^6 n)$ on the size of monotone Boolean circuits for this problem [2], as long as $q = O(\log \log n)$.

Jukna [5] gave a nice example of a Boolean function that admits polynomial size monotone 2-multilinear Boolean circuits but any monotone 1-multilinear Boolean circuit computing it has to have an exponential size. The function, termed $Isol_n$, verifies if an input $n \times n$ Boolean matrix has at least one Boolean 1 in each row and each column of the matrix. The lower bound follows from the fact that the shortest prime implicants of $Isol_n$ form exactly the set of prime implicants of the Boolean permanent combined with the exponential lower bound on the size of monotone multilinear Boolean circuits for the latter problem from [12] via arithmetization and the lower envelope argument from [3].

We generalize the function $Isol_n$ to include the function $Isol_{k,n}$ which verifies if the input k-dimensional Boolean matrix has a Boolean 1 in each line of the matrix (a generalization of a row or a column). We show that $Isol_{k,n}$ admits Π_2 monotone k-multilinear circuits of linear, i.e., $O(n^k)$, size. In a Π_2 monotone Boolean circuit, each output gate is an AND gate and on each path from an input gate to an output gate there is at most one block of OR gates and one block of AND gates. On the other hand, we demonstrate that any Π_2 monotone Boolean circuit for $Isol_{k,n}$ is at least k-multilinear. Also, we show under an additional assumption[2] that any Σ_3 monotone Boolean circuit for $Isol_{k,n}$ is not $(k-1)$-multilinear or it has an exponential in n size. In a Σ_3 monotone Boolean circuit, each output gate is an OR gate and on each path from an input gate to an output gate there is at most one block of AND gates and two blocks of OR gates.

2   Monotone Boolean Circuits and Functions

A *monotone Boolean circuit* is a finite directed acyclic graph with the following properties:

1. The indegree of each vertex (termed gate) is either 0 or 2.
2. The source vertices (i.e., vertices with indegree 0 called input gates) are labeled by variables or the Boolean constants 0, 1.
3. The vertices of indegree 2 are labeled by elements of the set $\{OR, AND\}$ and termed OR gates and AND gates, respectively.
4. A distinguished set of gates forms the set of output gates of the circuit.

For convenience, we shall denote also by g the function computed at a gate g of a monotone Boolean circuit. The *size* of Boolean circuit is the total number of its non-input gates.

A monotone Boolean circuit is *multilinear* if for any AND gate the two input Boolean functions have no variable in common. The *conjunction depth* of a monotone Boolean circuit is the maximum number of AND gates on a path from an input gate to an output gate. A monotone Boolean circuit has the *alternation depth* d iff d is the highest number of blocks of OR gates and blocks of AND

---

[2] Recently, the author has obtained an alternative proof, eliminating the need of the additional assumption. The new proof will be included in the journal version.

gates on paths from input gates to output gates. A Σ_d -circuit (respectively, Π_d-circuit) is a circuit with the alternation depth not exceeding d such that the output gates are OR gates (AND gates, respectively).

With each gate g of a monotone Boolean circuit, we shall associate a set $T(g)$ of terms in a natural way. Thus, with each input gate, we associate the singleton set consisting of the corresponding variable or constant. Next, with an OR gate, we associate the union of the sets associated with its direct predecessors. Finally, with an AND gate g, we associate the set of concatenations $t_1 t_2$ of all pairs of terms t_1, t_2, where $t_i \in T(g_i)$ and g_i stands for the i-th direct predecessor of g for $i = 1, 2$. The function computed at the gate g is the disjunction of the functions (called monoms) represented by the terms in $T(g)$. The monom $con(t)$ represented by a term t is obtained by replacing concatenations in t with conjunctions, respectively. A term in $T(g)$ is a *zero-term* if it contains the Boolean constant 0. Clearly, a zero-term represents the Boolean constant 0. By the definition of $T(g)$ and induction on the structure of the monotone Boolean circuit, $g = \bigvee_{t \in T(g)} con(t)$ holds. For a term $t \in T(g)$, the set of variables in t is denoted by $Var(t)$.

A Boolean form is a finite set of Boolean 0–1 functions. An *implicant* of a Boolean form F is a conjunction of some variables and/or Boolean constants (monom) such that there is a function belonging to F which is true whenever the conjunction is true. If the conjunction includes the Boolean 0 then it is a *trivial implicant* of F.

A non-trivial implicant of F that is minimal with respect to included variables is a *prime implicant* of F. The set of prime implicants of F is denoted by $PI(F)$.

A (monotone) *Boolean polynomial* is a disjunction of *monoms*, where each monom is a conjunction of some variables and Boolean constants. It is a *minimal Boolean polynomial* representing a given Boolean function if after the removal of any variable or constant occurrence, it does not represent this function.

A set F of monotone Boolean functions is a *semi-disjoint bilinear form* if it is defined on the set of variables $X \cup Y$ and the following properties hold.

1. For each minimal Boolean polynomial representing a Boolean function Q in F and each variable $z \in X \cup Y$, there is at most one monom of the polynomial containing z.
2. Each monom of a minimal Boolean polynomial representing a Boolean function in F consists of exactly one variable in X and one variable in Y.
3. The sets of monoms of minimal Boolean polynomials representing different Boolean functions in F are pairwise disjoint.

Boolean matrix product and Boolean vector convolution are the best known examples of semi-disjoint bilinear Boolean forms.

3   Monotone q-multilinear Boolean Circuits

Recall that a monotone Boolean circuit is multilinear if for any of its AND gates the two input functions do not share a variable.

The following lemma provides a characterization of the terms produced at the gates of a monotone multilinear circuit which lays ground to the generalization of the multilinearity to include the q-multilinearity. To specify the lemma, we need to introduce the following additional notation.

Let g stand for a gate of a monotone multilinear circuit. For two terms $t,\ t' \in T(g)$, the relationship $t' \le t$ holds if and only if for each variable x, the number of occurrences of x in t' does not exceed that in t. A *variable repetition* takes place in t if there is a variable which occur at least two times in t.

Lemma 1. *(companion lemma) Let g be a gate of a monotone multilinear Boolean circuit without the Boolean constants, and let $t \in T(g)$. There is $t' \in T(g)$ without variable repetitions such that $t' \le t$.*

Proof. The proof is by induction on the structure of the circuit in a bottom-up manner. If g is an input gate corresponding to a variable then $t' = t$. If g is an OR gate then the lemma for the gate immediately follows from the induction hypothesis. Suppose that g is an AND gate with two direct gate predecessors g_1 and g_2. Consider $t = t_1 t_2 \in T(g)$, where $t_i \in T(g_i)$ for $i = 1,\ 2$. By the induction hypothesis, there are (non-zero) terms $t'_i \in T(g_i)$ without variable repetitions such that $t'_i \le t_i$ for $i = 1,\ 2$. Let $t' = t'_1 t'_2$. It follows that $t' \le t$. If t' has a variable repetition then there exist a variable x and $j \in \{1, 2\}$ such that t'_j has an occurrence of the variable but the function g_j does not depend on x. Hence, there must exist a term $t''_j \in T(g_j)$ without an occurrence of x such that the monom represented by t''_j is implied by t'_j, i.e., $Var(t''_j) \subset Var(t'_j)$. We may assume without loss of generality that t''_j does not contain variable repetitions since otherwise we can replace it with a smaller term with respect to $\le$ without variable repetitions by the induction hypothesis. We may also assume without loss of generality that $j = 1$. Hence, $t''_1 t'_2 \le t$ and if $t''_1 t'_2$ is free from variable repetitions we are done. Otherwise, we repeat the procedure eliminating next variable on one of the sides. Because the number of variables is finite the process must eventually result in a term satisfying the lemma.   □

Fix a positive integer q. Roughly, a sufficient condition on a Boolean monotone circuit to call it q-multilinear mentioned in the introduction is that in the formal Boolean polynomials at the output gates of the circuit, no variable has multiplicity larger than q. More formally, in terms of our notation it can be rephrased as follows:

In each term in $T(g)$, where g is an output gate, each variable has at most q occurrences. We shall call a monotone Boolean circuit satisfying this condition *strictly q-multilinear*.

Our definition of a monotone q-multilinear Boolean circuit imposes a weaker condition and it corresponds to that of a monotone read-k Boolean circuit from [5].

A monotone Boolean circuit computing a monotone Boolean form F is said to be *q-multilinear* if for each prime implicant p of each function $f \in F$, there is a term $t \in T(g)$ representing p, where g is the output gate of the circuit computing f, such that no variable occurs more than q times in t.

By the companion lemma, a monotone multilinear Boolean circuit is 1-multilinear. For the reverse implication in terms of monotone read-1 Boolean circuits see Lemma 4 in [5]. Among other things because of the aforementioned equivalence, we believe that the name "k-multilinear" is more natural than the name "read-k" used in [5].

3.1 q-multilinearity Versus Bounded Conjunction Depth

Recall that a monotone Boolean circuit is of conjunction depth d if the maximum number of AND gates on any path from an input gate to an output gate in the circuit is d. A bounded conjunction depth yields a rather weak upper bound on the q-multilinearity of a monotone Boolean circuit.

Theorem 1. *Let F be a monotone Boolean form, and let k be the minimum number of variables forming a prime implicant of F. A monotone Boolean circuit of conjunction depth d computing F is strictly $(2^d - k + 1)$-multilinear.*

Proof. An AND gate can at most double the maximum length of the terms (i.e., the number of variable occurrences in the terms) produced by its direct predecessors. Hence, the output terms of a monotone Boolean circuit of conjunction depth d have length not exceeding 2^d. Consider a variable x occurring in an output term of a monotone Boolean circuit of conjunction depth d computing F. As the term represents an implicant of F, it has to contain at least $k - 1$ other variables. Hence, the maximum number of occurrences of x in the term is $2^d - k + 1$.   □

The reverse relationship is much stronger.

Theorem 2. *Let C be an optimal monotone q-multilinear Boolean circuit without the Boolean constants computing a monotone Boolean form F whose prime implicants are formed by at most k variables. The circuit has conjunction depth not exceeding $kq - 1$.*

Proof. Consider terms at the output gates of C representing prime implicants of F. We know that for each prime implicant p of F there is such a term t_p representing p, where each variable occurs at most q times. Consider the sub-dag C_p of the circuit generating the term t_p. Note that C_p includes the input gates corresponding to the at most k variables in t_p and for any OR gate included in C_p exactly one of the direct predecessors gates in C is included (such a sub-dag is termed *parse graph* in [19]). Let P be a path from an input gate labeled by a variable x to the output gate in the sub-dag having the maximum number of AND gates. Note that at each AND gate h on the path P, a subterm of t_p including x and belonging to $T(h)$ has to be larger at least by one variable occurrence than that belonging to the direct predecessor of h on the path. We conclude that there are at most $kq - 1$ AND gates on the path P.

Form the sub-dag C' of C that is the union of the sub-dags C_p, $p \in PI(F)$. Note that some OR gates in C' may have only one direct predecessor, we replace the missing one with the Boolean 0. Let g' be the output gate of C' corresponding to the output gate g of C. By the definition, $T(g')$ includes terms representing all prime implicants of F represented in $T(g)$. Consider a non-zero (i.e., not including 0) term $t \in T(g')$ that does not represent a prime implicant of F. Consider the sub-dag (parse-graph) C'_t of C' that generates exactly the term t. It also generates the term t in the original circuit C. We conclude that t is an implicant of F and consequently that C' computes F. By the definition, C' has conjunction depth bounded by $kq - 1$, size not exceeding that of C, and it is also q-multilinear. Since the Boolean constants can be eliminated from C' without increasing its size, we conclude that C' has the same size as C by the optimality of C and consequently that $C = C'$ by the construction of C'.   □

4   Lower Bounds for q-multilinear Boolean Circuits

4.1   Lower Bound Trade-Offs for Semi-disjoint Bilinear Forms

Our result in this subsection relies on Theorem 2 in [8] which in terms of our notation can be restricted and rephrased as follows.

Fact 1. *[8] Let C be a monotone Boolean circuit of conjunction depth at most d computing a semi-disjoint bilinear form F with p prime implicants. The circuit C has at least $\frac{p}{2^{4d}}(1 - \frac{1}{2^d})^{2^d - 2}$ AND gates.*

The following theorem is immediately implied by Fact 1 and Theorem 2.

Theorem 3. *Let C be a monotone q-multilinear Boolean circuit computing a semi-disjoint bilinear form F with p prime implicants. The circuit C has at least $\frac{p}{2^{8q-4}}(1 - \frac{1}{2^{2q-1}})^{2^{2q-1} - 2}$ gates.*

In particular, Theorem 3 yields the lower bound of $\Omega(\frac{n^2}{2^{8q-4}})$ on the number of gates in a monotone q-multilinear Boolean circuit computing the n-dimensional Boolean vector convolution. The latter bound subsumes the best known lower bound of $\Omega(n^2/\log^6 n)$ on the size of monotone Boolean circuits for this problem due to Grinchuk and Sergeev [2] as long as $q = o(\log \log n)$.

4.2   Lower Bounds for $Isol_{k,n}$

We generalize the monotone Boolean function $Isol_n$ defined on two-dimensional $n \times n$ Boolean matrices to the function $Isol_{k,n}$ defined on k-dimensional $n \times n...n$ Boolean matrices as follows.

Let $X = (x_{i,j,...,r})$ be an k-dimensional Boolean matrix, where the indices $i, j, ..., r$ are in $[n]$, where $[s]$ stands for the set of natural numbers not exceeding s. A *line* in X is any sequence of n variables in X, where $k - 1$ indices are fixed and the index on the remaining position varies from 1 to n. E.g., in the three-dimensional case, it can be $x_{7,1,5}, x_{7,2,5}, ..., x_{7,n,5}$. $Isol_{k,n}(X) = 1$ if and only if in each line in X there is at least one 1.

Jukna showed that $Isol_n$, i.e., $Isol_{2,n}$ in terms of our notation, admits a monotone 2-multilinear Boolean circuit with $\leq 2n^2$ gates [5]. We can easily generalize his result to include $Isol_{k,n}$.

Theorem 4. $Isol_{k,n}$ *admits a* Π_2 *monotone strictly* k-*multilinear Boolean circuit with* $kn^{k-1}(n-1)$ *OR gates and* $kn^{k-1} - 1$ *AND gates.*

Proof. Observe that there are kn^{k-1} lines in the input matrix. It is sufficient to compute for each line the disjunction of the variables in the line and then the conjunction of all the disjunctions. Since each variable occurs only in k lines, the strict k-multilinearity of the resulting circuit follows.   □

As we have mentioned in the introduction, Jukna observed an exponential gap between the size of monotone 2-multilinear Boolean circuits and the size of monotone multilinear (i.e., also 1-multilinear) circuits for $Isol_{2,n}$ [5]. In the following, we give an evidence that a similar gap holds between monotone k-multilinear Boolean circuits and monotone $(k-1)$-multilinear Boolean circuits for $Isol_{k,n}$, if we restrict the circuits to Σ_3 circuits.

Lemma 2. *Consider a gate computing a disjunction of variables in an optimal monotone Boolean circuit for* $Isol_{k,n}$. *All the variables in the disjunction belong to a common line in the input matrix.*

Proof. Suppose that there are two variables in the disjunction that do not share a line. Then, no variable occurrence from the disjunction is necessary to "guard" uniquely a line in any output term depending on the disjunction in order to make the term an implicant of $Isol_{k,n}$. Otherwise, the sibling output term resulting from replacing the variable by another one belonging to the disjunction but not lying on the line would not be an implicant of $Isol_{k,n}$. Hence, the gate can be replaced by the Boolean constant 1. Consequently, each pair of variables in the disjunction shares a line which implies that all variables in the disjunction occur on the same line of the matrix.   □

Theorem 5. *Any* Π_2 *monotone Boolean circuit for* $Isol_{k,n}$ *is not strictly* $(k-1)$-*multilinear.*

Proof. By Lemma 2, for each line there must be at least one OR gate or input variable gate computing a disjunction of some variables on the line that is a direct predecessor of an AND gate.

In fact, there are no two OR gates or input variable gates that are direct predecessors of AND gates and compute distinct disjunctions of variables on the same line. Otherwise, a term representing a shortest prime implicant including a single variable on this line belonging to the symmetric difference of the disjunctions could not occur at the output AND gate of the circuit. (Such a prime implicant can be formed by completing the single variable on the line with any minimum cardinality set of variables outside the line so all lines are guarded.) Therefore, for each line there must be at least one OR gate computing a disjunction of *all* variables on the line that is a direct predecessor of an AND gate. Otherwise, terms representing shortest prime implicants with a single variable x on the line missing in the disjunction could not occur at the output gate. (Such a prime implicant can be again formed by completing x with any minimum cardinality set of variables so all lines are guarded.)

Thus, for each entry of the input matrix, the variable corresponding to the entry has to appear at least in the k disjunctions corresponding to the k lines it occurs in, computed at k OR gates directly preceding AND gates. In effect, a term representing an implicant of $Isol_{k,n}$ with at least k occurrences of the variable will be created.   □

Note that the monotone strictly k-multilinear circuit for $Isol_{k,n}$, presented in Theorem 4, produces $n^{kn^{k-1}}$ output terms. For Σ_3 monotone Boolean circuit for $Isol_{k,n}$, we have the following conditional result.

Theorem 6. *Let $k > 2$. Consider an optimal Σ_3 monotone Boolean circuit for $Isol_{k,n}$ that has the smallest possible number of AND gates. Suppose that it produces at least $n^{kn^{k-1}} e^{-\delta n^{k-1}}$ terms, where $\delta < 1$. The circuit includes an exponential in n number of AND gates or it is not strictly $(k-1)$-multilinear.*

Proof. Suppose that the number of AND gates that are direct predecessors of the OR gates is at most t. Let $L \geq n^{kn^{k-1}} e^{-\delta n^{k-1}}$ be the total number of terms produced by the circuit at its output gate. Consequently, there must be an AND gate g among the direct predecessors of the OR gates on the top level such that the total number of terms in $T(g)$ is at least L/t.

Let C be the subcircuit with the gate g being an output gate. By Lemma 2, we may assume w.l.o.g. that each at least two variable disjunction computed at an OR gate that is a direct predecessor of an AND gate in the subcircuit is composed of variables lying on the same line. On the other hand, for each line, there must be at least one disjunction composed of some variables on the line, computed at an OR gate that is a direct predecessor of an AND gate in the subcircuit, in case the disjunction consists of a single variable on the line, the input gate corresponding to the variable may be a direct predecessor of an AND gate in the subcircuit. Otherwise, the terms in $T(g)$ would not represent implicants of $Isol_{k,n}$. Suppose that for a line, there are j disjunctions with at least two variables and all its variables lying on the line, computed at OR gates that are direct predecessors of AND gates in the subcircuit. Then, we can compute a disjunction of' these j disjunctions using $j - 1$ OR gates and use the resulting disjunction instead of the j disjunctions saving on $j - 1$ AND gates. The resulting circuit still computes $Isol_{k,n}$, it is optimal and it uses a smaller number of AND gates, a contradiction. So, from now on, we may assume that for each line there is at most one disjunction with at least two variables on the line and no variable outside it, representing the line.

For $i = 1, ..., kn^{k-1}$, let d_i, $1 \leq d_i \leq n$, be the number of variables in the disjunction representing the i-th line. In case, there is no at least two variable disjunction representing the i-th line, the line is represented by a single variable on the line at an input gate or an OR gate that is a direct predecessor of an AND gate. Such a single variable can represent up to k lines it belongs too. So, in this case, we set $d_i = 1$ for only one of the lines represented by the variable and for the remaining ones for which there are no at least two variable representatives, we set $d_i = 0$, alternatively, we can set d_i to a proportional fractional value. It follows that $\prod_{d_i \geq 2} d_i \geq L/t$.

Each tine some d_i is decreased by one, the product $\prod_{d_i \geq 2} d_i$ decreases at least by the factor of $1 - \frac{1}{n}$. Hence, the maximal number of kn^{k-1} variable occurrences in the disjunctions decreases at most by $n(\log_e(e^{\delta n^{k-1}}) + \log_e t)$ for sufficiently large n since L can be smaller by the factor of $e^{\delta n^{k-1}}$ than the maximal possible value of the product of the disjunctions and $\prod_{d_i \geq 2} d_i \geq L/t$ yields further possible decrease by at most t. We infer that $\sum_i d_i > kn^k - \delta n^k - n \log_e t$.

Now, it is sufficient to set t to an exponential in n number, e.g., e^n, so $kn^k - \delta n^k - n\log_e t > kn^k - \delta n^k - n^2 > (k-1)n^k$ by $\delta < 1$ and $k \geq 3$ for sufficiently large n. There are $\sum_i d_i > (k-1)n^k$ slots to be divided between at most n^k variables in the entries of the matrix, so at least one variable must be repeated at least k times, each time in a distinct disjunction.   □

Theorem 6 shows that the number of terms produced by a monotone Boolean circuit for $Isol_{k,n}$ is related to its q-multilinearity and size. In particular, if there was an optimal Σ_3 monotone $(k-1)$-multilinear Boolean circuit of polynomial size for $Isol_{k,n}$ then it would need to produce a substantially smaller number of terms than that k-multilinear one of Theorem 4.

Presumably, the lower bound assumption on the number of terms produced by the circuit in Theorem 6 can be substantially weakened to reach almost $n^{(k-1)n^k-1}$ by using more involved analysis.

On the other hand, note that the lowest possible number of produced output terms equal to the number of prime implicants of $Isol_{k,n}$ can be achieved by a monotone 1-multilinear Boolean circuit of exponential size computing just the (minimal) disjunctive normal form of $Isol_{k,n}$.

Acknowledgments. The author thanks Susanna de Rezende for bringing attention to the monotone Boolean circuit complexity of the Boolean permanent problem studied in [12,17] and valuable discussions. Thanks also go to Stasys Jukna for valuable comments. The research was supported by Swedish Research Council grant 621-2017-03750.

References

1. Alon, N., Boppana, R.: The monotone circuit complexity of Boolean functions. Combinatorica **7**(1), 1–22 (1987)
2. Grinchuk, M.I., Sergeev, I.S.: Thin circulant matrices and lower bounds on the complexity of some Boolean operations. Diskretn. Anal. Issled. Oper. 18, pp. 35–53, (2011). (See also CORR.abs/1701.08557 2017
3. Jerrum, M., Snir, M.: Some exact complexity results for straight-line computations over semirings. J. ACM **29**(3), 874–897 (1982)
4. Jukna, S.: Personnal communication, June (2022)
5. Jukna, S.: Notes on Boolean Read-k Circuits. Electron. Colloquium Comput. Complex. TR22-094 (2022)
6. Krieger, M.P.: On the incompressibility of monotone DNFs. Theor. Comput. Syst. **41**(2), 211–231 (2007)
7. Lingas, A.: A Note on Lower Bounds for Monotone Multilinear Boolean Circuits. Electron. Colloquium Comput. Complex. TR22-085 (2022)

8. Lingas, A.: Small normalized Boolean circuits for semi-disjoint bilinear forms require logarithmic conjunction-depth. Theoretical Computer Science 820, pp. 17–25 (2020) (prel. version Computational Complexity Conference (CCC) 2018)

9. Mehlhorn, K., Galil, Z.: Monotone switching circuits and Boolean matrix product. Computing **16**, 99–111 (1976)

10. Nešetřil, J., Poljak, S.: On the complexity of the subgraph problem. Commentationes Mathematicae Universitatis Carolinae **26**(2), 415–419 (1985)

11. Paterson, M.: Complexity of monotone networks for Boolean matrix product. Theor. Comput. Sci. **1**(1), 13–20 (1975)

12. Ponnuswami, A.K., Venkateswaran, H.: Monotone multilinear Boolean circuits for bipartite perfect matching require exponential size. In Proceedings of 24th International Conference on Foundations of Software Technology and Theoretical Computer Science (FST-TCS), pp. 16–18 (2004)

13. Pratt, R.: The power of negative thinking in multiplying Boolean matrices. SIAM J. Comput. **4**(3), 326–330 (1975)

14. Raz, R.: Multi-linear formulas for permanent and determinant are of super-polynomial size. Electron. Colloquium Comput. Complex. (067) (2003)

15. Raz, R., Wigderson, A.: Monotone circuits for matching require linear depth. J. ACM **39**(3), 736–744 (1992)

16. Razborov, A.A.: Lower bounds for the monotone complexity of some Boolean functions. Soviet Math. Dokl. **31**, 354–357 (1985)

17. Razborov, A.A.: Lower bounds on monotone complexity of the logical permanent. Math. Notes of the Acad. of Sci. of the USSR, 37(6), pp. 485–493 (1985)

18. Schnorr, C.P.: A lower bound on the number of additions in monotone computations. Theor. Comput. Sci. **2**(3), 305–315 (1976)

19. Sengupta, R., Venkateswaran, H.: Multilinearity can be exponentially restrictive (pre-liminary version). Technical Report GIT-CC-94-40, Georgia Institute of Technology. College of Computing (1994)

20. Shamir, E., Snir, M.: Lower bounds on the number of multiplications and the number of additions in monotone computations. Tech. Rep. RC 6757, IBM Thomas J. Watson Research Center, Yorktown Heights, NY, (1977)

A Faster Algorithm for Determining the Linear Feasibility of Systems of BTVPI Constraints

Piotr Wojciechowski and K. Subramani[✉]

LDCSEE, West Virginia University, Morgantown, WV, USA
{pwojciec,k.subramani}@mail.wvu.edu

Abstract. In this paper, we discuss an approach for determining the feasibility of a polyhedron defined by a system of Binary Two Variable Per Inequality (BTVPI) constraints. A constraint of the form: $a_i \cdot x_i + a_j \cdot x_j \geq b_{ij}$ is called a BTVPI constraint, if $a_i, a_j \in \{0, 1, -1, 2, -2\}$ and $b_{ij} \in \mathbb{Z}$. These constraints find applications in a number of domains, including scheduling and abstract interpretation. Our algorithm is based on a rewrite version of the well-known Fourier-Motzkin elimination procedure for linear programs. We show that our algorithm converges in polynomial time and is faster than all known algorithms for this class of problems.

1 Introduction

The focus of this paper is on the design and analysis of a fast algorithm for a specialized class of linear programs. This class of linear programs is defined by linear constraints having at most two non-zero coefficients. Furthermore, the non-zero coefficients must belong to the set $\{\pm 1, \pm 2\}$. These constraints are called Binary Two Variable Per Inequality (BTVPI) constraints and they subsume constraints such as difference constraints and UTVPI constraints (see Sect. 2). The more general class of TVPI constraints has been studied by others [6,14]. For this class, the best known algorithm runs in $O(m \cdot n^2 \cdot \log m)$ time [14], on a system with m constraints over n variables. Our algorithm runs in $O(m \cdot n^2)$ time and is thus asymptotically superior. However, our algorithm works for a smaller constraint class. We use the same idea as in [14], i.e., Fourier-Motzkin elimination. However, owing to the specialized nature of our constraints, we are able to incorporate substantially different pruning techniques, which simplify our algorithm and its analysis.

For the longest time, the Simplex algorithm discovered by Dantzig [9] was the algorithm of choice for linear programming problems. Since then, the linear programming problem was shown to be in **P** through the use of the ellipsoid algorithm [13]. Although provably polynomial time, the ellipsoid algorithm has not proven to be efficient in practice. The projective and affine scaling methods

This research was made possible by the NASA Established Program to Stimulate Competitive Research, Grant # 80NSSC22M0027.

for linear programming discussed in [15] are both theoretically and practically efficient. Research has also progressed along the direction of finding efficient algorithms for special classes of linear programs. For instance, Tardos [24] discusses a strongly polynomial time algorithm for combinatorial linear programs. Recall that a combinatorial linear program is one in which all the entries in the constraint matrix $\mathbf{A}$ are bounded by a polynomial in the dimensions of $\mathbf{A}$. Similarly, there exists a strongly polynomial time algorithm for Horn programs. Recall that in a Horn program, every entry in $\mathbf{A}$ is in the set $\{0, 1, -1\}$ and furthermore, there is at most one positive entry per constraint. Horn programs have several interesting properties which have been documented in [25] and [26]. In [5], an incremental approach to solve Horn programs with absolute constraints is discussed. This approach was distilled into a strongly polynomial time algorithm in [4].

Over the years, a number of approaches have been developed to handle TVPI systems and in particular, difference constraint systems, and UTVPI constraint systems. It is well-known that the $\mathbf{LF}$ problem in difference constraints is $\mathbf{AC^0}$-equivalent to the problem of finding shortest paths in a weighted, directed graph [7,27]. Consequently, strongly polynomial time algorithms such as Bellman-Ford [1] can be used to solve the same. It must be noted though that although Bellman-Ford is asymptotically the best choice for DCSs, there is documented evidence of other approaches being faster on real-world instances [12].

[16] and [18] showed that there exists an $\mathbf{AC^0}$-reduction from the $\mathbf{LF}$ problem in UCSs to a path problem in a specialized network. Similar reduction schemes with difference constraint networks have been discussed in [21] and [23].

For general TVPI constraints, Shostak provided the first $\mathbf{AC^0}$-reduction to transform the $\mathbf{LF}$ problem into a path problem in graphs. He defined the notion of *residues* in graph loops and identified negative loop residues with infeasibility of the underlying TVPI system. His approach is exponential in the worst case. This approach was improved by Nelson [19], who gave an $n^{O(\log n)}$ algorithm. The loop residue approach was further refined by Aspvall, et. al., in [3] and they produced the first polynomial time algorithm for linear feasibility in TVPI constraints. The main contribution in [3] was the efficient propagation of inequalities around loops. The running time of their algorithm is $O(m \cdot n^3 \cdot I)$, where I represents the number of bits required the represent the problem instance. In [14], Hochbaum and Naor provided the first strongly polynomial time algorithm for this problem. Their algorithm runs in time $O(m \cdot n^2 \cdot \log m)$ and does not depend upon the values in $\mathbf{b}$. The algorithm in [14] is based on the Fourier-Motzkin elimination technique used to solve systems of linear inequalities [11]. Yet another loop-following approach is proposed in [6]. Their approach runs in time $O(m \cdot n^2 \cdot (\log m + \log^2 n))$, which is slightly inferior to the approach in [14].

It is well-known that the linear feasibility problem is $\mathbf{P\text{-}complete}$ [2]. Thus, any problem in $\mathbf{P}$ can be reduced in $\mathbf{log\text{-}space}$ (or in $\mathbf{NC}$) to $\mathbf{LF}$. The advantage of such a reduction is that there are a number of highly optimized solvers such as CPLEX and Gurobi [20] that can be used to solve instances in these problem domains. From the parallel perspective, the $\mathbf{LF}$ problem in both DCSs and

UCSs is in **NC**. For arbitrary TVPI systems, Cohen and Meggido propose an algorithm that runs in $\tilde{O}(n)$ time using $O(m \cdot n)$ processors [6]. It is mentioned in [6] that the approach in [14] can be parallelized to run in $O(n^2 \cdot \log m)$ time. Our algorithm is superior to the current best algorithm for TVPI constraints. At the same time, we are considering only a subset of the constraints considered by [14] and [6]. Table 1 contains the best known results for solving the **LF** problem in DCSs, UCSs, BCSs, and systems of TVPI constraints.

Table 1. Best known algorithms

System	Linear feasibility
Difference	$O(m \cdot n)$ [7]
UTVPI	$O(m \cdot n)$ [16]
BTVPI	$O(m \cdot n^2)$ (This paper)
TVPI	$O(m \cdot n^2 \cdot \log m)$ [14]

The principal contributions of this paper are as follows: 1. A rewrite version of Fourier-Motzkin Elimination. 2. A combinatorial algorithm for the linear feasibility problem in BCSs.

2 Statement of Problems

In this section, we introduce the concepts examined in this paper and define the problems under consideration. In this paper, we examine the feasibility problem for a restricted form of linear programs.

Definition 1. *A* **Linear Program** *(LP) is a conjunction of constraints in which each constraint is an inequality of the form* $\mathbf{a_j} \cdot \mathbf{x} \geq b_j$ *where* $\mathbf{a_j} \in \mathbb{Z}^n$, $b_j \in \mathbb{Z}$, *and each variable* x_i *can take any real value.*

As mentioned before, an LP can be expressed in matrix form as: $\mathbf{A} \cdot \mathbf{x} \geq \mathbf{b}$. Furthermore, the **LF** problem is concerned with checking if a given **LP** is non-empty. In this paper, we focus on LPs where each constraint has at most two non-zero variables.

Definition 2. *A* **Two Variable Per Inequality** *(TVPI) constraint is a constraint with at most two non-zero variables.*

An LP in which every constraint is a TVPI constraint is called a TVPI Constraint System (TCS). We can further restrict constraints by limiting the values which the non-zero coefficients can take.

Definition 3. *A* **difference** *constraint is a TVPI constraint such that each non-zero variable has a coefficient belonging to the set* $\{\pm 1\}$, *at most one coefficient is 1, and at most one coefficient is* -1.

An LP in which every constraint is a difference constraint is known as a Difference Constraint System (DCS).

Definition 4. *A* **Unit Two Variable Per Inequality** *(UTVPI) constraint is a TVPI constraint such that each non-zero variable has a coefficient belonging to the set* $\{\pm 1\}$.

Note that every difference constraint is a UTVPI constraint. An LP in which every constraint is a UTVPI constraint is known as a UTVPI Constraint System (UCS).

Definition 5. *A* **Binary Two Variable Per Inequality** *(BTVPI) constraint is a TVPI constraint such that each non-zero variable has a coefficient belonging to the set* $\{\pm 1, \pm 2\}$.

Every UTVPI constraint is a BTVPI constraint and every BTVPI constraint is a TVPI constraint. An LP in which every constraint is a BTVPI constraint is known as a BTVPI Constraint System (BCS).

Definition 6. *A constraint with only one non-zero coefficient is known as an* **absolute** *constraint.*

It is well-known that absolute constraints can be rewritten as difference constraints using an auxiliary variable [7]. It follows that every absolute constraint is a difference constraint. At this juncture, we distinguish between binary constraints (constraints having two non-zero variables) and absolute constraints (constraints with only one non-zero variable).

From an input constraint system, new constraints can be derived using inference rules. For LPs, we use a single inference rule, viz., Rule (1).

$$\text{ADD} : \frac{\sum_{i=1}^{n} a_i \cdot x_i \geq b_1 \qquad \sum_{i=1}^{n} a_i' \cdot x_i \geq b_2}{\sum_{i=1}^{n} (a_i + a_i') \cdot x_i \geq b_1 + b_2} \tag{1}$$

Example 1. Consider the constraints $x_1 - x_2 \geq 1$ and $x_2 + 2 \cdot x_3 \geq -1$, applying the ADD rule to these constraints results in the constraint $x_1 + 2 \cdot x_3 \geq 0$.

We refer to Rule (1) as the **ADD rule**. It is easy to see that Rule (1) is **sound** since any assignment that satisfies the hypotheses also satisfies the consequent. Furthermore, as per Farkas' lemma (a theorem of the alternative relating the feasibility LPs), Rule (1) is **complete** as well. Thus, any infeasible LP has a refutation using the ADD rule.

In this paper, we use two specializations of the above-mentioned ADD rule. Fourier-Motzkin elimination – Fourier-Motzkin Elimination (FM) is a procedure to eliminate a single variable from a constraint system [10]. Consider a variable x_i. For each pair of constraints such that one constraint has x_i with positive coefficient and the other constraint has x_i with negative coefficient, the FM procedure derives a new constraint without x_i [10]. This can be accomplished through use of the ADD rule.

Example 2. Consider the constraints $x_1 - 2 \cdot x_2 \geq 2$ and $x_2 + 2 \cdot x_1 \geq 1$. Applying the FM procedure to eliminate x_2 results in the constraint $x_1 + 4 \cdot x_3 \geq 4$. This constraint can be derived using the ADD rule as follows:

1. Apply the ADD rule to the constraints $x_1 - 2 \cdot x_2 \geq 2$ and $x_2 + 2 \cdot x_1 \geq 1$ to get $x_1 - x_2 - 2 \cdot x_3 \geq 3$.
2. Apply the ADD rule to the constraints $x_1 - x_2 + 2 \cdot x_3 \geq 3$ and $x_2 + 2 \cdot x_1 \geq 1$ to get $x_1 + 4 \cdot x_3 \geq 4$.

To represent the FM procedure, we introduce a new rule called the FM Rule. The FM Rule takes two constraints which share a variable and derives the constraint that Fourier-Motzkin elimination would derive from those constraints. Rule (2) shows this when applied to TVPI constraints.

$$\text{FM} : \frac{a_i \cdot x_i + a_j \cdot x_j \geq b_1 \qquad a_i' \cdot x_i + a_k \cdot x_k \geq b_2 \qquad a_i \cdot a_i' < 0}{a_j \cdot |a_i'| \cdot x_j + a_k \cdot |a_i| \cdot x_k \geq b_1 \cdot |a_i'| + b_2 \cdot |a_i|} \quad (2)$$

Note that, if the constraints to which the FM rule is being applied are TVPI constraints, then so is the derived constraint. Observe that the application of the FM Rule in Rule (2) can be accomplished using $(|a_i| + |a_i'| - 1)$ applications of the ADD rule. Note that if $a_j = 0$, $a_k = 0$, or $x_j = x_k$, then the constraint derived by the FM Rule in Rule (2) rule is an absolute constraint.

Lifting – Lifting is a technique used in [4] to derive new absolute constraints by using existing absolute constraints to eliminate all but one variable from a constraint.

Example 3. Consider the constraint $x_1 - x_2 \geq 5$. If in addition, we have the absolute constraints $x_1 \geq 5$ and $x_2 \geq 6$, then we can derive (infer) the absolute constraint $x_1 \geq 11$. Note that this raises (or "lifts") the lower bound on x_1 from 5 to 11. This constraint can be derived using the ADD rule as follows: Apply the ADD rule to the constraints $x_1 - x_2 \geq 5$ and $x_2 \geq 6$ to get $x_1 \geq 11$.

The algorithm for linear feasibility in BCSs relies exclusively on the ADD operation and the two specializations described above.

3   A Rewrite Version of Fourier-Motzkin Elimination

In this section, we design a rewrite version of Fourier-Motzkin Elimination. In Sect. 4, we will utilize various properties of BCSs to show that, when applied to BCSs, this rewrite procedure can be implemented efficiently. We refer to this procedure as FM-REWRITE.

Fourier-Motzkin elimination is a well known procedure for determining if a system of linear constraints is linearly feasible [10]. In Fourier-Motzkin Elimination, variables are eliminated one-by-one from the system until only one variable remains. The bounds on that variable are then used to determine if the original system is feasible. It is important to note that Fourier-Motzkin Elimination is both sound and complete [10].

Let **L** be a system of linear constraints. Throughout this section, we refer to useless constraints.

Definition 7. *A constraint* $l : \sum_{i=1}^{n} a_i \cdot x_i \geq b$ *is* **useless** *with respect to* **L***, if there exists a constraint* $l' : \sum_{i=1}^{n} a'_i \cdot x_i \geq b'$ *in* **L** *such that: 1. l and l' use the same variables. 2. There exists a positive constant c, such that $c \cdot a_i = a'_i$ for $i = 1 \ldots n$ and $c \cdot b \leq b'$.*

Any **x** that satisfies l' also satisfies l. Thus, we do not need to add the constraint l to **L**, i.e., it is **redundant**. Thus, we only keep the tightest version of each constraint, breaking ties arbitrarily. Note that each constraint in **L** is trivially useless with respect to **L**. This means that no additional copies of those constraints need to be added to **L**. Any constraint which is not useless with respect to **L** is called useful with respect to **L**. There are three special types of systems that we want to consider. These are as follows:

1. Systems that contain a constraint of the form $0 \geq b$ where $b > 0$ (Type 1).
2. Non-Type 1 systems in which no constraint that is useful with respect to **L** can be derived by a single application of the FM Rule (Type 2).
3. Non-Type 1 systems in which a constraint that is useful with respect to **L** can be derived by a single application of the FM Rule (Type 3).

Note that Type 1 systems are trivially infeasible. We now show that Type 2 systems are feasible.

Theorem 1. *Let* **L** *be a system of linear constraints. If* **L** *is a Type 2 system, then* **L** *is feasible.*

Proof. Let **L** be a Type 2 system. Assume, for the sake of contradiction, that **L** is infeasible. Note that Fourier-Motzkin elimination is a complete refutation system [10]. Thus, Fourier-Motzkin elimination, when applied to **L** will eventually produce a constraint l^* of the form $0 \geq b$ where $b > 0$. Thus, the constraint l^* is derivable from **L** by a sequence of applications of the FM Rule.

Since **L** does not contain a constraint of the form $0 \geq b$ where $b > 0$, the constraint l^* is useful with respect to **L**. Consider the sequence of applications of the FM Rule used to derive l^* and let l' be the first constraint derived by this sequence that is useful with respect to **L**. Since l^* is useful with respect to **L**, such a constraint must exist.

Let l'_1 and l'_2 be the constraints used to derive l'. Note that l'_1 was either derived earlier by the sequence of applications of the FM Rule, or $l'_1 \in$ **L**. In either case, l'_1 is useless with respect to **L**. The same holds for l'_2.

Let $l_1 \in$ **L** and $l_2 \in$ **L** be the constraints that make l'_1 and l'_2 useless, respectively. Consider the constraint l_3 derived from l_1 and l_2 by the FM rule. Since **L** is a Type 2 system, l_3 is useless with respect to **L**. Thus, there is a constraint $l'_3 \in$ **L** that makes l_3 useless. However, l'_3 also makes l' useless. This is a contradiction. Thus, **L** must be feasible.   □

The rewrite version of Fourier-Motzkin Elimination classifies LPs by rewriting them until they become either a Type 1 system or a Type 2 system. Those that become Type 1 systems will be declared infeasible. Those that become Type

2 systems will be declared feasible. This rewriting will be accomplished through use of the FM Rule.

Let **L** be a Type 3 system. Let l be a constraint that is useful with respect to **L** and is derivable from **L** by an application of the FM Rule. Note that if no such constraint exists, then **L** would be a Type 2 system. We rewrite **L** as **L** ∪ {l} and then repeat this procedure until **L** becomes a Type 1 system or a Type 2 system. The flowchart in Fig. 1 illustrates this procedure:

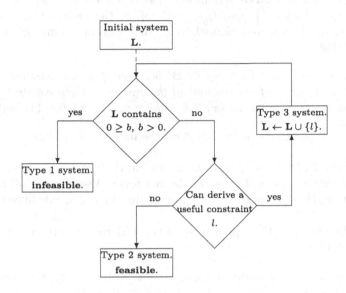

Fig. 1. Flow of the FM-REWRITE algorithm.

We now show that FM-REWRITE correctly determines the feasibility of a linear program.

Theorem 2. *Let* **L** *be an LP.* FM-REWRITE*(***L***) returns* **feasible** *if and only if* **L** *is linearly feasible.*

Proof. If **L** is a Type 1 system, then **L** is infeasible and FM-REWRITE(**L**) correctly returns **infeasible**. If **L** is a Type 2 system, then, by Theorem 1, **L** is feasible and FM-REWRITE(**L**) correctly returns **feasible**.

If **L** is a Type 3 system, then there is a constraint l that is useful with respect to **L** and is derivable from **L** by a single application of the FM rule. Since Fourier-Motzkin Elimination is a sound refutation system [10], **L** is feasible if and only if **L** ∪ {l} is feasible. In this case, FM-REWRITE(**L**) correctly returns the same value as FM-REWRITE(**L** ∪ {l}). □

In the worst case, for a system **L** with m constraints over n variables, Fourier-Motzkin Elimination produces $4 \cdot (\frac{m}{4})^{(2^n)}$ constraints [10]. Thus, FM-REWRITE(**L**) runs in time $\Omega(m^{(2^n)})$ with a naive implementation.

4   The BCS Linear Programming Algorithm

In this section, we show that FM-REWRITE can be implemented efficiently for BCSs.

Let $\mathbf{B}$ be a BCS with m constraints over n variables and let $\mathbf{B}^*$ be a set of constraints derivable from $\mathbf{B}$ by a sequence of applications of the FM rule that are useful with respect to $\mathbf{B}$. First, we establish that the constraints in $\mathbf{B}^*$ take a specific form. In fact, each constraint in $\mathbf{B}^*$ is of the form $a_i \cdot 2^{l_i} \cdot x_i + a_j \cdot 2^{l_j} \cdot x_j \geq b_{ij}$ where $a_i, a_j \in \{1, 0, -1\}$, and $l_i, l_j \geq 0$. Since the order of x_i and x_j does not matter, the proofs and algorithms in this section assume without loss of generality that $l_i \geq l_j$.

Theorem 3. *Let $\mathbf{B}$ be a BCS and let $\mathbf{B}^*$ be a set of useful constraints derivable from $\mathbf{B}$ by a sequence of applications of the FM rule. Each constraint in $\mathbf{B}^*$ is of the form $a_i \cdot 2^{l_i} \cdot x_i + a_j \cdot 2^{l_j} \cdot x_j \geq b_{ij}$ where $a_i, a_j \in \{1, 0, -1\}$, and $l_i, l_j \geq 0$.*

Proof. We will prove this by induction on the number of applications of the FM rule.

Before the FM rule is applied, there are no derived constraints. Since $\mathbf{B}$ is a BCS, all constraints in $\mathbf{B}$ have the desired form. Assume that all constraints derivable from $\mathbf{B}$ by a sequence of h applications of the FM rule have the desired form.

Consider the $(h + 1)^{th}$ application of the FM rule. There are two cases we need to consider.

1. We derive a new constraint using $a_i \cdot 2^{l_i} \cdot x_i + a_j \cdot 2^{l_j} \cdot x_j \geq b_{ij}$ and $-a_i \cdot 2^{l_i'} \cdot x_i + a_k \cdot 2^{l_k} \cdot x_k \geq b_{ik}$ where $l_i \geq l_i'$. In this case, the constraint derived by the FM Rule is $a_j \cdot 2^{l_j} \cdot x_j + a_k \cdot 2^{l_k + l_i - l_i'} \cdot x_k \geq b_{ij} + 2^{l_i - l_i'} \cdot b_{ik}$.
2. We derive a new constraint using $a_i \cdot 2^{l_i} \cdot x_i + a_j \cdot 2^{l_j} \cdot x_j \geq b_{ij}$ and $-a_i \cdot 2^{l_i'} \cdot x_i + a_k \cdot 2^{l_k} \cdot x_k \geq b_{ik}$ where $l_i' \geq l_i$. In this case, the constraint derived by the FM rule is $a_j \cdot 2^{l_j + l_i' - l_i} \cdot x_j + a_k \cdot 2^{l_k} \cdot x_k \geq 2^{l_i' - l_i} \cdot b_{ij} + b_{ik}$.

Note that in each case, the derived constraint is of the desired form. Note that if $x_j = x_k$, then the derived constraint is an absolute constraint. □

We further reduce the number of constraints in $\mathbf{B}^*$ by looking at the value of $(l_i - l_j)$ for each constraint in $\mathbf{B}^*$.

Theorem 4. *Let $\mathbf{B}$ be a BCS and let $\mathbf{B}^*$ be a set of useful constraints derivable from $\mathbf{B}$. For each $1 \leq i, j \leq n$, $a_i, a_j \in \{1, -1\}$, and $l \in \mathbb{Z}$, we can assume without loss of generality that $\mathbf{B}^*$ contains at most one constraint of the form $a_i \cdot 2^{l_i} \cdot x_i + a_j \cdot 2^{l_j} \cdot x_j \geq b_{ij}$ such that $l_i - l_j = l$.*

Proof. Let $a_i \cdot 2^{l_i} \cdot x_i + a_j \cdot 2^{l_j} \cdot x_j \geq b_{ij}$ and $a_i \cdot 2^{l_i'} \cdot x_i + a_j \cdot 2^{l_j'} \cdot x_j \geq b_{ij}'$ be two constraints derivable from $\mathbf{B}$ such that $l = l_i - l_j = l_i' - l_j'$. Since $l_i - l_j = l_i' - l_j'$, $l_i - l_i' = l_j - l_j'$

Assume without loss of generality that $l_i \geq l_i'$. In this case, the constraint $a_i \cdot 2^{l_i'} \cdot x_i + a_j \cdot 2^{l_j'} \cdot x_j \geq b_{ij}'$ is equivalent to the constraint $a_i \cdot 2^{l_i} \cdot x_i + a_j \cdot 2^{l_j' + l_i - l_i'} \cdot x_j \geq b_{ij}' \cdot 2^{l_i - l_i'}$. Since $l_i - l_i' = l_j - l_j'$, this is equivalent to the constraint $a_i \cdot 2^{l_i} \cdot x_i + a_j \cdot 2^{l_j} \cdot x_j \geq b_{ij}' \cdot 2^{l_i - l_i'}$.

If $b_{ij} > b_{ij}' \cdot 2^{l_i - l_i'}$, then the constraint $a_i \cdot 2^{l_i} \cdot x_i + a_j \cdot 2^{l_j} \cdot x_j \geq b_{ij}$ is stronger than the constraint $a_i \cdot 2^{l_i'} \cdot x_i + a_j \cdot 2^{l_j'} \cdot x_j \geq b_{ij}'$. Thus, the constraint $a_i \cdot 2^{l_i'} \cdot x_i + a_j \cdot 2^{l_j'} \cdot x_j \geq b_{ij}'$ is useless with respect to $\mathbf{B}$.

Similarly, if $b_{ij}' \cdot 2^{l_i - l_i'} > b_{ij}$, then the constraint $a_i \cdot 2^{l_i'} \cdot x_i + a_j \cdot 2^{l_j'} \cdot x_j \geq b_{ij}'$ is stronger than the constraint $a_i \cdot 2^{l_i} \cdot x_i + a_j \cdot 2^{l_j} \cdot x_j \geq b_{ij}$. Thus, the constraint $a_i \cdot 2^{l_i} \cdot x_i + a_j \cdot 2^{l_j} \cdot x_j \geq b_{ij}$ is useless with respect to $\mathbf{B}$. $\square$

From Theorem 4, $\mathbf{B}^*$ only needs to contain at most one constraint for a given i, j, a_i, a_j and $l = l_i - l_j$. Thus, we can store $\mathbf{B}^*$ as a 5-dimensional array in which $\mathbf{B}^*[i, j, a_i, a_j, l]$ is a constraint of the form $a_i \cdot 2^{l_i} \cdot x_i + a_j \cdot 2^{l_j} \cdot x_j \geq b_{ij}$ such that $l_i - l_j = l$.

Algorithm 4.1, uses the technique described in the proof of Theorem 4 to eliminate useless constraints. This algorithm takes as input a new constraint and the array of derived constraints. The algorithm checks to see if an already derived constraint makes the new constraint useless with respect to $\mathbf{B} \cup \mathbf{B}^*$. If no constraint does, the new constraint is added to the array of derived constraints.

Input: Array $\mathbf{B}^*$ and BTVPI constraint c.

Output: None.

1: **procedure** ADD-CONSTRAINT-LIN($\mathbf{B}^*$, c)
2:   **If** ($\mathbf{B}^*$ contains a constraint c' whose left hand side is a multiple of c's left-hand side) **then**
3:     **if** (c makes c' useless) **then**
4:       Replace c' with c in $\mathbf{B}^*$.
5:   **else**
6:     Add c to $\mathbf{B}^*$.

Algorithm 4.1: Add the constraint c to $\mathbf{B}^*$ if it is useful.

Note that Algorithm 4.1 runs in constant time. From Theorem 4, the constraint c is added to $\mathbf{B}^*$ by Algorithm 4.1 if and only if it is useful with respect to $\mathbf{B} \cup \mathbf{B}^*$.

We will later establish that at any point in the derivation procedure $l \leq n$. Thus, we only need to keep at most $4 \cdot n$ constraints using any pair of variables. Thus $\mathbf{B}^*$ contains at most $4 \cdot n^3$ non-absolute constraints.

We explain how to handle absolute constraints. Algorithm 4.1 handles the addition of non-absolute constraints. We handle the derivation of new absolute constraints by using a technique called lifting. This technique utilizes existing absolute constraints to derive new absolute constraints. Theorem 5 shows how new absolute constraints are generated through lifting.

Theorem 5. *Let* $\mathbf{B}$ *be a BCS and let* $a_i \cdot 2^{l_i} \cdot x_i + a_j \cdot 2^{l_j} \cdot x_j \geq b_{ij}$ *be a constraint derivable from* $\mathbf{B}$. *If* $\mathbf{B}$ *contains the absolute constraint* $-a_j \cdot x_j \geq b_j$, *then the constraint* $a_i \cdot 2^{l_i} \cdot x_i \geq b_{ij} + 2^{l_j} \cdot b_j$ *is derivable from* $\mathbf{B}$.

Proof. We can use the constraint $-a_j \cdot x_j \geq b_j$ to eliminate the term $a_j \cdot 2^{l_j} \cdot x_j$ from the constraint $a_i \cdot 2^{l_i} \cdot x_i + a_j \cdot 2^{l_j} \cdot x_j \geq b_{ij}$. This results in the constraint $a_i \cdot 2^{l_i} \cdot x_i \geq b_{ij} + 2^{l_j} \cdot b_j$ as desired.   □

In addition to using $\mathbf{B}^*$ to store non-absolute constraints, the absolute constraints derivable from $\mathbf{B}$ can be represented by a two dimensional array $\mathbf{D}$, where $\mathbf{D}[i, a_i]$ stores the tightest bound on the term $a_i \cdot x_i$ derived so far. Thus, at any point we know that $-\mathbf{D}[i, -1] \geq x_i \geq \mathbf{D}[i, 1]$.

Algorithm 4.2 uses the technique described in the proof of Theorem 5 to generate new absolute constraints.

Input: Array $\mathbf{D}$ and BTVPI constraints c.

Output: None.

1: **procedure** UPDATE-BOUNDS-LIN($\mathbf{D}$, c)

2:     Assume that c is of the form $a_i \cdot 2^{l_i} \cdot x_i + a_j \cdot 2^{l_j} \cdot x_j \geq b$.

3:     **if** ($\frac{b + 2^{l_j} \cdot \mathbf{D}[j, -a_j]}{2^{l_i}} > \mathbf{D}[i, a_i]$) **then**

4:         $\mathbf{D}[i, a_i] \leftarrow \frac{b + 2^{l_j} \cdot \mathbf{D}[j, -a_j]}{2^{l_i}}$.   ▷ $a_i \cdot 2^{l_i} \cdot x_i \geq b + 2^{l_j} \cdot \mathbf{D}[j, -a_j]$ is a derivable absolute constraint.

5:     **if** ($\frac{b + 2^{l_i} \cdot \mathbf{D}[i, -a_i]}{2^{l_j}} > \mathbf{D}[j, a_j]$) **then**

6:         $\mathbf{D}[j, a_j] \leftarrow \frac{b + 2^{l_i} \cdot \mathbf{D}[i, -a_i]}{2^{l_j}}$.   ▷ $a_j \cdot 2^{l_j} \cdot x_j \geq b + 2^{l_i} \cdot \mathbf{D}[i, -a_i]$ is a derivable absolute constraint.

Algorithm 4.2: Update the bounds stored in $\mathbf{D}$ to account for the constraint c.

Note that Algorithm 4.2 runs in constant time. From Theorem 5, the new bounds derived by Algorithm 4.2 are all derivable from constraint c and the existing bounds stored in $\mathbf{D}$.

We utilize Algorithms 4.1 and 4.2 to solve the linear feasibility problem for BCSs. Algorithm 4.3 represents our approach. A flowchart for this algorithm is depicted in Fig. 2.

Observe that each application of the FM rule uses a constraint in $\mathbf{B}$. Thus, each application of the FM rule is done to constraints of the form $a_i \cdot 2^{l_i} \cdot x_i + a_j \cdot 2^{l_j} \cdot x_j \geq b_{ij}$ and $-a_i \cdot 2^{l'_i} \cdot x_i + a_k \cdot 2^{l'_k} \cdot x_k \geq b_{ik}$ where $l'_i, l'_k \in \{0, 1\}$. Thus, each iteration of the **for** loop on Line 9 of Algorithm 4.3 increases the maximum value of $l = l_i - l_j$ by at most 1. Thus, $l \leq n$ as desired.

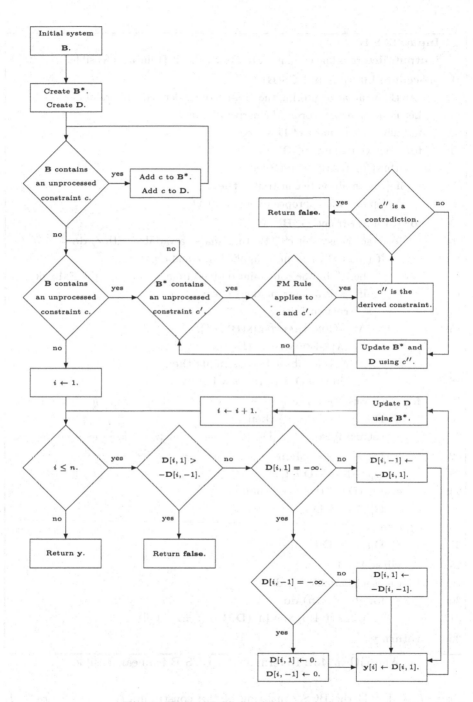

Fig. 2. Flow of BCS feasibility algorithm.

Input: BCS **B**.

Output: Return a linear solution to **B** or **false** if **B** linear infeasible.

1: **procedure** LIN-BINARY-BCS(**B**)
2:    Let **B*** be an array storing the non-absolute derived constraints.
3:    Let **D** be an array storing the derived bounds.
4:    Initially, each element of **D** is $-\infty$.
5:    **for** (each constraint $c \in$ **B**) **do**
6:        ADD-CONSTRAINT-LIN(**B***, c).
7:        **if** (c is an absolute constraint) **then**
8:            Update the appropriate bound in **D**.
9:    **for** (each constraint $c \in$ **B**) **do**
10:        **for** (each constraint $c' \in$ **B*** that shares a variable with c) **do**
11:            **if** (the FM rule can be applied to c and c') **then**
12:                Let c'' be the constraint derived from c and c' by the FM rule.
13:                **if** (c'' is a contradiction) **then**
14:                    **return false.**
15:                ADD-CONSTRAINT-LIN(**B***, c'').
16:                UPDATE-BOUNDS-LIN(**D**, c'').
17:                **if** (c'' is an absolute constraint) **then**
18:                    Update **D** with the new bound.
19:    **for** (each variable x_i) **do**   ▷ Construct a feasible solution.
20:        **if** (**D**$[i, 1] > -$**D**$[i, -1]$) **then**
21:            **return false.**   ▷ The lower bound on x_i exceeds the upper bound.
22:        **if** (**D**$[i, 1] \neq -\infty$) **then**
23:            **D**$[i, -1] \leftarrow -$**D**$[i, 1]$.
24:        **else if** (**D**$[i, -1] \neq -\infty$) **then**
25:            **D**$[i, 1] \leftarrow -$**D**$[i, -1]$.
26:        **else**   ▷ $-\infty \leq x_i \leq \infty$.
27:            **D**$[i, 1] \leftarrow$ **D**$[i, -1] \leftarrow 0$.
28:        **y**$[i] \leftarrow$ **D**$[i, 1]$.
29:        **for** (each variable x_j) **do**
30:            **for** ($l = 0 \ldots n$) **do**
31:                UPDATE-BOUNDS-LIN(**D**,**B***$[i, j, \pm 1, \pm 1, l]$).
32:    **return y.**

Algorithm 4.3: Determine if a BCS **B** is linear feasible.

Example 4. Let **B** the BCS consisting of the constraints $l_1 : x_1 + 2 \cdot x_2 \geq 1$, $l_2 : -x_2 + 2 \cdot x_3 \geq 1$, $l_3 : 2 \cdot x_4 - x_1 \geq 2$, $l_4 : -x_1 \geq -2$, and $l_5 : x_1 \geq 1$. Initially, we have **B***$[1, 2, 1, 1, 1] = 1$, **B***$[2, 3, -1, 1, 1] = 1$, **B***$[1, 4, -1, 1, 1] = 2$, **D**$[1, -1] = -2$, and **D**$[1, 1] = 1$.

First, we process the constraint l_1. l_1 can be used together with the constraint l_2, the constraint l_3, or the constraint l_4 to derive a new constraint. When used with the constraint l_2, the constraint derived by the FM Rule is $l_6 : x_1 + 4 \cdot x_3 \geq 3$. This constraint is stored by setting $\mathbf{B}^*[1, 3, 1, 1, 2] = 3$. When used with the constraint l_3, the constraint derived by the FM Rule is $l_7 : 2 \cdot x_2 + 2 \cdot x_4 \geq 3$. This constraint is stored by setting $\mathbf{B}^*[2, 4, 1, 1, 0] = 2$. When used with the constraint l_4, the constraint derived by The FM Rule is $l_8 : 2 \cdot x_2 \geq -1$. This constraint is stored by setting $\mathbf{D}[2, 1] = -\frac{1}{2}$.

Next, we process the constraint l_2. l_2 can be used together with the constraint l_1 or the constraint l_8 to derive a new constraint. When used with the constraint l_1, the constraint derived by the FM Rule is l_6 which was already generated. When used with the constraint l_8, the constraint derived by Algorithm 4.2 is $l_9 : 4 \cdot x_3 \geq 1$. This constraint is stored by setting $\mathbf{D}[3, 1] = \frac{1}{4}$.

Next, we process the constraint l_3. l_3 can be used together with the constraint l_1, the constraint l_5, or the constraint l_6 to derive a new constraint. When used with the constraint l_1, the constraint derived by the FM Rule is l_7 which was already generated. When used with the constraint l_5, the constraint derived by Algorithm 4.2 is $l_{10} : 2 \cdot x_4 \geq 3$. This constraint is stored by setting $\mathbf{D}[4, 1] = \frac{3}{2}$. When used with the constraint l_6, the constraint derived by the FM Rule is $l_{11} : 2 \cdot x_4 + 4 \cdot x_3 \geq 5$. This constraint is stored by setting $\mathbf{B}^*[4, 3, 1, 1, 1] = 5$.

When constructing the solution, we set $x_1 = \mathbf{D}[1, 1] = 1$. At this point, l_1 is used to set $\mathbf{D}[2, 1] = 0$ and l_6 is used to set $\mathbf{D}[3, 1] = \frac{1}{2}$. Then, we set $x_2 = \mathbf{D}[2, 1] = 0$. At this point, no further updates to $\mathbf{D}$ are made. Then, we set $x_3 = \mathbf{D}[3, 1] = \frac{1}{2}$. At this point, no further updates to $\mathbf{D}$ are made. Finally, we set $x_4 = \mathbf{D}[4, 1] = \frac{3}{2}$. This gives a solution of $\mathbf{x} = (1, 0, \frac{1}{2}, \frac{3}{2})$. It is easy to see that this solution satisfies every constraint.

4.1   Analysis

Note that the the runtime of Algorithm 4.3 is dominated by the **for** loop between Line 9 and Line 18. This loop iterates over all m constraints, all n variables, and all n possible values of l. In each iteration, a fixed number of constant time operations are performed. Thus, the **for** loop on Line 9 of Algorithm 4.3 runs in time $O(m \cdot n^2)$. Consequently, Algorithm 4.3 runs in time $O(m \cdot n^2)$.

The linear solution $\mathbf{y}$ is constructed by the **for** loop between Line 19 and Line 31. This loop assigns each x_i a valid value and then updates the bounds of the remaining variables based on that assignment. To update the bounds, each constraint in $\mathbf{B}^*$ that uses x_i is processed. There are $O(n^2)$ such constraints. Thus, producing $\mathbf{y}$ takes $O(n^3)$ time. We now prove that Algorithm 4.3 always returns a valid linear assignment to BCS $\mathbf{B}$ if one exists.

Theorem 6. *Let $\mathbf{B}$ be a BCS. Algorithm 4.3 returns a valid linear assignment to $\mathbf{B}$, if and only if $\mathbf{B}$ is linear feasible.*

Proof. First assume that Algorithm 4.3 returns **false**. Thus, there is a variable x_i such that $\mathbf{D}[i, 1] > -\mathbf{D}[i, -1]$. From Theorem 5, the constraints $x_i \geq \mathbf{D}[i, 1]$ and

$-x_i \geq \mathbf{D}[i, -1]$ are both derivable from $\mathbf{B}$. Thus, $x_i \geq \mathbf{D}[i, 1] > -\mathbf{D}[i, -1] \geq x_i$. This is a contradiction. Consequently, $\mathbf{B}$ is infeasible.

Now assume that Algorithm 4.3 returns a vector $\mathbf{y}$. Assume that there is a constraint B_k that is violated by $\mathbf{y}$. From Theorems 3, 4, and 5, when the **for** loop on Line 9 processes constraint B_k the bounds in $\mathbf{D}$ and $\mathbf{B}^*$ are adjusted so that this cannot happen. Thus, the constraint B_k is satisfied by $\mathbf{y}$. Consequently, $\mathbf{B}$ is feasible. □

5   Conclusion

In this paper, we introduced the BTVPI constraint system and studied the linear feasibility problem in the same. As mentioned before, BTVPI constraint systems find applicability in a wide range of domains, including program verification [22] and operations research [17]. It follows that there is a need for efficient algorithms which exploit their structure. We designed and analyzed a combinatorial algorithm for the same. The running time of our algorithm improves on the heretofore best running time for this problem [6,14]. Additionally, our algorithms are simple and based on well-understood refutation rules. In fact our algorithm uses a single refutation rule, i.e., the ADD rule and its variants. We plan on implementing the algorithm discussed in this paper and contrasting its performance with a general-purpose solvers such as CPLEX [8]. It would be interesting to see if our specialized algorithm outperforms an optimized solver.

References

1. Ahuja, R.K., Magnanti, T.L., Orlin, J.B.: Network Flows: Theory. Prentice-Hall, Algorithms and Applications (1993)
2. Àlvarez, C., Greenlaw, R.: A compendium of problems complete for symmetric logarithmic space. Electronic Colloquium on Computational Complexity (ECCC), vol. 3, no. 39 (1996). https://doi.org/10.1007/PL00001603
3. Aspvall, B., Shiloach, Y.: A fast algorithm for solving systems of linear equations with two variables per equation. Linear Algebra Appl. **34**, 117–124 (1980)
4. Chandrasekaran, R., Subramani, K.: A combinatorial algorithm for Horn programs. Discret. Optim. **10**, 85–101 (2013)
5. Chandru, V., Hooker, J.N.: Optimization methods for logical inference. Series in Discrete Mathematics and Optimization. John Wiley & Sons Inc. (1999)
6. Cohen, E., Meggido, N.: Improved algorithms for linear inequalities with two variables per inequality. SIAM J. Comput. **23**(6), 1313–1347 (1994)
7. Cormen, T.H., Leiserson, C.E., Rivest, R.L., Stein, C.: Introduction to Algorithms, 3rd edn. The MIT Press, Cambridge, MA (2009)
8. CPLEX Manual. IBM ILOG CPLEX Optimization Studio V12.8.0 Documentation (2017)
9. Dantzig, G.B.: Linear Programming and Extensions. Princeton University Press, Princeton, NJ (1963)
10. Dantzig, G.B., Eaves, B.C.: Fourier-motzkin elimination and its dual. J. Comb. Theor. (A) **14**, 288–297 (1973)

11. Fiduccia, C.M., Mattheyses, R.M.: A linear time heuristics for improving network partitions. In: Proceedings of the 19th Design Automation Conference, pp. 175–181 (1982)
12. Gallo, G., Pallottino, S.: Shortest path algorithms. Ann. Oper. Res. **13**, 1–79 (1988)
13. Grötschel, M., Lovász, L., Schrijver, A.: The ellipsoid method and its consequences in combinatorial optimization. Combinatorica **1**(2), 169–197 (1981)
14. Hochbaum, D.S., (Seffi) Naor, J.: Simple and fast algorithms for linear and integer programs with two variables per inequality. SIAM J. Comput. **23**(6), 1179–1192 (1994)
15. Karmarkar, N.K.: A new polynomial-time algorithm for linear programming. In: Proceedings of the 16th Annual ACM Symposium on Theory of Computing, pp. 302–311 (1984)
16. Lahiri, S.K., Musuvathi, M.: An efficient decision procedure for UTVPI constraints. In: Gramlich, B. (ed.) FroCoS 2005. LNCS (LNAI), vol. 3717, pp. 168–183. Springer, Heidelberg (2005). https://doi.org/10.1007/11559306_9
17. Luyo, L.E.F., Agra, A., Figueiredo, R., Anaya, E.O.: Mixed integer formulations for a routing problem with information collection in wireless networks. Eur. J. Oper. Res. **280**(2), 621–638 (2020)
18. Miné, A.: The octagon abstract domain. Higher-Order Symbolic Comput. **19**(1), 31–100 (2006)
19. Nelson, C.G.: An $n^o(\log n)$ algorithm for the two-variable-per-constraint linear program satisfiability problem. Technical Report Technical Note STA, Stanford University, Computer Science Department, pp. 78–689 (1978)
20. Gurobi Optimization. Gurobi optimizer reference manual4. http://www.gurobi.com (2014)
21. Revesz, P.Z.: Tightened transitive closure of integer addition constraints. In: Symposium on Abstraction, Reformulation, and Approximation (SARA), pp. 136–142 (2009)
22. Singh, G., Gehr, T., Püschel, M., Vechev, M.T.: An abstract domain for certifying neural networks. PACMPL **3**(POPL), 1–30 (2019)
23. Subramani, K., Wojciechowski, P.J.: A combinatorial certifying algorithm for linear feasibility in UTVPI constraints. Algorithmica **78**(1), 166–208 (2017)
24. Tardos, E.: A strongly polynomial algorithm to solve combinatorial linear programs. Oper. Res. **34**(2), 250–256 (1986)
25. Veinott, A.F., Dantzig, G.B.: Integral extreme points. SIAM Rev. **10**, 371–372 (1968)
26. Veinott, A.F., LiCalzi, M.: Subextremal functions and lattice programming, unpublished manuscript (1992)
27. Vollmer, H.: Introduction to Circuit Complexity. Springer (1999). https://doi.org/10.1007/978-3-662-03927-4

Quantum Complexity for Vector Domination Problem

Andris Ambainis and Ansis Zvirbulis[✉]

Center for Quantum Computer Science, Faculty of Computing,
University of Latvia, Riga, Latvia
`ansis.z@inbox.lv`

Abstract. In this paper we investigate quantum query complexity of two vector problems: *vector domination* and *minimum inner product*. We believe that these problems are interesting because they are closely related to more complex 1-dimensional dynamic programming problems. For the general case, the quantum complexity of vector domination is $\Theta(n^{1-o(1)})$, similarly to the more known *orthogonal vector* problem (OV). We prove a $\tilde{O}(n^{2/3})$ upper bound and a $\Omega(n^{2/3})$ lower bound for special case of *vector domination* where vectors are from $\{1, \ldots, W\}^d$ and number of dimensions d is a constant and $W \in O(poly\ n)$. We also prove a $\Omega(n^{2/3})$ lower bound for *minimum inner product* with the same constraints. To prove bounds we use reductions from the *element distinctness* problem as well as a classical data structure - *Fenwick trees*.

Keywords: Quantum query · Quantum query complexity · LWS · Element distinctness · QSETH · Fenwick trees

1 Introduction

Dynamic programming is one of among the most widely used methods for algorithm design. It has been used to solve many well known problems. For example, Dijkstra's algorithm for finding shortest paths in graphs, the $O(n^2)$ algorithm for edit distance [17,21], the fastest exponential time algorithm for the travelling salesman problem by Held and Karp [11] are all based on dynamic programming and there are many more applications of it.

Given the widespread applicability of dynamic programming, it is natural to ask whether it can be combined with quantum effects to produce quantum speedups for problems whose best classical solutions involve dynamic programming. Up to now, this question has been considered for two types of dynamic programming algorithms.

First, it was studied for exponential-time algorithms that use dynamic programming over subsets of an n-element set. The most famous algorithm of this type is the Held-Karp algorithm for Travelling Salesman problem (TSP) [11] which solves it in time $O(2^n n^2)$ and is still the fastest classical algorithm for this problem. This algorithm can be combined with quantum search, resulting in a quantum algorithm that runs in time $O(1.728...^n)$ [2]. This was followed by

L. Gąsieniec (Ed.): SOFSEM 2023, LNCS 13878, pp. 328–341, 2023.
https://doi.org/10.1007/978-3-031-23101-8_22

other quantum speedups for exponential-time dynamic programming algorithms based on similar combinatorial structures [15, 20].

Second, possible quantum speedups for dynamic programming have been studied in the context of the Edit Distance problem whose best classical algorithm uses dynamic programming over a 2-dimensional array. Classically, the best algorithm for this problem runs in time $O(n^2)$. Despite a considerable effort, no quantum speedup has been discovered. An $\tilde{O}(n^{1.5})$ quantum lower bound has been shown, either in the quantum query model for an abstract version of the problem [3] or in the time complexity model under a plausible complexity assumption [6].

Another class of dynamic programming algorithms is based on computing values for elements of a 1-dimensional array [16]. A prototypical example of a problem that can be solved in this way is the Least Weight Subsequence (LWS) problem. In this problem, one is given a sequence of n items together with weights for every pair of items and has to determine a subsequence S which minimizes the sum of weights of pairs that are adjacent in S. As described in [16], this problem can serve as a model for a variety of situations. Classically, various cases of LWS are subquadratically equivalent to much simpler problems, such as Orthogonal Vectors (OV), Vector Domination (VD) and Minimum Inner Product (MIP), called *core problems* in [16].

The goal of this paper is to understand the complexity of these core problems in the quantum setting, with the perspective of finding quantum lower and upper bounds for problems that are classically solved via dynamic problem.

Orthogonal vector problem (OV) is rather well known and researched from the perspective of both classical and quantum computing [1]. In this problem we are given 2 sets of the same size A, B of d-dimensional vectors with binary coordinates and we must find whether there are 2 vectors perpendicular to each other, where one belongs to set A and the other - to set B.

The main focus of this paper is another similar problem, called Vector domination problem (VD) and is defined as follows: we are given 2 sets of the same size A, B of d-dimensional vectors, with integer coordinates from $\{-W, ..., W\}$. We must find whether there are 2 vectors $u \in A, v \in B$ such that $u \leq v \iff \forall i : u[i] \leq v[i]$. We can notice that the problem OV is actually a special case of vector domination problem, because with binary coordinates: $\langle u|v \rangle = 0 \Leftrightarrow u \leq \bar{v}$. Vector $\bar{v}$ is vector v with inverted coordinates.

Natural question might arise - why this problem is interesting to look at? One reason is that although general case is as at least as complex as OV problem, this problem has more special cases than OV and we can prove some new things about them. Another reason is that this problem is core problem for *nested boxes problem* [16]. Finding faster algorithm than currently known for VD will immediately provide faster algorithm for *nested boxes* problem. At the moment, we have not seen or found any faster quantum algorithm for this problem than $O(n^{1.5})$ - there are trivial and more complex algorithms, however both yield the same complexity. [24]

In this paper we will also look into problem *minimum inner product* (MIP) [16]. This problem is also somewhat similar - we are given 2 sets of the same

A, B size of d-dimensional vectors, with integer coordinates from $\{-W, ..., W\}$ and integer r. We must find whether there are 2 vectors $a \in A, b \in B$, where $\langle a|b \rangle \le r$.

This problem is also interesting because of the same reason - it is also a core problem of the dynamic programming problem *LowRankLWS* [16].

1.1 Prior Work

All three problems - OV, VD and MIP have been researched quite a lot from the perspective of classical computing, but at the moment we have found quantum results only for the OV problem.

Problem OV was researched in [1] where author proved conditional lower bounds for OV under *QSETH*. Strangely enough, for general case of this problem, there have not been found better quantum algorithm than just running Grover's search over all pairs of vectors. And if *QSETH* is true then it also is the best algorithm.

Results of [1] for the OV problem can be seen in the following table:

Dimension	Algorithm	Lower bounds	Upper bounds
$\Theta(1)$	Classical	$\Omega(n)$	$O(n)$
	Quantum	$\Omega(\sqrt{n})$	$O(\sqrt{n})$
$poly \log n$	Classical	$n^{2-o(1)}$ under SETH	$n^{2-o(1)}$
	Quantum	$n^{1-o(1)}$ under QSETH	$\tilde{O}(n)$

1.2 Our Results

Our main results are for special cases of the vector domination problem - we prove quantum lower and upper bounds for some of these special cases. We either use already proven bounds for OV or use results for *element distinctness* (ED) problem and use Radix and Fenwick tree classical data structures. We also prove quantum lower bounds for special case of MIP where number of dimensions is constant. We prove that to solve MIP in such case at least $\Omega(n^{2/3})$ queries will be necessary (*Th. 5*). Results for VD can be seen in the following table.

Constraints	Lower Bounds	Upper Bounds
$d \in O(poly \log n)$	$n^{1-o(1)}$ under QSETH *(Th. 2)*	$\tilde{O}(n)$ *(Th. 2)*
$d \in \Theta(1), W \in O(poly\, n)$	$\Omega(n^{\frac{2}{3}})$ *(Th. 3)*	$\tilde{O}(n^{\frac{2}{3}})$ *(Th. 4)*
$d = c \log n, W \le 4$	$\Omega(n^{\frac{2}{3}})$ *(Th. 3)*	$\tilde{O}(n^{\frac{2}{3}+c})$ *(Th. 4)*
$d, W \in \Theta(1)$	$\Omega(\sqrt{n})$ *(sec. 3.2)*	$O(\sqrt{n})$ *(sec. 3.2)*

2 Preliminaries

2.1 Problem Definitions

In this section we define problems considered in this paper in a more formal way.

Definition 1 (Orthogonal vectors, OV). *Given two sets A, B of n vectors in $\{0,1\}^d$, find whether there are 2 vectors $a \in A$ and $b \in B$ such that $\langle a|b \rangle = 0$.*

Definition 2 (Vector domination, VD). *Given two sets A, B of n vectors in $\{-W, \ldots, W\}^d \subset \mathbb{Z}^d$, find whether there are two vectors $a \in A$ and $b \in B$ such that $a \leq b$. $a \leq b \iff \forall i : a[i] \leq b[i]$, where with $u[i]$ we denote coordinate i of vector u.*

Definition 3 (Minimum inner product, MIP). *Given two sets A, B of n vectors in $\{-W, \ldots, W\}^d \subset \mathbb{Z}^d$ and $r \in \mathbb{Z}$ find whether there are 2 vectors $a \in A$ and $b \in B$ such that $\langle a|b \rangle \leq r$.*

2.2 Quantum Query Model

We consider the quantum query model in this work [4]. Main parts of this model are a quantum register with finite number of base states (typically depends on the size of input), a quantum query transformation Q also called *oracle* and unitary operations $U_0, U_1, \ldots, U_t$.

In the beginning the quantum register is in a starting state $|\psi_{start}\rangle$. Then query transformations and unitary operations are applied to the starting state in an alternating way. In the end we get final state $|\psi_{final}\rangle$. If algorithm contains T query transformations, its final state is

$$|\psi_{final}\rangle = U_T Q U_{T-1} Q \ldots U_1 Q U_0 |\psi_{start}\rangle$$

In the end we measure the final state. The starting state and the unitary transformations U_i do not depend on input data - only on its length.

Oracle Q. Oracle is one of the most important components of the model. It is used to change quantum state depending on the input $(x_1, \ldots, x_N)$. In general case we can define it in following way:

$$Q|i\rangle|a\rangle = |i\rangle|a \oplus x_i\rangle$$

where typically $i \in \{1, 2, \ldots, N\}$ is index of input variable x_i. Here, N denotes the length of the input.

We say that algorithm solves a particular problem if it gives correct answer on all possible inputs with probability at least $\frac{2}{3}$. When we calculate complexity of the algorithm we count the number of oracle calls.

2.3   Techniques

In this section we describe techniques used in our proofs.

QSETH. QSETH is quantum version of classical SETH (*strong exponential time hypothesis*). It is used to prove conditional lower bounds for OV problem in [1] and we use it when we reduce VD to OV.

Hypothesis 1. *For every $\epsilon > 0$ there exists $k = k(\epsilon) \in \mathbb{N}$ such that there is no quantum algorithm which could solve the k-SAT problem in time $O(2^{(\frac{1}{2}-\epsilon)n})$, where n is the number of variables in the formula. Additionally this holds true even if the number of clauses is no more than $c(\epsilon)n$ where $c(\epsilon)$ denotes some constant, which depends entirely on value of ϵ.*

Algorithm for Element Distinctness Problem . This algorithm allows us to find out whether there are at least 2 equal elements in the input or more generally - whether there are at least 2 elements with some common property or relation. This generalization is called 2-Subset Finding or Claw finding when relation is equality and the first of the two elements has to be from the first half of the input and the second element from the second half of the input.

Theorem 1. *Our input contains 2 lists $A, B \subseteq \{1, 2, \ldots, n\}$, where $|A| = |B| = n$ and we have a function $f : A \times B \longrightarrow \{0, 1\}$. We need to find out whether there are 2 different input numbers $a \in A, b \in B$ such that $f(a, b) = 1$. There exists a quantum algorithm that solves this problem with $O(n^{2/3})$ queries and this is optimal - there is also a $\Omega(n^{2/3})$ lower bound on the number of queries. ([7, 19, 23]).*

In this paper, we will also be concerned about the time complexity of quantum algorithms. The element distinctness algorithm uses a quantum memory that stores values a_i, $i \in S$ for a set $S : |S| \in \{m, m + 1\}$, $m = O(n^{2/3})$. The algorithm uses the following resources:

1. $O(n^{2/3})$ queries to oracle $(x_1, ..., x_n)$;
2. $O(n^{2/3})$ update operations that add or remove an element from a data structure storing the set S and values a_i, $i \in S$;
3. $O(n^{2/3})$ steps of a quantum walk, each of which can be implemented in time $O(\log n)$ in the QRAM model with an appropriate data structure storing the set S;
4. $O(n^{1/3})$ checking operations that check whether the current set S contains $x, y : f(x, y) = 1$.

In the case of element distinctness (when $f(x, y) = 1$ if $x = y$), there is a data structure for which both update operations and checking operations can be performed in time $O(\log^c n)$ in the QRAM model of computation. This leads to an implementation of the whole quantum algorithm in time $\tilde{O}(n^{2/3})$.

For other choices of $f(x, y)$, to construct a time-efficient implementation of the quantum algorithm, a data structure with similar properties needs to be provided. We note that one can reuse the implementation for the steps of the quantum walk because this part does not depend on the choice of $f(x, y)$. Thus, one only needs to describe an additional data structure which ensures that the checking operations can be performed efficiently.

To be usable within a quantum algorithm, the data structure must be *history-independent*. If S (and values $x_i, i \in S$) is stored as a quantum state $|\psi_S\rangle$, then $|\psi_S\rangle$ must only depend on S and not on the order in which elements have been added and removed to S.

Fenwick Tree. Fenwick tree also called BIT (binary indexed tree) is a simple yet powerful classical data structure, which allows to quickly find prefix sums and max/min values in an array ([8,18]). This data structure will turn out to be very important when proving upper bounds for the vector domination problem.

In the simplest version we are given an array $T[x], x \in [W]$ with natural numbers and we want the following:

(a) Find $T[1] + T[2] + \cdots + T[i], 1 \leq i \leq W$
(b) Change value $T[i]$ by δ.

With Fenwick trees we can accomplish these operations in $O(\log W)$ time. We can also generalize Fenwick trees to d dimensional data. We are given d-dimensional array $T[x_1, x_2, \ldots, x_d], x_1, \ldots, x_d \in [W]$ and we want very similar operations:

(a) Find $\sum_{x_1, x_2, \ldots, x_d} T[x_1, x_2, \ldots, x_d]$, where $\forall i : 1 \leq x_i < a_i$, where $a_1, \ldots, a_d$ are given. Sometimes it is convenient to denote this sum sum as $T[(1, \ldots, 1) : (a_1, \ldots, a_d)]$.
(b) Change value $T[a_1, a_2, \ldots, a_d]$ by some δ.

In this d-dimensional case we can execute these operations in $O(\log^d W)$ time.

A simple implementation of a Fenwick tree would require $O(W^d)$ memory cells, with $O(\log^d W)$ cells accessed in each update or lookup. If the number of performed updates is bounded by $T << W^d$, we can use a dictionary of size $S = O(T \log^d W)$ to simulate the Fenwick tree. Namely, we use dictionary to store the memory cells of the Fenwick tree that have been set to a value that is different from its initial value. To read the value from a memory cell, we first use the dictionary to find it. If a memory cell is not in the dictionary, we conclude that it has not been changed away from its initial value.

If dictionary is implemented efficiently, $O(\log S)$ time will be sufficient to access a cell, where S is number of keys in dictionary. Then, each update or lookup for the Fenwick tree takes $O(\log^d W \log S)$ time, as it involves accessing $O(\log^d W)$ memory cells.

If we use Fenwick tree as a part of a quantum algorithm (for example, to implement the element distinctness algorithm), there is a very important detail - the implementation must be history independent, that is, it must be stored in the memory in a way that depends only on the current data in the set and not on any operations done previously. This causes an issue - how to implement a dictionary. Many popular implementations of efficient dictionaries (such as AVL, R&B tree, Treap) are history dependent and, therefore, cannot be used.

We can solve this issue by implementing dictionary with a quantum radix tree [1,5,14] - a quantum version of the radix tree (a binary tree, where all values are stored in leaves). This data structure achieves history-independence by randomizing memory layout of the tree and using the superposition over all possible memory layouts to store the dictionary. As shown in [1,5,14]), quantum radix tree can be implemented so that elements can be added and deleted from the data structure in $O(l)$ time, where l is maximum length of binary key string. In the cases considered in this paper, $l = d * \log W$, since the key is made of d numbers each from $[W]$.

With this layer, the overall complexity of each operation in a quantum implementation of a Fenwick tree is $O(\log^d W * l) = O(\log^d W * d * \log W) = O(d * \log^{d+1} W)$.

3   Vector Domination

We first analyze the general case of the problem, where $d \in O(poly \log n)$ and coordinates are integers from $\{-W, \ldots, W\}$. We can get results for it directly from previous results for the OV problem.

Theorem 2. *If the input to VD is $a_1, a_2, \ldots, a_n, b_1, b_2, \ldots, b_n \in \{-W, \ldots, W\}^d$ and $d \in O(poly \log n)$, then we can solve this problem in $\tilde{O}(n)$ time. Furthermore, if QSETH is true, there is no quantum algorithm that can solve VD in $O(n^{1-o(1)})$ time.*

Proof. Firstly we can notice, that the orthogonal vector problem is special case of the vector domination problem - VD is equivalent to OV when all coordinates are binary. To be more precise if $a, b, \in \{0, 1\}^d$, then $\langle a|b \rangle = 0 \Leftrightarrow a \leq \bar{b}$, where $\bar{b}$ is vector b with inverted coordinates. Therefore, VD is at least as hard as OV - we get a conditional lower bound of $\Omega(n^{1-o(1)})$ under *QSETH*.

We can also solve VD in the same way OV is solved - we run Grover search ([10]) over all pairs a, b and check whether $a \leq b$, achieving an $\tilde{O}(n)$ upper bound.

Now, let us move on to the special case where the number of dimensions is constant.

3.1   Lower Bounds

In this section we show $\Omega(n^{2/3})$ lower bounds by making a reduction from the element distinctness problem.

Theorem 3. *If VD input is $a_1, a_2, \ldots, a_n, b_1, b_2, \ldots, b_n \in \{1, \ldots, n\}^d$ and $d \in \Theta(1)$, then any quantum algorithm will require at least $\Omega(n^{2/3})$ queries to solve this problem.*

Proof. We consider the Claw Finding problem where we are given two lists A, B and we must find if there are $a \in A, b \in B$ such that $a = b$ [23]). By Theorem 1, solving this problem requires $\Omega(n^{2/3})$ queries, even if all elements are natural numbers from $\{1, \ldots, n\}$.

Let $Q_{CF(k,i)}$ denote an oracle answering queries about an input to Claw Finding (CF). For where $k \in \{1, 2\}$ and $i \in \{1, \ldots, n\}$, $Q_{CF}(k, i)$ returns the ith number from the kth list (the list A if $k = 1$, the list B if $k = 2$).

Then we define oracle for VD: $Q_{VD}(k, i, j)$. k denotes the choice between A and B, $i \in \{1, \ldots, n\}$ - index of the vector and $j \in \{1, 2\}$ coordinate of the vector.

$$Q_{VD}(k, i, j) = \begin{cases} Q_{CF}(k, i) & \text{if } j = 1 \\ n + 1 - Q_{CF}(k, i) & \text{if } j = 2 \end{cases}$$

We claim that if an algorithm solves the VD problem for the oracle input Q_{VD} with t queries, it also solves the CF problem for the input Q_{CF} with the same number of queries.

Let us look at two vectors $a \in A, b \in B$ which correspond to input variables $u \in A, v \in B$ for the CF problem.

$$a = \begin{pmatrix} u \\ n + 1 - u \end{pmatrix}, b = \begin{pmatrix} v \\ n + 1 - v \end{pmatrix}$$

Then there are 3 possible cases:

1) $u < v \implies n + 1 - u > n + 1 - v \implies a \not\leq b \wedge a \not\geq b$
2) $u > v \implies n + 1 - u < n + 1 - v \implies a \not\leq b \wedge a \not\geq b$
3) $u = v \implies a \leq b \wedge a \geq b$

To put it informally - there are 2 comparable vectors if and only if there are 2 equal numbers.

Since CF requires $\Omega(n^{2/3})$ queries and, in the above reduction, one query to VD involves one query to CF, the lower bound for VD is also $\Omega(n^{2/3})$.

3.2   Upper Bounds

In this section we describe the main result of this work - quantum algorithm for a special case of the vector domination problem. The algorithm is based on the element distinctness algorithm. It also uses such classical data structures as radix trees and Fenwick trees. The complexity of this algorithm is $\tilde{O}(n^{\frac{2}{3}} \log^{d+1} W)$. Although the algorithm can be used in general case, its complexity gets worse than trivial Grover search, therefore it is useful only in special cases, e.g. with constant dimensions.

Theorem 4. *If the input to VD is $a_1, a_2, \ldots, a_n, b_1, b_2, \ldots, b_n \in \{1, \ldots, W\}^d$, then we can solve this problem in $\tilde{O}(n^{\frac{2}{3}} \log^{d+1} W)$ quantum time, provided that $d \in O(poly \log n)$.*

Proof. To solve this problem we use the element distinctness algorithm *(Theorem 1)* with an appropriate data structure for the set S and $x_i, i \in S$. As described in Sect. 2.3, we can reuse the part of the element distinctness algorithm that implements the quantum walk. Thus, we only need to provide a data structure that provides fast updates to S and fast checking if S contains $x, y : f(x, y) = 1$.

We say that there is an edge $u \to v \Leftrightarrow u \leq v$. To implement S we define two d-dimensional Fenwick trees that will be stored using dictionaries (see Sect. 2.3):

- $F_{in} \underbrace{[W, \ldots, W]}_{d}$ - to find the number of incoming edges to vector u

- $F_{out} \underbrace{[W, \ldots, W]}_{d}$ - to find the number of outgoing edges from vector u.

We define $F_{in}[u] = F_{out}[w + 1 - u] = $ *number of vectors* u *in* S, where $w + 1 = \begin{pmatrix} W+1 \\ \vdots \\ W+1 \end{pmatrix}$. Then:

- the number of edges going into u is equal to the sum:

$$F_{in}[(1, \ldots, 1) : (x_1, \ldots, x_d)]$$

- the number of edges going out of u is equal to the sum:

$$F_{out}[(1, \ldots, 1) : (W + 1 - x_1, \ldots, W + 1 - x_d)]$$

If $v \leq u$, then it will be counted in the F_{in} interval, else if $v \geq u$ it will be counted in the F_{out} interval, because $v \geq u \Leftrightarrow w + 1 - u \geq w + 1 - v$. In other cases the vector will not be counted. Additionally we create counter *edgeCnt* in which we store number of edges or comparable pairs of vectors.

Addition of Vector u to the Set

1) Find the number of outgoing edges (including edges $u \to u$):

$$outCnt = F_{out}[(1, \ldots, 1) : (W + 1 - x_1, \ldots, W + 1 - x_d)]$$

2) Find the number of incoming edges (not including edges $u \to u$ because those were already counted in the previous step):

$$inCnt^* = F_{in}[(1, \ldots, 1) : (x_1 + 1, \ldots, x_i, \ldots, x_d)] + \cdots +$$
$$+ F_{in}[(1, \ldots, 1) : (x_1, \ldots, x_i + 1, \ldots, x_d)] + \cdots +$$
$$+ F_{in}[(1, \ldots, 1) : (x_1, \ldots, x_i, \ldots, x_d + 1)].$$

3) $edgeCnt = edgeCnt + outCnt + inCnt^*$
4) $F_{in}[(x_1, \ldots, x_d)] = F_{in}[(x_1, \ldots, x_d)] + 1$
5) $F_{out}[(W + 1 - x_1, \ldots, W + 1 - x_d)] = F_{out}[(W + 1 - x_1, \ldots, W + 1 - x_d)] + 1$

Deletion of Vector u from the Set

1) Find the number of outgoing edges (including edges $u \to u$):

$$outCnt = F_{out}[(1,\ldots,1) : (W+1-x_1,\ldots,W+1-x_d)] - 1$$

We need to subtract 1, to ignore the edge from itself.

2) Find the number of incoming edges (not including edges $u \to u$ because those were already counted in the previous step):

$$inCnt^* = F_{in}[(1,\ldots,1) : (x_1+1,\ldots,x_i,\ldots,x_d)] + \cdots +$$
$$+ F_{in}[(1,\ldots,1) : (x_1,\ldots,x_i+1,\ldots,x_d)] + \cdots +$$
$$+ F_{in}[(1,\ldots,1) : (x_1,\ldots,x_i,\ldots,x_d+1)]$$

(the same as in addition of vector).

3) $edgeCnt = edgeCnt - outCnt - inCnt^*$
4) $F_{in}[(x_1,\ldots,x_d)] = F_{in}[(x_1,\ldots,x_d)] - 1$
5) $F_{out}[(W+1-x_1,\ldots,W+1-x_d)] = F_{out}[(W+1-x_1,\ldots,W+1-x_d)] - 1$

Time Complexity. The time for each addition or deletion is $O(d*\log^{d+1} W)$, as described in Sect. 2.3. The time for checking if the current set S contains an edge can be done in constant time, by examining $edgeCnt$. Since element distinctness algorithm involves $O(n^{2/3})$ updates and $O(n^{1/3})$ checking operations, the overall complexity is $\tilde{O}(n^{2/3} \log^{d+1} W)$.

Some More Special Cases. In the following special cases, the complexity bound from Theorem 4 is close to $O(n^{2/3})$:

1. If $d \in \Theta(1), W \in O(poly\ n)$, we have $\tilde{O}(n^{2/3} * \log^{d+1} W) = \tilde{O}(n^{2/3})$.
2. If $d = c * \log n, W \le 4$, then we have $\tilde{O}(n^{2/3} * \log^{d+1} W) = \tilde{O}(n^{\frac{2}{3}+c})$.

Constant Number of Dimensions and Values. If vectors are from the set $\{1, 2, \ldots, W\}^d$ where $d, W \in O(1)$ we can generalize OV algorithm to get optimal algorithm:

1) Generate all possible vectors and put them into the set $V, |V| = W^d$.
2) $\forall a, b \in V$: if $a \le b$ search with Grover's algorithm vector a in A and b in B. If both are found - we can give the answer yes.
3) If no comparable vectors were found, give the answer no.

Time complexity of the algorithm is $O(W^d) + O(W^d * 2\sqrt{n}) = O(\sqrt{n})$, which is in fact optimal (follows from Grover's search algorithm [10]).

4   Minimum Inner Product

Firstly we must note, that orthogonal vector problem is a special case of this problem as well - if vectors are from $\{0,1\}^d$ and $r = 0$ MIP is equivalent to OV. Secondly, there does not seem to be a simple relationship between MIP and VD, therefore we cannot reuse VD results.

Theorem 5. *Solving MIP with constant number of dimensions ($d \geq 4$) requires $\Omega(n^{2/3})$ queries.*

Proof. The proof is via a reduction from the Claw finding variant of the element distinctness problem (Theorem 1):

- We are given 2 arrays $a_1, \ldots, a_n$ and $b_1, \ldots, b_n$ with n number each. Each number is from $\{1, \ldots, n\} = [n]$.
- We must find if there are i, j such that $a_i = b_j$.

Usually the numbers would be from larger set $[M]$, but it has been proven that this version is just as hard - the lower bound for this problem is still $\Omega(n^{2/3})$ [19].

4.1   Idea

We show a reduction from ED to MIP with 4 dimensional vectors. For this reduction we need to do some preprocessing. First we choose n vectors from $\{1, 2, \ldots, n\}^4$, with all n vectors being of the same length l. We then define

$1 : 1$ correspondence $f : [n] \rightarrow \begin{pmatrix} u_1 \\ u_2 \\ u_3 \\ u_4 \end{pmatrix}$ between $[n]$ and the chosen vectors. This

preprocessing will depend only on the length of the input.

Then in the reduction, for each ED variable a_i we take the vector $f(\vec{a_i})$ and for each b_j, we take the vector $-f(\vec{b_j})$. Then all inner products of these vectors will be in the interval $[-l^2; 0)$ and, for any i, j, $\left\langle f(\vec{a_i})| - f(\vec{b_j}) \right\rangle = -l^2 \Leftrightarrow a_i = b_j$. Next we show how to construct such n vectors that are of length l for this correspondence f.

4.2   Preprocessing

We use the four square theorem [12]:

$$r_4(S) = 8 * \sum_{m \mid S} m, S \text{ is odd}$$

Here $r_4(S)$ denotes number of ways it is possible to write natural number S as sum of 4 integer squares. Order is important - $1 + 0 + 0 + 0$ and $0 + 0 + 1 + 0$ are 2 different ways how to write 1.

We note that $r_4(S)$ above includes quadruples $(x, y, z, t) : x^2 + y^2 + z^2 + t^2 = S$ in which some of x, y, z, t are negative. For our purposes, we would like to use only (x, y, z, t) in which all numbers are non-negative.

To count the number of such (x, y, z, t), we note that, from each (x, y, z, t) with non-negative elements, we can obtain at most 15 (x, y, z, t) in which some elements are negative by changing signs of some of x, y, z, t (possibly less, because $0 = -0$). Then we can claim, that $r_4^+(S) \geq \frac{r_4(S)}{16}$, where r_4^+ counts only non-negative integer square sums.

We can notice that $r_4(S) \geq S$. This implies $r_4^+(S) \geq \frac{r_4(S)}{16} \geq \frac{8S}{16} = \frac{S}{2}$. If we choose $S = 2n + 1$:

$$r_4^+(2n + 1) \geq \left\lceil \frac{2n + 1}{2} \right\rceil = n$$

Then if we choose $l = \sqrt{S} = \sqrt{2n + 1}$ there will be at least n vectors with length l and natural number coordinates. Such vectors can be found with the following algorithm:

1) Generate and save all $i, j \in [l]$ and also sum of their squares. The number of such pairs are l^2. Then sort these pairs by their sum of squares.
2) Traverse all saved sums. For each sum $i^2 + j^2 = s$: search for a square sum $2n + 1 - s = x^2 + y^2$.
   (a) If n vectors already found - exit.
   (b) If found - save the vector in the array f: $\begin{pmatrix} i \\ j \\ x \\ y \end{pmatrix}$.
   (c) If not - move on to the next sum

Time complexity of the algorithm is $O(l^2 \log l^2) = O(n \log n)$. This is preparation algorithm that depends only on the length of problem's input. It does not make any queries - there is no effect on the query complexity.

4.3   Reduction

We assume we are given ED oracle $Q_{ED}(k, i)$ that returns a_i if $k = 1$ and b_i if $k = 2$ (the same as in VD). We define MIP oracle $Q_{MIP}(k, i, j)$, where

- k denotes whether the query is to A or B;
- i denotes the index of the queried vector from A or B;
- j denotes the queried coordinate.

$Q_{MIP}(k, i, j)$ can be calculated by a following procedure that uses 1 query to Q_{ED}:

1) $ind = Q_{ED}(k, i)$
2) $u = f(\vec{ind})$

3) Return $\begin{cases} u_j, & \text{if } k = 1 \\ -u_j, & \text{else} \end{cases}$

Then we can solve ED with 4-dimensional MIP $\implies$ if $d \geq 4$ MIP lower bounds are $\Omega(n^{2/3})$.

5 Open Questions

The results in this paper help to understand the complexity of VD and MIP problems in the quantum case but there are many questions that remain open. Firstly, we did not find better than trivial linear upper bounds for MIP special case. It seems that Fenwick trees do not help here, but maybe we can use something similar to technique used for *closest pair* problem in [1]. It seems reasonable to believe that special case can be solved faster.

Secondly, it might be interesting to look at other more general special cases of VD and find whether anything interesting can be proven. Maybe quantum algorithms can improve the classical algorithm in [13].

And, thirdly, the most ambitious question is: can we solve VD or MIP (and also OV) faster than in linear time and break **QSETH**? However, this is a question, that is unlikely to be answered.

Acknowledgment. This research was supported by ESF project 1.1.1.5/18/A/020 "Quantum algorithms: from complexity theory to experiment".

References

1. Aaronson, S., Chia, N.H., Lin, H.H., Wang, C., Zhang., R.: On the quantum complexity of closest pair and related problems. In: Proceedings of the 35th Computational Complexity Conference (CCC 2020), vol. 16, pp. 1–43 (2020)
2. Ambainis A., Balodis K., Iraids J., Kokainis M., Prusis K., Vihrovs J.: Quantum speedups for exponential-time dynamic programming algorithms. In: Proceedings of the Thirtieth Annual ACM-SIAM Symposium on Discrete Algorithms, pp. 1783–1793 (2019) Society for Industrial and Applied Mathematics. https://doi.org/10.1137/1.9781611975482.107
3. Ambainis A., et al.: Quantum lower and upper bounds for 2D-grid and dyck language. In: 45th International Symposium on Mathematical Foundations of Computer Science, vol. 170, pp. 1–14. Prague, Czech Republic (2020)
4. Ambainis, A., Yeung, D.: CS860 Quantum algorithms and complexity, lecture 07 (2006)
5. Bernstein, J.D., Jeffery, S., Lange, T., Meurer, A.: Quantum algorithms for the subset-sum problem. In: International Workshop on Post-Quantum Cryptography, pp. 16–33 (2013)
6. Buhrman, H., Patro, S., Speelman, F.: A framework of quantum strong exponential-time hypotheses. In: 38th International Symposium on Theoretical Aspects of Computer Science (STACS 2021). Schloss Dagstuhl-Leibniz-Zentrum für Informatik (2021)
7. Childs, A.: Lecture Notes on Quantum Algorithms University of Maryland (2017)

8. Fenwick, P.M.: A new data structure for cumulative frequency tables. Softw. Pract. Experience **24**(3), 327–336 (1994)
9. Gall F.L., Seddighin, S.: Quantum meets fine-grained complexity: sublinear time quantum algorithms for string problems. In: 13th Innovations in Theoretical Computer Science Conference (ITCS 2022), vol. 215, pp. 1–23 (2022)
10. Grover, L. K.: A fast quantum mechanical algorithm for database search. In: Proceedings of the Twenty-Eighth Annual ACM Symposium on Theory of Computing, pp. 212–219 (1996)
11. Held, M., Karp, R.M.: The traveling-salesman problem and minimum spanning trees. Oper. Res. **18**(6), 1138–1162 (1970)
12. Hirschhorn, M.D., McGowan, J.A.: Algebraic consequences of Jacobi's two-and four-square theorems. In: Symbolic Computation, Number Theory, Special Functions, Physics and Combinatorics, pp. 107–132. Springer, Boston, MA (2001). https://doi.org/10.1007/978-1-4613-0257-5_7
13. Impagliazzo, R., Lovett, S., Paturi, R., Schneider, S.: 0–1 Integer linear programming with a linear number of constraints. Electronic Colloquium on Computational Complexity, p. 24 (2014)
14. Jeffery, S.: Frameworks for quantum algorithms, Ph. D. thesis, University of Waterloo (2014)
15. Klevickis, V., Prusis, K., Vihrovs, J.: Quantum speedups for treewidth. arXiv preprint arXiv:2202.08186 (2022)
16. Künnemann, M., Paturi, R., Schneider, S.: On the fine-grained complexity of one-dimensional dynamic programming. ICALP **80**, 1–15 (2017). https://doi.org/10.4230/LIPIcs.ICALP.2017.21
17. Levenshtein, V.I.: Binary codes capable of correcting deletions, insertions, and reversals. Sov. Phys. Dokl. **10**(8), 707–710 (1966)
18. Mishra, P.: A new algorithm for updating and querying sub-arrays of multidimensional arrays. arXiv preprint arXiv:1311.6093 (2013)
19. Rosmanis, A.: Quantum adversary lower bound for element distinctness with small range. Chic. J. Theor. Comput. Sci. **4**, 2014 (2014)
20. Shimizu, K., Mori, R.: Exponential-time quantum algorithms for graph coloring problems. In: Kohayakawa, Y., Miyazawa, F.K. (eds.) LATIN 2021. LNCS, vol. 12118, pp. 387–398. Springer, Cham (2020). https://doi.org/10.1007/978-3-030-61792-9_31
21. Wagner, R., Fisher, M.: The string-to-string correction problem. J. JACM **21**, 1 (1974)
22. Williams, R.: A new algorithm for optimal 2-constraint satisfaction and its implications. Theoret. Comput. Sci. **348**(2–3), 357–365 (2005)
23. Zhang, S.: Promised and distributed quantum search. In: Wang, L. (ed.) COCOON 2005. LNCS, vol. 3595, pp. 430–439. Springer, Heidelberg (2005). https://doi.org/10.1007/11533719_44
24. Zvirbulis, A.: A quantum query complexity for dynamic programming problems. [Unpublished master's thesis]. Faculty of Computing, University of Latvia (2022)

Learning Through Imitation by Using Formal Verification

Avraham Raviv, Eliya Bronshtein, Or Reginiano, Michelle Aluf-Medina,
and Hillel Kugler$^{(\boxtimes)}$

Faculty of Engineering, Bar-Ilan University, Ramat Gan, Israel
{ravivav1,hillelk}@biu.ac.il

Abstract. Reinforcement-Learning-based solutions have achieved many successes in numerous complex tasks. However, their training process may be unstable, and achieving convergence can be difficult, expensive, and in some instances impossible. We propose herein an approach that enables the integration of strong formal verification methods in order to improve the learning process as well as prove convergence. During the learning process, formal methods serve as experts to identify weaknesses in the learned model, improve it, and even lead it to converge. By evaluating our approach on several common problems, which have already been studied and solved by classical methods, we demonstrate the strength and potential of our core idea of incorporating formal methods into the training process of Reinforcement Learning methods.

Keywords: Reinforcement learning · Q-learning · Formal verification · Model checking

1 Introduction

Reinforcement Learning (RL) [1] is a paradigm of machine learning focused on training intelligent agents leading to a strategy of selecting actions in an environment to maximize their cumulative reward. RL involves an agent, a set of states ($\mathcal{S}$), and a set of actions per state ($\mathcal{A}$). The transition of an agent from state to state is effected by performing an action $a \in \mathcal{A}$. Each (s, a) pair, i.e. action execution in a specific state, provides the agent with a reward (r), represented by a numerical score.

The agent's goal is to maximize total reward over a complete trajectory, a set of N steps defined by $\tau = \{a_0, s_0, r_0, ..., a_n, s_n, r_n\}$. Formally, the goal is to find the steps that maximize the following expression:

$$\sum_{t=1}^{N} r_t$$

Link to our code: https://github.com/eliyabron/Formal_verification_with_RL.

L. Gąsieniec (Ed.): SOFSEM 2023, LNCS 13878, pp. 342–355, 2023.
https://doi.org/10.1007/978-3-031-23101-8_23

The numerous RL algorithms that deal with this expression and its variants, can be divided into two groups – model-based methods, those which use known/learned models, and model-free methods, those that do not. Model-free algorithms focus on control, searching for the best policy, and prediction, evaluating the future when given a policy.

Despite its success in solving real-world problems, such as autonomous driving [2], natural language processing [3], genetic algorithms [4], and more, RL faces several significant challenges. The algorithms do not use pure labeled data, so they have to explore it themselves. A major challenge with RL solutions is that the state distribution changes as policies change. Using some exploration policy, sampling many states and actions, and then using that model to improve the policy will result in a revised distribution over states. The model may have been fairly accurate for the previous distribution, but there is no guarantee that it will be accurate for the updated distribution as well.

Rather than trying to learn only from sparse rewards or manually specifying a reward function, Imitation Learning [5] provides us with a set of demonstrations from an expert (typically a human). Following and imitating the expert's decisions, the agent attempts to learn the optimal policy. Behavioral cloning [6] is the simplest type of imitation learning, which uses supervised learning to mimic the expert's policy. Using imitation learning methods tackles the mentioned challenge since learning from an expert reduces the changes in the environmental distribution and allows for more stable and accurate learning. It is unfortunate that experts can be expensive, and sometimes they are simply not available. One can use Inverse Learning [7] to reduce the dependencies on experts, but it still requires prior data.

In this paper, we propose a method of learning through imitation in which formal verification tools serve as experts. It includes two separate innovations: The first is to formulate RL solutions as a transition system, essential for applying formal verification tools; the second is to use these tools' output to re-input the learned model as if it were a human expert's trajectories. In particular, we evaluate the model as a transition system and formulate a specification for checking. Model checkers then take both the translated transition system and a specification and check for counter examples, that the model failed to handle, and use them as expert trajectories with appropriate rewards. When model checkers return true, i.e. they didn't find a counter example, the model has converged. This is another strength of our method, as RL solutions are difficult to prove in terms of convergence. A high-level illustration of the entire mechanism is shown in Fig. 1.

To demonstrate the effectiveness of our approach, we examine several common tasks in the field of RL, each of which contains an environment of potentially hundreds of thousands of possibilities. This highlights how formal methods can play a crucial role in the learning process. Our contribution can be expressed in several ways: reducing the number of epochs to reach convergence, proving convergence of a given model, exploring states not seen during training, and assisting model convergence when getting stuck in non-terminal states by exploring new trajectories.

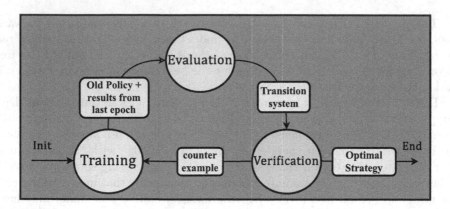

Fig. 1. An overview of the approach. In the training phase, we pass the evaluated model to the Formal Verification tool, and use the output as re-input for the next epoch. By doing so, convergence can both be reached and proved.

In summary, our work has three main contributions:

- Demonstrates a new simple way of bridging two distinct fields — Reinforcement Learning and Formal Verification — by translating RL models into transition systems and using model checker outputs as an expert's trajectories.
- Exhibits the ability of this combination to prove convergence where standard RL algorithms may have difficulty.
- Shows this combination's effectiveness using applications to different problematic scenarios such as getting stuck in non-terminal states, and too many epochs until convergence.

2 Related Work

The field of Artificial Intelligence [8,9] suggests many approaches toward safe reinforcement learning, though most do not take advantage of formal verification. García et al. [10] divide many of these algorithms that do not utilize formal verification into two groups by their objective goal—the optimality criterion (the discounted horizon), and the exploration process.

In recent years, several works have used formal methods for RL. In [11], formal verification and run-time monitoring are combined in order to ensure system safety. Their technique, called Justified Speculative Control (JSC), provides a set of proofs that transfer computer-checked safety proofs for hybrid dynamical systems to policies derived by common reinforcement learning algorithms.

Another approach, called *Verily* [12], verifies deep-RL-based systems by introducing deep neural networks (DNNs) as Satisfiability (SAT) problems, and then solving them via a SAT solver. This method is based on algorithms for DNN verification and it expands them so that they are suitable for RL systems based on DNNs as well. Yang et al. [13] use RL methods to synthesize DNN controllers for nonlinear systems subject to safety constraints. In [14], a fully integrated system that allows both training and verification is proposed. It enables Deep RL systems to be trained with an abstraction-based approach, resulting in DNNs with finite input states that can be directly verified using model checking. As these methods require the algorithm being verified to be based on a DNN, they cannot be applied to other RL algorithms.

Zhu et al. [15] suggested a counter example driven technique, in which states and actions that are heavily weighted could be found directly and more efficiently. The search of safety moves is followed by checking trajectories that satisfy safety properties by forcing the agent to take certain actions in key states to avoid property violations. During the epochs, the policy is improved by utilizing only formal methods that lead the agent to avoid bad actions and choose actions with higher weights.

3 Method

In order to implement formal verification methods, the problem should be presented as a finite-state machine (FSM), and then solved using an RL algorithm whose solution can be represented by an FSM. We also require the RL algorithm to have a structure and characteristic that allows creation of a robust communication infrastructure between the algorithm and the model checker. The RL field includes many algorithms that offer different approaches to solutions for the same problem and could potentially use our approach.

To demonstrate our approach, we consider several common tasks and apply our method to Q-learning algorithms, though it can naturally be extended to other algorithms as well. In Sect. 3.1 we give a brief overview on Q-learning and in Sect. 3.2, we explain how to formulate the model as a transition system which enables the model to be verified by a model checker. In addition, we describe the output of the model checker, along with how it is used.

3.1 Q-Learning

Q-learning, one of the most popular model-free algorithms, is a family of methods that attempts to assess the quality of an action taken in order to move to a given state. The algorithm calculates the quality of each (s, a) pair:

$$Q : \mathcal{S} \times \mathcal{A} \to \mathbb{R} \tag{1}$$

Before learning begins, Q is initialized to an arbitrary fixed value, and is then updated at each time t with action a_t and reward r_t. The updated equation takes into account the previous value of Q along with the new information for the current reward, as follows:

$$Q^{new}(s_t, a_t) = Q(s_t, a_t) + \alpha \cdot \left(r_t + \gamma \cdot \max_a Q(s_{t+1}, a) - Q(s_t, a_t) \right), 0 \le \alpha \le 1 \tag{2}$$

This equation can be rearranged into a more easily interpreted form:

$$\mathcal{Q}^{new}(s_t, a_t) = (1 - \alpha)\mathcal{Q}(s_t, a_t) + \alpha\, r_t + \alpha\gamma \max_a \mathcal{Q}(s_{t+1}, a) \tag{3}$$

This form has three elements:

1. The current value, weighted by the learning rate α. The higher the learning rate, the faster Q changes.
2. The reward, weighted by the learning rate.
3. The maximum reward that can be obtained from state s_{t+1}, weighted by the learning rate and a discount factor γ.

This equation is the basis of the learning process, where at each step the value of a specific state is updated, based on a number of factors—the previous value, the reward and the possible reward of the next state.

The learning process creates a table, called a Q-table, which represents a strategy in which each (s_t, a_t) pair has a value. This table guides the agent to take an action from the current state, based on the expression $a = \text{argmax}_a(s_t, a_t)$, thereby moving the agent from s_t to s_{t+1}. To balance between exploration and exploitation we use the ϵ-greedy method: before choosing an action we sample a random number from a uniform distribution $\mathcal{U}[0, 1]$. If the number is below the chosen epsilon, a random action is selected, otherwise the action with the highest value in the Q-table is selected.

Although this algorithm can reach convergence [16], if the largest difference between the current and updated Q-table is less than the given ϵ, it is not always a simple task and is often so computationally expensive that it is impractical. Bearing in mind the possible implications exploration-exploitation might have on the duration until convergence is reached, the proposed approach adds an external tool as an add-on to the RL algorithm to speed-up, ensure and verify reaching the optimal policy.

3.2   Using a Model Checker on the Q-learning Model

RL solutions are formulated as transition systems, which can be read by model checkers. In order to implement the solution as an FSM, we describe the updated Q-table as a transition system:

$$T = \langle \mathcal{V}, \theta, \rho \rangle \tag{4}$$

where $\mathcal{V}$ are the system variables, θ is the initial condition and ρ is the transition relation. In general, the variables $\mathcal{V}$ encode the state space $\mathcal{S}$, ρ represents the actions from each state based on the the current policy $\pi^{\mathcal{Q}}$ and θ includes the initial position(s). Once the Q-table is encoded as an FSM, it can be checked by model checkers using temporal logic queries. A query will return one of two options through model checkers—True, or False with a counter example. True means the Q-table holds for the query, which in the current context means an optimal strategy has been identified. On the other hand, whenever False is

Algorithm 1. Convergence of Q-table

```
procedure CONV(Q-table, task)          ▷ calculate optimal Q-table for a given task
    Q-table ← 0
    ans ← False
    while Q-table did not converge do
        x epoch of Q-learning
        ans ← Check LTL\CTL spec
        if ans is False then                                           ▷ There is a path
            Path ← model checker counter example
            update Q-table using Path
        else                                                        ▷ There is no path
            break
    return                                                   ▷ The strategy is optimal
```

returned and a counter example is found, the counter example can then be used to improve the model. To find the optimal solution of a specific task, we use the model checker repeatedly, update the results through counter examples until reaching convergence. The algorithm is depicted in Algorithm 1.

A number of critical scenarios can be greatly improved by applying this algorithm which uses the model checker as an expert. Situations where the state space is large and the agent takes a long time to reach convergence, or more seriously, gets stuck in non-terminal states, as well as when it is unknown whether convergence has been reached, can all benefit from using model checkers.

The abstraction of the Q-table as a transition system can be easily converted to a standard model checker's input, which is made up of three main parts. The first initializes all possible vectors for the defined task derived from its characteristics. The second describes the selected move for the agent according to a given state, for which the next selected move is either all the possible moves for this state or the one with the highest score in the Q-table. If values in the Q-table are equal, there may be more than one possible next move. In the third and final part, a temporal logic specification is used to check if the model is successful under all circumstances. This specification enables encoding of formulae about the system's future path, and applying a condition or rule on the system behavior.

4 Model Checker as an Expert—Applications

Here, we demonstrate our approach and its effectiveness in several applications. We will show that even on small and easy problems, classical Q-learning may not always work properly and could benefit greatly from the incorporation of model checkers. Specifically, RL results are improved in a number ways by the implementation of model checkers as an expert.

4.1   Convergence with Fewer Epochs

We use the well-investigated game of Cops and Robbers [17], also termed the pursuit-evasion game [18], in order to demonstrate how we can reduce the number of required epochs to reach convergence. Typically the game is presented as a graph [17,18], but can also be viewed as a two-dimensional board, as illustrated in Fig. 2. There are two kinds of players in a grid, cops and robbers, where each game must have at least one cop and one robber. Every turn, only one player can move in one of four directions—up, down, left or right—but not outside of the border. As soon as the cops catch the robbers, the cops win, and when the maximum number of turns is over, the robbers win.

In RL terminology, the board is the environment, agents are the cops and the goal is to converge in the smallest number of games (i.e. an epoch) to an optimal policy, if possible. An optimal policy ensures the agent wins from every initial position. Alongside the standard training procedure, we use model checkers to indicate, at each game, whether or not the optimal policy has been reached. In the case where the model checker returns False (optimal policy hasn't yet been reached), a counter example is provided, and can be used as an initial state for the next epoch. By classic machine learning terms, the model checker holds the loss function derived at each game's end, thus returning the gradient's direction.

In this system, each state is defined by the current position of cops and robbers on the board using x, y coordinates, i.e. 1,3 indicates the third cell from left in the first row from the top. A letter indicating the team taking the current turn ($\mathcal{C}$ for cops, $\mathcal{R}$ for robbers) is attached to the state as well. Possible moves for the next turn emerge from each state, up/down/left/right, for each player in the team, excluding illegal moves (e.g. moving to coordinate 0).

As shown in Eq. 4, using a model checker is made possible by formulating the RL solution as a transition system—a tuple $\langle \mathcal{V}, \theta, \rho \rangle$. The variables encode the position of the agent in the network and its direction of movement, $\mathcal{V} = \{vec, turn\}$. The variable vec represents the cops and robbers positions. The positions, together with the team carrying out the next move, are saved each turn as a state vector. State vectors are modified by a number that holds the cops and robbers positions (x, y coordinates). The second variable, $turn \in \{\mathcal{C}, \mathcal{R}\}$, identifies the team making the next move.

The initial condition, θ, is chosen from all vec options by the model checker, as explained below. The transition relation ρ specifies how variables are updated based on the current state:

$$(turn' = \mathcal{C} \wedge turn = \mathcal{R}) \vee (turn' = \mathcal{R} \wedge turn = \mathcal{C}) \tag{5}$$

$$vec' = \text{Choose Action(Available Positions)} \tag{6}$$

"Available Positions" relate to the set of all possible positions for the next state. "Choose Action" is a function of these "Available Positions". This function may differ for each team. Cops select from the available positions by the learned Q-table, meaning they choose from the actions with the highest value in the table.

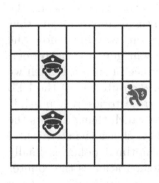

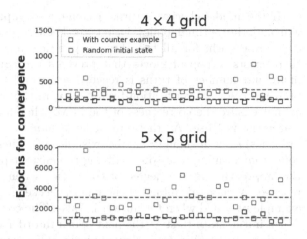

Fig. 2. Cops and Robbers. (Left) 5 × 5 Cops and Robbers grid with two cops (blue) and one robber (red). (Right) For each experiment we run 50 games, 25 with the model checker (blue squares) and 25 without (red squares). Median is marked with dotted line. Number of epochs required for convergence, as well as variance, with model checker is much lower than without, suggesting a more stable learning method. (Color figure online)

The robber can choose from all the possible actions for its current state. The last part given to the model checker is a temporal logic statement, which requires that in each possible FSM path, the cops eventually win. When the model checker answers False, it accompanies this with a counter example by which the initial position of the following game in the epoch is derived. This initial position is selected from the loop detected in the counter example (i.e. a repeated set of steps the robber can force the cops to take without being caught). Starting from such an initial position, which can result in a loss under the current strategy, forces the model to improve the current strategy for this position.

The first game runs on the initial Q-learning table (i.e. all values set to 0), therefore it uses an arbitrary policy. The game is finished when a winner is declared, and the reward of 100 and −100 points is divided according to the rules previously defined. The reward per state is back-propagating, so the overall reward is divided unequally across all states by the modified hyperparameter. The Q-table that describes the policy is saved after each game, thus implementing the "experience" from previous games. Communication with the model checker is established at the end of each game. The state-vectors list is fed into the model checker, in effect defining an FSM by which the connection between the Q-learning algorithm and formal verification is created.

The FSM created at each game's end is used as the input into the model checker together with the CTL or LTL statement. The CTL and LTL statements ask the model checker if, for all branches in the FSM, the cops win. Formally:

$$\text{LTLSPEC} := F\ (vec = 0), \quad \text{CTLSPEC} := AF\ (vec = 0) \tag{7}$$

If the model checker returns a counter example it means the robber can win and the optimal policy hasn't yet been reached, otherwise it means the cops always win for all branches. In practice, a counter example means that there exists a loop of moves that prevents the cops from winning before the maximum number of turns is reached, therefore the win goes to the robber. Shifting to formal verification terminology, in the presence of such a branch we cannot ensure the correctness of the FSM. Therefore, the outcome of the LTL statement is False. As there are a large number of trajectories (i.e. possible branches) even in small grids, the existence of a single branch that prevents the cops from winning, thus disqualifying the current policy, shows the strength of this approach and the degree of confidence when an optimal policy is finally reached. A main advantage of the model checker implementation is the counter example provided when the answer is False. The counter example indicates from which initial state to start the next epoch in order to prevent the robbers from reaching this specific branch, maximizing policy convergence.

In essence, the model checker returns the loss function's gradient to the optimal policy. The option to play without utilizing the counter example was also evaluated, meaning that the model checker returns only the True or False answer, while the initial position for next game is decided randomly. While such an answer from the external tool to the Q-learning algorithm can be helpful toward convergence efforts, the added value of the counter example's contribution to these efforts (together with the True-False outcome) is non-negligible.

Experiments have been run with different exploration rates, discount factors, grid sizes, and cop numbers, all leading to the same result. By using a model checker as an expert, the number of epochs required to reach convergence is significantly reduced. Additionally, variance is much lower, resulting in much more stable learning. A few examples of small-size boards are shown in Fig. 2, while the trends are the same for large-sized boards with different parameters.

4.2   Help Convergence by Avoiding Sub-optimal Solutions

Unlike the previous example, where the model was able to reach convergence without the help of a model checker though under higher number of epochs, sometimes the model does not reach optimal convergence at all. Integrating the model checker is even more critical in these cases. We use Frozen Lake, a game consisting of a grid containing four types of squares (start, finish, frozen and hole), as seen in Fig. 3 (upper left), for demonstration. Possible actions are the same as in Cops and Robbers, though if the agent takes a step into a hole, it loses the game, and if it reaches the goal, it wins.

As seen in Fig. 3 (bottom), several boards are difficult to learn as they have many holes and few paths to the goal. As seen in Fig. 3 (bottom), several boards are difficult to learn as they have many holes and few paths to the goal. We train the model using Q-learning until the difference between the Q-tables of two consecutive epochs is smaller than epsilon. Even when using a very small epsilon ($\epsilon = 10^{-30}$), the algorithm fails to converge to the optimal policy on these boards.

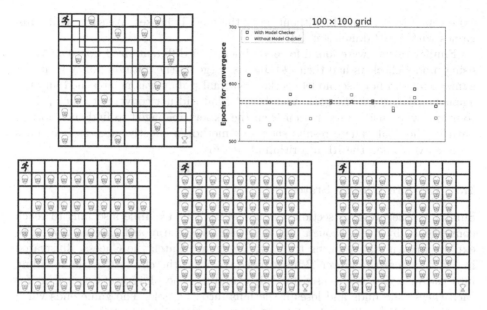

Fig. 3. Frozen Lake. (Upper Left) 10 × 10 Frozen Lake grid. (Upper Right) For a 100 × 100 grid, we run 20 games, 10 with the model checker (blue squares) and 10 without (red squares). Median is marked with dotted line. Though the number of epochs to convergence is similar, the policy reached with the model checker is optimal, while the policy reached without is sub-optimal. (Bottom) Three difficult scenarios that fail reaching optimal policy without the model checker. (Color figure online)

We overcome this problem by utilizing an expert to assist Q-learning by supplying a possible route. The expert is also able to declare if there are no possible routes to the goal in advance. Similar to the previous task, we define the possible states as $\mathcal{V}$, while the transition relation ρ is derived from the possible actions for each state. Every environment is initially defined by the placement of holes in the grid, with the agent starting at the top left, setting the initial state θ. Once the system has been defined and formulated, LTL statements are added asking whether the agent will always reach the goal according to the current strategy:

$$\text{LTLSPEC} := \text{!F (currentPosition} = \text{Goal)} \tag{8}$$

As the goal of the game is to reach the last square, we ask the model checker if "there is no route by which the agent can reach the goal". Whenever a counter example is returned, there is a route that can be used to help the Q-table converge. For every x epochs of updating the Q-table, the model checker is used to find a possible path with which to update the Q-table. Eventually, the Q-table converges to the optimal policy.

Experiments conducted on various grids that are hard for Q-learning to learn, all showed that the model was unable to converge without routes from the model checker. For each board 20 games are run, 10 using the model checker for the expert's trajectory and 10 without using the model checker. We successfully ran games with a 10^5 dimension Q-table.

Similar results were found for experiments on different sized boards. Games using model checkers had their Q-table converge to the optimal policy while for games that did not use model checkers optimal policy was not reached and the agent failed to reach the goal. Even boards that did not converge to the optimal solution, when using information from the expert managed to find the shortest path to the goal. These results show our method's effectiveness, as it enables success even when the RL algorithm alone fails.

4.3   Explore Unseen States

Similar to the previous section, the model for this example also fails to reach convergence by itself, though, unlike frozen lake, the model checker is used to explore unseen states during training. The game Catch the Cheese, illustrated in Fig. 4, where the player $\mathcal{P}$ must move to catch the cheese $\mathcal{C}$ without falling into the hole $\mathcal{O}$, will be used for this example. The player gets one point for each cheese she finds and loses if she falls into the hole. The game ends when the player either collects the cheese $\mathcal{K}$ times, or falls into the hole.

Each state of the transition system is defined by the location of the player. As in the previous games, the agent can move to another square in the grid only by legal actions (right or left, except for the borders where there is only one action). In this game, the number of steps the player can take is deliberately limited so she won't be able to reach all the possible states, and thus the goal (the cheese). This is a simple example for the problem where the run-time of each epoch is limited. The expert starts from the most advanced state the player was able to reach and teaches the Q-learning algorithm the states it was not able to see due to the steps limit.

Similar to the previous tasks, we define the possible states as $\mathcal{V}$ while the transition relation ρ is derived from the possible actions for each state. Each environment is initially defined by the placement of hole in the grid, with the agent starting somewhere in the middle of the grid, thus setting the initial state θ. Once the system has been defined and formulated, we add the following LTL:

$$\text{LTLSPEC} := !F \ (score \geq \mathcal{K}) \tag{9}$$

As the goal of the game is to reach the cheese at least $\mathcal{K}$ times, the LTL asks if "the agent cannot reach the cheese more than $\mathcal{K}$ times". If a counter example is returned, it is used to teach the Q-learning algorithm.

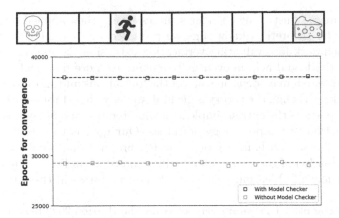

Fig. 4. Catch the Cheese. (Upper) 1×10 Catch the Cheese grid. (Lower) For a 1×100 grid, we run 20 games, 10 with the model checker (blue squares) and 10 without (red squares). Median is marked with dotted line. Even though number of epochs to convergence is on average higher, the policy reached with the model checker is optimal, and the policy reached without is sub-optimal. (Color figure online)

In this algorithm, for every x epochs (determined by the user) of updating the Q-table, we check for convergence. Whenever a model is not convergent, we input the counter example path into the Q-table. This procedure is repeated until the Q-table converges to the optimal policy.

In order to test the impact of the model checker, we performed a large number of experiments, while for each setup we ran 20 games—10 using the model checker for information on unseen states and 10 without. When using the expert, the player managed to achieve the desired score, while the player failed to converge to a Q-table containing the optimal solution when not using the expert. This shows the value of using formal verification tools as experts for unseen states, as a reduced number of steps per epoch still achieves optimal results. This approach is also helpful when it is hard to estimate the number of steps needed, as you can choose an approximate number of steps, and use the model checker to explore paths beyond the limits of the agent.

5 Discussion

We present a method combining formal verification as an expert for reinforcement learning algorithms. This enables both obtaining formal guarantees on the correctness of the learned strategy, as well as speeding up the learning process by utilizing counter examples produced by the model checker. Using well-studied tasks, we demonstrate the feasibility and effectiveness of the approach on a Q-learning algorithm that solves finite-length games on a grid, and show empirically that we can identify when the Q-learning algorithm has converged on an optimal policy, and can both speed up this convergence, and direct the player in scenarios in which the policy fails. Our tasks included tens of thousands of scenarios and possibilities, illustrating the idea's success even when applied to non trivial problems. The theoretical knowledge we proposed and examined can serve as a good benchmark for evaluating verification extensions to more reinforcement learning methods and help in eventually moving to more practical applications.

Future research may focus on some of the limitations and shortcomings of our current work. Although the theory is general, we only show tabular-RL problems, and several parts of the current implementation require manual engineering, e.g. the construction of temporal logic formulas. Our goal is to extend our infrastructure so that it can handle large-scale RL problems automatically. We also omit an evaluation of the overhead of verification tools and the comparison with modern versions of Q-learning, which we hope to address in the future.

Acknowledgments. This work is supported by the Horizon 2020 research and innovation programme for the Bio4Comp project under grant agreement number 732482 and by the ISRAEL SCIENCE FOUNDATION (Grant No. 190/19). We would like to thank Assaf Grundman and Shlomi Mamman for their work and feedback on an early version of this project, and the Data Science Institute at Bar-Ilan University.

References

1. Sutton, R.S., Barto, A.G.: Reinforcement learning: an introduction. MIT press (2018)
2. Kiran, B.R., et al.: Deep reinforcement learning for autonomous driving: a survey. IEEE Trans. Intell. Transp. Sys.**23**, 4909–4926 (2021)
3. Xiong, C., Zhong, V., Socher, R.: DCN+: mixed objective and deep residual coattention for question answering. arXiv preprint arXiv:1711.00106 (2017)
4. Weile, D.S., Michielssen, E.: Genetic algorithm optimization applied to electromagnetics: a review. IEEE Trans. Antennas Propag. **45**(3), 343–353 (1997)
5. Hussein, A., Gaber, M.M., Elyan, E., Jayne, C.: Imitation learning: a survey of learning methods. ACM Comput. Surv. (CSUR) **50**(2), 1–35 (2017)
6. Pomerleau, D.A.: Alvinn: an autonomous land vehicle in a neural network. In: Advances in Neural Information Processing Systems 1 (1988)
7. Abbeel, P., Ng, A.Y.: Apprenticeship learning via inverse reinforcement learning. In: Proceedings of the Twenty-First International Conference on Machine Learning, p. 1 (2004)
8. Menzies, T., Pecheur, C.: Verification and validation and artificial intelligence **65**, 153–201 (2005)
9. Seshia, S.A., Sadigh, D., Sastry, S.S.: Towards verified artificial intelligence. ArXiv e-prints (2016)
10. García, J., Fernández, F.: A comprehensive survey on safe reinforcement learning. J. Mach. Learn. Res. **16**(1), 1437–1480 (2015)
11. Fulton, N., Platzer, A.: Safe reinforcement learning via formal methods: toward safe control through proof and learning. In: AAAI (2018)
12. Kazak, Y., Barrett, C., Katz, G., Schapira, M.: Verifying deep-rl-driven systems. In: Proceedings of the 2019 Workshop on Network Meets AI & ML, NetAI'19, pp. 83–89 New York, NY, USA (2019). Association for Computing Machinery

13. Yang, Z., et al.: An iterative scheme of safe reinforcement learning for nonlinear systems via barrier certificate generation. In: Silva, A., Leino, K.R.M. (eds.) CAV 2021. LNCS, vol. 12759, pp. 467–490. Springer, Cham (2021). https://doi.org/10.1007/978-3-030-81685-8_22

14. Jin, P., Zhang, M., Li, J., Han, L., Wen, X.: Learning on abstract domains: a new approach for verifiable guarantee in reinforcement learning. CoRR. arXiv:2106.06931 (2021)

15. Zhu, H., Magill, S.: Systems support for hardware anti-rop. Technical report, Galois Inc (2017). https://galois.com/reports/formal-methods-for-reinforcement-learning/

16. Watkins, C.J.C.H., Dayan, P.: Q-learning. In: Machine Learning, vol. 8, pp. 279–292 (1992). https://doi.org/10.1007/BF00992698

17. Aigner, M., Fromme, M.: A game of cops and robbers. Discret. Appl. Math. **8**(1), 1–12 (1984)

18. Parsons, T.D.: Pursuit-evasion in a graph. In: Theory and Applications of Graphs, pp. 426–441. Springer (1978). https://doi.org/10.1007/BFb0070400

Robots and Strings

Delivery to Safety with Two Cooperating Robots

Jared Coleman[2]([✉]), Evangelos Kranakis[3], Danny Krizanc[4],
and Oscar Morales-Ponce[1]

[1] Department of Computer Engineering and Computer Science,
California State University, Long Beach, CA, USA
Oscar.MoralesPonce@csulb.edu
[2] Department of Computer Science, University of Southern California,
Los Angeles, CA, USA
jaredcol@usc.edu
[3] School of Computer Science, Carleton University, Ottawa, ON, Canada
kranakis@scs.carleton.ca
[4] Department of Mathematics and Computer Science, Wesleyan University,
Middletown, CT, USA
dkrizanc@wesleyan.edu

Abstract. Two cooperating, autonomous mobile robots with arbitrary
nonzero max speeds are placed at arbitrary initial positions in the plane.
A remotely detonated bomb is discovered at some source location and
must be moved to a safe distance away from its initial location as quickly
as possible. In the *Bomb Squad* problem, the robots cooperate by communicating face-to-face in order to pick up the bomb from the source
and carry it away to the boundary of a disk centered at the source in the
shortest possible time. The goal is to specify trajectories which define
the robots' paths from start to finish and their meeting points which
enable face-to-face collaboration by exchanging information and passing
the bomb from robot to robot.

We design algorithms reflecting the robots' knowledge about orientation and each other's speed and location. In the offline case, we design
an optimal algorithm. For the limited knowledge cases, we provide online
algorithms which consider robots' level of agreement on orientation as
per `OneAxis` and `NoAxis` models, and knowledge of the boundary as per
`Visible`, `Discoverable`, and `Invisible`. In all cases, we provide upper
and lower bounds for the competitive ratios of the online problems.

Keywords: Boundary · Mobile robots · Delivery · Cooperative ·
Competitive ratio

1 Introduction

A remotely detonated bomb is located at the center of some critical zone. Since
the time of detonation is unknown, the bomb must be removed as quickly as

Research supported in part by NSERC Discovery grant.

possible from the critical zone by two autonomous mobile robots. How can these robots, each with their own speeds and initial location, collaborate to carry the bomb out of the critical zone as quickly as possible? We assume the bomb is initially located at a point S (the source) and must be transported at least distance D (called the *critical distance*) away from the source. The critical distance defines a disk centered at S of radius D. Each robot has its own maximum speed and the bomb can be passed from robot to robot in a face-to-face communication exchange. We refer to this as the *Bomb Squad* problem. The perimeter of the disk centered at S and radius D is also called the *boundary* and encloses the critical zone which must be rid of the bomb.

In the sequel, we study various versions of the Bomb Squad problem which depend on what knowledge the collaborating robots have regarding the location of the other robot and the boundary. We are interested in designing both offline (full knowledge) and online (limited knowledge) algorithms that describe the trajectories and collaboration of the participating robots.

1.1   Model, Notation, and Preliminaries

There are two autonomous mobile robots, r_1 and r_2, with maximum speeds v_1 and v_2 initially placed in the plane distances d_1 and d_2 from the source, respectively. We use the standard mobility model for the robots. At any time, they may stop and start, change direction/speed, and carry the bomb when they decide to do so. A robot trajectory is a continuous function $f : [0, T] \to \mathbb{R}^2$ such that $f(t)$ is the location of the robot at time t and T is the duration of a robot's trajectory. If the robot's speed cannot exceed v then $\|f(t) - f(t')\|_2 \leq v|t - t'|$, for all $0 \leq t, t' \leq T$, where $\| \cdot \|_2$ denotes the Euclidean norm in the plane $\mathbb{R}^2$.

Robots may collect information as they traverse their trajectories. Moreover, they may exchange information only when they are collocated (also known as F2F model). When collocated, they may compare their speeds and decide which robot is faster. They can recognize the bomb initially placed at location S and can carry it around and pass it from robot to robot without their speed being affected.

We assume robots have a common unit of distance. We consider both the offline and online settings. In the offline setting, all information regarding the robots (their initial positions and speeds) is available and an algorithm provides robot trajectories and a sequence of robot meetings that relay the bomb from the source to the boundary in optimal time.

In the online setting, a robot has limited knowledge of the other robot's location and the critical distance. We consider both OneAxis and NoAxis (or Disoriented) models (see [12]). In the OneAxis model robots agree on a single axis and direction (i.e. North). In the NoAxis model, we say robots are *disoriented* and do not agree on any axis or direction. With respect to knowledge of the critical distance, we consider three models:

1. VisibleBoundary: the boundary is always visible and thus the critical distance D is known by all robots.

2. DiscoverableBoundary: the boundary (and thus the critical distance) is not known ahead of time but is "discoverable". Robots can *discover* the boundary (and the critical distance) by visiting any point on the boundary or by encountering another robot which has already discovered it.

3. InvisibleBoundary: the boundary is completely invisible and robots have no knowledge of whether or not they've already visited a point on the boundary.

Each of these models has intuitive inspiration from the bomb-squad scenario. The VisibleBoundary model considers the situation where a safe distance is known ahead of time, while the DiscoverableBoundary model considers a situation where a boundary—physical (i.e. a fence or a wall) or abstract (i.e. a border, patrol line, maximum communication distance)—must be discovered by the robots. Finally, the InvisibleBoundary model considers the situation where a safe distance is not known by the robots (i.e. they don't know the detonation radius of the bomb). In this case, the goal is to deliver the bomb to an unknown radius as quickly as possible. Interestingly, each of these models yields unique algorithms with different competitive ratios.

In all of our algorithms, both robots start at the same time from arbitrary locations in the plane. The delivery time $T_A(I)$ of an algorithm A solving the Bomb Squad problem is the time it takes the algorithm A to deliver the bomb to the boundary for an instance I of the problem (a source location S, critical distance D, and robots' initial positions and maximum speeds). If $T_{opt}(I)$ is the optimal time of an offline algorithm for the same instance I, then the competitive ratio of an online algorithm A is defined by the ratio $CR_A := \sup_I T_A(I)/T_{opt}(I)$. If $\mathcal{A}$ is a class of algorithms solving an online version of the Bomb Squad problem, then its competitive ratio is defined by $CR_{\mathcal{A}} := \inf_{A \in \mathcal{A}} CR_A$. Usually, the subscripts will be omitted since the online version of the problem will be easily understood from the context.

In proving upper bounds on the competitive ratio, if the faster robot cannot arrive at S before the slow robot then we may restrict our attention to the case where the slow robot starts at the source. We state this useful claim as a lemma.

Lemma 1.1. *Consider any online algorithm solving an online version of the Bomb Squad problem. Assume that the faster robot cannot arrive at S before the slow robot does. If c is an upper bound on the competitive ratio of the algorithm for all instances in which the slow robot starts at the source, then c is also an upper bound on the competitive ratio for that algorithm.*

1.2   Related Work

The *Bomb Squad* problem is closely related to the message delivery problem with a set of robots. In that problem, the source and destination are predefined and robots jointly work to deliver the message. Two different objective functions have been studied. The first assumes that the robots have limited battery and consequently the objective function is to minimize the maximum movement (minmax). The second is to minimize the time to deliver the message. Anaya et al. [7] study a more general minmax problem where the message must be delivered to many destinations. The authors show that the decision problem is NP-hard and provide a 2 approximation algorithm.

Chalopin et al. [8] study the minmax problem on a line and show that the decision problem is NP-Complete for instances where all input values are integers. The authors also provide an algorithm for delivering the message that runs in $O(d^2 \cdot n^{1+4\log d})$ time where d is the distance between the source and destination for the general case. Coleman et al. [10] study the broadcast and unicast versions of the problem on a line and present optimal offline algorithms and online algorithms with optimal constant competitive ratio. In [11] Czyzowicz et al. study the problem in a weighted graph and show that the problem is NP-Complete. They also show that by allowing robots to exchange energy, the problem can be solved in polynomial time. Carvalho et al. [6] also study the problem in weighted graphs. They provide an offline algorithm that runs in $O(kn \log n + km)$ time where k is the number of robots, n is the number of nodes, and m the number of edges.

More recently, Coleman et al. [9] studied the point-to-point delivery problem on the plane and gave an optimal offline algorithm for two robots as well as approximation offline algorithms and online algorithms with constant competitive ratio. The delivery problem differs significantly from the problem studied in our current paper, where the goal is to reach any point on a given boundary (namely the perimeter of a disk centered at the source) as opposed to a specific destination.

The delivery problem studied in our paper focuses on the knowledge the robots have about each other as well as the environment. To this end we design algorithms for the OneAxis and NoAxis models. In particular, in the latter model and based on the knowledge the robots have in Subsect. 4.3 one has to design a search algorithm that makes the robots perform a "zigzag" procedure in order to collect appropriate information and pass the bomb to the faster robot, if feasible, that will eventually deliver the bomb to the boundary. This has similarities to the well-known linear search algorithms proposed by Baeza-Yates et al. [3], Beck [4] and Bellman [5], Ahlswede et al. [1], as well as Alpern et al. [1,2]. However, search in the previously given research works is based only on one robot while in our case we have two collaborating robots with incomplete information about the environment.

1.3   Outline and Results of the Paper

In this paper, we design and analyze algorithms for the Bomb Squad problem with two cooperating robots. In the offline case, we design an optimal algorithm that assumes robots have knowledge of their own and each other's location but does not require knowledge of each other's speed. For the online case Table 1 displays upper and lower bounds on the competitive ratio for the OneAxis and NoAxis models, and for Visible, Discoverable, and Invisible Boundary as well as the specific (sub)section where the results are proved. Section 2 presents an optimal offline algorithm, Sect. 3 presents an online algorithm for the OneAxis

Table 1. Upper and lower bounds on the competitive ratio of online algorithms for two cooperating robots in the `OneAxis` and `NoAxis` models for `Visible`, `Discoverable`, and `Invisible` Boundary:

Axis Model	Boundary Model	Upper Bound	Lower Bound	Section
`OneAxis`	All	$\frac{1}{7}\left(5 + 4\sqrt{2}\right) \approx 1.5224$	1.48102	3
`NoAxis`	`Visible`	$1 + \sqrt{2}$	$1 + \sqrt{2}$	4.1
`NoAxis`	`Discoverable`	$\frac{15}{4}$	3	4.2
`NoAxis`	`Invisible`	$\frac{7+\sqrt{17}}{2}$	$2 + \sqrt{5}$	4.3

model, while Sect. 4 includes the results of the three Subsections for the `NoAxis` model. There are many interesting open problems and in Sect. 5 we summarize the results and discuss potential extensions and alternatives. The full version of this paper with all missing proofs can be found on arXiv [11].

2 Optimal Offline Algorithm

Our problem may be solved optimally using Algorithm 1.

Algorithm 1. Offline Delivery Algorithm for Two Robots

1: move toward S
2: **if** arrived at S **then**
3:     pick up the bomb
4:     move in direction of other robot
5: **else if** encountered other robot with bomb and other robot is slower **then**
6:     take the bomb from other robot
7:     move away from S toward boundary

Theorem 2.1. *For any two robots r_1, r_2 such that $v_1 \leq v_2$, the offline Algorithm 1 is optimal in that it delivers the bomb to the perimeter of the circle centered at S with radius D in minimum time*

$$\min\left(\frac{d_1 + D}{v_1}, \; \frac{d_2 + D}{v_2}, \; \frac{D - d_2}{v_2} + 2\frac{d_1 + d_2}{v_1 + v_2}\right) \tag{1}$$

where S is the initial location of the bomb and d_1, d_2 are the starting distances (from S) of the r_1, r_2, respectively.

Proof. First, observe that the cases where the fast robot can reach S first or where the slow robot can deliver the bomb before the fast robot can get within a distance D of S are trivial and justify the first two arguments of the min term in (1). In each case the robot which reaches S first simply completes the delivery by itself and the algorithm is optimal. In all other cases, the slow robot reaches S first and must hand the bomb over to the fast robot at some point M which then delivers it to the boundary. Observe that the trajectory of the bomb itself must be a straight line since the closest point from M to the perimeter of the circle must be along SM (by the definition of a circle).

Consider all the candidate trajectories of the bomb. Since it must travel a total distance of exactly D, the trajectory which minimizes the delivery time is clearly that which involves the faster of the two robots carrying the bomb for the greatest portion of this distance. In other words, if s is the distance the slow robot carries the bomb before handing it over to the fast robot (Fig. 1), then the delivery time is

$$\frac{s}{v_1} + \frac{D-s}{v_2} = \frac{s(v_2 - v_1) + Dv_1}{v_1 v_2}$$

which is clearly minimized when s is minimum since $v_1 \leq v_2$. Intuitively, this means the slow robot should carry the bomb as short a distance as possible. Clearly, s is minimum when the slow robot moves directly toward the fast robot.

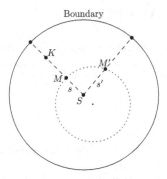

Fig. 1. Two candidate trajectories given by robot meeting points M and M', where the slow robot starts at S and the fast robot at K. Clearly, M is superior since $s < s'$ and the faster robot spends a larger portion of the bomb's trajectory carrying it. In other words, the bomb is moving at the faster speed v_2 for a larger portion of its trip to the boundary.

The delivery time for this case can then be easily written as the sum of the time for the robots to meet and the time for the fast robot to travel back to the boundary for delivery:

$$\frac{d_1 + d_2}{v_1 + v_2} + \frac{1}{v_2}\left(D - \left(\left(\frac{d_1 + d_2}{v_1 + v_2} - \frac{d_1}{v_1}\right)v_1\right)\right) = \frac{D - d_2}{v_2} + 2\frac{d_1 + d_2}{v_1 + v_2}.$$

$\square$

3   Online Algorithm for the OneAxis Model

Here we assume the robots agree on a single axis and direction and can therefore choose to move along the same radius emanating from S. We start by proving a lower bound.

Theorem 3.1. *Any online algorithm for the* OneAxis *model has a competitive ratio of at least* 1.48102.

Now, we propose the following Algorithm 2 and prove its competitive ratio is at most $\frac{1}{7}\left(5 + 4\sqrt{2}\right)$.

Algorithm 2. Online Delivery Algorithm for the OneAxis Model

1: move toward S, taking the bomb from any slower robot encountered
2: upon reaching S, move along common axis/direction away from S, taking the bomb from any slower robot encountered
---

Theorem 3.2. *Algorithm 2 has competitive ratio* $\frac{1}{7}\left(5 + 4\sqrt{2}\right)$.

Proof. If the fast robot arrives at the center first, then clearly the algorithm is optimal (it completes the delivery entirely by itself). Similarly, if the optimal algorithm involves only the slow robot (i.e. the fast robot is too far away to help), the algorithm is also optimal. Thus, we may consider only the case where the slow robot arrives first and where an optimal offline algorithm involves cooperation between the two robots.

Unlike in the optimal algorithm, the slow robot will not move directly toward the fast robot, since it doesn't know where it is. Rather, the slow robot will move along the shared axis in a previously agreed-upon direction (i.e. North). The fast robot will continue to move toward the source and, seeing the bomb is no longer there, begin to move along the shared axis. If the fast robot is fast enough, it will catch the slow robot, take the bomb, and complete the delivery. Otherwise, the slow robot will deliver the bomb.

Let d_1 and d_2 be the initial distances of the slow and fast robots to the source, respectively. Without loss of generality, suppose $D = 1$ and the fast robot moves at speed $v_2 = 1$. By Lemma 1.1, setting $d_1 = 0$ cannot decrease the competitive ratio, and so a bound on the competitive ratio can be written as

$$\frac{\min\left\{\frac{1}{v_1}, d_2 + 1\right\}}{\frac{d_2}{v_1+1} + \left(1 - \frac{d_2}{v_1+1}v_1\right)} = \frac{\min\left\{\frac{1}{v_1}, d_2 + 1\right\}}{d_2\frac{1-v_1}{1+v_1} + 1}.$$

For the first case, where $\frac{1}{v_1} \leq d_2 + 1$, we can write an upper bound by substituting $1/(d_2 + 1)$ for v_1 since $1/v_1$ decreases w.r.t v_1 and $\frac{1-v_1}{1+v_1}$ increases w.r.t v_1:

$$\frac{\frac{1}{v_1}}{d_2\frac{1-v_1}{1+v_1} + 1} \leq 1 + \frac{2d_2}{2 + d_2 + d_2^2}$$

which is maximized when $d_2 = \sqrt{2}$ (giving a value of $\frac{1}{7}\left(5 + 4\sqrt{2}\right)$). For the second case, when $\frac{1}{v_1} > d_2 + 1$, observe:

$$\frac{d_2 + 1}{\frac{d_2}{v_1 + 1}(1 - v_1) + 1} = \frac{(1 + v_1) + d_2(1 + v_1)}{(1 + v_1) + d_2(1 - v_1)} \le \frac{1 + v_1}{1 + v_1(2v_1 - 1)}$$

which is maximized when $v_1 = \sqrt{2} - 1$ (giving a value of $\frac{1}{7}\left(5 + 4\sqrt{2}\right)$).   □

Remark 1. This algorithm makes no use of the critical distance and thus applies to all three boundary-knowledge models (`VisibleBoundary`, `DiscoverableBoundary`, and `InvisibleBoundary`).

4   Online Algorithms for the `NoAxis` Model

The previous algorithms made use of a common axis and direction between the two robots. Now we consider an even weaker model where robots are disoriented (they have no common axis or sense of direction). We consider the three boundary-knowledge models introduced in Sect. 1.

We begin with the following lemma which will be useful for the analysis of online algorithms.

Lemma 4.1. *Assume at the start the slow robot is at S. Any online algorithm that involves the robots meeting at any point other than S cannot have constant competitive ratio.*

4.1 `VisibleBoundary` Model

First, we study the model where the critical distance D is known. Clearly, the optimally competitive algorithm for the `OneAxis` is not feasible in this model, since robots cannot decide on a common axis or direction to move along and potentially meet for a handover. By Lemma 4.1, the robots only hope for cooperation is by meeting at S. So if robots are going to collaborate, at least one of the robots will need to wait at S for the other robot to arrive. Clearly, it cannot do this forever though—it may be the case that the other robot is so far away or slow that the first robot may as well have delivered the bomb by itself! On the other hand, if the first robot simply commits to delivering the bomb by itself without waiting for the other robot to arrive, it may miss an opportunity to deliver the bomb very quickly if the other robot arrives at S shortly after and is very fast. It would seem, then, that an optimal algorithm must balance the cost of waiting for the other robot to arrive and completing the delivery without collaboration.

The main result of this section is the following theorem:

Theorem 4.1. *There exists an algorithm in the `NoAxis`/`VisibleBoundary` model with optimal competitive ratio $1 + \sqrt{2}$.*

We'll start by proving that there exists no algorithm with a better competitive ratio:

Lemma 4.2. *Any online algorithm for the* NoAxis/VisibleBoundary *model has competitive ratio at least* $1 + \sqrt{2}$.

Now, we propose Algorithm 3 and then prove its competitive ratio to be at most $1 + \sqrt{2}$.

Algorithm 3. Online Algorithm for NoAxis model for robot with speed v

---
1: move to S and wait for D/v time
2: **if** other robot arrives within D/v time **then**
3:     faster of two robots picks up bomb and moves toward boundary for delivery
4: **else**
5:     pick up bomb and move toward boundary for delivery

Lemma 4.3. *Algorithm 3 has a competitive ratio of at most* $1 + \sqrt{2}$.

4.2 DiscoverableBoundary Model

In this section, we consider the scenario where the robots are disoriented (NoAxis model) and do not know the distance D of the boundary from the source, but *can* discover it by passing through a point on the boundary or by encountering another robot which has previously discovered it.

Theorem 4.2. *Any algorithm for the* NoAxis / DiscoverableBoundary *model has competitive ratio at least* 3.

Now, we present Algorithm 4 with competitive ratio of 15/4.

Algorithm 4. Online Algorithm for DiscoverableBoundary Model for robot with speed v

---
1: move to S
2: **if** discovered boundary on the way to S **then**
3:     wait time D/v;
4: **else**
5:     move away from S (without the bomb) until arriving at the boundary
6:     return to S
7: **if** bomb is still at S **then**
8:     take bomb to the boundary

Theorem 4.3. *Algorithm 4 has a competitive ratio of exactly* 15/4.

4.3 InvisibleBoundary Model

Finally, we analyze an online algorithm under much stricter conditions. Robots cannot perceive the boundary at any time and therefore can never know the critical distance. It follows, then, that any valid algorithm must involve robots carrying the bomb away from the source without knowing how far they must take it in order to terminate.

Assume the slow robot starts at the source. By Lemma 4.1 there is no online algorithm with bounded competitive ratio unless the two robots have a meeting at the source. Further, it is easy to see that if the slow robot leaves the source without the bomb then unless it returns to the source there can be no online algorithm with bounded competitive ratio.

Lemma 4.4. *There exists no algorithm with constant competitive ratio for any instance of the problem under the* NoAxis/InvisibleBoundary *model in which one robot starts at S and no lower bound on D is known to the robots.*

In order to provide an online algorithm with constant competitive ratio, we make the necessary assumption (by Lemma 4.4) that the critical distance $D \geq 1$. In the sequel, we provide an algorithm that involves the first robot arriving at S taking the bomb a certain distance away from S and then returning (without the bomb) to see if a faster robot has arrived. If a faster robot *has* arrived, it shares information about the distance and direction of the bomb and allows the faster robot to complete the delivery. Otherwise, it travels back to where it left the bomb and carries it a bit further, expanding the distance each time. Formally, we present Algorithm 5 below.

We now prove a theorem that gives an upper bound on the competitive ratio of Algorithm 5. Note that Algorithm 5 uses the as yet unspecified expansion factor $a > 1$. The optimal selection of a will turn out to be $a = \frac{3+\sqrt{17}}{4}$ and this will be determined in the course of the proof of the following theorem.

Theorem 4.4. *For two robots, Algorithm 5 with $a = \frac{3+\sqrt{17}}{4}$ delivers the bomb to the boundary in at most $\frac{7+\sqrt{17}}{2}$ times the optimal offline time.*

Before proceeding to show a lower bound of $2 + \sqrt{5}$ for any online algorithm for the NoAxis/InvisibleBoundary model, we introduce a few basic concepts and ideas. Assume two robots r_1, r_2 with $v_1 = 1 < v_2$ and a source S. Recall that each robot knows the location of the source and its own location and speed but not the speed and location of the other robot. If the two robots meet, they can exchange information and determine which of the two is faster. If a robot knows it is faster than the other robot then if/when it acquires the bomb, it should simply move away from S forever to guarantee eventual delivery. If a robot holds the bomb and is "searching" for the perimeter but does not know whether it is the faster robot then it must return to the source (without the bomb) to check whether or not the other robot is waiting there. If it is, then it will share the direction of the bomb so that the fast robot can complete the delivery. Therefore

---

Algorithm 5. Online Algorithm for the `InvisibleBoundary` Model for a robot speed v and expansion factor a

---

1: $i \leftarrow 0$
2: **while** never encountered another robot **do**
3:    move to S
4:    **if** faster robot is at S **then**
5:       share direction of bomb with faster robot and stay at S forever
6:    **else if** slower robot is at S **then**
7:       get direction of bomb from slow robot (if not already known)
8:       move to bomb, pick it up, and continue moving away from S forever
9:    **else if** bomb is at S **then**
10:       Wait for another robot for at most time $2/v$
11:       **if** faster robot has not arrived **then**
12:          pick up bomb
13:          move away from S for a distance a
14:          set the bomb down
15:       **else**
16:          stay at S forever
17:    **else if** bomb is not at S but its location is known **then**
18:       move toward the bomb a distance of a^i distance away from S
19:       pick up bomb
20:       move another a^i distance away from S
21:       set the bomb down, marking its location
22:    **else**
23:       wait for other robot to return
24:       **if** other robot is slower **then**
25:          get direction of the bomb from other robot
26:    $i \leftarrow i + 1$

---

the slow robot is forced to execute a zigzag strategy as defined below, otherwise, the adversary will make the competitive ratio arbitrarily large.

A general algorithm is encapsulated by a strategy in which the robot starts at the source and executes Algorithm 6 implying a search at a distance x_k in the k-th round of the algorithm, for each $k \geq 1$. The algorithm is parameterized by an infinite ordered sequence of positive distances $X = \{x_1, x_2, \ldots, x_k, \ldots\}$ measured from the source that specifies the turning points that a moving robot will make. In the argument below we assume that given a strategy X the adversary has the power to choose the speed of the fast robot and its initial distance from the source S.

To ensure progress in searching, each trip away from the source should explore farther towards the perimeter than in the previous trip: this is formalized by the requirement that $x_k < x_{k+1}$ for all $k \geq 1$. Moreover, $\lim_{k\to\infty} x_k = +\infty$ (if not, the strategy could not solve all instances of the problem).

Consider a strategy X. Let the perimeter be at an unknown distance D. In each round k for which the perimeter is not found the robot must return to the source and will therefore cover a length $2x_k$. The total length covered up to and

---

Algorithm 6. Zig-Zag Delivery Algorithm (X)

1: **Input:** Infinite sequence $X = \{x_1, x_2, \ldots, x_k, \ldots\}$ with $0 < x_k < x_{k+1}$ for all $k \geq 1$;
2: **for** $k \leftarrow 1, 2, 3, \ldots$ **do**
3:     **if** $k = 1$ **then**
4:         move distance x_k away from S (in any direction)
5:     **else**
6:         move distance x_k away from S in diretion of bomb, picking it up on the way
7:     set down bomb
8:     return to source and $k \leftarrow k + 1$

---

including round k will be equal to $2\sum_{i=1}^{k} x_i$. If the perimeter is found during the next round the total distance covered by the robot will be $D + 2\sum_{i=1}^{k} x_i$. The resulting competitive ratio will be equal to

$$\frac{D + 2\sum_{i=1}^{k} x_i}{D} = 1 + \frac{2\sum_{i=1}^{k} x_i}{D}$$

Since the perimeter can be placed arbitrarily close to x_k by an adversary it follows that the highest lower bound on the competitive ratio for this step will be equal to

$$1 + \sup_{D > x_k} \frac{2\sum_{i=1}^{k} x_i}{D} = 1 + \frac{2\sum_{i=1}^{k} x_i}{x_k} = 3 + \frac{2\sum_{i=1}^{k-1} x_i}{x_k}$$

It follows from the previous discussion that the resulting competitive ratio of the strategy X will satisfy

$$CR_X = 3 + \frac{2\sup_{k \geq 1} \sum_{i=1}^{k-1} x_i}{x_k}. \tag{2}$$

Observe that the lower bound obtained in Eq. (2) is valid for two robots provided the adversary can force the slow robot to execute the zigzag strategy. So we consider an optimal strategy $X = \{x_1, x_2, \ldots, x_k, \ldots\}$. Let x_k be the last move of this strategy with which we reach the destination perimeter.

Now we are ready to complete our analysis in the `InvisibleBoundary` model by proving the following:

Theorem 4.5. *The competitive ratio of every strategy X solving the bomb squad problem in the* `InvisibleBoundary` *model must satisfy* $CR_X \geq 2 + \sqrt{5}$.

5   Conclusion

The main focus of the paper was to investigate algorithms for delivering a bomb to a safe location and compare the performance of online algorithms under several models which describe the knowledge the two robots have about each other and the environment (in this case the boundary). There are many interesting and challenging open problems remaining. For the two-robot case studied in the present paper, one can see the gaps remaining by glancing at the results displayed in Table 1. An interesting class of problems arises in the multi-robot (more than two robots) case, where, generally, it is much harder to give tight performance bounds. Finally, it would be interesting to investigate algorithms that are resilient to faults that arise either from robot miscommunication or faults caused by the planar environment on which the robots operate (e.g. based on visibility obstructions, distance constraints, etc.) and/or are sensitive to energy consumption limitations.

References

1. Ahlswede, R., Wegener, I.: Search problems. Wiley-Interscience (1987)
2. Alpern, S., Gal, S.: The theory of search games and rendezvous. Springer, New York (2003). https://doi.org/10.1007/b100809
3. Baeza Yates, R., Culberson, J., Rawlins, G.: Searching in the plane. Inf. Comput. **106**(2), 234–252 (1993)
4. Beck, A.: On the linear search problem. Israel J. Math. **2**(4), 221–228 (1964)
5. Bellman, R.: An optimal search. SIAM Rev. **5**(3), 274–274 (1963)
6. Carvalho, I.A., Erlebach, T., Papadopoulos, K.: On the fast delivery problem with one or two packages. J. Comput. Syst. Sci. **115**, 246–263 (2021)
7. Chalopin, J., Das, S., Mihal'ák, M., Penna, P., Widmayer, P.: Data delivery by energy-constrained mobile agents. In: Flocchini, P., Gao, J., Kranakis, E., Meyer auf der Heide, F. (eds.) ALGOSENSORS 2013. LNCS, vol. 8243, pp. 111–122. Springer, Heidelberg (2014). https://doi.org/10.1007/978-3-642-45346-5_9
8. Chalopin, J., Jacob, R., Mihalák, M., Widmayer, P.: Data delivery by energy-constrained mobile agents on a line. In: Esparza, J., Fraigniaud, P., Husfeldt, T., Koutsoupias, E. (eds.) ICALP 2014. LNCS, vol. 8573, pp. 423–434. Springer, Heidelberg (2014). https://doi.org/10.1007/978-3-662-43951-7_36
9. Coleman, J., Kranakis, E., Krizanc, D., Ponce, O.M.: Message delivery in the plane by robots with different speeds. In: Johnen, C., Schiller, E.M., Schmid, S. (eds.) SSS 2021. LNCS, vol. 13046, pp. 305–319. Springer, Cham (2021). https://doi.org/10.1007/978-3-030-91081-5_20
10. Coleman, J., Kranakis, E., Krizanc, D., Morales-Ponce, O.: The pony express communication problem. In: Flocchini, P., Moura, L. (eds.) IWOCA 2021. LNCS, vol. 12757, pp. 208–222. Springer, Cham (2021). https://doi.org/10.1007/978-3-030-79987-8_15
11. Czyzowicz, J., Diks, K., Moussi, J., Rytter, W.: Communication problems for mobile agents exchanging energy. In: Suomela, J. (ed.) SIROCCO 2016. LNCS, vol. 9988, pp. 275–288. Springer, Cham (2016). https://doi.org/10.1007/978-3-319-48314-6_18
12. Flocchini, P., Prencipe, G., Santoro, N.: Distributed computing by mobile entities. Current Research in Moving and Computing 11340 (2019). https://doi.org/10.1007/978-3-030-11072-7

Space-Efficient STR-IC-LCS Computation

Yuuki Yonemoto[1], Yuto Nakashima[2(✉)] (iD), Shunsuke Inenaga[2,3] (iD),
and Hideo Bannai[4] (iD)

[1] Department of Information Science and Technology, Kyushu University,
Fukuoka, Japan
yonemoto.yuuki.240@s.kyushu-u.ac.jp
[2] Department of Informatics, Kyushu University, Fukuoka, Japan
nakashima.yuto.003@m.kyushu-u.ac.jp, inenaga@inf.kyushu-u.ac.jp
[3] PRESTO, Japan Science and Technology Agency, Kawaguchi, Japan
[4] M&D Data Science Center, Tokyo Medical and Dental University, Tokyo, Japan
hdbn.dsc@tmd.ac.jp

Abstract. One of the most fundamental methods for comparing two
given strings A and B is the *longest common subsequence* (LCS), where
the task is to find (the length) of the longest common subsequence. In
this paper, we address the *STR-IC-LCS* problem which is one of the
constrained LCS problems proposed by Chen and Chao [J. Comb. Optim,
2011]. A string Z is said to be an STR-IC-LCS of three given strings A,
B, and P, if Z is one of the longest common subsequences of A and
B that contains P as a substring. We present a space efficient solution
for the STR-IC-LCS problem. Our algorithm computes the length of an
STR-IC-LCS in $O(n^2)$ time and $O((\ell + 1)(n - \ell + 1))$ space where ℓ is
the length of a longest common subsequence of A and B of length n.
When $\ell = O(1)$ or $n - \ell = O(1)$, then our algorithm uses only linear
$O(n)$ space.

Keywords: String algorithm · Constrained longest common
subsequence · Dynamic programming

1 Introduction

Comparison of two given strings (sequences) has been a central task in Theoretical Computer Science, since it has many applications including alignments of biological sequences, spelling corrections, and similarity searches.

One of the most fundamental method for comparing two given strings A and B is the *longest common subsequence LCS*, where the task is to find (the length of) a common subsequence L that can be obtained by removing zero or more characters from both A and B, and no such common subsequence longer than L exists. A classical dynamic programming (DP) algorithm is able to compute an LCS of A and B in quadratic $O(n^2)$ time with $O(n^2)$ working space, where n

L. Gąsieniec (Ed.): SOFSEM 2023, LNCS 13878, pp. 372–384, 2023.
https://doi.org/10.1007/978-3-031-23101-8_25

is the length of the input strings [12]. In the word RAM model with ω machine word size, the so-called "Four-Russian" method allows one to compute the length of an LCS of two given strings in $O(n^2/k + n)$ time, for any $k \leq \omega$, in the case of constant-size alphabets [9]. Under a common assumption that $\omega = \log_2 n$, this method leads to weakly sub-quadratic $O(n^2/\log^2 n)$ time solution for constant alphabets. In the case of general alphabets, the state-of-the-art algorithm computes the length of an LCS in $O(n^2 \log^2 k/k^2 + n)$ time [2], which is weakly sub-quadratic $O(n^2(\log \log n)^2/\log^2 n)$ time for $k \leq \omega = \log_2 n$. It is widely believed that such "log-shaving" improvements would be the best possible one can hope, since an $O(n^{2-\epsilon})$-time LCS computation for any constant $\epsilon > 0$ refutes the famous strong exponential time hypothesis (SETH) [1].

Recall however that this conditional lower-bound under the SETH does not enforce us to use (strongly) quadratic *space* in LCS computation. Indeed, a simple modification to the DP method permits us to compute the length of an LCS in $O(n^2)$ time with $O(n)$ working space. There also exists an algorithm that computes an LCS string in $O(n^2)$ time with only $O(n)$ working space [6]. The aforementioned log-shaving methods [2,9] use only $O(2^k + n)$ space, which is $O(n)$ for $k \leq \omega = \log_2 n$.

In this paper, we follow a line of research called the *Constrained* LCS problems, in which a pattern P that represents a-priori knowledge of a user is given as a third input, and the task is to compute the longest common subsequence of A and B that meets the condition w.r.t. P [3–5,7,8,11]. The variant we consider here is the *STR-IC-LCS* problem of computing a longest string Z which satisfies that (1) Z *includes* P as a *substring* and (2) Z is a common subsequence of A and B. We present a space-efficient algorithm for the STR-IC-LCS problem in $O(n^2)$ time with $O((\ell+1)(n-\ell+1))$ working space, where $\ell = \mathsf{lcs}(A, B)$ denotes the length of an LCS of A and B. Our solution improves on the state-of-the-art STR-IC-LCS algorithm of Deorowicz [5] that uses $\Theta(n^2)$ time and $\Theta(n^2)$ working space, since $O((\ell+1)(n-\ell+1)) \subseteq O(n^2)$ always holds. Our method requires only sub-quadratic $o(n^2)$ space whenever $\ell = o(n)$. In particular, when $\ell = O(1)$ or $n - \ell = O(1)$, which can happen when we compare very different strings or very similar strings, respectively, then our algorithm uses only linear $O(n)$ space.

Our method is built on a non-trivial extension of the LCS computation algorithm by Nakatsu et al. [10] that runs in $O(n(n - \ell + 1))$ time with $O((\ell + 1)(n - \ell + 1))$ working space. We remark that the $O(n^{2-\epsilon})$-time conditional lower-bound for LCS also applies to our case since STR-IC-LCS with the pattern P being the empty string is equal to LCS, and thus, our solution is almost time optimal (except for log-shaving, which is left for future work).

Related Work. There exists four variants of the *Constrained* LCS problems, STR-IC-LCS/SEQ-IC-LCS/STR-EC-LCS/SEQ-EC-LCS, each of which is to compute a longest string Z such that (1) Z includes/excludes the constraint pattern P as a substring/subsequence and (2) Z is a common subsequence of the two target strings A and B [3–5,7,8,11]. Yamada et al. [13] proposed an $O(n\sigma + (\ell'+1)(n-\ell'+1)r)$-time and space algorithm for the STR-EC-LCS problem, which is also based on the method by Nakatsu et al. [10], where σ is the alphabet size, ℓ' is the length of an STR-EC-LCS and r is the length of P. However, the design of our solution to STR-IC-LCS is quite different from that of Yamada et al.'s solution to STR-EC-LCS.

2   Preliminaries

2.1   Strings

Let Σ be an *alphabet*. An element of Σ^* is called a *string*. The length of a string S is denoted by $|S|$. The empty string ε is a string of length 0. For a string $S = uvw$, u, v and w are called a *prefix*, *substring*, and *suffix* of S, respectively.

The i-th character of a string S is denoted by $S[i]$, where $1 \leq i \leq |S|$. For a string S and two integers $1 \leq i \leq j \leq |S|$, let $S[i..j]$ denote the substring of S that begins at position i and ends at position j, namely, $S[i..j] = S[i] \cdots S[j]$. For convenience, let $S[i..j] = \varepsilon$ when $i > j$. S^R denotes the reversed string of S, i.e., $S^R = S[|S|] \cdots S[1]$. A non-empty string Z is called a *subsequence* of another string S if there exist increasing positions $1 \leq i_1 < \cdots < i_{|Z|} \leq |S|$ in S such that $Z = S[i_1] \cdots S[i_{|Z|}]$. The empty string ε is a subsequence of any string. A string that is a subsequence of two strings A and B is called a *common subsequence* of A and B.

2.2   STR-IC-LCS

Let A, B, and P be strings. A string Z is said to be *an STR-IC-LCS* of two target strings A and B *including* the pattern P if Z is a longest string such that (1) P is a substring of Z and (2) Z is a common subsequence of A and B.

For ease of exposition, we assume that $n = |A| = |B|$, but our algorithm to follow can deal with the general case where $|A| \neq |B|$. We can also assume that $|P| \leq n$, since otherwise there clearly is no solution. In this paper, we present a space-efficient algorithm that computes an STR-IC-LCS in $O(n^2)$ time and $O((\ell+1)(n-\ell+1))$ space, where $\ell = \mathsf{lcs}(A, B)$ is the longest common subsequence length of A and B. In case where there is no solution, we use a convention that $Z = \perp$ and its length $|\perp|$ is -1. We remark that $\ell \geq |Z|$ always holds.

3   Space-efficient Solution for STR-IC-LCS Problem

In this section, we propose a space-efficient solution for the STR-IC-LCS problem.

Problem 1 (STR-IC-LCS problem). For any given strings A, B of length n and P, compute an STR-IC-LCS of A, B, and P.

Theorem 1. *The STR-IC-LCS problem can be solved in $O(n^2)$ time and $O((\ell+ 1)(n-\ell+1))$ space where ℓ is the length of LCS of A and B.*

$$A = \text{b c a } \mathbf{a} \text{ } \mathbf{b} \text{ a } \mathbf{b} \text{ c b}$$

$$B = \text{c b a c b } \mathbf{a} \text{ } \mathbf{b} \text{ } \mathbf{b} \text{ c}$$

Fig. 1. Let $A = \text{bcdababcb}$, $B = \text{cbacbaaba}$, and $P = \text{abb}$. The length of an STR-IC-LCS of these strings is 6. One of such strings can be obtained by minimal intervals $[4..7]$ over A and $[6..8]$ over B because $\mathsf{lcs}(\text{bca}, \text{cbacb}) = 2$, $|P| = 3$, and $\mathsf{lcs}(\text{cb}, \text{c}) = 1$.

In Sect. 3.1, we explain an overview of our algorithm. In Sect. 3.2, we show a central technique for our space-efficient solution and Sect. 3.3 concludes with the detailed algorithm.

3.1 Overview of Our Solution

Our algorithm uses an algorithm for the STR-IC-LCS problem which was proposed by Deorowicz [5]. Firstly, we explain an outline of the algorithm. Let I_A be the set of minimal intervals over A which have P as a subsequence. Remark that I_A is linear-size since each interval cannot contain any other intervals. There exists a pair of minimal intervals $[b_A, e_A]$ over A and $[b_B, e_B]$ over B such that the length of an STR-IC-LCS is equal to the sum of the three values $\mathsf{lcs}(A[1..b_A - 1], B[1..b_B - 1])$, $|P|$, and $\mathsf{lcs}(A[e_A + 1..n], B[e_B + 1..n])$ (see also Fig. 1 for an example). First, the algorithm computes I_A and I_B and computes the sum of three values for any pair of intervals. If we have an LCS table d of size $n \times n$ such that $d(i, j)$ stores $\mathsf{lcs}(A[1..i], B[1..j])$ for any integers $i, j \in [1..n]$, we can check any LCS value between prefixes of A and B in constant time. It is known that this table can be computed in $O(n^2)$ time by using a simple dynamic programming. Since the LCS tables for prefixes and suffixes requires $O(n^2)$ space, the algorithm also requires $O(n^2)$ space.

Our algorithm uses a space-efficient LCS table by Nakatsu et al. [10] instead of the table d for computing LCSs of prefixes (suffixes) of A and B. The algorithm by Nakatsu et al. also computes a table by dynamic programming, but the table does not gives $\mathsf{lcs}(A[1..i], B[1..j])$ for several pairs (i, j). In the next part, we show a way to resolve this problem.

3.2 Space-efficient Prefix LCS

First, we explain a dynamic programming solution by Nakatsu et al. for computing an LCS of given strings A and B. We give a slightly modified description in order to describe our algorithm. For any integers $i, s \in [1..n]$, let $f_A(s, i)$ be the length of the shortest prefix $B[1..f_A(s, i)]$ of B such that the length of the longest common subsequence of $A[1..i]$ and $B[1..f_A(s, i)]$ is s. For convenience, $f_A(s, i) = \infty$ if no such prefix exists. The values $f_A(s, i)$ will be computed using dynamic programming as follows:

$$f_A(s, i) = \min\{f_A(s, i - 1), j_{s,i}\},$$

where $j_{s,i}$ is the index of the leftmost occurrence of $A[i]$ in $B[f_A(s-1,i-1)+1..n]$. Let s' be the largest value such that $f_A(s',i) < \infty$ for some i, i.e., the s'-th row is the lowest row which has an integer value in the table f_A. We can see that the length of the longest common subsequence of A and B is s' (i.e., $\ell = \mathsf{lcs}(A,B) = s'$). See Fig. 2 for an instance of f_A. Due to the algorithm, we do not need to compute all the values in the table f_A for obtaining the length of an LCS. Let F_A be the sub-table of f_A such that $F_A(s,i)$ stores a value $f_A(s,i)$ if $f_A(s,i)$ is computed in the algorithm of Nakatsu et al. Intuitively, F_A stores the first $n - l + 1$ diagonals of length at most l. Let $\langle i \rangle$ be the set of pairs in the i-th diagonal line ($1 \le i \le n$) of the table f_A:

$$\langle i \rangle = \{(s, i + s - 1) \mid 1 \le s \le n - i + 1\}.$$

Formally, $F_A(s,i) = \mathsf{undefined}$ if

1. $s > i$,
2. $(s,i) \in \langle j \rangle$ $(j > n - \ell + 1)$, or
3. $F_A(s - 1, i - 1) = \infty$ or undefined.

Any other $F_A(s,i)$ stores the value $f_A(s,i)$. Since the lowest row number of each diagonal line $\langle j \rangle$ $(j > n - \ell + 1)$ is less than ℓ, we do not need to compute values which is described by the second item. Actually, we do not need to compute the values in $\langle n - \ell + 1 \rangle$ for computing the LCS since the maximum row number in the last diagonal line is also ℓ. However, we need the values on the last line in our algorithm. Hence the table F_A uses $O((\ell + 1)(n - \ell + 1))$ space (subtable which need to compute is parallelogram-shaped of height ℓ and base $n - \ell$). See Fig. 3 for an instance of F_A.

Now we describe a main part of our algorithm. Recall that a basic idea is to compute $\mathsf{lcs}(A[1..i], B[1..j])$ from F_A. If we have all the values on the table f_A, we can check the length $\mathsf{lcs}(A[1..i], B[1..j])$ as follows.

Observation 1. *The length of an LCS of $A[1..i]$ and $B[1..j]$ for any $i, j \in [1..n]$ is the largest s such that $f_A(s,i) \le j$. If no such s exists, $A[1..i]$ and $B[1..j]$ have no common subsequence of length s.*

However, F_A does not store several integer values w.r.t. the second condition of undefined for some i and j. See also Fig. 3 for an example of the fact. In this example, we can see that $\mathsf{lcs}(A[1..7], B[1..4]) = f_A(3,7) = 3$ from the table f_A, but $F_A(3,7) = \mathsf{undefined}$ in F_A. In order to resolve this problem, we also define F_B (and f_B). Formally, for any integers $j, s \in [1..n]$, let $f_B(s,j)$ be the length of the shortest prefix $A[1..f_B(s,j)]$ of A such that the length of the longest common subsequence of $B[1..j]$ and $A[1..f_B(s,j)]$ is s. Our algorithm accesses the length of an LCS of $A[1..i]$ and $B[1..j]$ for any given i and j by using two tables F_A and F_B. The following lemma shows a key property for the solution.

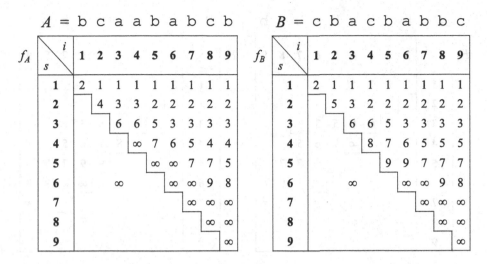

Fig. 2. The LCS-table f_A which is defined by Nakatsu et al. of $A =$ bcdababcb. This figure also illustrates the table f_B of $B =$ cbacbaaba.

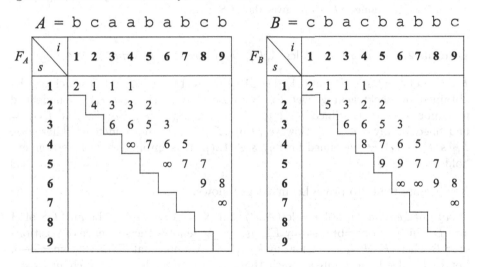

Fig. 3. A sparse table F_A of f_A for $A =$ bcdababcb and $B =$ cbacbaaba does not give $\mathsf{lcs}(A[1..i], B[1..j])$ for some (i, j).

Lemma 1. *Let s be the length of an LCS of $A[1..i]$ and $B[1..j]$. If $F_A(s,i) =$* undefined *then $F_B(s,j) \neq$* undefined.

This lemma implies that the length of an LCS of $A[1..i]$ and $B[1..j]$ can be obtained if we have the two sparse tables (see also Fig. 4). Before we prove this lemma, we show the following property. Let U_{F_A} be the set of pairs (s, i) of integers such that $F_B(s,j) \neq$ undefined where $F_A(s,i) = j$.

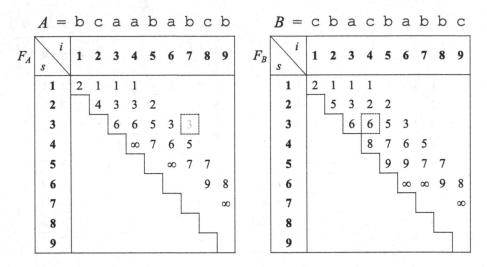

Fig. 4. Due to Observation 1, $f_A(3,7)$ gives the fact that $\mathsf{lcs}(A[1..7], B[1..4]) = 3$. However, $F_A(3,7) = \mathsf{undefined}$. Then we can obtain the fact that $\mathsf{lcs}(A[1..7], B[1..4]) = 3$ by using F_B. Namely, $F_B(3,4)$ gives the LCS value.

Lemma 2. *For any $1 \le s \le \mathsf{lcs}(A, B)$, there exists i such that $(s, i) \in U_{F_A}$.*

Proof. Let $\ell = \mathsf{lcs}(A, B)$ and $A[i_1] \cdots A[i_\ell]$ be an LCS of A and B which can be obtained by backtracking over F_A. Suppose that $F_B(s, F_A(s, i_s)) = \mathsf{undefined}$ for some $s \in [1..\ell]$. Since $F_A(1, i_1) < \cdots < F_A(\ell, i_\ell)$, $F_B(s', F_A(s', i_{s'})) = \mathsf{undefined}$ for any $s' \in [s..\ell]$. However, $F_B(\ell, F_A(\ell, i_\ell))$ is not $\mathsf{undefined}$. Therefore, $F_B(s, F_A(s, i_s)) \ne \mathsf{undefined}$ for any $s \in [1..\ell]$. This implies that the statement holds.   □

Now we are ready to prove Lemma 1 as follows.

Proof (of Lemma 1). Let $\ell = \mathsf{lcs}(A, B)$ and $X = A[i_1] \cdots A[i_\ell]$ be an LCS of A and B which can be obtained by F_A. $j_1, \ldots, j_\ell$ denotes the sequence of positions over B where $F_A(k, i_k) = j_k$ for any $k \in [1..\ell]$. Assume that $F_A(s, i) = \mathsf{undefined}$. Let m be the largest integer such that $i_{s+m} \le i$ holds. If no such m exists, namely $i < i_1$, the statement holds since $F_A(s, i) \ne \mathsf{undefined}$. Due to Lemma 1, $F_A(s + m, i) > j$. Thus $j < j_{s+m}$ holds. On the other hand, we consider the table F_B (and f_B). Let $i' = f_B(s, j)$ and $i'' = f_B(s + m, j)$. Due to Lemma 1, $i' \le i < i''$ holds. From Lemma 2, $F_B(s + m, j_{s+m}) \ne \mathsf{undefined}$. This implies that $F_B(s + m, j) (= i'')$ is not $\mathsf{undefined}$. By the definition of X, $j_{s+m} - j \ge m - 1$. Notice that (s, j) is in $(j - s + 1)$-th diagonal line. These facts imply that $F_B(s, j) \ne \mathsf{undefined}$. See also Fig. 5 for an illustration.   □

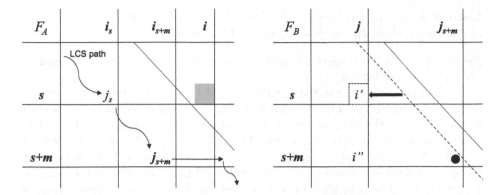

Fig. 5. This figure shows an illustration for the proof of Lemma 1. The length s of an LCS of $A[1..i]$ and $B[1..j]$ cannot be obtained over F_A because $F_A(i,s) = $ undefined (the highlighted cell). However, the length can be obtained by $F_B(s,j)$ over F_B. The existence of $F_B(s+m, j_{s+m})$ from an LCS path guarantees the fact that $F_B(s,j) \neq$ undefined.

3.3   Algorithm

First, our algorithm computes sets of minimal intervals I_A and I_B (similar to the algorithm by Deorowicz [5]). Second, compute the tables F_A and F_B for computing LCSs of prefixes, and the tables F_{A^R} and F_{B^R} for computing LCSs of suffixes (similar to the algorithm by Nakatsu et al. [10]). Third, for any pairs of intervals in I_A and I_B, compute the length of an LCS of corresponding pre-fixes/suffixes and obtain a candidate of the length of an STR-IC-LCS. As stated above, the first and the second steps are similar to the previous work. Here, we describe a way to compute the length of an LCS of prefixes on F_A and F_B in the third step. We can also compute the length of an LCS of suffixes on F_{A^R} and F_{B^R} by using a similar way.

We assume that I_A and I_B are sorted in increasing order of the beginning positions. Let $[\mathsf{b}_A(x)..\mathsf{e}_A(x)]$ and $[\mathsf{b}_B(y)..\mathsf{e}_B(y)]$ be a x-th interval in I_A and a y-th interval in I_B, respectively. We process $O(n^2)$-queries in increasing order of the beginning position of the intervals in I_A. For each interval $[\mathsf{b}_A(x)..\mathsf{e}_A(x)]$ in I_A, we want to obtain the length of an LCS of $A[1..\mathsf{b}_A(x) - 1]$ and $B[1..\mathsf{b}_B(1) - 1]$. For convenience, let $i_x = \mathsf{b}_A(x) - 1$ and $j_y = \mathsf{b}_B(y) - 1$. In the rest of this section, we use a pair (x, y) of integers to denote a prefix-LCS query (computing $\mathsf{lcs}(A[1..i_x], B[1..i_y])$). We will find the LCS by using Observation 1. Here, we describe how to compute prefix-LCS queries $(i_x, j_1), \ldots, (i_x, j_{|I_B|})$ in this order for a fixed i_x.

Lemma 3. *All required prefix-LCS values for an interval $[b_A(x)..e_A(x)]$ in I_A and all intervals in I_B can be computed in $O(n)$ time.*

Proof. There exist two cases for each i_x. Formally, (1) $F_A[i_x, 1] \neq$ undefined or (2) $F_A[i_x, 1] =$ undefined.

In the first case, we scan the i_x-th column of F_A from the top to the bottom in order to find the maximum value which is less than or equal to j_1. If such a value exists in the column, then the row number s_1 is the length of an LCS. After that, we are given the next prefix-LCS query (i_x, j_2). It is easy to see that $s_0 = \mathsf{lcs}(A[1..i_x], B[1..j_1]) \leq \mathsf{lcs}(A[1..i_x], B[1..j_2])$ since $j_1 < j_2$. This implies that the next LCS value is equal to s_0 or that is placed in a lower row in the column. This means that we can start to scan the column from the s_0-th row. Thus we can answer all prefix-LCSs for a fixed i_x in $O(n)$ time (that is linear in the size of I_B).

In the second case, we start to scan the column from the top $F_A[i_x, i_x - n - \ell + 1]$ (the first $i_x - n - \ell$ rows are undefined). If $F_A[i_x, i_x - n - \ell + 1] \leq j_1$, then the length of an LCS for the first query (i_x, j_1) can be found in the table (similar to the first case) and any other queries $(i_x, j_2), \ldots, (i_x, j_{|I_B|})$ can be also answered in the similar way. Otherwise (if $F_A[i_x, i_x - n - \ell + 1] > j_1$), the length which we want may be in the "undefined" domain. Then we use the other table F_B. We scan the j_1-th column in F_B from the top to the bottom in order to find the maximum value which is less than or equal to i_x. By Lemma 1, such a value must exist in the column (if $\mathsf{lcs}(A[1..i_x], B[1..j_1]) > 0$ holds) and the row number s' is the length of an LCS. After that, we are given the next query (i_x, j_2). If $F_A[i_x, i_x - n - \ell + 1] \leq j_2$, then the length can be found in the table (similar to the first case). Otherwise (if $F_A[i_x, i_x - n - \ell + 1] > j_2$), the length must be also in the "undefined" domain. Since such a value must exist in the j_2-th column in F_B by Lemma 1, we scan the column in F_B. It is easy to see that $s' = \mathsf{lcs}(A[1..i_x], B[1..j_1]) \leq \mathsf{lcs}(A[1..i_x], B[1..j_2])$. This implies that the length of an LCS that we want to find is in lower row. Thus it is enough to scan the j_2-th column from the s'-th row to the bottom. Then we can answer the second query (i_x, j_2). Hence we can compute all LCSs for a fixed i_x in $O(n + \ell)$ time (that is linear in the size of I_B or the number of rows in the table F_B).

Therefore we can compute all prefix-LCSs for each interval in I_A in $O(n)$ time (since $n \geq \ell$).   □

On the other hand, we can compute all required suffix-LCS values with computing prefix-LCS values. We want a suffix-LCS value of $A[e_A(x) + 1..n]$ and $B[e_B(y) + 1..n]$ $(1 \leq y \leq |I_B|)$ when we compute the length of an LCS of $A[1..b_A(x) - 1]$ and $B[1..b_B(y) - 1]$. Recall that we process all intervals of I_B in increasing order of the beginning positions when computing prefix-LCS values with a fixed interval of I_A. This means that we need to process all intervals of I_B in "decreasing order" when computing suffix-LCS values with a fixed interval of I_A. We can do that by using an almost similar way on F_{A^R} and F_{B^R}. The most significant difference is that we scan the $|A[e_A(x) + 1..n]|$-th column of F_{A^R} from the ℓ-th row to the first row.

Overall, we can obtain the length of an STR-IC-LCS in $O(n^2)$ time in total. Also this algorithm requires space for storing all minimal intervals and tables, namely, requiring $O(n + (\ell + 1)(n - \ell + 1)) = O((\ell + 1)(n - \ell + 1))$ space in the worst case. Finally, we can obtain Theorem 1.

Algorithm 1. Algorithm for computing the length of STR-IC-LCS

Input: A, B, P $(|A| = n, |B| = n, |P| = r)$
Output: l, C

1: compute I_A and I_B
2: compute F_A, F_B, F_{A^R}, and F_{B^R}
3: $\ell \leftarrow \mathsf{lcs}(A, B)$;
4: $l \leftarrow 0$;
5: **for** $i = 1$ to $|I_A|$ **do**
6:    $k_1^A \leftarrow 1;\ k_1^B \leftarrow 1;\ k_2^A \leftarrow \ell;\ k_2^B \leftarrow \ell$;
7:    **for** $j = 1$ to $|I_B|$ **do**
8:       $k_1 \leftarrow 0;\ k_2 \leftarrow 0$;
9:       compute $\mathsf{lcs}(A[1..b_A(i) - 1], B[1..b_B(j) - 1])$ // as k_1 by Algorithm 2
10:     compute $\mathsf{lcs}(A[e_A(i) + 1..n], B[e_B(j) + 1..n])$ // as k_2 by Algorithm 3
11:     **if** $k_1 + k_2 + r > l$ **then**
12:       $l \leftarrow k_1 + k_2 + r$
13:     **end if**
14:    **end for**
15: **end for**
16: **return** l

In addition, we can also compute an STR-IC-LCS (as a string), if we store a pair of minimal intervals which produce the length of an STR-IC-LCS. Namely, we can find a cell which gives the prefix-LCS value over F_A or F_B. Then we can obtain a prefix-LCS string by a simple backtracking (a suffix-LCS can be also obtained by backtracking on F_{A^R} or F_{B^R}). On the other hand, we can also use an algorithm that computes an LCS string in $O(n^2)$ time and $O(n)$ space by Hirschberg [6]. We conclude with supplemental pseudocodes of our algorithm (see Algorithms 1,2, and 3).

Algorithm 2. Computing $\mathsf{lcs}(A[1..\mathsf{b}_A(i) - 1], B[1..\mathsf{b}_B(j) - 1])$

```
 1: for k ← k₁ᴬ to ℓ do
 2:     if F_A[b_A(i) − 1, k] ≤ b_B(j) − 1 then
 3:         if F_A[b_A(i) − 1, k + 1] > b_B(j) − 1 then
 4:             k₁ ← k
 5:             k₁ᴬ ← k
 6:             break
 7:         end if
 8:     else if F_A[b_A(i) − 1, k] > b_B(j) − 1 then
 9:         if F_A[b_A(i) − 1, k − 1] = undefined then
10:             k₁ᴬ ← k
11:             for k′ = k₁ᴮ to ℓ do
12:                 if F_B[b_B(j) − 1, k′] > b_A(i) − 1 then
13:                     k₁ ← 0
14:                     k₁ᴮ ← k′
15:                     break
16:                 else if F_B[b_B(j) − 1, k′ + 1] > b_A(i) − 1 then
17:                     k₁ ← k′
18:                     k₁ᴮ ← k′
19:                     break
20:                 end if
21:             end for
22:         else
23:             k₁ ← 0
24:             k₁ᴬ ← k
25:             break
26:         end if
27:     end if
28: end for
```

Algorithm 3. Computing $\mathsf{lcs}(A[\mathsf{e}_A(i) + 1..n], B[\mathsf{e}_B(j) + 1..n])$

```
 1: for k = k₂ᴬ to 1 do
 2:     if F_{AR}[n − e_A(i), k] ≤ n − e_B(j) then
 3:         k₂ ← k
 4:         k₂ᴬ ← k
 5:         break
 6:     else if F_{AR}[n − e_A(i), k] > n − e_B(j) then
 7:         if F_{AR}[n − e_A(i), k − 1] = undefined then
 8:             k₂ᴬ ← k
 9:             for k′ = k₂ᴮ to 1 do
10:                 if F_{BR}[n − e_B(j), k′] ≤ n − e_A(i) then
11:                     k₂ ← k′
12:                     k₂ᴮ ← k′
13:                     break
14:                 else if F_{BR}[n − e_B(j), k′ − 1] = undefined then
15:                     k₂ ← 0
16:                     k₂ᴮ ← k′
17:                     break
18:                 end if
19:             end for
20:         end if
21:     end if
22: end for
```

4   Conclusions and Future Work

We have presented a space-efficient algorithm that finds an STR-IC-LCS of two given strings A and B of length n in $O(n^2)$ time with $O((\ell+1)(n-\ell+1))$ working space, where ℓ is the length of an LCS of A and B. Our method improves on the space requirements of the algorithm by Deorowicz [5] that uses $\Theta(n^2)$ space, irrespective of the value of ℓ.

Our future work for STR-IC-LCS includes improvement of the $O(n^2)$-time bound to, say, $O(n(n - \ell + 1))$. We note that the algorithm by Nakatsu et al. [10] for finding (standard) LCS runs in $O(n(n - \ell + 1))$ time. There also exists an $O(n\sigma + (\ell' + 1)(n - \ell' + 1)r)$-time solution for the STR-EC-LCS problem that runs fast when the length ℓ' of the solution is small [13], where $r = |P|$.

Acknowledgments. This work was supported by JSPS KAKENHI Grant Numbers JP21K17705 (YN), JP22H03551 (SI), JP20H04141 (HB), and by JST PRESTO Grant Number JPMJPR1922 (SI).

References

1. Abboud, A., Backurs, A., Williams, V.V.: Tight hardness results for LCS and other sequence similarity measures. In: FOCS 2015, pp. 59–78 (2015)
2. Bille, P., Farach-Colton, M.: Fast and compact regular expression matching. Theor. Comput. Sci. **409**(3), 486–496 (2008)
3. Chen, Y.C., Chao, K.M.: On the generalized constrained longest common subsequence problems. J. Comb. Optim. **21**(3), 383–392 (2011). https://doi.org/10.1007/s10878-009-9262-5
4. Chin, F.Y., Santis, A.D., Ferrara, A.L., Ho, N., Kim, S.: A simple algorithm for the constrained sequence problems. Inf. Process. Lett. 90(4), 175–179 (2004). https://doi.org/10.1016/j.ipl.2004.02.008, http://www.sciencedirect.com/science/article/pii/S0020019004000614
5. Deorowicz, S.: Quadratic-time algorithm for a string constrained LCS problem. Inf. Process. Lett. 112(11), 423–426 (2012). https://doi.org/10.1016/j.ipl.2012.02.007, http://www.sciencedirect.com/science/article/pii/S0020019012000567
6. Hirschberg, D.S.: A linear space algorithm for computing maximal common subsequences. Commun. ACM **18**(6), 341–343 (1975)
7. Kuboi, K., Fujishige, Y., Inenaga, S., Bannai, H., Takeda, M.: Faster STR-IC-LCS computation via RLE. In: Kärkkäinen, J., Radoszewski, J., Rytter, W. (eds.) 28th Annual Symposium on Combinatorial Pattern Matching, CPM 2017, July 4–6, 2017, Warsaw, Poland. LIPIcs, vol. 78, pp. 20:1–20:12. Schloss Dagstuhl - Leibniz-Zentrum fuer Informatik (2017). https://doi.org/10.4230/LIPIcs.CPM.2017.20
8. Liu, J.J., Wang, Y.L., Chiu, Y.S.: Constrained Longest Common Subsequences with Run-Length-Encoded Strings. Comput. J. **58**(5), 1074–1084 (2014). https://doi.org/10.1093/comjnl/bxu012
9. Masek, W.J., Paterson, M.: A faster algorithm computing string edit distances. J. Comput. Syst. Sci. **20**(1), 18–31 (1980)
10. Nakatsu, N., Kambayashi, Y., Yajima, S.: A longest common subsequence algorithm suitable for similar text strings. Acta Inf. **18**, 171–179 (1982). https://doi.org/10.1007/BF00264437

11. Tsai, Y.T.: The constrained longest common subsequence problem. Inf. Process. Lett. 88(4), 173–176 (2003). https://doi.org/10.1016/j.ipl.2003.07.001, http://www.sciencedirect.com/science/article/pii/S002001900300406X

12. Wagner, R.A., Fischer, M.J.: The string-to-string correction problem. J. ACM 21(1), 168–173 (1974). https://doi.org/10.1145/321796.321811, http://doi.acm.org/10.1145/321796.321811

13. Yamada, K., Nakashima, Y., Inenaga, S., Bannai, H., Takeda, M.: Faster STR-EC-LCS computation. In: Chatzigeorgiou, A., et al. (eds.) SOFSEM 2020. LNCS, vol. 12011, pp. 125–135. Springer, Cham (2020). https://doi.org/10.1007/978-3-030-38919-2_11

The k-Centre Problem for Classes
of Cyclic Words

Duncan Adamson[1]([✉]), Argyrios Deligkas[2], Vladimir V. Gusev[3,4],
and Igor Potapov[4]

[1] Department of Computer Science, Reykjavik University, Reykjavik, Iceland
`duncana@ru.is`
[2] Department of Computer Science, Royal Holloway, University of London,
Egham, UK
`argyrios.deligkas@rhul.ac.uk`
[3] Materials Innovation Factory, University of Liverpool, Liverpool, UK
`Vladimir.Gusev@liverpool.ac.uk`
[4] Department of Computer Science, University of Liverpool, Liverpool, UK
`potapov@liverpool.ac.uk`

Abstract. The problem of finding k uniformly spaced points (centres) within a metric space is well known as the k-centre selection problem. In this paper, we introduce the challenge of k-centre selection on a class of objects of exponential size and study it for the class of combinatorial necklaces, known as cyclic words. The interest in words under translational symmetry is motivated by various applications in algebra, coding theory, crystal structures and other physical models with periodic boundary conditions. We provide solutions for the centre selection problem for both one-dimensional necklaces and largely unexplored objects in combinatorics on words - multidimensional combinatorial necklaces. The problem is highly non-trivial as even verifying a solution to the k-centre problem for necklaces can not be done in polynomial time relative to the length of the cyclic words and the alphabet size unless $P = NP$. Despite this challenge, we develop a technique of centre selection for a class of necklaces based on de-Bruijn Sequences and provide the first polynomial $O(k \cdot n)$ time approximation algorithm for selecting k centres in the set of 1D necklaces of length n over an alphabet of size q with an approximation factor of $O\left(1 + \frac{\log_q (k \cdot n)}{n - \log_q (k \cdot n)}\right)$. For the set of multidimensional necklaces of size $n_1 \times n_2 \times \ldots \times n_d$ we develop an $O(k \cdot N^2)$ time algorithm with an approximation factor of $O\left(1 + \frac{\log_q (k \cdot N)}{N - \log_q (k \cdot N)}\right)$ in $O(k \cdot N^2)$ time, where $N = n_1 \cdot n_2 \cdot \ldots \cdot n_d$ by approximating de Bruijn hypertori technique.

1 Introduction

The problem of finding k uniformly spaced points (centres) within a metric space is well known as the k-centre selection problem. So far, the problem has been

I. Potapov—Partially supported by ESPRC grant (EP/R018472/1).

L. Gąsieniec (Ed.): SOFSEM 2023, LNCS 13878, pp. 385–400, 2023.
https://doi.org/10.1007/978-3-031-23101-8_26

intensely studied for finite, and explicitly given inputs like the k-centre problem for graphs, grids, or a set of strings, which are essential in the context of facility location and distribution [9,16,33].

The k-centre problem is also a tool in state space exploration, where cluster centres or equally spaced centres need to be selected to guarantee effective coverage of the configuration space. For algebraic and combinatorial structures with a state space of exponential size, sampling techniques have been used to generate equally probable objects [7]. However, while such sampling techniques can give uniform probability to the selection of any given object, there is a substantial gap in the problem of ensuring that k samples are representative. The k-centre problem is a natural means of modelling this objective, with the goal of ensuring that no object is significantly far from the set of samples under a distance based on some similarity metric. However, if the explicit representation of the whole class of objects is infeasible to store and process due to its exponential size, the k-centre selection problem requires new solutions and approaches.

In this work, we consider the class of combinatorial necklaces (also known as cyclic words). The study of 1D necklaces has been motivated by applications in the coding theory, free Lie algebras, and Hall sets [2,3,5,17,23,24,30]. Moreover, 2D necklaces have been recently studied for counting the number of toroidal codes in [6] and can be used in the construction of 2D Gray codes [8]. Algorithms for multidimensional combinatorial necklaces have remained a largely unexplored area in combinatorics on words [2,27,32]. A multidimensional necklace is an equivalence class of multidimensional words under *translational symmetry*, which is the natural generalisation of the shift operation in 1D, see Fig. 1.

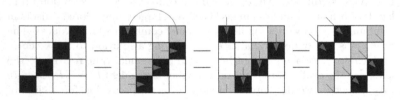

Fig. 1. An illustration of translational symmetry for a 4×4 word. Note that all 4 words presented here (out of a total of 16) correspond to the same necklace and can be reached from one another through some two-dimensional translation denoted (g_1, g_2). In red, the translation from the starting word to the new word is highlighted, with the original word overlaid in grey. (Color figure online)

One natural use of multidimensional necklaces up to dimension three is the combinatorial representation of crystal structures. In computational chemistry, crystals are represented by periodic motives known as "unit cells". Informally, translational symmetry can be thought of as the equivalence of two crystals under translation in space. Intuitively this symmetry makes sense in the context of real structures, where two different "snapshots" of a unit cell both represent the same periodic and infinite global structure, see Fig. 2.

Crystal Structure Prediction (CSP) is one of the most central and challenging problems in materials science and computational chemistry [1,4]. The objective is to find the "best" periodic three-dimensional structure of ions that yields the lowest interatomic potential energy. The aim of our k-centre selection algorithms for multidimensional necklaces is to replace the currently-used random generation approaches of unit cells [12] that often lead to identical crystal structures during the process of configuration-space exploration.

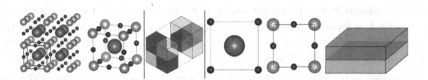

Fig. 2. The crystal of $SrTiO_3$ (left) and its 3D (middle) and 1D (right) necklace representations.

Most of the existing methods for CSP require the exploration of different configurations of periodic structures that combine local exploration and selection of new random locations in unexplored configuration space. The number of unit cells of size $n_1 \times n_2 \times \ldots \times n_d$ in the d-dimensional integer lattice and considering them up to translational symmetry is exponential and it is larger than $\frac{q^N}{N}$, where q is the number of different ions and $N = \prod_{i=1}^{d} n_i$. The size of such an object makes it infeasible to represent this set explicitly in the form of a weighted graph. By extension, applying existing centre selection algorithms will lead to an EXPSPACE solution and therefore require new techniques for operating on implicitly represented combinatorial objects. The same problem exists when it may be required to construct k equally spaced code words from a set of multidimensional cyclic words.

Even the original k-centre problem on graphs is non-trivial. The k-centre problem is both NP-hard with respect to the size of a graph and is APX-hard [18], making a PTAS unlikely. Additionally, the k-centre problem is unlikely to be fixed-parameter tractable in the context of the most natural parameter k [13]. A different form of the k-centre problem appears in stringology with important applications in computational biology; for example, to find the approximate gene clusters for a set of words over the DNA alphabet [14,25,26].

This paper introduces the challenge of k-centre selection for implicitly represented sets. Notably, we aim for polynomial time algorithms in the length of the output rather than in the size of the graph. The length of the output corresponds to a logarithmic factor relative to the size of the graph, multiplied by some function on the number of centres. The k-centre problem for strings or words can be defined over various distance functions. In this paper we focus on the *overlap distance*, based on the *overlap coefficient* (well known in linguistic processing [11,28,29]). The overlap coefficient measures the similarity of two words relative to the number of common subwords. This measure can, in turn, be used as a proxy for the closeness of potential energy in crystals. However, it is not critical for our algorithmic results; all results could be reformulated using other distance functions at the cost of slightly different approximation bounds.

In particular we develop a technique of centre selection based on de-Bruijn Sequences and provide the first polynomial $O(k \cdot n)$ time approximation algorithm for selecting k centres in the set of 1D necklaces of length n over an alphabet of size q with an approximation factor of $O\left(1 + \frac{\log_q(k \cdot n)}{n - \log_q(k \cdot n)}\right)$. In the multidimensional case, the results on generating de Bruijn tori are highly limited, so we developed a technique to select centres by approximating de Bruijn hypertori. We present an algorithm that generates k centres for the set of multidimensional necklaces of size $n_1 \times n_2 \times \ldots \times n_d$ with an approximation of $O\left(1 + \frac{\log_q(k \cdot N)}{N - \log_q(k \cdot N)}\right)$ in $O(k \cdot N^2)$ time, where $N = n_1 \cdot n_2 \cdot \ldots \cdot n_d$. Moreover, we show that verifying a solution to the k-centre problem for necklaces can not be done in polynomial time relative to the length of the cyclic words and the alphabet size unless $P = NP$, indicating that the k-centre problem itself is likely to be at least NP-hard.

2 Preliminaries

Let Σ be a finite alphabet of size q. In this paper, we assume that Σ is linearly ordered. We denote by Σ^* the set of all words over Σ and by Σ^n the set of all words of length n. The length of a word $u \in \Sigma^*$ is denoted by $|u|$. We use u_i, for any $i \in \{1, \ldots, |u|\}$, to denote the ith symbol of u.

Let $[n]$ return the ordered set of integers from 1 to n inclusive. Given 2 words $u, v \in \Sigma^*$ where $|u| = |v|$, $u = v$ if and only if $u_i = v_i$ for every $i \in [|u|]$. A word u is *lexicographically smaller* than v if there exists an $i \in [|u|]$ such that $u_1 u_2 \ldots u_{i-1} = v_1 v_2 \ldots v_{i-1}$ and $u_i < v_i$ or $|u| < |v|$ and $u_1 u_2 \ldots u_{|u|} = v_1 v_2 \ldots v_{|u|}$. For example, given the alphabet $\Sigma = \{a, b\}$ where $a < b$, the word $aaaba$ is smaller than $aabaa$ as the first 2 symbols are the same and a is smaller than b. For a given set of words $\mathbf{S}$, the rank of v with respect to $\mathbf{S}$ is the number of words in $\mathbf{S}$ that are smaller than v.

The *translation (cyclic shift)* of a word $w = w_1 w_2 \ldots w_n$ by $r \in [n]$ returns the word $w_{r+1} \ldots w_n w_1 \ldots w_r$, and is denoted by $\langle w \rangle_r$, i.e. $\langle w_1 w_2 \ldots w_n \rangle_r = w_{r+1} \ldots w_n w_1 \ldots w_r$. Under the translation operation, u is equivalent to v if $v = \langle w \rangle_r$ for some $r \in [n]$. The tth power of a word w, denoted by w^t, is equal to w repeated t times. For example $(aab)^3 = aabaabaab$. A word w is *periodic* if there is some word u and integer $t \geq 2$ such that $u^t = w$. Equivalently, word w is *periodic* if there exists some translation $r \in [|w| - 1]$ where $w = \langle w \rangle_r$. A word is *aperiodic* if it is not periodic. The *period* of a word w is the length of the smallest word u for which there exists some value t for which $w = u^t$.

A *necklace* is an equivalence class of words under the translation operation. An aperiodic necklace is called a *Lyndon word*. For notation, a word w is written as $\tilde{\mathbf{w}}$ when treated as a necklace. Given a necklace $\tilde{\mathbf{w}}$, the *canonical form* of $\tilde{\mathbf{w}}$

is the lexicographically smallest element of the set of words in the equivalence class $\tilde{w}$. The canonical form of $\tilde{w}$ is denoted by $\langle \tilde{w} \rangle$, and the rth shift of the canonical form is denoted by $\langle \tilde{w} \rangle_r$. Given a word w, $\langle w \rangle$ denotes the canonical form of the necklace containing w.

A *subword* of the necklace w, denoted by $w_{[i,j]}$ is the word u of length $|w| + j - i - 1 \bmod |w|$ such that $u_a = w_{i-1+a \bmod |w|}$ for every $a \in |w|$. For notation, $u \sqsubseteq w$ denotes that u is a subword of w. Further, $u \sqsubseteq_i w$ denotes that u is a subword of w of length i. If a word u is a subword of w, then u is also a subword of the necklace $\langle w \rangle$. We denote that u is a subword of some necklace $\tilde{w}$ by $u \sqsubseteq \tilde{w}$, and that u is a subword of $\tilde{w}$ of length i by $u \sqsubseteq_i \tilde{w}$.

If $w = uv$, then u is a prefix and v is a suffix. For notation, the tuple $\mathbf{S}(v, \ell)$ is defined as the set of all subwords of v of length ℓ. Formally let $\mathbf{S}(v, \ell) = \{s \sqsubseteq v : |s| = \ell\}$. Further, $\mathbf{S}(v, \ell)$ is assumed to be in lexicographic order, i.e. $\mathbf{S}(v, \ell)_1 \geq \mathbf{S}(v, \ell)_2 \geq \ldots \mathbf{S}(v, \ell)_{|v|}$, where $\mathbf{S}(v, \ell)_i$ denotes the ith entry of $\mathbf{S}(v, \ell)$. The set of necklaces of length n over an alphabet of size q is denoted by $\mathcal{N}_q^n$.

Multidimensional Words and Necklaces: In order to establish multidimensional necklaces, notation for *multidimensional words* must first be introduced. A d-*dimensional word* over Σ is an array of size $\vec{n} = (n_1, n_2, \ldots, n_d)$ of elements from Σ. In this work we tacitly assume that $n_1 \leq n_2 \leq \ldots \leq n_d$ unless otherwise stated. Let $|w|$ denote the vector of length d defining the size of the multidimensional word w. Given a size vector $\vec{n} = (n_1, n_2, \ldots, n_d)$, $\Sigma^{\vec{n}}$ is used to denote the set of all words of size $\vec{n}$ over Σ. Given a vector $\vec{n} = (n_1, n_2, \ldots, n_d)$ where every $n_i \geq 0$, $[\vec{n}]$ is used to denote the set $\{(x_1, x_2, \ldots, x_d) \in \mathbb{N}^d | \forall i \in [d], x_i \leq n_i\}$. Similarly $[\vec{m}, \vec{n}]$ is used to denote the set $\{(x_1, x_2, \ldots, x_d) \in \mathbb{N}^d \mid \forall i \in [d], m_i \leq x_i \leq n_i\}$.

For a d-dimensional word w, the notation $w_{(p_1, p_2, \ldots, p_d)}$ is used to refer to the symbol at position $(p_1, p_2, \ldots, p_d)$ in the array. Given two d-dimensional words w, u such that $|w| = (n_1, n_2, \ldots, n_{d-1}, a)$ and $|u| = (n_1, n_2, \ldots, n_{d-1}, b)$, the concatenation wu is performed along the last dimension, returning the word v of size $(n_1, n_2, \ldots, n_{d-1}, a + b)$ such that $v_{\vec{p}} = w_{\vec{p}}$ if $p_d \leq a$ and $v_{\vec{p}} = u_{(p_1, p_2, \ldots, p_{d-1}, p_d - a)}$ if $p_d > a$.

A *multidimensional subword* of w of size $\vec{m}$ is denoted by $v \sqsubseteq_{\vec{m}} w$. As in the 1D case, a subword is defined by a start and an end position within the original word w. Let $w[\vec{i}, \vec{j}]$ for $\vec{i}, \vec{j} \in [\vec{n}]$ denote the subword u of size $(j_1 - i_1 + 1, j_2 - i_2 + 1, \ldots, j_d - i_d + 1)$. The symbol at position $\vec{p}$ of u equals the symbol at position $(i_1 + p_1, i_2 + p_2, \ldots, i_d + p_d)$ of w, i.e. $u_{\vec{p}} = w_{(i_1 + p_1, i_2 + p_2, \ldots, i_d + p_d)}$.

A d-*dimensional translation* r is defined by a d-tuple $r = (r_1, r_2, \ldots, r_d)$. The translation of the word $w \in \Sigma^{\vec{n}}$ by r, denoted by $\langle w \rangle_r$, returns the word $v \in \Sigma^{\vec{n}}$ such that $v_{\vec{p}} = w_{\vec{j}}$ for every position $\vec{p} \in [\vec{n}]$ where $\vec{j} = (p_1 + r_1 \bmod n_1, p_2 + r_2 \bmod n_2, \ldots, p_d + r_d \bmod n_d)$. It is assumed that $r_i \in [0, n_i - 1]$, so the set of translations $Z_{\vec{n}}$ is equivalent to the direct product of the cyclic groups, giving $Z_{\vec{n}} = Z_{n_1} \times Z_{n_2} \times \ldots \times Z_{n_d}$. Given two translations $r = (r_1, r_2, \ldots, r_d)$ and $t = (t_1, t_2, \ldots, t_d)$ in $Z_{\vec{n}}$, $t + r$ is used to denote the translation $(r_1 + t_1 \bmod n_1, r_2 + t_2 \bmod n_2, \ldots, r_d + t_d \bmod n_d)$.

Definition 1. *A **multidimensional necklace** $\tilde{w}$ is an equivalence class of multidimensional words under the translation operation. The set of multidimensional necklaces over an alphabet of size q of size $n_1 \times n_2 \times \ldots \times n_d$ is denoted by $\mathcal{N}_q^{\vec{n}}$ where $\vec{n} = (n_1, n_2, \ldots, n_d)$.*

Proposition 1. *The number of multidimensional necklaces of size $n_1 \times n_2 \times \ldots \times n_d$ over an alphabet of size q is bounded by $\frac{q^N}{N} \leq |\mathcal{N}_q^{\vec{n}}|$, where $N = \prod_{i=1}^{d} n_i$.*

Proof. Given any word $w \in \Sigma^{\vec{n}}$, there are at most $N - 1$ words equivalent to w under the translation operation, giving $|\mathcal{N}_q^{\vec{n}}| \geq \frac{|\Sigma^{\vec{n}}|}{N} = \frac{q^N}{N}$. $\square$

3   The k-centre Problem for Necklaces

In this section, we formally define the k-centre problem for a set of necklaces. The input to our problem is some positive integer k, an alphabet Σ, and positive integer length n. The goal is to choose a set $\mathbf{S}$ of k centres from the implicitly defined set of necklaces such that the maximum distance between any member of the input set and the set of centres $\mathbf{S}$ is minimised. For example, when the problem is defined over the set $\mathcal{N}_q^n$ of q-ary necklaces of length n, the problem is to select some subset $\mathbf{S} \subseteq \mathcal{N}_q^n$ such that $|\mathbf{S}| = k$ and the distance between each necklace $\tilde{w} \in \mathcal{N}_q^n$ and the necklace $\tilde{u} \in \mathbf{S}$ that is closest to $\tilde{w}$ is minimised.

The remainder of this section formalises the k centre problem for necklaces. Section 3.1 defines the *overlap distance* between necklaces. At a high-level, the overlap distance between two necklaces is the inverse of the *overlap coefficient* between them, in this case, 1 minus the overlap coefficient. This distance can be seen as a natural distance based on "bag-of-words" techniques used in machine learning [15]. Section 3.2 uses the overlap distance to define the k-centre problem for classes of necklaces. Along with a problem definition, we provide preliminary results on the complexity of the k-centre problem for necklaces, as well as theoretical lower bounds on the optimal solution in the necklace setting.

3.1   The Overlap Distance and the k-centre Problem

Our definition of the overlap distance depends on the well-studied *overlap coefficient*, defined for a pair of sets A and B as $\mathfrak{C}(A, B) = \frac{|A \cap B|}{\min(|A|, |B|)}$. In the context of necklaces $\mathfrak{C}(\tilde{w}, \tilde{v})$ is defined as the overlap coefficient between the multisets of all subwords of $\tilde{w}$ and $\tilde{v}$. For some necklace $\tilde{w}$ of size $\vec{n}$, the multiset of subwords of size $\vec{\ell}$ contains all $u \sqsubseteq_{\vec{\ell}} w$. For each subword u appearing m times in $\tilde{w}$, m copies of u are added to the multiset. This gives a total of N subwords of size $\vec{\ell}$ for any $\vec{\ell}$, where $N = n_1 \cdot n_2 \cdot \ldots \cdot n_d$. For example, given the necklace represented by $aaab$, the multiset of subwords of length 2 are $\{aa, aa, ab, ba\} = \{aa \times 2, ab, ba\}$. The multiset of all subwords is the union of the multisets of the subwords for every size vector, with a total of N^2 subwords; see Fig. 3.

To use the overlap coefficient as a distance between $\tilde{w}$ and $\tilde{v}$, the overlap coefficient is inverted so that a value of 1 means $\tilde{w}$ and $\tilde{v}$ share no common subwords while a value of 0 means $\tilde{w} = \tilde{v}$. The overlap distance (see example in Fig. 3) between two necklaces $\tilde{w}$ and $\tilde{v}$ is $\mathfrak{O}(\tilde{w}, \tilde{v}) = 1 - \mathfrak{C}(\tilde{w}, \tilde{v})$. Proposition 2 shows that this distance is a metric distance.

	word $ababab$	word $abbabb$	Intersection
1	$a \times 3, b \times 3$	$a \times 2, b \times 4$	5
2	$ab \times 3, ba \times 3$	$ab \times 2, bb \times 2, ba \times 2$	4
3	$aba \times 3, bab \times 3$	$abb \times 2, bba \times 2, bab \times 2$	2
4	$abab \times 3, baba \times 3$	$abba \times 2, bbab \times 2, babb \times 2$	0
5	$ababa \times 3, babab \times 3$	$abbab \times 2, bbabb \times 2, babba \times 2$	0
6	$ababab \times 3, bababa \times 3$	$abbabb \times 2, bbabba \times 2, babbab \times 2$	0
Total			11

Fig. 3. Example of the overlap coefficient calculation for a pair of words $ababab$ and $abbabb$. There are 11 common subwords out of the total number of 36 subwords of length from 1 till 6, so $\mathfrak{C}(ababab, abbabb) = \frac{11}{36}$ and $\mathfrak{O}(ababab, abbabb) = \frac{25}{36}$.

A $aaaa$ B $aaab$ C $aabb$
D $abab$ E $abbb$ F $bbbb$

$\tilde{w}\backslash\tilde{v}$	A	B	C	D	E	F
A	0	$\frac{10}{16}$	$\frac{13}{16}$	$\frac{14}{16}$	$\frac{15}{16}$	1
B	$\frac{10}{16}$	0	$\frac{9}{16}$	$\frac{10}{16}$	$\frac{12}{16}$	$\frac{15}{16}$
C	$\frac{13}{16}$	$\frac{9}{16}$	0	$\frac{10}{16}$	$\frac{8}{16}$	$\frac{13}{16}$
D	$\frac{14}{16}$	$\frac{10}{16}$	$\frac{10}{16}$	0	$\frac{6}{16}$	$\frac{14}{16}$
E	$\frac{15}{16}$	$\frac{12}{16}$	$\frac{8}{16}$	$\frac{6}{16}$	0	$\frac{8}{16}$
F	1	$\frac{15}{16}$	$\frac{13}{16}$	$\frac{14}{16}$	$\frac{8}{16}$	0

Fig. 4. Example of the overlap distance $\mathfrak{O}(\langle\tilde{w}\rangle, \langle\tilde{v}\rangle)$ for all necklaces in $\mathcal{N}_2^4$.

Proposition 2. *The overlap distance for necklaces is a metric distance.*

Proof. Let $\tilde{a}, \tilde{b}, \tilde{c} \in \mathcal{N}_q^{\vec{n}}$, for some arbitrary vector $\vec{n} \in \mathbb{N}^d$ and $q \in \mathbb{N}$. The overlap distance is metric if and only if $\mathfrak{O}(\tilde{a}, \tilde{b}) \leq \mathfrak{O}(\tilde{a}, \tilde{c}) + \mathfrak{O}(\tilde{b}, \tilde{c})$. Rewriting this gives $1 - \mathfrak{C}(\tilde{a}, \tilde{b}) \leq 2 - \mathfrak{C}(\tilde{a}, \tilde{b}) - \mathfrak{C}(\tilde{b}, \tilde{c})$ which can be rewritten in turn as $\mathfrak{C}(\tilde{a}, \tilde{b}) + \mathfrak{C}(\tilde{b}, \tilde{c}) \leq 1 + \mathfrak{C}(\tilde{a}, \tilde{b})$. Observe that if $\mathfrak{C}(\tilde{a}, \tilde{c}) + \mathfrak{C}(\tilde{b}, \tilde{c}) > 1$ then $\frac{|\tilde{a}\cap\tilde{c}|}{N^2} + \frac{|\tilde{b}\cap\tilde{c}|}{N^2} > 1$, meaning that $|\tilde{a} \cap \tilde{c}| + |\tilde{b} \cap \tilde{c}| > N^2$. This implies that $\tilde{a}$ and $\tilde{b}$ share at least $|\tilde{a} \cap \tilde{c}| + |\tilde{b} \cap \tilde{c}| - N^2$ subwords. Therefore $\mathfrak{C}(\tilde{a}, \tilde{n})$ must be at least $\mathfrak{C}(\tilde{a}, \tilde{n}) + \mathfrak{C}(\tilde{b}, \tilde{c}) - 1$. Hence $\mathfrak{O}(\tilde{a}, \tilde{b}) \leq \mathfrak{O}(\tilde{a}, \tilde{c}) + \mathfrak{O}(\tilde{b}, \tilde{c})$. $\square$

3.2   The k-centre Problem

The goal of the k-centre problem for necklaces is to select a set of k necklaces of size $\vec{n}$ over an alphabet of size q that are "central" within the set of necklaces $\mathcal{N}_q^{\vec{n}}$. Formally the goal is to choose a set **S** of k necklaces such that the maximum distance between any necklace $\tilde{w} \in \mathcal{N}_q^{\vec{n}}$ and the nearest member of **S** is minimised. Given a set of necklaces $\mathbf{S} \subset \mathcal{N}_q^{\vec{n}}$, we use $\mathfrak{D}(\mathbf{S}, \mathcal{N}_q^{\vec{n}})$ to denote the maximum overlap distance between any necklace in $\mathcal{N}_q^{\vec{n}}$ and its closest necklace in $\mathbf{S}$. Formally $\mathfrak{D}(\mathbf{S}, \mathcal{N}_q^{\vec{n}}) = \max_{\tilde{\mathbf{w}} \in \mathcal{N}_q^{\vec{n}}} \min_{\tilde{\mathbf{v}} \in \mathbf{S}} \mathfrak{D}(\tilde{\mathbf{w}}, \tilde{\mathbf{v}})$.

Problem 1. k-centre problem for necklaces.

Input: A size vector of d-dimensions $\vec{n} \in \mathbb{N}^d$, an alphabet of size q, and an integer $k \in \mathbb{N}$.

Question: What is the set $\mathbf{S} \subseteq \mathcal{N}_q^{\vec{n}}$ of size k minimising $\mathfrak{D}(\mathbf{S}, \mathcal{N}_q^{\vec{n}})$?

There are two significant challenges we have to overcome in order to solve Problem 1: the exponential size of $\mathcal{N}_q^{\vec{n}}$, and the lack of structural, algorithmic, and combinatorial results for multidimensional necklaces. We show that the conceptually more straightforward problem of verifying whether a set of necklaces is a solution for Problem 2 is NP-hard for any dimension d.

Problem 2. The k-centre verification problem for necklaces.

Input: A d-dimensional size vector $\vec{n} \in \mathbb{N}$, an alphabet of size q, a rational distance $\ell \in \mathbb{Q}$, and a subset $\mathbf{S} \subseteq \mathcal{N}_q^{\vec{n}}$.

Question: Does there exist a necklace $\tilde{\mathbf{w}} \in \mathcal{N}_q^{\vec{n}}$ such that $\mathfrak{D}(\tilde{\mathbf{w}}, \mathbf{S}) \geq \ell$?

Theorem 1. *Given a set $\mathbf{S} \subseteq \mathcal{N}_q^{\vec{n}}$ and a distance ℓ, it is NP-hard to determine if there exists some necklace $\tilde{\mathbf{v}} \in \mathcal{N}_q^{\vec{n}}$ such that $\mathfrak{D}(\tilde{\mathbf{s}}, \tilde{\mathbf{v}}) > \ell$ for every $\tilde{\mathbf{s}} \in \mathbf{S}$.*

Proof. This claim is proven via a reduction from the Hamiltonian cycle problem on bipartite graphs to Problem 2 in 1D. Note that if the problem is hard in the 1D case, then it is also hard in any dimension $d \geq 1$ by using the same reduction for necklaces of size $(n_1, 1, 1, \ldots, 1)$. Let $G = (V, E)$ be a bipartite graph containing an even number $n \geq 6$ of vertices. The alphabet Σ is constructed with size n such that there is a one to one correspondence between each vertex in V and symbol in Σ. Using Σ a set $\mathbf{S}$ of necklaces is constructed as follows. For every pair of vertices $u, v \in V$ where $(u, v) \notin E$, the necklace corresponding to the word $(uv)^{n/2}$ is added to the set of centres $\mathbf{S}$. Further the word v^n, for every $v \in V$, is added to the set $\mathbf{S}$.

For the set $\mathbf{S}$, we ask if there exists any necklace in $\mathcal{N}_q^n$ that is further than a distance of $1 - \frac{3}{n^2}$. For the sake of contradiction, assume that there is no Hamiltonian cycle in G, and further that there exists a necklace $\tilde{\mathbf{w}} \in \mathcal{N}_q^{\vec{n}}$ such that the distance between $\tilde{\mathbf{w}}$ and every necklace $\tilde{\mathbf{v}} \in \mathbf{S}$ is greater than $1 - \frac{3}{n^2}$. If $\tilde{\mathbf{w}}$ shares a subword of length 2 with any necklace in $\mathbf{S}$ then $\tilde{\mathbf{w}}$ would be at a distance of no less than $1 - \frac{3}{n^2}$ from $\mathbf{S}$. Therefore, as every subword of length 2 in $\mathbf{S}$ corresponds to a edge that is not a member of E, every subword of length 2 in $\tilde{\mathbf{w}}$ must correspond to a valid edge.

As $\tilde{\mathbf{w}}$ can not correspond to a Hamiltonian cycle, there must be at least one vertex v for which the corresponding symbol appears at least 2 times in $\tilde{\mathbf{w}}$. As G is bipartite, if any cycle represented by $\tilde{\mathbf{w}}$ has length greater than 2, there must exist at least one vertex u such that $(v, u) \notin E$. Therefore, the necklace $(uv)^{n/2}$ is at a distance of no more than $1 - \frac{3}{n^2}$ from $\tilde{\mathbf{w}}$. Alternatively, if every cycle represented by $\tilde{\mathbf{w}}$ has length 2, there must be some vertex v that is represented at least 3 times in $\tilde{\mathbf{w}}$. Hence in this case $\tilde{\mathbf{w}}$ is at a distance of no more than $1 - \frac{3}{n^2}$ from the word $v^n \in \mathbf{S}$. Therefore, there exists a necklace at a distance of greater than $1 - \frac{3}{n^2}$ if and only if there exists a Hamiltonian cycle in the graph G. Therefore, it is NP-hard to verify if there exists any necklace at a distance greater than l for some set $\mathbf{S}$. □

The combination of this negative result with the exponential size of $\mathcal{N}_q^{\vec{n}}$ relative to $\vec{n}$ and q makes finding an optimal solution for Problem 1 exceedingly unlikely.

Lemma 1. *Let* $\mathbf{S} \subseteq \mathcal{N}_q^{\vec{n}}$ *be an optimal set of* k *centres minimising* $\mathfrak{D}(\mathbf{S}, \mathcal{N}_q^{\vec{n}})$ *then* $\mathfrak{D}(\mathbf{S}, \mathcal{N}_q^{\vec{n}}) \geq 1 - \frac{\log_q(k \cdot N)}{N}$.

Proof. Recall that the distance between the furthest necklace $\tilde{\mathbf{w}} \in \mathcal{N}_q^n$ and the optimal set $\mathbf{S}$ is bounded from bellow by determining an upper bound on the number of shared subwords between $\tilde{\mathbf{w}}$ and the words in $\mathbf{S}$. For the remainder of this proof let $\tilde{\mathbf{w}}$ to be the necklace furthest from the optimal set $\mathbf{S}$. Further for the sake of determining an upper bound, the set $\mathbf{S}$ is treated as a single necklace $\tilde{\mathbf{S}}$ of length $n \cdot k$. As the distance between $\tilde{\mathbf{w}}$ and $\tilde{\mathbf{S}}$ is no more than the distance between $\tilde{\mathbf{w}}$ and any $\tilde{\mathbf{v}} \in \mathbf{S}$, the distance between $\tilde{\mathbf{w}}$ and $\tilde{\mathbf{S}}$ provides a lower bound on the distance between $\tilde{\mathbf{w}}$ and $\mathbf{S}$.

In order to determine the number of subwords shared by $\tilde{\mathbf{w}}$ and $\tilde{\mathbf{S}}$, consider first the subwords of length 1. In order to guarantee that $\tilde{\mathbf{w}}$ shares at least one subword of length 1, $\tilde{\mathbf{S}}$ must contain each symbol in Σ, requiring the length of $\tilde{\mathbf{S}}$ to be at least q. Similarly, in order to ensure that $\tilde{\mathbf{w}}$ shares two subwords of length 1 with $\tilde{\mathbf{S}}$, $\tilde{\mathbf{S}}$ must contain 2 copies of every symbol on Σ, requiring the length of $\tilde{\mathbf{S}}$ to be at least $2q$. More generally for $\tilde{\mathbf{S}}$ to share i subwords of length 1 with $\tilde{\mathbf{w}}$, $\tilde{\mathbf{S}}$ must contain i copies of each symbol in Σ, requiring the length of $\tilde{\mathbf{S}}$ to be at least $i \cdot q$. Hence the maximum number of subwords of length 1 that $\tilde{\mathbf{w}}$ can share with $\tilde{\mathbf{S}}$ is either $\lfloor \frac{n \cdot k}{q} \rfloor$, if $\lfloor \frac{n \cdot k}{q} \rfloor \leq n$, or n otherwise.

For subwords of length 2, the problem becomes more complicated. In order to share a single word of length 2, it is not necessary to have every subword of length 2 appear as a subword of $\tilde{\mathbf{w}}$. Instead, it is sufficient to use only the prefixes of the canonical representations of each necklace. For example, given the binary alphabet $\{a, b\}$, every necklace has either aa, ab or bb as the prefix of length 2. Note that any necklace of length 2 followed by the largest symbol q in the alphabet $n - 2$ times belongs to the set N_q^n. As such, a simple lower bound on the number of prefixes of the canonical representation of necklaces is the number of necklaces of length 2, which in turn is bounded by $\frac{q^2}{2}$. Noting that the prefixes in $\tilde{\mathbf{S}}$ may overlap, to ensure that $\tilde{\mathbf{S}}$ and $\tilde{\mathbf{w}}$ share at least one subword of length 2, the length of $\tilde{\mathbf{S}}$ must be at least $\frac{q^2}{2}$. Similarly, for $\tilde{\mathbf{S}}$ and $\tilde{\mathbf{w}}$ to share i subwords of length 2, the length of $\tilde{\mathbf{S}}$ must be at least $\frac{i \cdot q^2}{2}$. Hence the maximum number of subwords of length 2 that $\tilde{\mathbf{S}}$ and $\tilde{\mathbf{w}}$ can share is either $\lfloor \frac{2n \cdot k}{q^2} \rfloor$, if $\lfloor \frac{2n \cdot k}{q^2} \rfloor \leq n$, or

n otherwise. In order for $\tilde{\mathbf{S}}$ to share at least one subword of length j with $\tilde{\mathbf{w}}$, the length of $\tilde{\mathbf{S}}$ must be at least $\frac{q^j}{j}$. Further the maximum number of subwords of length j that $\tilde{\mathbf{S}}$ and $\tilde{\mathbf{w}}$ can share is either $\lfloor \frac{j \cdot n \cdot k}{q^j} \rfloor$, if $\lfloor \frac{j \cdot n \cdot k}{q^j} \rfloor \le n$ or n otherwise.

The maximum length of a common subword that $\tilde{\mathbf{w}}$ can share with $\tilde{\mathbf{S}}$ is the largest value l such that $\frac{q^l}{l} \le n \cdot k$. By noting that $\frac{q^l}{l} \ge \frac{q^l}{n}$, a upper bound on l can be derived by rewriting the inequality $\frac{q^l}{n} \le n \cdot k$ as $l = 2\log_q(n \cdot k)$. Note further that, for any value $l' > l$, there must be at least one necklace that does not share any subword of length l' with $\tilde{\mathbf{S}}$ as $\tilde{\mathbf{S}}$ can not contain enough subwords to ensure that this is the case. This bound allows an upper bound number of shared subwords between $\tilde{\mathbf{w}}$ and $\tilde{\mathbf{S}}$ to be given by the summation

$$\sum_{i=1}^{2\log_q(n \cdot k)} \min(\lfloor \tfrac{i \cdot n \cdot k}{q^i} \rfloor, n) \le n \cdot \log_q(n \cdot k) + \frac{\log_q(k \cdot n)}{q-1} \approx \frac{q \cdot n \log_q(k \cdot n)}{q-1} \approx n\log_q(k \cdot n).$$

Using this bound, the distance between $\tilde{\mathbf{w}}$ and $\tilde{\mathbf{S}}$ must be no less than $1 - \frac{\log_q(k \cdot n)}{n}$.

In the multidimensional case, let $\vec{m} = (m_1, m_2, \ldots, m_d)$ be a size vector of d-dimensions such that $M = m_1 \cdot m_2 \cdot \ldots \cdot m_d$. The largest value of M such that $\tilde{\mathbf{S}}$ can contain every subword with M positions is $2\log_q(n \cdot k)$. From Proposition 1, the lower bound on the number of necklaces of size $\vec{m}$ is $\frac{q^M}{M}$. The maximum number of shared subwords between $\tilde{\mathbf{w}}$ and $\tilde{\mathbf{S}}$ is $\sum_{i=1}^{M} i \cdot \frac{N \cdot k}{q^i} \le \log_q(k \cdot N)$. Hence the distance between $\tilde{\mathbf{w}}$ and $\tilde{\mathbf{s}}$ is at most $1 - \frac{\log_q(k \cdot N)}{N}$. $\square$

The key idea behind our algorithms for approximating the k-centre problem on necklaces is to find the largest vector $\vec{\ell} = (l_1, l_2, \ldots, l_d)$ such that every word of size $\vec{\ell}$ appears as a subword within the set of centres. In this setting $\vec{m}$ is larger than $\vec{\ell}$ if $m_1 \cdot m_2 \cdot \ldots \cdot m_d > l_1 \cdot l_2 \cdot \ldots \cdot l_d$. This is motivated by observing that if two necklaces share a subword of length l, they must also share 2 subwords of length $l - 1$, 3 of length $l - 2$, and so on.

Lemma 2. *Given* $\tilde{\mathbf{w}}, \tilde{\mathbf{v}} \in \mathcal{N}_q^{\vec{n}}$ *sharing a common subword* a *of size* $\vec{m}$, *let* $x_i = n_i \cdot m_i$ *if* $n_i = m_i$, *and* $x_i = \frac{m_i(m_i+1)}{2}$ *otherwise. The distance between* w *and* v *is bounded by* $\mathfrak{O}(w,v) \le 1 - \frac{\prod_{i=1}^{d} x_i}{N^2} \le 1 - \frac{M^2}{2N^2}$ *where* $N = \prod_{i \in [d]} n_i$ *and* $M = \prod_{i \in [d]} m_i$.

Proof. Note that the minimum intersection between $\tilde{\mathbf{w}}$ and $\tilde{\mathbf{v}}$ is the number of subwords of a, including the word a itself. To compute the number of subwords of a, consider the number of subwords starting at some position $\vec{j} \in [|a|]$. Assuming that $|a|_i < n_i$ for every $i \in [d]$, the number of subwords starting at $\vec{j}$ corresponds to the size of the set $[\vec{j}, |a|]$, equal to $\prod_{i=1}^{d} m_i - |a|_i$. This gives the number of shared subwords as being at least $\sum_{\vec{j} \in [|a|]} \prod_{i \in [d]} m_i - |a|_i \ge \sum_{j \in [M]} j \ge \frac{M^2}{2}$. Therefore, the distance between $\tilde{\mathbf{w}}$ and $\tilde{\mathbf{v}}$ is no more than $1 - \frac{M^2}{2N^2}$. $\square$

4 Approximating the k-centre Problem for necklaces

In this section we provide our approximation algorithms. The main idea is to determine the longest de-Bruijn sequence that can fit into the set of k-centres. As the de Bruijn sequence of order l contains every word in Σ^l as a subword, by representing the de Bruijn sequence of order l in the set of centres we ensure that every necklace shares a subword of length l with the set of k-centres.

Definition 2. *A de Bruijn hypertorus of order $\vec{n}$ is a cyclic d-dimensional word T containing as a subword every word in $\Sigma^{\vec{n}}$ exactly once.*

Lemma 3. *There exists an $O(n \cdot k)$ time algorithm for the k-centre problem on $\mathcal{N}_q^n$ returning a set of centres $\mathbf{S}$ such that $\mathfrak{D}(\mathbf{S}, \mathcal{N}_q^n) \leq 1 - \frac{\log_q^2(k \cdot n)}{2n^2}$.*

Proof. Our algorithm operates by partitioning a de Bruijn sequence S of order λ into a set of k centres of size $n - \lambda + 1$, with the final $\lambda - 1$ symbols of the ith centre being shared with the $(i + 1)^{th}$ centre. In this manner, the first centre is generated by taking the first n symbols of the de Bruijn sequence. To ensure that every subword of length λ occurs, the first $\lambda - 1$ symbols of the second centre is the same as the last $\lambda - 1$ symbol of the first centre. Repeating this, the ith centre is the subword of length n starting at position $i(n - \lambda + 1) + 1$ in the de Bruijn sequence. An example of this is given in Fig. 5.

Sequence:	0000001000011000101000111001001011001101001111010101110110111111
Centre	Word
1	000000100001100010100
2	101000111001001011001
3	110011010011110101011
4	000000 0101110110111111

Fig. 5. Example of how to split the de Bruijn sequence of order 6 between 4 centres. Highlighted parts are the shared subwords between two centres.

This leaves the problem of determining the largest value of λ such that $q^\lambda \leq k \cdot (n - \lambda + 1)$. Rearranging $q^\lambda \leq k \cdot (n - \lambda + 1)$ in terms of λ gives $\lambda \leq \log_q(k \cdot (n + 1) - k \cdot \lambda)$. Noting that $\lambda \leq \log_q(k \cdot n)$, this upper bound on the value of λ can be rewritten as $\log_q(k \cdot (n + 1 - \log_q(k \cdot n))) \approx \log_q(k \cdot n)$. Using Lemma 2, along with $\log_q(k \cdot n)$ as an approximate value of λ gives an upper bound on the distance between each necklace in $\mathcal{N}_q^n$ and the set of centres of $1 - \frac{\log_q^2(kn)}{2n^2}$. As the corresponding de Bruijn sequence can be computed in no more than $O(k \cdot n)$ time [31], the total complexity is at most $O(k \cdot n)$. $\square$

Theorem 2. *The k-centre problem for $\mathcal{N}_q^n$ can be approximated in $O(n \cdot k)$ time with an approximation factor of $1 + \frac{\log_q(k \cdot n)}{n - \log_q(k \cdot n)} - \frac{\log_q^2(k \cdot n)}{2n(n - \log_q(k \cdot n))}$.*

Proof. Using the lower bound of $1 - \frac{\log_q^2(kn)}{2n^2}$ given by Lemma 3 gives $\frac{1 - \frac{\log_q^2(kn)}{2n^2}}{1 - \frac{\log_q(k \cdot n)}{n}}$

$= \frac{2n^2 - \log_q^2(kn)}{2n^2 - 2n \log_q(kn)} = 1 + \frac{2n \log_q(kn) - \log_q^2(kn)}{2n^2 - 2n \log_q(kn)} = 1 + \frac{\log_q(kn)}{n - \log_q(kn)} - \frac{\log_q^2(kn)}{2n(n - \log_q(kn))}.$ □

Theorem 3. *Let T be a d-dimensional de Bruijn hypertorus of size $(x, x, \ldots, x)$. There exist k subwords of T that form a solution to the k-centre problem for $\mathcal{N}_q^{(y,y,\ldots,y)}$ with an approximation factor of $1 + \frac{\log_q(kN)}{N - \log_q(k \cdot N)} - \frac{\log_q^2(k \cdot N)}{2N(N - \log_q(k \cdot N))}$ where $y^d = N$ and $x^d = \log_N(y)$.*

Proof. Recall from Lemma 1 that the lower bound on the distance between the centre and every necklace in $\mathcal{N}_q^{\vec{n}}$ is $1 - \frac{\log_q(k \cdot N)}{N}$. As in Theorem 2, the goal is to find the largest de Bruijn torus that can "fit" into the centres. To simplify the reasoning, the de Bruijn hyper tori here is limited to those corresponding to the word where the length of each dimension is the same. Formally, the de Bruijn hypertori are restricted to be of the size $m_1 = m_2 = \ldots = m_j = \sqrt[j]{N}$ for some $j \in [d]$, giving the total number of positions in the tori as M. Similarly, the centres is assumed to have size $n_1 = n_2 = \ldots = n_d = \sqrt[d]{N}$, giving N total positions.

Observe that the largest torus that can be represented in the set of centres has M positions such that $q^M \leq k \cdot N^{(d-j)/d}(\sqrt[d]{N} - \sqrt[j]{M} + 1)^j$. This can be rewritten to give $M \leq \log_q(k \cdot N^{(d-j)/d}(\sqrt[d]{N} - \sqrt[j]{M} + 1)^j)$. Noting that M is of logarithmic size relative to N, this is approximately equal to $M \leq \log_q(k \cdot N)$. Using Lemma 2, the minimum distance between any necklace in $\mathcal{N}_q^{\vec{n}}$ is $1 - \frac{\log_q^2(kN)}{2N^2}$. Following the arguments from Theorem 2 gives a ratio of $1 + \frac{2 \cdot N \log_q(k \cdot N) - \log_q^2(k \cdot N)}{2 \cdot N^2 - 2 \cdot N \cdot \log_q(k \cdot N)} = 1 + \frac{\log_q(kN)}{N - \log_q(kN)} - \frac{\log_q^2(kN)}{2N(N - \log_q(kN))}.$ □

While this provides a good starting point for solving the k-centre problem for $\mathcal{N}_q^{\vec{n}}$, this work is restricted by the limited results on generating de Bruijn hypertori, particularly in higher dimensions [10,19–22]. As such, we present an alternative approach below. The high-level idea is to reduce the problem from the multi-dimensional setting to the 1D problem, which we can approximate well using Theorem 2. Given a size vector $\vec{n}$, integer k and alphabet Σ our approach can be thought of as finding a set of $k \cdot n_1 \cdot \ldots \cdot n_{d-1}$ centres of length n_d over Σ, taking advantage of the added number of centres to increase the length of shared subwords.

Case 1, $q^{n_d} \geq k \cdot \frac{N}{n_d}$: In this case the set of centres is constructed by using $k' = \frac{k \cdot N}{n_d}$ centres of $\mathcal{N}_q^{n_d}$. The motivation behind this approach is to optimise the length of the 1D subwords that are shared by the centre and every other necklace in $\mathcal{N}_q^{\vec{n}}$. Let $\mathbf{S} \subseteq \mathcal{N}_q^{n_d}$ be a set of centres $k \cdot \frac{N}{n_d}$ from $\mathcal{N}_q^{n_d}$ constructed following the algorithm outlined in Lemma 3. Following the arguments from Lemma 3, every necklace in $\mathcal{N}_q^{n_d}$ must share a subword of length $\log_q(k \cdot N)$ with at least one centre in $\mathbf{S}$. As every subword of size $(1, 1, \ldots, 1, n_d)$ of any necklace in $\mathcal{N}_q^{\vec{q}}$ belongs to a necklace $\tilde{w} \in \mathcal{N}_q^{n_d}$, by ensuring that every necklace in **S** appears as a subword in the centre $\mathbf{S}' \subseteq \mathcal{N}_q^{\vec{n}}$ it is ensured that $\tilde{w}$ shares at least one subword of length $\log_q(k \cdot N)$ with some necklace in $\mathbf{S}'$. This can be done by simply splitting **S** into k sets of $\frac{N}{n_d}$ centres, each of which can be made into a word of size $\vec{n}$ through concatenation. From Lemma 2, the maximum distance between any necklace in **S**$'$ and necklace in $\mathcal{N}_q^{\vec{n}}$ is $1 - \frac{\log_q^2(k \cdot N)}{2N^2}$. This equals the bound given by Lemma 3, giving the same approximation ratio.

Case 2, $q^{n_d} < k \cdot \frac{N}{n_d}$: Following the process outlined above, it is possible to represent every word of length n_d over Σ with some redundancy. In order to reduce the redundancy an alternative reduction from the 1D setting is constructed. The high-level idea is to construct a new alphabet such that each symbol corresponds to some word in $\Sigma^{\vec{m}}$ for some size vector $\vec{m}$.

The first problem is determining the size vector allowing for this reduction. Let $\Sigma(\vec{m})$ denote the alphabet of size $q^{m_1 \cdot m_2 \cdots \cdot m_d}$ such that each symbol in $\Sigma(\vec{m})$ corresponds to some word in $\Sigma^{\vec{m}}$. Given a word $w \in \Sigma(\vec{m})^{n_1/m_1, n_2/m_2, \ldots, n_d/m_d}$ a word $v \in \Sigma^{\vec{n}}$ can be constructed by replacing each symbol in w with the corresponding word in $\Sigma^{\vec{m}}$. Note that the largest value of $\vec{m}$ such that every symbol in $\Sigma(\vec{m})$ can be represented in k words from $\Sigma(\vec{m})^{n_1/m_1, n_2/m_2, \ldots, n_d/m_d}$ is bounded by the inequality $q^{m_1 \cdot m_2 \cdots \cdot m_d} \leq k \cdot \lfloor \frac{n_1}{m_1} \rfloor \cdot \lfloor \frac{n_2}{m_2} \rfloor \cdots \cdot \lfloor \frac{n_d}{m_d} \rfloor$. Letting $M = m_1 \cdot m_2 \cdots \cdot m_d$, this inequality can be rewritten as approximately $q^M \leq k \cdot \frac{N}{M}$. Treating M as being approximately N gives $M \leq \log_q(k)$.

Using this bound on M let $\vec{m}$ be some set of vectors such that $M = m_1 \cdot m_2 \cdots \cdot m_d$. We may assume without loss of generality that $m_d = 1$. The centres for $\mathcal{N}_q^{\vec{n}}$ are constructed by making a set **S** of $k\frac{N}{M \cdot n_d}$ centres for $\mathcal{N}_{q^M}^{n_d}$. Following the arguments from Lemma 3, every necklace in $\mathcal{N}_{q^M}^{n_d}$ must share a subword of length at least $\log_{q^M}(k \cdot \frac{N}{M}) = \frac{\log_q(k \cdot \frac{N}{M})}{M} = \frac{\log_q\left(k \cdot \frac{N}{\log_q(k)}\right)}{\log_q(k)}$. Note further that, as each symbol in $\Sigma(\vec{m})$ corresponds to a word in $\Sigma^{\vec{m}}$, converting each word in **S** to a word of size $(m_1, m_2, \ldots, m_{d-1}, n_1)$ provides a set of centres such that every necklace in $\mathcal{N}_q^{(m_1, m_2, \ldots, m_{d-1}, n_1)}$ shares a subword of size

$$\left(m_1, m_2, \ldots, m_{d-1}, \frac{\log_q\left(k \cdot \frac{N}{\log_q(k)}\right)}{\log_q(k)}\right)$$ with some centre. Converting this new set

of centres into a set $\mathbf{S}' \subseteq \mathcal{N}_q^{\vec{n}}$ maintains the same size of shared subwords. From Lemma 2, the furthest distance between $\mathbf{S}'$ and any necklace in $\mathcal{N}_q^{\vec{n}}$ is bounded

from above by $1 - \dfrac{\log_q^2(k) \cdot \frac{\log_q^2\left(k \cdot \frac{N}{\log_q(k)}\right)}{\log_q^2(k)}}{2N^2} = 1 - \dfrac{\log_q^2\left(k \cdot \frac{N}{\log_q(k)}\right)}{2N^2} \approx 1 - \dfrac{\log_q^2(k \cdot N)}{2N^2}$.

Theorem 4. *The k-centre problem for $\mathcal{N}_q^{\vec{n}}$ can be approximated in $O(N^2k)$ time within a factor of $1 + \dfrac{\log_q(kN)}{N - \log_q(kN)} - \dfrac{\log_q^2(kN)}{2N(N - \log_q(kN))}$, where $N = \prod_{i=1}^d n_i$.*

Proof. Following the above construction, note that in both cases the distance between the set of centres $\mathbf{S}$ and the necklaces $\mathcal{N}_q^{\vec{n}}$ is bounded from above by $1 - \frac{\log_q^2(k \cdot N)}{2N^2}$. The approximation ratio of $1 + \frac{\log_q(kN)}{N - \log_q(kN)} - \frac{\log_q^2(kN)}{2N(N - \log_q(kN))}$ is derived using the same arguments as in Theorem 2. Regarding time complexity, in the first case the problem can be solved in $O(k \cdot N)$ time using Theorem 2. In the second case, a brute force approach to find to best value of $\vec{m}$ can be done in $O(N)$ additional time steps giving a total complexity of $O(k \cdot N^2)$. $\qquad\square$

Acknowledgements. The authors thank the Leverhulme Trust via the Leverhulme Research Centre for Functional Materials Design at the University of Liverpool for their support.

References

1. Adamson, D., Deligkas, A., Gusev, V.V., Potapov, I.: On the hardness of energy minimisation for crystal structure prediction. In: Chatzigeorgiou, A., Dondi, R., Herodotou, H., Kapoutsis, C., Manolopoulos, Y., Papadopoulos, G.A., Sikora, F. (eds.) SOFSEM 2020. LNCS, vol. 12011, pp. 587–596. Springer, Cham (2020). https://doi.org/10.1007/978-3-030-38919-2_48
2. Adamson, D., Deligkas, A., Gusev, V.V., Potapov, I.: Combinatorial algorithms for multidimensional necklaces (2021). https://arxiv.org/abs/2108.01990, https://doi.org/10.48550/ARXIV.2108.01990
3. Adamson, D.: Ranking binary unlabelled necklaces in polynomial time. In: Han, Y.S., Vaszil, G. (eds.) Descriptional Complexity of Formal Systems, DCFS 2022. LNCS, vol. 13439, pp. 15–29. Springer, Cham (2022). https://doi.org/10.1007/978-3-031-13257-5_2
4. Adamson, D., Deligkas, A., Gusev, V.V., Potapov, I.: The complexity of periodic energy minimisation. In: Szeider, S., Ganian, R., Silva, A. (eds.) 47th International Symposium on Mathematical Foundations of Computer Science (MFCS 2022). Leibniz International Proceedings in Informatics (LIPIcs), vol. 241, pp. 8:1–8:15, Dagstuhl, Germany (2022). Schloss Dagstuhl - Leibniz-Zentrum für Informatik. https://drops.dagstuhl.de/opus/volltexte/2022/16806, https://doi.org/10.4230/LIPIcs.MFCS.2022.8
5. Adamson, D., Gusev, V.V., Potapov, I., Deligkas, A.: Ranking bracelets in polynomial time. In: Gawrychowski, P., Starikovskaya, T. (eds.) 32nd Annual Symposium on Combinatorial Pattern Matching (CPM 2021). Leibniz International Proceedings in Informatics (LIPIcs), vol. 191, pp. 4:1–4:17. Dagstuhl, Germany (2021). Schloss Dagstuhl - Leibniz-Zentrum für Informatik. https://drops.dagstuhl.de/opus/volltexte/2021/13955, https://doi.org/10.4230/LIPIcs.CPM.2021.4
6. Anselmo, M., Madonia, M., Selmi, C.: Toroidal codes and conjugate pictures. In: Martín-Vide, C., Okhotin, A., Shapira, D. (eds.) LATA 2019. LNCS, vol. 11417, pp. 288–301. Springer, Cham (2019). https://doi.org/10.1007/978-3-030-13435-8_21
7. Babai, L.: Local expansion of vertex-transitive graphs and random generation in finite groups. In: Proceedings of the Twenty-Third Annual ACM Symposium on Theory of Computing, STOC 1991, New York, NY, USA, pp. 164–174. Association for Computing Machinery (1991). https://doi.org/10.1145/103418.103440

8. Bae, M.M., Bose, B.: Gray codes for torus and edge disjoint Hamiltonian cycles. In: Proceedings of the 14th International Parallel and Distributed Processing Symposium, IPDPS 2000, pp. 365–370 (2000). https://doi.org/10.1109/IPDPS.2000.846007

9. Chakrabarty, D., Goyal, P., Krishnaswamy, R.: The non-uniform k-center problem. ACM Trans. Algorithms 16(4) (2020). https://doi.org/10.1145/3392720

10. Chung, F., Diaconis, P., Graham, R.: Universal cycles for combinatorial structures. Discret. Math. 110(1–3), 43–59 (1992)

11. Cohen, W.W., Ravikumar, P., Fienberg, S.E., et al.: A comparison of string distance metrics for name-matching tasks. IIWeb 2003, 73–78 (2003)

12. Collins, C., et al.: Accelerated discovery of two crystal structure types in a complex inorganic phase field. Nature 546(7657), 280 (2017)

13. Feldmann, A.E., Marx, D.: The parameterized hardness of the k-center problem in transportation networks. Algorithmica 82(7), 1989–2005 (2020)

14. Frances, M., Litman, A.: On covering problems of codes. Theory Comput. Syst. 30(2), 113–119 (1997)

15. Gärtner, T.: A survey of kernels for structured data. ACM SIGKDD Explor. Newsl. 5(1), 49–58 (2003)

16. Gasieniec, L., Jansson, J., Lingas, A.: Efficient approximation algorithms for the Hamming Center Problem. In: SODA 1999, pp. 905–906 (1999)

17. Graham, R.L., Knuth, D.E., Patashnik, O.: Concrete Mathematics?: A Foundation for Computer Science. Addison-Wesley, Upper Saddle River (1994)

18. Hochbaum, D.S.: Approximation algorithms for NP-hard problems. In: Various Notions of Approximations: Good, Better, Best and More (1997)

19. Horan, V., Stevens, B.: Locating patterns in the de Bruijn Torus. Discret. Math. 339(4), 1274–1282 (2016)

20. Hurlbert, G., Isaak, G.: On the de Bruijn Torus problem. J. Comb. Theory Ser. A 64(1), 50–62 (1993)

21. Hurlbert, G., Isaak, G.: New constructions for de Bruijn Tori. Des. Codes Crypt. 6(1), 47–56 (1995)

22. Hurlbert, G.H., Mitchell, C.J., Paterson, K.G.: On the existence of de Bruijn Tori with two by two windows. J. Comb. Theory Ser. A 76(2), 213–230 (1996)

23. Kociumaka, T., Radoszewski, J., Rytter, W.: Computing k-th Lyndon word and decoding lexicographically minimal de Bruijn sequence. In: Kulikov, A.S., Kuznetsov, S.O., Pevzner, P. (eds.) CPM 2014. LNCS, vol. 8486, pp. 202–211. Springer, Cham (2014). https://doi.org/10.1007/978-3-319-07566-2_21

24. Kopparty, S., Kumar, M., Saks, M.: Efficient indexing of necklaces and irreducible polynomials over finite fields. Theory Comput. 12(1), 1–27 (2016)

25. Lanctot, J.K., Li, M., Ma, B., Wang, S., Zhang, L.: Distinguishing string selection problems. Inf. Comput. 185(1), 41–55 (2003)

26. Li, M., Ma, B., Wang, L.: On the closest string and substring problems. J. ACM 49(2), 157–171 (2002)

27. Lothaire, M.: Combinatorics on Words. Cambridge Mathematical Library, 2nd edn. Cambridge University Press, Cambridge (1997). https://doi.org/10.1017/CBO9780511566097

28. Piskorski, J., Sydow, M., Wieloch, K.: Comparison of string distance metrics for lemmatisation of named entities in polish. In: Language and Technology Conference, pp. 413–427 (2007)

29. Recchia, G., Louwerse, M.M.: A comparison of string similarity measures for toponym matching. In: SIGSPATIAL 2013, pp. 54–61 (2013)

30. Ruskey, F., Savage, C., Min Yih Wang, T.: Generating necklaces. J. Algorithms **13**(3), 414–430 (1992)
31. Ruskey, F., Sawada, J.: Generating Necklaces and Strings with Forbidden Substrings. In: Du, DZ., Eades, P., Estivill-Castro, V., Lin, X., Sharma, A. (eds) Computing and Combinatorics, COCOON 2000. LNCS, vol 1858, pp. 330–339. Springer, Heidelberg (2000). https://doi.org/10.1007/3-540-44968-X_33
32. Siromoney, G., Siromoney, R., Robinson, T.: Kahbi kolam and cycle grammars, pp. 267–300. Springer-Verlag (1987). https://www.worldscientific.com/doi/abs/10.1142/9789814368452_0017, https://doi.org/10.1142/9789814368452_0017
33. Thorup, M.: Quick k-median, k-center, and facility location for sparse graphs. SIAM J. Comput. **34**(2), 405–432 (2005). http://arxiv.org/abs/https://doi.org/10.1137/S0097539701388884, https://doi.org/10.1137/S0097539701388884

Author Index

Printed in the United States
by Baker & Taylor Publisher Services

Printed in the United States
by Baker & Taylor Publisher Services